Green Energy and Technology

Climate change, environmental impact and the limited natural resources urge scientific research and novel technical solutions. The monograph series Green Energy and Technology serves as a publishing platform for scientific and technological approaches to “green”—i.e. environmentally friendly and sustainable—technologies. While a focus lies on energy and power supply, it also covers "green" solutions in industrial engineering and engineering design. Green Energy and Technology addresses researchers, advanced students, technical consultants as well as decision makers in industries and politics. Hence, the level of presentation spans from instructional to highly technical.
Indexed in Scopus.
Indexed in Ei Compendex.

Solomon Workneh Fanta
Mesfin Wogayehu Tenagashaw
Muluken Temsgen Tigabu
Ephrem Yetbarek Gedilu
Editors

Sustainable Development Research in Manufacturing, Process Engineering, Green Infrastructure, and Water Resources

Advancement of Science and Technology

Springer

Editors
Solomon Workneh Fanta
Faculty of Chemical and Food Engineering
Bahir Dar Institute of Technology
Bahir Dar University
Bahir Dar, Ethiopia

Muluken Temsgen Tigabu
Bahir Dar Institute of Technology
Bahir Dar University
Bahir Dar, Ethiopia

Mesfin Wogayehu Tenagashaw
Bahir Dar Institute of Technology
Bahir Dar University
Bahir Dar, Ethiopia

Ephrem Yetbarek Gedilu
Bahir Dar Institute of Technology
Bahir Dar University
Bahir Dar, Ethiopia

ISSN 1865-3529 ISSN 1865-3537 (electronic)
Green Energy and Technology
ISBN 978-3-031-77338-9 ISBN 978-3-031-77339-6 (eBook)
https://doi.org/10.1007/978-3-031-77339-6

This Springer imprint is published by the registered company Springer Nature Switzerland AG
The registered company address is: Gewerbestrasse 11, 6330 Cham, Switzerland

Contents

1 **Impact of Manufacturing Strategies on the Effectiveness of Mass Customization** 1
Mehari Kumilachew Bezuneh, Assefa Asmare Tsegaw, Bereket Haile Woldegiorgis, Teshome Mulatie Bogale, Matthias Brossog, and Jörg Franke
1.1 Introduction 1
1.2 Review of Literature on Mass Customization 2
1.2.1 Capabilities of Mass Customization 3
1.3 Enablers of MC Capabilities 3
1.3.1 Lean Manufacturing 4
1.3.2 Agile Manufacturing 4
1.3.3 Manufacturing Flexibility 5
1.3.4 Cellular Manufacturing (Group Technology) 5
1.3.5 Customer-Driven Design and Manufacturing 5
1.4 Methodology 6
1.4.1 Data Collection and Analysis 6
1.4.2 Reliability of Questionnaire 10
1.5 Results and Discussion 10
1.6 Conclusion and Recommendation 13
References 14

2 **Detecting Misalignments of Drilled Holes Using Machine Vision** 17
Getachew Ambaye, Enkhsaikhan Boldsaikhan, and Krishna Krishnan
2.1 Introduction 17
2.2 Modeling and Methods 19
2.3 Results and Discussions 21

2.3.1 Edge Detection 22
2.3.2 Mean Square Error (MSE) 25
2.4 Conclusion 27
References 28

3 Tractive Performances of Traction Devices of Two-Wheeled Tractors in Dry Field Conditions: A Review 31
Samuel Getachew Chekol, Smegnew Moges Mintesinot, Berihun Abebaw Mebratie, and Addisu Alamirew Meku
3.1 Introduction 31
3.1.1 Two-Wheeled Tractors 31
3.1.2 Traction Devices in 2WTs 33
3.2 Tractive Performance of Traction Devices 35
3.2.1 Pneumatic Tires 36
3.2.2 Cage Wheels 37
3.2.3 Rubber Tracks 37
3.3 Discussion 38
3.3.1 Maximum Pull Developed By Tractive Devices 38
3.3.2 Effects of Weight on Tractive Performance 40
3.3.3 Effects of Design Parameters on Tractive Performance 41
3.4 Conclusion 42
3.5 Future Studies 43
References 43

4 Recent Advances of Additive Manufacturing for Aerospace Industries: Methods, Materials, Challenges, and Future Outlooks 47
Melkamu Tadesse Getachew, Menberu Zeleke Shiferaw, and Bekalu Sintayehu Ayele
4.1 Introduction 47
4.2 Additive Manufacturing Methods (Techniques) for Aerospace Components 49
4.2.1 Powder Bed Fusion Techniques 50
4.2.2 Directed Energy Deposition (DED) Techniques 55
4.2.3 Material Extrusion Techniques 59
4.2.4 Comparison of Additive Manufacturing Methods with Their Advantages and Limitations in Aerospace Applications 62
4.3 Materials for Aerospace Additive Manufacturing 63
4.3.1 Polymers 63
4.3.2 Ceramics 64
4.3.3 Metal and Alloys 65
4.3.4 Carbon Fiber-Reinforced Polymers (CFRPs) 69
4.4 Aerospace Components Manufactured Using Additive Manufacturing 71
4.4.1 Engine Components 71

4.4.2 Interior Components . . . 71
4.4.3 Aircraft Structures . . . 73
4.4.4 Converging-Diverging Rocket Nozzles . . . 74
4.4.5 Anti-icing Systems . . . 74
4.4.6 Lattice Heat Sinks . . . 75
4.5 Challenges and Limitations of Additive Manufacturing in Aerospace Application . . . 75
4.6 Future of Additive Manufacturing in the Aerospace Industry . . . 76
4.7 Conclusion . . . 77
References . . . 78

5 Rotating Machinery Fault Diagnosis Using One-Dimensional Convolutional Neural Network . . . 83
Fasikaw Kibrete, Dereje Engida Woldemichael, and Hailu Shimels Gebremedhen
5.1 Introduction . . . 83
5.2 Convolutional Neural Network . . . 85
5.3 The Proposed Method . . . 87
5.3.1 Network Architecture . . . 87
5.3.2 Fault Diagnosis Process . . . 88
5.4 Experimental Results and Discussion . . . 89
5.4.1 Data Description . . . 89
5.4.2 Setting Parameters of 1D-CNN . . . 89
5.4.3 Diagnosis Results and Discussion . . . 90
5.5 Conclusion . . . 92
References . . . 93

6 Development and Evaluation of Metering Mechanism for Sowing of Berseem (*Trifolium alexandrinum L.*) and Lucerne (Medicagosativa) Crops . . . 95
Smegnew Moges Mintesinot and Rohinish Khurana
6.1 Introduction . . . 95
6.2 Materials and Methods . . . 96
6.2.1 Development Consideration for Seed-Metering Mechanism of Berseem and Lucerne Crop . . . 96
6.2.2 Development of Seed-Metering Mechanism . . . 97
6.2.3 Laboratory Evaluation of Metering Mechanisms . . . 98
6.2.4 Selection of Independent and Dependent Parameters . . . 100
6.2.5 Data Analysis . . . 102
6.3 Results and Discussion . . . 103
6.3.1 Effect of the Metering Device and Forward Speed on the Seed Rate . . . 103
6.3.2 Effect of Metering Device and Forward Speed on Seed-to-Seed Distance . . . 104
6.3.3 Effect of Metering Device and Forward Speed on Coefficient of Seed Distribution Variation . . . 105

6.3.4 Effect of Metering Device and Forward Speed on Seed Damage 106
6.3.5 Suitable Combination of Independent Parameters 108
6.4 Conclusions 108
References 109

7 **Effects of Crop Parameters and Feed Rates on Performances of Axial Rice Thresher** 111
Solomon Tekeste Hailemesikel, Mersha Alebachew, Mulugeta Admasu Delele, Addisu Negash Ali, Solomon Workneh Fanta, and Geta Kidanemariam
7.1 Introduction 111
7.2 Materials and Method 113
7.2.1 Description of the Test Thresher 113
7.2.2 Rice Cultivars Used in This Study 114
7.2.3 Experimental Design 115
7.2.4 Performance Evaluation of the Thresher 115
7.2.5 Fuel Consumption and Specific Energy Consumed 117
7.2.6 Data Analysis 117
7.3 Result and Discussion 118
7.3.1 Crop Parameters 118
7.3.2 Threshing Performances 118
7.4 Conclusion 122
References 123

8 **Enhancing Mass Customization Using a Digital Twin: Review and Model Development** 125
Mehari Kumilachew Bezuneh, Assefa Asmare Tsegaw, Bereket Haile Woldegiorgis, Teshome Mulatie Bogale, Matthias Brossog, and Jörg Franke
8.1 Introduction 125
8.2 Methodology 126
8.3 Related Literature on Mass Customization 127
8.4 Review on Digital Twin 129
8.4.1 Digital Twin Concepts 129
8.4.2 Application of Digital Twin in the Manufacturing Industry 131
8.5 Application of the Digital Twin in Mass Customization 132
8.5.1 Digital Twin for DTO Customization 134
8.5.2 Digital Twin-Based MTO Customization 134
8.5.3 Digital Twin-Based ATO Customization 136
8.6 Discussion, Conclusion, and Recommendation 137
8.6.1 Discussion 137
8.6.2 Conclusion and Recommendation 137
References 139

9 Physical Properties of Berseem (*Trifolium alexandrinum*) and Lucerne (*Medicago sativa*) Seeds 143
Smegnew Moges Mintesinot, Rohinish Khurana, Samuel Getachew Chekol, and Mequanint Demeke Aynalem
9.1 Introduction 143
9.2 Materials and Methods 144
9.2.1 Sample Collection, Preparation, and Materials Used 144
9.2.2 Determination of Size and Shape 145
9.2.3 Determination of Thousand-Grain Weight (TGW) 146
9.2.4 Determination of Bulk Density and True Density 146
9.2.5 Determination of Angle of Repose and Coefficient of Static Friction 147
9.3 Result and Discussion 148
9.3.1 Size and Shape of Seeds 148
9.3.2 Thousand-Grain Weight (TGW) of Seeds 150
9.3.3 Bulk Density and True Density of Seeds 150
9.3.4 Angle of Repose and Coefficient of Static Friction of Seeds 151
9.4 Conclusion 151
References 152

10 Recent Developments in Biomechanics-Based Prediction of Musculoskeletal Disorders: A Review 155
Alehegn Melesse Semegn, Bereket Haile Woldegiorgis, and Zerihun Wondimu Lemessa
10.1 Introduction 155
10.2 Methodology 156
10.3 Results and Discussion 158
10.3.1 Biomechanical Models 158
10.3.2 Wearable Devices Advancement in Biomechanical-Based Risk Assessment 159
10.3.3 Prediction Algorithms in Biomechanics to Predict MSDs 162
10.4 Conclusion and Future Research Opportunities 164
10.4.1 Conclusion 164
10.4.2 Future Research Opportunities 165
References 165

11 Opportunities and Challenges of E-SCM in Developing Countries 169
Zerihun Wondimu Lemessa, Bereket Haile Woldegiorgis, and Alehegn Melesse Semegn
11.1 Introduction 169
11.2 Theoretical Background 170
11.2.1 Introduction to E-SCM 171
11.2.2 e-Supply Chain Management and E-Commerce 171

11.3 Methodology ... 172
11.3.1 Paper Location ... 172
11.3.2 Paper Selection ... 172
11.3.3 Paper Analysis ... 174
11.4 Results and Discussion ... 174
11.4.1 E-Commerce Implementation in Developing Countries ... 175
11.4.2 Barriers to Successful E-SC Adopting in Developing Countries ... 179
11.4.3 Benefits of Adopting E-SCM ... 180
11.5 Conclusion and Future Work ... 180
References ... 181

12 Utilization, Nutritional, and Microbiological Quality of *Awaze*: Fermented Domestic Spicy Red Pepper Paste of North Western Part of Amhara, Ethiopia ... 185
Biadge Kefale Anteneh, Mulugeta Admasu, Solomon Workneh Fanta, and Solomon Abate
12.1 Introduction ... 185
12.2 Materials and Methods ... 186
12.2.1 Study Site ... 186
12.2.2 Study Design and Sample Size ... 186
12.2.3 Sampling ... 188
12.2.4 Data Collection Method ... 188
12.2.5 Physicochemical Analysis ... 188
12.2.6 Data Analysis ... 191
12.3 Result and Discussion ... 191
12.3.1 Socio-demographic Characteristics of Households ... 191
12.3.2 Utilization of *Awaze* ... 192
12.3.3 Adjusted Associations ... 194
12.3.4 Materials Used by Women for *Awaze* Storage ... 195
12.3.5 Antioxidant Activity, Physicochemical, and Microbiological Quality of Collected *Awaze* Samples ... 195
12.4 Conclusion ... 199
References ... 200

13 Microwave-Assisted Hydrothermal Pretreatment of Coffee Husk: Process Parameter Optimization Using Response Surface Methodology ... 203
Gashaw Muluken Tessera, Nigus Gabbiye Habtu, Metadel Kassahun Abera, Kefyalew Dagnew Addisu, and Fantahun Woldesenbet Misganaw
13.1 Introduction ... 203
13.2 Materials and Method ... 205
13.2.1 Sample Collection and Preparation ... 205

13.2.2 Pretreatment of Coffee Husk Biomass 205
13.2.3 Experimental Design 206
13.2.4 Characterization of Coffee Husk Biomass 207
13.2.5 Statistical Analysis 209
13.3 Results and Discussion 209
13.3.1 Compositional Analysis of Coffee Husk 209
13.3.2 Structural Changes in the Coffee Husk Biomass 210
13.3.3 Model Fittingness Analysis 212
13.3.4 Statistical Analysis of the Experimental Design 213
13.3.5 Effect of Process Factors 216
13.3.6 Validation of Optimal Data 219
13.3.7 Evaluation of Pretreatment Efficiency 219
13.4 Conclusion 220
Annex: Statistical Tables 220
References 221

14 Formulation, Extraction, and Characterization of Biopesticides from Birbira (*Milletia ferruginea*) Seed and Neem (*Azadirachta indica*) Leaf 225
Mequanint Demeke Aynalem, Surafel Aregahegn Agdew, Ayalneh Adnew Alemayehu, Menber Berihun Nigatu, Addisu Getahun Adugna, Asmare Tezera Admassie, Kindie Bitew Worku, and Alebel Abebaw Teshager
14.1 Introduction 225
14.2 Materials and Methods 227
14.2.1 Equipment and Instrument 227
14.2.2 Chemicals and Reagents 227
14.2.3 Experimental Setup and Description 227
14.2.4 Methods 228
14.2.5 Formulations of Biopesticides 228
14.2.6 Extraction and Recovery of Biopesticides 229
14.3 Raw Materials and Product Characterization 230
14.3.1 Determination of Moisture Content 230
14.3.2 Determination of Percentage Yield 230
14.3.3 Determination of Specific Gravity 230
14.3.4 Determination of P^HValues 231
14.3.5 Thin Layer Chromatography (TLC) Analysis of Biopesticide 231
14.3.6 FTIR Analysis of Biopesticide 231
14.3.7 Determination of Total Phenolic Content (TPC) 231
14.3.8 Determination of Total Flavonoids Content (TFC) 232
14.3.9 Efficacy of Biopesticides on Worker Termites 232
14.4 Results and Discussion 233
14.4.1 Raw Material Characterization 233
14.5 Physicochemical Characteristics of Biopesticides 233
14.5.1 Determination of Biopesticide P^H Values 233

14.5.2 Specific Gravity of Organic Biopesticide. 234
14.5.3 Percentage Yield of Biopesticide 235
14.5.4 Thin Layer Chromatography (TLC) Biopesticide Analysis. 235
14.5.5 Fourier-Transform Infrared Spectroscopy Analysis of Biopesticide 236
14.5.6 Determination of Phenolic Content 237
14.5.7 Determination of Flavonoid Content 238
14.5.8 Efficacy of Biopesticides on Worker Termites. 238
14.6 Conclusions 239
References. 240

15 Formulation and Development of Organic Hair Shampoo from Discharged Beer in Filler Machine (Brewery Industry) and Aloe Vera Plant Gel 243
Mequanint Demeke Aynalem, Tegen Dagnew Tessema, Surafel Aregahegn Agdew, Tessafa Abrham Ashagrie, Belete Adane Mandie, Asmare Tezera Admassie, Smegnew Moges Mintesinot, and Desalegn Abera Mersha
15.1 Introduction 243
15.2 Materials and Methods. 245
15.2.1 Equipment and Instrument. 245
15.2.2 Chemicals and Reagents 245
15.2.3 Experimental Setup and Description 245
15.3 Methods 246
15.3.1 Collection and Preparation of Raw Materials 246
15.3.2 Preparation of Aloe Vera Gel/Oil 246
15.3.3 Organic Shampoo Formulation 247
15.4 Product Characterization 247
15.4.1 Physical Appearance (Visual Inspection). 247
15.4.2 Determination of P^H 248
15.4.3 Determination of Percentage Solid Contents. 248
15.4.4 Dirt Dispersion Test 248
15.4.5 Evaluation of Foaming Ability and Stability 248
15.4.6 Microbial Analysis of Organic Shampoo. 248
15.5 Results and Discussion 249
15.5.1 Aloe Vera Gel Characterization 249
15.5.2 Determination of Shampoo P^H Values 250
15.5.3 Foaming Ability and Foam Height. 251
15.5.4 Percentage of Solid Content. 252
15.5.5 Dirt Dispersion. 253
15.5.6 Physical Appearance/Visual Inspection 253
15.5.7 Microbial Analysis of Organic Shampoo. 254
15.6 Conclusion 254
References. 255

16 **Formulation and experimental evaluation of antidandruff shampoo from leaf-extracted herbs of *Aloe vera* (Eret), *Ocimum sanctum* (Damakase), and *Withania somnifera* (Gizewa): An outlook as a herbal medicine** 259
Tegen Dagnew Tessema, Yohannes Shetawn Wendante, Asmare Tezera Admassie, Mequanint Demeke Aynalem, Asnakew Baye Workie, Desalegn Abera Mersha, Alemu Engashu Wubie, Haile Mariam Mulusaw Minalu, Abraham Sintayehu Siyum, Bereded Gedamu Eshetie, and Behailu Bisenebt Mussie
16.1 Introduction 259
16.1.1 Background of the study and knowledge gaps. 259
16.2 Materials and Methods. 261
16.2.1 Materials, Reagents, and Equipment Used 261
16.2.2 Equipments and Apparatus Used 261
16.2.3 Collection of Plant Materials and Preparation for Extraction 262
16.2.4 Extraction of Plant Materials 262
16.2.5 Formulation of Herbal Shampoo 262
16.2.6 Strains and Media Preparation 263
16.2.7 Experimental Design 263
16.2.8 Evaluation of Antidandruff Herbal Shampoo. 265
16.3 Results and Discussion 267
16.3.1 Physical Appearance and pH 267
16.3.2 Evaluation of Shampoo for Foamability and Foam Stability 267
16.3.3 Dirt Dispersion and Percentage of Solid Content 268
16.3.4 Viscosity. 268
16.3.5 Antidandruff Activity Evaluations 269
16.3.6 Statistical analysis in One-Way ANOVA 270
16.3.7 FT-IR Analysis of Oil. 271
16.4 Concluding Remarks 275
References. 276

17 **Recent Advances in Microalgae Biomass Drying, Pretreatment, and Extraction of Polyunsaturated Fatty Acids (PUFAs): A Review**. 279
Tsegaye Mekuria Ayele, Metadel Kassahun Abera, Tsehayneh Geremew Yohannes, Gizachew Muluneh Amera, and Amha Belay Sahlu
17.1 Introduction 279
17.2 Prospects of Microalgae as Alternative Source of PUFAs. 280
17.2.1 Microalgae with Potential of PUFA Production 281
17.2.2 Microalgae Used at Industrial Scale Production of PUFAs 281

17.3 Methods of Drying Microalgae Biomass.................... 283
17.4 Biomass Pretreatment Techniques in PUFA Extraction.......... 283
17.4.1 Bead Milling.................................. 284
17.4.2 A High-Speed Homogenizer (HSH) 285
17.4.3 High-Pressure Homogenization (HPH) 285
17.4.4 Ultrasonication................................ 285
17.4.5 Microwaves 285
17.4.6 Pulsed Electric Field (PEF) 286
17.4.7 Enzymatic Lysis................................ 286
17.4.8 Osmotic Shock................................ 286
17.5 Extraction of PUFAs from Microalgae Biomass.............. 287
17.5.1 Conventional PUFA Extraction Methods and Their Drawbacks............................. 288
17.5.2 Alternative Solvents............................. 288
17.6 Conclusions 292
References... 293

18 Effect of Spontaneous Sedimentation on the Chemical Composition of Compressed Water Hyacinth Juice as a Pretreatment for Anaerobic Digestion: Comparison of Different Growing Areas 297
Pranshu Bhatia, Anas Hijazi, Shinichi Akizuki, Nigus Gabbiye Habtu, Shinjiro Sato, and Tatsuki Toda
18.1 Introduction 297
18.2 Material and Methods 299
18.2.1 Substrate Collection............................. 299
18.2.2 Sedimentation Pretreatment...................... 299
18.2.3 Analytical Methods 300
18.2.4 Calculations 301
18.3 Results and Discussions................................ 302
18.3.1 Effect of Sedimentation Pretreatment on the Suspended Solids 302
18.3.2 Effect of Sedimentation Pretreatment on the Chemical Composition of Different Types of WHJ..... 305
18.3.3 Comparison of Different Types of Water Hyacinth Juice 311
18.4 Conclusion.. 311
References... 312

19 Treatment of Water Hyacinth Juice in an Upflow Anaerobic Digester: Process Stability and Quality of Digestate's Macronutrients 317
Shinichi Akizuki, Pranshu Bhatia, Nigus Gabbiye Habtu, Tatsuki Toda, and Shinjiro Sato
19.1 Introduction 317

19.2 Materials and Methods. 319
19.2.1 Preparation of Substrate and Seed Sludge 319
19.2.2 Reactor Operation and Experimental Description 319
19.2.3 Analytical Parameters 320
19.2.4 Calculations 321
19.3 Results and Discussion 322
19.3.1 Process Stability and Reactor Performance 322
19.3.2 Changes in the Macronutrient Concentration After Passing Through the Reactor 324
19.3.3 Evaluation of the Digestate Quality as an Alternative to the Microalgae Cultivation Medium 327
19.4 Conclusions 328
References. 329

20 Sustainable Energy Catalyzed Rural Entrepreneurship for Smallholder Farmers in Ethiopia Through a Living Lab Approach 333
Tena Tegbar
20.1 Introduction 333
20.2 Research Methodology 335
20.3 Results and Discussion 336
20.4 Conclusions and Recommendations 337
References. 339

21 Contract Time Determination and Work Item Scheduling in Public Building Projects: Case Study in Bahir Dar 341
Befekadu C. Gashe, Asregidew K. Woldesenbet, Solomon M. Belay, and Habtamu E. Endalew
21.1 Introduction 341
21.2 Methodology 343
21.2.1 Study Design 343
21.2.2 Study Area 344
21.2.3 Sample Size/Research Population 344
21.3 Results and Discussion 345
21.3.1 Current Practice of Contract Time Determination Procedures 345
21.3.2 Accuracy of Contract Time 346
21.3.3 Factors Affecting Contract Time 347
21.3.4 Linear Regression Models for Productivity Rate. 347
21.3.5 Model Mean Validation 350
21.4 Case Study 352
21.4.1 General Overview of Selected Projects 352
21.5 Conclusion 356
References. 358

22 Assessment of Selected Physicochemical Parameters of Selamco Dam Water, South Gondar, Ethiopia . . . 359
Menber Berihun Nigatu, Mengistu Berihun Nigatu, Minaleshewa Atlabachew, Mequanint Demeke Aynalem, and Kindie Bitew Worku
22.1 Introduction . . . 359
22.2 Methodology . . . 360
22.2.1 Study Area Description . . . 360
22.2.2 Apparatus and Chemicals . . . 361
22.2.3 Sample Collection . . . 361
22.2.4 Sample Preparation . . . 362
22.2.5 Method Validation . . . 363
22.2.6 Instrumentation . . . 363
22.2.7 Data Analysis . . . 364
22.3 Analysis of Metals . . . 364
22.3.1 Validation of Experimental Procedures . . . 364
22.4 Result and Discussion . . . 368
22.4.1 Physical Parameters . . . 368
22.4.2 Analyzed Concentration of Metals . . . 371
22.4.3 Discussions for the Heavy Metal Results . . . 371
22.5 Conclusion . . . 374
References . . . 374

23 Evaluating Flood Vulnerability in Bahir Dar City Through an Indicator-Based Method . . . 377
Elias S. Leggesse, Walelign A. Derseh, Afework A. Simegn, and Mengiste A. Meshesha
23.1 Introduction . . . 377
23.2 Description of The Project Area . . . 378
23.3 Methodology . . . 380
23.3.1 Vulnerability Dimensions and Indicators . . . 382
23.3.2 Quantification of Vulnerability Dimensions and Indicators . . . 383
23.3.3 Assessment Framework . . . 385
23.3.4 Spatial Pattern of Vulnerability Indices . . . 387
23.4 Result and Discussion . . . 388
23.4.1 Physical Vulnerability (PhyVI) . . . 388
23.4.2 Social Vulnerability (SoVI) . . . 389
23.4.3 Economical Vulnerability (EcoVI) . . . 389
23.4.4 Combined Vulnerability (ComVI) . . . 389
23.5 Conclusions and Recommendations . . . 390
References . . . 391

24 Hydrological Impact of Khat Cultivation in Aba Gerima Watershed, Lake Tana Basin, Ethiopia 395
Kindie Bitew Worku, Dagnenet Sultan, Nigussie Haregeweyn, Mulatu Liyew Berihun, Menber Berihun Nigatu, and Mequanint Demeke Aynalem
24.1 Introduction 395
24.2 Methods 396
24.2.1 Study Area Description 396
24.2.2 Experimental Setup, Data Collection, and Analysis 397
24.2.3 Rainfall Measurement and Climate Data 399
24.2.4 Soil Data Analysis 399
24.2.5 Plot Water balance Determination 400
24.2.6 Rainfall, Runoff, and Soil Loss Measurement 400
24.2.7 Moisture Data Measurement 402
24.2.8 Reference Evapotranspiration (ETO) 402
24.3 Results and Discussions 403
24.3.1 Effect of Khat on Seasonal Runoff, Runoff Coefficient, and Soil Loss 403
24.3.2 Soil Moisture Availability 407
24.3.3 Water Balance in the Khat Field 409
24.4 Conclusions 410
References 411

25 Surface Water Quality Assessment of Chilanchil Abay Watershed: The Case of Bahir-Dar City Waste Disposal Site 413
Dargie Haile and Nigus Gabbiye
25.1 Introduction 413
25.2 Materials and Methods 414
25.2.1 Descriptions of the Study Area 414
25.2.2 Sample Collection, Preservation, and Laboratory Analyses 415
25.3 Results and Discussions 416
25.3.1 Spatial and Temporal Variation of Surface Water Quality 416
25.3.2 Spatial and Temporal Variation in the Measured Values of Nutrients and Microbiological Parameters of Surface Water Quality 426
25.3.3 Spatial and Temporal Variation in the Concentrations of Heavy Metals of Surface Water Quality 430
25.4 Conclusion 432

Appendices 433
Appendix A: Graphs of Spatial and Seasonal Variation of Physical Parameters of Surface Water 433
Appendix B: Graphs of Spatial and Seasonal Variation of Chemical Parameters of Surface Water 434
Appendix C: Graphs of Spatial and Seasonal Variation of Nutrients and Biological Parameters of Surface Water 435
Appendix D: Graphs of Spatial and Seasonal Variation of Heavy Metal Concentration in Surface Water 436
References. 436

Index. 439

Chapter 1
Impact of Manufacturing Strategies on the Effectiveness of Mass Customization

Mehari Kumilachew Bezuneh, Assefa Asmare Tsegaw, Bereket Haile Woldegiorgis, Teshome Mulatie Bogale, Matthias Brossog, and Jörg Franke

1.1 Introduction

The manufacturing industry is currently confronting significant computational issues due to globalization, market dynamics, and variations in customers' interests. As a result, the sector is shifting from mass production to mass customization (MC) [1]. In the modern industrial environment, MC is being viewed as an emerging technique to increase the competitiveness of the manufacturing industry by developing customized products with additional value to meet the customers' individual needs [2, 3]. Due to its ability to provide customized products efficiently and effectively, MC has gained increasing importance in recent years, and manufacturing companies are continuously searching for ways to develop their MC abilities [4].

Mass customization enables manufacturers to achieve multiple competing priorities simultaneously and resolve trade-offs between customization and costs [5, 6]. Indeed, researchers report that the implementation of MC improves various performance outcomes, such as value to customers, customer satisfaction, and operational performance in terms of product cost, product quality, delivery time, flexibility, product innovation, and firm performance [5, 7].

For the success of MC, according to da Silveira, customer demand for variety must exist, market conditions must be appropriate, value chains should be ready, technology must be available, products should be customizable, and knowledge must be shared [8]. Manufacturing strategies or systems, like manufacturing processes and methodologies, are the main elements needed to make the MC strategy successful. Mainly, agile manufacturing, supply chain management, lean

M. K. Bezuneh (✉) · A. A. Tsegaw · B. H. Woldegiorgis · T. M. Bogale
Bahir Dar Institute of Technology, Bahir Dar University, Bahir Dar, Ethiopia

M. Brossog · J. Franke
Friedrich-Alexander-Universität Erlangen-Nürnberg, Erlangen, Germany

S. W. Fanta et al. (eds.), *Sustainable Development Research in Manufacturing, Process Engineering, Green Infrastructure, and Water Resources*, Green Energy and Technology, https://doi.org/10.1007/978-3-031-77339-6_1

manufacturing, customer-driven design and manufacturing, flexible manufacturing systems [9, 10], and cellular manufacturing (group technology) are the basic enablers of MC [11].

Despite the fact that the MC strategy enables achieving multiple competing priorities simultaneously, like cost, quality, delivery time, variety, and product volume, by using manufacturing strategy enabling factors, the literature does not clearly define the relative importance of each enabling factor on each capability to maximize the efficiency and effectiveness of MC. As a result, this research attempted to identify the enabling factors in the manufacturing strategy or system and analyze the degree to which each enabling factor impacts each MC's capabilities using the fuzzy Delphi approach.

This paper is structured as follows: Sect. 1.2 reviews work on MC; Sect. 1.3 deals with MC capabilities; Sect. 1.4 focuses on research methodology; Sect. 1.5 contains results and discussions; and Sect. 1.6 deals with conclusions and recommendations.

1.2 Review of Literature on Mass Customization

Different scholars define the concept of "mass customization"as follows: The term "mass customization" was first defined by Stan Davis in the 1987 book *Future Perfect* as "the same large number of customers can be reached as in mass markets, while simultaneously they can be treated individually as in customized markets" [12]. Pine defined MC as "the ability to produce customized items at a low cost, high quality, and large volume of individually personalized products" [13]. Tseng and Jiao defined MC as "the technologies and systems to deliver goods and services that meet individual customers' needs with near mass production efficiency" [14]. MC refers to the capability to reliably offer a high volume of different product options in a relatively large market that demands customization without substantial trade-offs in cost, delivery, and quality [15]. MC is the ability to quickly produce customized products and services in large volumes at a low cost, high quality, and minimum delivery time comparable to that achieved by mass production [16]. According to Tu and Sandrin, MC consists of five elements: high-volume customization, customization cost-effectiveness, customization responsiveness, reliable quality, and customization quality [17, 18].

Three primary concepts serve as the foundation of mass customization systems. First, production systems can deliver a greater variety at a reduced cost using flexible manufacturing and information technology. Second, there is an increasing demand for product variety and customization (even segmented markets are now too broad to permit the development of niche strategies). Finally, the shortening of product life cycles and expanding industrial competition have led to the breakdown of many mass industries, increasing the need for production strategies focused on individual customers [19].

Today, mass customization is widespread across different industries to better meet customer requirements. There are many examples of industries that are pursuing a mass customization strategy: electronics, automotive, healthcare, housing construction [20, 21], apparel [22], mobile phones and computers, etc. [23].

1.2.1 Capabilities of Mass Customization

The capability of MC can be defined as its effectiveness, which addresses how well MC is implemented to achieve multiple competing priorities simultaneously [15, 24, 25]. MC capability can help firms quickly and cheaply manufacture a variety of products and services that meet specific customer needs [26]. According to da Silveira, MC capability is measured by the ability of the firm to achieve more variety, high volume, low cost, high quality, and fast delivery of products simultaneously [8]. Some other literature [17, 27] defines MC capability as the capability to reliably produce a high volume of different product options for a relatively large market and rapidly adjust product and process design in response to customers' needs without significant cost, delivery, or quality trade-offs. Sheng described the MC capabilities of a manufacturing firm in terms of producing quality customized products on a large scale in a time, and cost-efficient manner [26]. According to [17, 28], MC capability can be defined as the ability to consistently produce a high volume of a variety of product options for a relatively large market and to quickly adjust product and process design according to customer demand, without substantial trade-offs in cost, delivery, or quality.

The five basic potential dimensions of MC capabilities, as stated by [29], are as follows:

Volume customization: Customizing large batches of products based on individual customer needs.

Customization cost efficiency: Producing products without significantly increasing their costs compared to mass production.

Customization responsiveness: The ability to satisfy customized needs with a proper total lead time.

Customization of quality: The quality assurance of the products when the variety of products suddenly increases.

Customization of variety: The ability to produce a variety of products in high volume without increasing their cost.

1.3 Enablers of MC Capabilities

The primary success criteria for MC are the existence of customer desire for variety and customization, the right market conditions, the readiness of value chains, the availability of technology, the ability to customize products, and the sharing of information [9]. The following four areas are identified as enablers for MC, namely, methodologies (agile and lean), manufacturing processes (design, manufacturing, and supply chain) [30], manufacturing technologies (advanced manufacturing technologies like 3D), and information technologies for MC [31]. Agile manufacturing, supply chain management, customer-driven design and manufacture, lean manufacturing, cellular manufacturing (group technology) [32], and flexible manufacturing systems are basic types of manufacturing processes and methodologies that enable the factors of MC [33]. Standardization, modularity, and commonality are also other types of MC enablers [34]. For this research, we focused on analyzing the impact or relative importance of process and methodological enablers, specifically lean manufacturing, agile manufacturing, flexible manufacturing, cellular manufacturing, and customer-driven manufacturing, to improve MC capabilities, namely, product variety, delivery time, cost, quality, and volume.

1.3.1 Lean Manufacturing

Lean manufacturing can be defined as waste disposal in any manufacturing sector, which is an effective and popular tool to tackle non-valued activities and waste [35]. Originally, it was focused on the elimination of wastes as defects, such as those requiring rework, unnecessary processing steps, movement of materials or people, waiting time, excess inventory, and overproduction [36]. The ultimate goal of implementing lean production is to enhance product quality, increase productivity, reduce production costs, and be first to market and quick to respond to customers' needs, which are the basic factors that determine the competitiveness of the manufacturing industry [37].

1.3.2 Agile Manufacturing

Agile manufacturing is the ability to adjust to quickly changing markets that are driven by customer-based product valuation in order to thrive and prosper in a competitive environment of unpredictable change [29, 38]. The implementation of agile manufacturing has some benefits for firms, including quick and efficient reactions to changing market demands, the ability to customize products and services delivered to customers, the capability to manufacture and deliver new products in a cost-efficient model, decreased production costs, enhanced customer satisfaction, the elimination of non-value-added activities, and increased competitiveness [39, 40].

1.3.3 Manufacturing Flexibility

Manufacturing flexibility is the capacity of a manufacturing system to adapt successfully to changing environmental conditions as well as changing product and process requirements. It is the ability of the manufacturing function to make the adjustments needed to react to environmental changes without significant sacrifices to firm performance [41, 42]. Flexibility provides the manufacturing plant with the ability to maintain customer satisfaction and profitability under conditions of change and uncertainty [42]. According to Newman, flexibility aims to provide a balance between the requirements of customers in terms of customized products or services and the ability of operations processes and suppliers to deliver such customized products or services within the required levels of quality, cost, time, variety, and volume [43, 44].

1.3.4 Cellular Manufacturing (Group Technology)

Cellular manufacturing is an application of group technology in manufacturing and involves the processing of a collection of similar parts (part families) on dedicated clusters of machines or manufacturing processes (cells) [45]. The main advantages of implementing cellular manufacturing or group technology are that work in process and flow time are reduced, improving market response, reducing setup times, reducing tooling costs, reducing space utilization, and simplifying production planning and control. Other advantages include the simplification of handling materials, reducing inventory, and shortening throughput times [38, 46]. Cellular manufacturing is important for the effectiveness of product quality, variety, cost, volume, and responsiveness [47, 48].

1.3.5 Customer-Driven Design and Manufacturing

Customer-driven design and manufacturing practices consider individual customer requirements during the design and manufacturing of a product [49, 50]. Lack of customer knowledge about the product and customer ignorance about his or her real individual needs are the key challenges related to the uncertainty encountered by customers when they intend to customize their products. Customer-driven manufacturing promises to the customer must always be kept (time, quality, and cost); there is a need for speed and flexibility; the order book shrinks while delivery times shorten; there is outsourcing; there are supply and value networks; and there are productivity improvements from supply chain management and coordination [51].

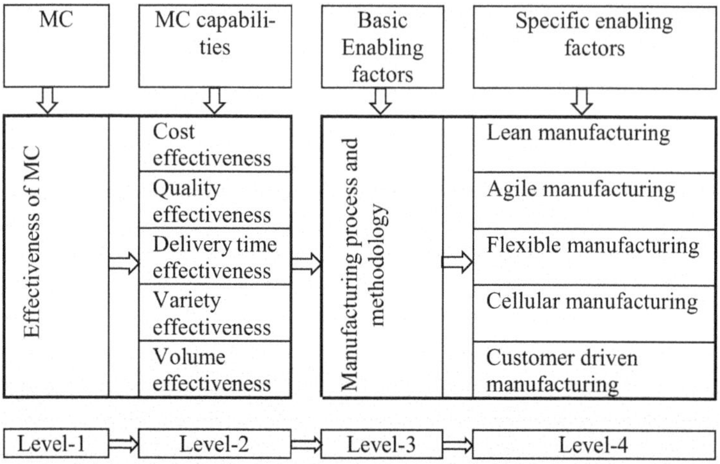

Fig. 1.1 The integration between the effectiveness of MC, MC capabilities, and enabling factors

From the above literature, it can be concluded that the efficiency and effectiveness of MC are measured by the system's ability to achieve the five potential dimensions of MC capabilities, namely, cost-effectiveness, quality effectiveness, delivery time effectiveness, variety effectiveness, and volume effectiveness. The effectiveness of MC capabilities is directly or indirectly affected by the manufacturing process and methodological enabling factors. Most literature agrees that lean manufacturing, agile manufacturing, flexible manufacturing, cellular manufacturing or group technology, and customer-driven design and manufacturing systems are specific enabling factors for MC, which are classified under manufacturing strategy enabling factors. Therefore, it can be concluded that these specific enabling factors can directly affect the effectiveness of MC as a whole. The integration of the effectiveness of MC, MC capabilities, basic enabling factors, and specific enabling factors are shown in Fig. 1.1.

1.4 Methodology

1.4.1 Data Collection and Analysis

The fuzzy Delphi method is used for this research to compare the impact level or relative importance of manufacturing processes and methodology enabling factors on MC capabilities. This study used an expert-based questionnaire survey to collect the opinions of the experts. The questionnaire was prepared based on five enabling factors for each of the five MC capabilities, which totals 25 questionnaires. We conducted 25 interviews, of which 22 experts properly completed the questionnaire, and the response rate was 88%. The expert profile includes their job, academic qualifications, and work experience (see Table 1.1).

Table 1.1 Expert profiles include the expert's job, qualifications, and work experience

Expert's job	No	Percentage	Qualification	No	Percentage	Experience	No	Percentage
University Professors.	10	45.5	Ph.D. (professors)	4	18.2	>20	7	31.82
Researcher	5	22.7	M.Sc	18	81.8	15–20	3	13.64
Ind. Manager	3	13.6	B.Sc	0		10–14	7	31.82
Ind. Expertise	4	18.2				5–9	3	13.64
						<5	2	9.09
Total	22	100	Total	22	100	Total	22	100

Table 1.2 Linguistic variables with their corresponding linguistic values [52]

Linguistic variables with importance weight	Linguistics variables with their effects	Triangular fuzzy numbers (TFNS)
Extremely low	Extremely low importance (EL)	(0.0, 0.0, 0.1)
Very low	Very low importance (VL)	(0.0, 0.1, 0.3)
Low	Low importance (L)	(0.1, 0.3, 0.5)
Moderate	Moderate importance (M)	(0.3, 0.5, 0.7)
High	High importance (H)	(0.5, 0.7, 0.9)
Very high	Very high importance (VH)	(0.7, 0.9, 1)
Extremely high	Extremely high importance (EH)	(0.9, 1, 1)

We asked the experts to rate the impact level of each manufacturing process and methodology enabling factor on MC capabilities using a seven-point Likert-type scale with their respective linguistic terms [52] (see Table 1.2).

Because they are intuitive, easily applicable, and useful in promoting representation and information processing in a fuzzy environment, this study applied linguistic variables with their respective triangular fuzzy numbers [53]. The fuzzy Delphi approach was used in this study to get fuzzy group opinions from experts on each indication. The detailed process is presented in Fig. 1.2.

This research applied the similarity and difference-based approaches equally, or the consistency aggregation method, to determine the group consensus through the following algorithm using [54].

Step 1: For each expert *Ei* ($i = 1, 2, \ldots, n$), construct a triangular fuzzy number to represent each opinion to the alternative under a given criterion, i.e., $EOi = (ai, bi, ci)$, which are the minimum (a), optimum (b), and maximum (c), where, n is the total number of decision-makers or experts participating in the survey and EOi is the *i*th expert's opinion.

Step 2: Determine the similarity $S\,(EOi, EOj)$ between experts using Eq. (1.1) [55]:

$$S(EOi,EOj) = 1 - \left[\left(/ai - aj/ + /bi - bj/ + /ci - cj/\right)/3\right] \tag{1.1}$$

Step 3: Calculate the distance difference $D\,(EOi, EOj)$ and the normalized distance $d\,(EOi, EOj)$ between each pair of experts by using Eqs. (1.2 and 1.3):

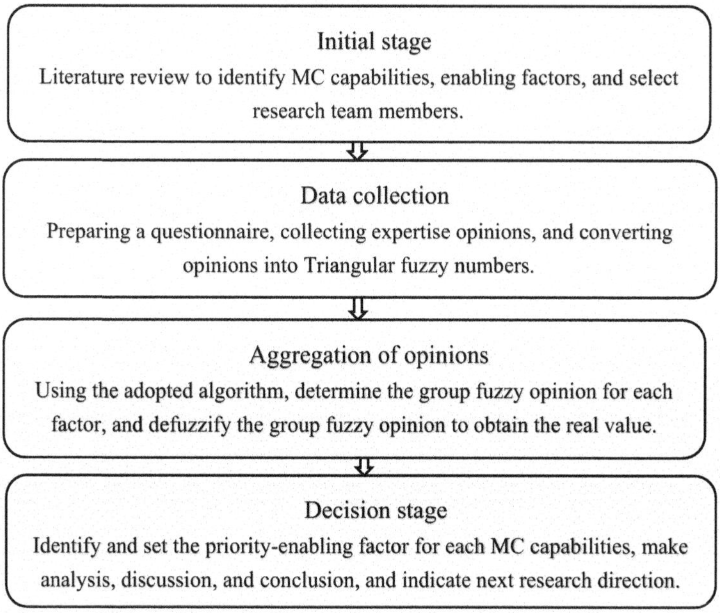

Fig. 1.2 Research flowchart using fuzzy Delphi method

$$D(EOi,EOj) = \frac{1}{2}\{\max(/ai - aj/, /ci - cj/) + /bi - bj/\} \tag{1.2}$$

Using the result from Eq. (1.3), the normalized distance (*d*) can be determined as follows:

$$d(E0i,E0j) = D(EOi,EOj) / \max i,j\, D(EOi,EOj) \tag{1.3}$$

Step 4: Calculate the consistency degree *r* (*EOi*, *EOj*) using Eq. (1.4). The calculated consistency degree is shown in Table 1.3.

$$r(EOi,EOj) = \alpha S(EOi,EOj) + (1-\alpha)[1 - d(EOi,EOj)] \tag{1.4}$$

Step 5: Determine the level of significance of each expert by considering their work experience and academic qualifications.

Step 6: Calculate the weighted consistency degree *C* (*Ei*) using Eq. (1.5):

$$C(Ei) = \sum_{j=1}^{n} r(EOi,EOj)\, ej \tag{1.5}$$

Table 1.3 The consistency degree between each expert's opinions

Experts	EO1	EO2	EO3	EO4	EO5	EO6	EO7
EO1	1	0.87	0.68	0.48	0.27	0.11	0.04
EO2	0.87	1	0.81	0.61	0.40	0.24	0.11
EO3	0.68	0.81	1	0.80	0.59	0.40	0.27
EO4	0.48	0.61	0.80	1	0.80	0.61	0.48
EO5	0.27	0.40	0.59	0.80	1	0.81	0.63
EO6	0.11	0.24	0.40	0.61	0.81	1	0.82
EO7	0.04	0.11	0.27	0.48	0.63	0.82	1

Table 1.4 Experts' degree of importance, consistency degree, and combined weight

Experts	Degree of importance	Consistency deg. C (Ei)	Combined weight w (Ei)	Experts	Degree of importance	Consistency Deg. C (Ei)	Combined weight w (Ei)
E1	0.0541	10.2441	0.0472	E12	0.0270	10.5476	0.0486
E2	0.0676	10.0979	0.0465	E13	0.0541	10.2441	0.0472
E3	0.0405	10.3914	0.0478	E14	0.0270	10.5376	0.0485
E4	0.0338	10.4639	0.0482	E15	0.0270	10.5376	0.0485
E5	0.0541	10.2441	0.0472	E16	0.0473	10.3177	0.0475
E6	0.0405	10.3914	0.0478	E17	0.0338	10.4639	0.0482
E7	0.0541	10.2441	0.0472	E18	0.0473	10.3177	0.0475
E8	0.0405	10.3914	0.0478	E19	0.0473	10.3177	0.0475
E9	0.0405	10.3914	0.0478	E20	0.0405	10.3914	0.0478
E10	0.0473	10.3177	0.0475	E21	0.0541	10.2441	0.0472
E11	0.0676	10.0979	0.0465	E22	0.0541	10.2441	0.0472

Step 7: Determine the aggregate weight $w\,(Ei)$ of the decision-maker using Eq. (1.6).

$$w(Ei) = \frac{C(Ei)}{\sum_{i=1}^{n} C(Ei)} \tag{1.6}$$

The results of experts' degree of importance, consistency degree, and combined weight are shown in Table 1.4.

Step 8: Aggregate each fuzzy opinion for each indicator into a group fuzzy opinion R_I using Eq. (1.7) [56].

$$R1 = \sum_{i=0}^{n} w(Ei).)EOi \tag{1.7}$$

$$R_I = \left\{ \begin{pmatrix} w(E1)a1 + w(E2)a2 + \ldots + w(En)an, \begin{pmatrix} w(E1)b1 + w(E2)b2 + \ldots \\ +w(En)bn \end{pmatrix} \\ ,(w(E1)c1) + w(E2)c2 \end{pmatrix} \right.$$

$$+\ldots + W(En)cn)\}$$

$R_I = (aI, bI, cI)$ *For* $I = 1, 2, 3..., 22$

Step 9: Transform the fuzzy group opinion value into real values (i.e., defuzzification) using Eq. (1.8). The summary of the results is shown in Table 1.5.

$$\text{Defuzzified} = \frac{\text{a1} + \text{b1} + \text{c1}}{3} \tag{1.8}$$

1.4.2 Reliability of Questionnaire

The reliability of the questionnaire is used to determine to what extent the data are consistent. The most frequently used reliability measure is the Cronbach alpha coefficient [56]. For this purpose, the study determined Cronbach's alpha for a sample of 22 experts and 25 items (questions) by using SPSS software. The value of Cronbach's alpha obtained was 0.832, which is greater than the commonly used value (0.7).

1.5 Results and Discussion

This study aimed to identify the manufacturing process and methodological enabling factors that improve, directly or indirectly, the basic potential of MC capabilities. Additionally, it aimed to determine the relative importance of the enabling factors using the data analysis of questionnaires completed by experts. Five enabling factors and five MC capabilities were identified, and 25 questionnaires were prepared. The number of experts who participated in the survey was 25. Of the 25 experts, 22 responded correctly and their responses were used in the analysis. The study applied expert-based judgment using the fuzzy Delphi method. An algorithm that combines the aggregation of similarity and difference-based expert opinions into a group consensus was applied in the analysis. Table 1.5 presents the results depicting the relative importance of each enabling factor on various MC capabilities.

A detailed description that shows the relative importance of each enabling factor on each capability is shown in Fig. 1.3.

As the result shown in Table 1.5 and Fig. 1.3

(a) Lean manufacturing: From the five MC capability indicators, applying lean manufacturing to the MC strategy has the highest value (0.8897) on product cost and a lower value (0.6820) on product variety. This indicates that applying lean manufacturing in MC has a relatively high effect on minimizing the cost of the product, but on the other side, lean manufacturing has a relatively lower effect on maximizing the variety of the product.

Table 1.5 Combined group opinions of enabling factors on MC capabilities

Basic enabling factor (Ef)	Specific enabling factor (Ef)	MC capability indicators (CI)	Effect of Ef on CI (EC)	Group fuzzy opinion			
				Minimum	Optimum	Maximum	Defuzzified value
Manufacturing methodology and process enabling factors	Lean manufacturing (Ef1)	Cost-effectiveness (CI1),	(EC11)	0.7324	0.9182	1.0185	0.8897
			(EC12)	0.5716	0.7572	0.8903	0.7397
			(EC13)	0.5530	0.7576	0.7900	0.7002
			(EC14)	0.4853	0.6899	0.8707	0.6820
			(EC15)	0.5714	0.7712	0.8055	0.7161
	Agile manufacturing (Ef2)	Quality effectiveness (CI2),	(EC21)	0.4761	0.6760	0.8519	0.6680
			(EC22)	0.5521	0.7565	0.9042	0.7376
			(EC23)	0.5908	0.7811	0.8949	0.7556
			(EC24)	0.7051	0.8812	0.9952	0.8605
			(EC25)	0.5146	0.7097	0.7877	0.6707
	Flexible manufacturing (Ef3)	Time effectiveness (CI3),	(EC31)	0.4425	0.6424	0.8238	0.6363
			(EC32)	0.4942	0.7130	0.8847	0.6973
			(EC33)	0.5906	0.7809	0.8805	0.7506
			(EC34)	0.7623	0.9426	1.0329	0.9126
			(EC35)	0.6193	0.8096	0.8117	0.7469
	Cellular manufacturing (Ef4)	Variety effectiveness (CI4),	(EC41)	0.5574	0.7429	0.9045	0.7349
			(EC42)	0.5707	0.7751	0.9040	0.7499
			(EC43)	0.6282	0.8233	0.9183	0.7899
			(EC44)	0.5044	0.7043	0.8805	0.6964
			(EC45)	0.6186	0.8137	0.7879	0.7823
	Customer-driven manufacturing (Ef5)	Volume effectiveness (CI5).	(EC51)	0.4186	0.6232	0.8186	0.6201
			(EC52)	0.6767	0.8716	0.9810	0.8431
			(EC53)	0.4431	0.6429	0.7854	0.6238
			(EC54)	0.7808	0.9468	1.0279	0.9185
			(EC55)	0.4233	0.6184	0.7582	0.6000

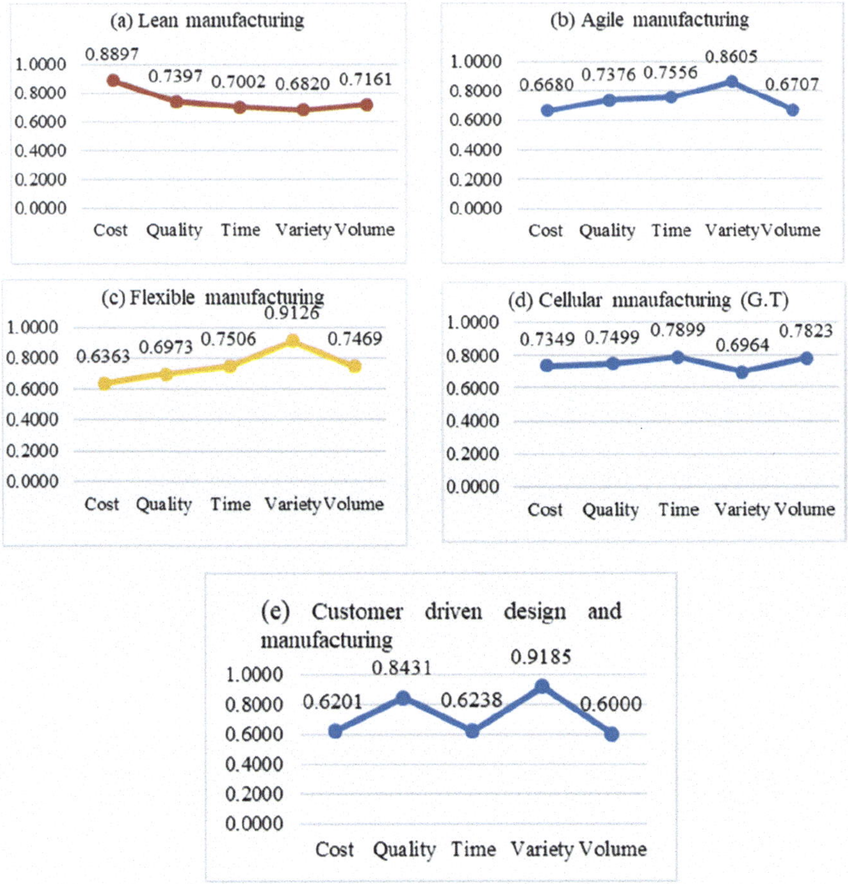

Fig. 1.3 Relative effect of each of the enabling factors on each of the MC capabilities

(b) Agile manufacturing: Applying agile manufacturing to the MC strategy has a higher value on product variety (0.8605) and a lower value on product cost (0.6680) out of the five MC capability indicators. This shows that implementing agile manufacturing in MC manufacturing has a relatively large impact on maximizing product variety, but a relatively smaller effect on minimizing product cost.

(c) Flexible manufacturing: Using flexible manufacturing in an MC strategy scores higher on product variety (0.7899) and lower on product cost (0.6363). This demonstrates that applying flexible manufacturing to the MC strategy has a relatively high influence on increasing product variety but a slight impact on reducing product cost.

(d) Cellular manufacturing or (G.T): Utilizing cellular manufacturing or group technology as an MC strategy results in a higher value for product volume (0.9126) and a lower value for product variety (0.6964). This illustrates that integrating cellular manufacturing or group technology into MC manufacturing strategy has a modest impact on lowering product variety while having a relatively high impact on boosting product volume.

(e) Customer-driven design and manufacturing: Applying customer-driven design and manufacturing to the MC strategy has the best value (0.9185) on product variety but a lower value (0.6000) on increasing production volume. This shows that implementing a customer-driven design and manufacturing system in the MC strategy has a comparatively high effect on maximizing product variety, while on the other hand, implementing a customer-driven design, and manufacturing system has a relatively low effect on increasing product volume.

1.6 Conclusion and Recommendation

Achieving an effective MC manufacturing strategy is a complex problem that can be measured by MC capability indicators. The identification of manufacturing process and methodology enabling factors as relatively important to maximize the effectiveness of the capability indicators represents an effort to minimize the complexity of MC and increase its effectiveness. In this study, we identified five MC capability indicators (effectiveness of product cost, delivery time, product quality, product variety, and product volume) and five manufacturing process and methodology enabling factors (lean manufacturing, agile manufacturing, flexible manufacturing, cellular manufacturing or group technology, and customer design and manufacturing). Using fuzzy Delphi, a questionnaire based on linguistic variables was designed to collect experts' opinions on the relative effect of each enabling factor on the MC capability indicators in order to maximize the effectiveness of MC. The consistency of the data was examined using Cronbach's alpha, and in this study, Cronbach's alpha of the data was 0.832, which is greater than the minimum acceptable value (0.7).

From the results discussed earlier, it can be concluded that all the identified manufacturing strategy-enabling factors have their own effect on MC capabilities. Nevertheless, it is clear that each of the enabling factors has different relative effects on different capabilities. This indicates that selecting the manufacturing strategies enabling factor depends on the capabilities of MC, on which the manufacturing industry focuses.

This study has a contribution to make in improving the effectiveness of MC capabilities to maximize the overall MC manufacturing strategy. In addition, the result of the study will guide the manufacturing industries to apply different manufacturing processes and methodologies, which will improve directly and indirectly the manufacturing strategy.

As a recommendation, the current study focused on identifying enabling factors, which are included only in the manufacturing process and methodology. For future research directions, it would be interesting to focus on other MC enabling factors that have direct or indirect effects on the capability indicators, such as technological factors and others.

References

1. Hu, S. J. (2013). Evolving paradigms of manufacturing: From mass production to mass customization and personalization. *Procedia CIRP, 7*, 3–8.
2. Zhang, C., Chen, D., Tao, F., & Liu, A. (2019). Data driven smart customization. *Procedia CIRP, 81*, 564–569.
3. Jain, P., Garg, S., & Kansal, G. (2022). Implementation of mass customization for competitive advantage in Indian industries: An empirical investigation. *International Journal of Advanced Manufacturing Technology, 121*, 737–752.
4. Wiengarten, F., Singh, P. J., Fynes, B., & Nazarpour, A. (2017). Impact of mass customization on cost and flexiblity performances: The role of social capital. *Operations Management Research, 10*, 137–147.
5. Zhang, M., Qi, Y., Zhao, X., & Duray, R. (2015). Mass customisation systems: Complementarities and performance consequences. *International Journal of Logistics Research and Applications, 18*, 459–475.
6. Mahamood, R. M., & Akinlabi, E. T. (2016). Achieving mass customization through additive manufacturing. In C. Schlick & S. Trzcieliński (Eds.), *Advances in ergonomics of manufacturing: Managing the enterprise of the future* (pp. 385–390). Springer. https://doi.org/10.1007/978-3-319-41697-7_34
7. Zhang, M., Lettice, F., & Zhao, X. (2015). The impact of social capital on mass customisation and product innovation capabilities. *International Journal of Production Research, 53*, 5251–5264.
8. Da Silveira, G., Borenstein, D., & Fogliatto, F. S. (2001). Mass customization: Literature review and research directions. *International Journal of Production Economics, 72*, 1–13.
9. Fogliatto, F. S., da Silveira, G. J. C., & Borenstein, D. (2012). The mass customization decade: An updated review of the literature. *International Journal of Production Economics, 138*, 14–25.
10. Smith, S., Smith, G. C., Jiao, R., & Chu, C.-H. (2013). Mass customization in the product life cycle. *Journal of Intelligent Manufacturing, 24*, 877–885. https://doi.org/10.1007/s10845-012-0691-0
11. Maalouf, E., Daaboul, J., Le Duigou, J., & Hussein, B. (2022). Production management for mass customization and smart cellular manufacturing system: NSGAII and SMPSO for factory-level planning. *International Journal of Advanced Manufacturing Technology, 120*, 6833–6854.
12. Schwartz, H. S. (1989). "Future perfect", by Stanley M. Davis (book review). *Human Resource Management, 28*, 145.
13. Christy, D. P. (1992). Mass customisation: The new frontier in business competition by B. Joseph Pine II, 1993 (Boston, Harvard Business School Press) US$29.95, hardback, pp.xxi + 333. *Australian Journal of Management, 13*.
14. Tseng, M. M., & Jiao, J. (2001). Mass customization, handbook of industrial engineering. *Technology and Operation Management*.
15. Liu, G., Shah, R., & Schroeder, R. G. (2006). Linking work design to mass customization: A sociotechnical systems perspective. *Decision Sciences, 37*, 519–545.
16. MacCarthy, B., Brabazon, P. G., & Bramham, J. (2003). Fundamental modes of operation for mass customization. *International Journal of Production Economics, 85*, 289–304.
17. Liu, G., Shah, R., & Babakus, E. (2012). When to mass customize: The impact of environmental uncertainty. *Decision Sciences, 43*, 851–887.
18. Lai, F., Zhang, M., Lee, D. M., & Zhao, X. (2012). The impact of supply chain integration on mass customization capability: An extended resource-based view. *IEEE Transactions on Engineering Management, 59*, 443–456.
19. Mourtzis, D. (2016). Challenges and future perspectives for the life cycle of manufacturing networks in the mass customisation era. *Logistics Research, 9*, 2.

20. Gandhi, A., Magar, C., & Roberts, R. (2013). How technology can drive the next wave of mass customization, p. 8.
21. Shin, Y., An, S.-H., Cho, H.-H., Kim, G.-H., & Kang, K.-I. (2008). Application of information technology for mass customization in the housing construction industry in Korea. *Automation in Construction, 17*, 831–838.
22. Liu, N., Chow, P.-S., & Zhao, H. (2020). Challenges and critical successful factors for apparel mass customization operations: Recent development and case study. *Annals of Operations Research, 291*, 531–563.
23. Wang, Y., Ma, H.-S., Yang, J.-H., & Wang, K.-S. (2017). Industry 4.0: A way from mass customization to mass personalization production. *Advanced Manufacturing, 5*, 311–320.
24. Huang, X., Kristal, M. M., & Schroeder, R. G. (2008). Linking learning and effective process implementation to mass customization capability. *Journal of Operations Management, 26*, 714–729.
25. Kortmann, S., Gelhard, C., Zimmermann, C., & Piller, F. T. (2014). Linking strategic flexibility and operational efficiency: The mediating role of ambidextrous operational capabilities. *Journal of Operations Management, 32*, 475–490.
26. Sheng, H., Feng, T., & Chen, L. (2020). Motives and performance outcomes of mass customization capability: Evidence from Chinese manufacturers. *Journal of Manufacturing Technology Management.*
27. Trentin, A., Forza, C., & Perin, E. (2015). Embeddedness and path dependence of organizational capabilities for mass customization and green management: A longitudinal case study in the machinery industry. *International Journal of Production Economics, 169*, 253–276.
28. Zhang, M., Zhao, X., & Qi, Y. (2014). The effects of organizational flatness, coordination, and product modularity on mass customization capability. *International Journal of Production Economics, 158*, 145–155.
29. Qi, Y., Mao, Z., Zhang, M., & Guo, H. (2020). Manufacturing practices and servitization: The role of mass customization and product innovation capabilities. *International Journal of Production Economics, 228*, 107747.
30. Gosling, J., & Naim, M. M. (2009). Engineer-to-order supply chain management: A literature review and research agenda. *International Journal of Production Economics, 122*, 741–754.
31. Daanen, H., & Hong, S. (2008). Made-to-measure pattern development based on 3D whole body scans. *International Journal of Clothing Science and Technology, 20*, 15–25.
32. Duray, R., Ward, P. T., Milligan, G. W., & Berry, W. L. (2000). Approaches to mass customization: Configurations and empirical validation. *Journal of Operations Management, 18*, 605–625.
33. Suzić, N., Forza, C., Trentin, A., & Anišić, Z. (2018). Implementation guidelines for mass customization: Current characteristics and suggestions for improvement. *Production Planning & Control, 29*, 856–871.
34. Steiner, F., & Hergenröther, I. (2014). Modular product architectures as an enabler of the simultaneous application of a mass customisation strategy and efficient ramp-up management. *International Journal of Product Development, 19*, 231–253.
35. Rahman, N. A. A., Sharif, S. M., & Esa, M. M. (2013). Lean manufacturing case study with Kanban system implementation. *Procedia Economics and Finance, 7*, 174–180.
36. Mrugalska, B., & Wyrwicka, M. K. (2017). Towards lean production in industry 4.0. *Procedia Engineering, 182*, 466–473.
37. Bevilacqua, M., Ciarapica, F. E., & De Sanctis, I. (2017). Lean practices implementation and their relationships with operational responsiveness and company performance: An Italian study. *International Journal of Production Research, 55*, 769–794.
38. Renna, P., & Ambrico, M. (2011). Evaluation of cellular manufacturing configurations in dynamic conditions using simulation. *The International Journal of Advanced Manufacturing Technology, 56*, 1235–1251.
39. Leite, M., & Braz, V. (2016). Agile manufacturing practices for new product development: Industrial case studies. *Journal of Manufacturing Technology Management, 27*, 560–576.

40. Gunasekaran, A., Yusuf, Y. Y., Adeleye, E. O., & Papadopoulos, T. (2018). Agile manufacturing practices: The role of big data and business analytics with multiple case studies. *International Journal of Production Research, 56*, 385–397.
41. D'Souza, D. E., & Williams, F. P. (2000). Toward a taxonomy of manufacturing flexibility dimensions. *Journal of Operations Management, 18*, 577–593.
42. Swamidass, P. M. (2000). Manufacturing flexibility. In P. M. Swamidass (Ed.), *Innovations in competitive manufacturing* (pp. 117–136). Springer.
43. Brettel, M., Klein, M., & Friederichsen, N. (2016). The relevance of manufacturing flexibility in the context of Industrie 4.0. *Procedia CIRP, 41*, 105–110.
44. Ojha, D., White, R. E., & Rogers, P. P. (2013). Managing demand variability using requisite variety for improved workflow and operational performance: The role of manufacturing flexibility. *International Journal of Production Research, 51*, 2915–2934.
45. Javadian, N., Aghajani, A., Rezaeian, J., & Ghaneian Sebdani, M. J. (2011). A multi-objective integrated cellular manufacturing systems design with dynamic system reconfiguration. *The International Journal of Advanced Manufacturing Technology, 56*, 307–317.
46. Askin, R. G. (2013). Contributions to the design and analysis of cellular manufacturing systems. *International Journal of Production Research, 51*, 6778–6787.
47. Tu, Q., Vonderembse, M. A., & Ragu-Nathan, T. S. (2004). Manufacturing practices: Antecedents to mass customization. *Production Planning & Control, 15*, 373–380.
48. Um, J., Lyons, A., Lam, H. K. S., Cheng, T. C. E., & Dominguez-Pery, C. (2017). Product variety management and supply chain performance: A capability perspective on their relationships and competitiveness implications. *International Journal of Production Economics, 187*, 15–26.
49. Cristiano, J. J., Liker, J. K., & White, C. C., III. (2000). Customer-driven product development through quality function deployment in the US and Japan. *Journal of Product Innovation Management, 17*, 286–308.
50. Liu, C., Ramirez-Serrano, A., & Yin, G. (2011). Customer-driven product design and evaluation method for collaborative design environments. *Journal of Intelligent Manufacturing, 22*, 751–764.
51. Heilala, J., Montonen, J., Järvinen, P., & Kivikunnas, S. (2010). Decision support using simulation for customer-driven manufacturing system design and operations planning. In *Advances in decision support systems* (pp. 235–260).
52. Habibi, A., Jahantigh, F. F., & Sarafrazi, A. (2015). Fuzzy Delphi technique for forecasting and screening items. *Asian Journal of Research in Business Economics and Management, 5*, 130–143.
53. Zhang, L., Xu, X., & Tao, L. (2013). Some similarity measures for triangular fuzzy number and their applications in multiple criteria group decision-making. *Journal of Applied Mathematics, 2013*.
54. Lu, C., Lan, J., & Wang, Z. (2006). Aggregation of fuzzy opinions under group decision-making based on similarity and distance. *Journal of Systems Science and Complexity, 19*, 63–71.
55. Chen, S.-M. (2010). Aggregating fuzzy opinions in the group decision-making environment. *Cybernetics & Systems*.
56. Sanchez-Lezama, A. P., Cavazos-Arroyo, J., & Albavera-Hernandez, C. (2014). Applying the Fuzzy Delphi method for determining socio-ecological factors that influence adherence to mammography screening in rural areas of Mexico. *Cad Saúde Pública, 30*, 245–258.

Chapter 2
Detecting Misalignments of Drilled Holes Using Machine Vision

Getachew Ambaye, Enkhsaikhan Boldsaikhan, and Krishna Krishnan

2.1 Introduction

The manufacturing industry specializes in producing quality parts, ranging from small components to large equipment. As a result, rigorous inspections of these products are imperative for the success of the trade. Inspection ensures the quality of parts at every stage of production, helping to avoid recalls and reputational damage while optimizing the quality control budget.

Different inspection methods have been implemented, particularly in the Industry 4.0 era where quality is of utmost importance. Manufacturing industries focus on producing products and services of superior quality to maintain competitiveness among end-users. Machine vision [1, 2] and robots [3] play a significant role in ensuring product quality. They are involved in direct inspections by utilizing inspection devices such as laser scanners and machine vision, or in picking and placing parts along the production line.

Inspection is an organized exercise that involves measurements, tests, and gauges applied to specific characteristics, which are then compared with standards or preset values. This inspection can be either destructive (DTE) or nondestructive (NDT). In the former, physical parts undergo physical damage during inspection or testing procedures, while the latter enables the inspection of parts without causing any physical damage. Various methods can be employed for detecting defective parts, including machine vision, ultrasonic testing, filtering, and others.

Drilling holes to join parts is a prevalent machining activity in many industries. The types of holes required, such as blind holes, through holes, interrupted holes,

G. Ambaye (✉) · E. Boldsaikhan · K. Krishnan
Department of Industrial, Systems and Manufacturing Engineering, Wichita State University, Wichita, KS, USA
e-mail: gaambaye@shockers.wichita.edu

S. W. Fanta et al. (eds.), *Sustainable Development Research in Manufacturing, Process Engineering, Green Infrastructure, and Water Resources*, Green Energy and Technology, https://doi.org/10.1007/978-3-031-77339-6_2

counter-bore holes, and others, depend on the specific application. These holes can be created using contact methods, utilizing rotor machines with a drill bit, or non-contact methods like laser drilling. Both tools can be mounted on the robot's tool center point and operated by robots. Various techniques are employed to drill holes in sheet metal or plates, aiming to enhance hole quality. Zhang et al. [4] focused on improving the surface condition of laser-drilled holes.

The location and shape of the holes serve as indicators of product quality and the precision of drilling machines. In industries such as aerospace, where millions of holes are required for riveted joints, robots assist in drilling these holes [5]. The quality of the holes depends on parameters such as cutting force, tool geometry, workpiece material, and coating. Krol et al. [6] investigated the influence of feed rate and material type, which are drilling conditions, on hole coordinate deviations in wood-based materials. Their findings revealed that drill deviation increases with feed rate and material hardness. The slenderness of the hole and difficulties during the hole initiation process have a less significant influence. Apart from the hole's center location, quality characteristics include hole size, circularity, burrs, and surface roughness. The inherent vibration from moving machine components significantly affects hole quality [7]. Vibration impacts product quality in two ways: it can displace the workpiece or cutting tool from the intended position, and it can cause wear on the cutting tool, consequently affecting hole quality [8]. Geometric deviations in holes are influenced by three elements of the drilling process: locators, the workpiece datum surface, and the machine tool. Polini and Corrado [9] explored a general method for estimating location deviations in drilling. Additionally, Shen et al. [10] proposed an improved hole position accuracy in robotic drilling processes.

Numerous methodologies have been implemented to detect and ensure product quality. For instance, Wang et al. [11] utilized laser edge detection to measure the bending deflection of a screw rod. Edge detection, a crucial process in the modern world, involves identifying and locating sharp discontinuities in an image, representing sudden changes in pixel concentration. It is employed to distinguish object boundaries in a scene.

Various edge detection methods have been employed for different applications. For instance, Luo et al. [12] used the Sobel method for workpiece defect detection, Basavaraju et al. [13] utilized the Laplacian of Gaussian method for text detection and segmentation, and He et al. [14] employed the Canny method for monitoring running trains, among others. Edge detection finds application in a wide range of areas, including vehicle paint defect detection [15–18], obstacle detection for drones [19], aircraft skin detection [20], lane detection in construction [21], and bridge crack detection [22]. The state-of-the-art research on defect detection based on machine vision is extensively reviewed in the following papers: Ren et al. [23] and Luo et al. [24].

He et al. [25] utilized a convolutional neural network (CNN) to investigate surface defects on steel plates, achieving an accuracy of 99.67% for defect classification. Tabernik et al. [26] also applied a deep learning approach for surface defect detection on plates, achieving good accuracy. Tan et al. [27] achieved a 99.95% accuracy in die defect edge detection. The application of edge detection systems finds use in numerous industries, including ceramic crack detection [28–30], liquid level detection in the food and beverage industry [31], stability analysis of injections in casting by Zhou et al. [32], turning tool wear [33], self-assembly robots [34], blade edge measurement [35], surface defect detection [36], and component identification such as bolts and nuts [37]. Additionally, Du et al. [38] employed the Canny-Zernike sub-pixel edge detection method to investigate the roundness of shaft end faces.

Holistically-nested edge detection (HED) is also the most efficient method of edge detection, which combines the fully CNN with deep supervision to learn image edges effectively. It is employed for chair ergonomic detection [39], 3D pavement crack detection [40], and detection of the wet/dry shoreline at fish pass [41]. However, this method requires training and computational resources due to its deep learning approach. On the other hand, the Canny edge detection method offers simplicity, fast computation, and good results for general-purpose edge detection tasks. Moreover, the Canny method can be employed for real-time edge detection, which is particularly important in certain applications.

Nowadays, computer-aided design (CAD) and computer-aided engineering (CAE) [42, 43] models are increasingly being utilized for developing datasets in image processing and machine learning applications. Tang et al. [44] employed CAD models for object detection using edge detection techniques. In their study, datasets were created using CATIA V5 software and preprocessed to minimize image size. In addition, in our recent study, the responses obtained by simulation using ABAQUS have been analyzed to detect the crack location, size, and orientation of a robot link [45]. This approach opens up possibilities for detecting deviations in the center of holes during the machining process. Moreover, it can be integrated into a direct production system by implementing a tolerable difference method. This would involve making decisions based on image processing and utilizing robots to identify faulty parts on the production line.

The mean difference of the models can be detected by observing pixel variations. However, by defining an acceptable tolerance of deviation and using a threshold value, such deviations can be deemed acceptable. In practical situations, the color of the workpiece may vary due to inefficient coating, dust and dirt, and variations in surface finish and edge quality. Consequently, numerous image processing operations need to be performed to identify differences solely based on edges. One approach to address this is by selecting a suitable perfect model that closely resembles the model to be tested. This ensures that the evaluation of the models is conducted using a model with a similar color and surface finish as the one being tested.

2.2 Modeling and Methods

A 400 mm by 250 mm plate with a thickness of 5 mm was modeled using CATIA V5. The reference drilled hole center is positioned at a distance of 40 mm from both the horizontal and vertical directions, as illustrated in Fig. 2.1a. The remaining 14

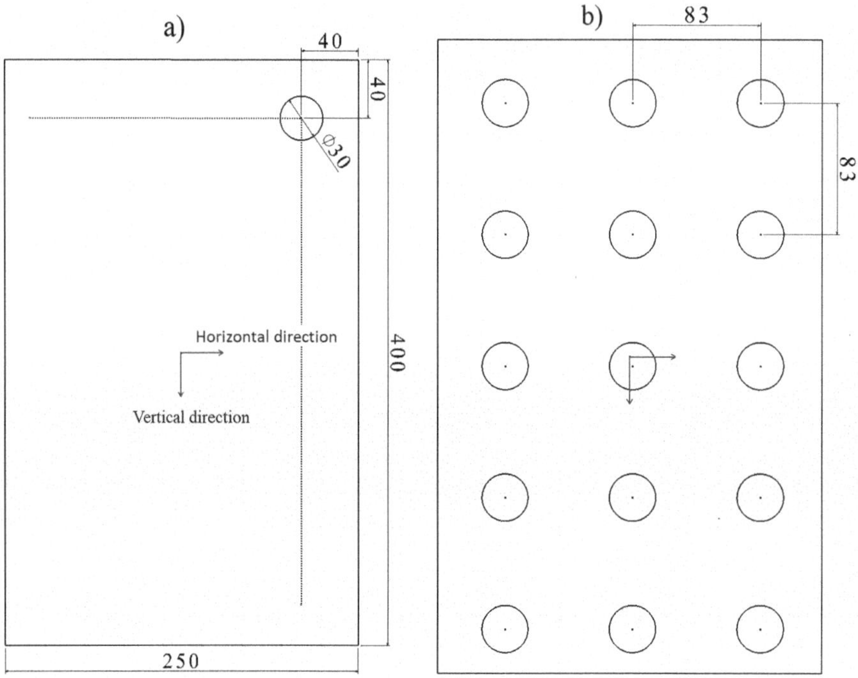

Fig. 2.1 Basic model: (**a**) with reference hole, (**b**) perfect model with a maximum spacing

holes will be drilled using a rectangular pattern, with spacing distances ranging from 80 mm to 83 mm in increments of 0.1 mm. This results in a total of 91 datasets, considering the variation in hole spacing relative to the reference drilled hole.

The minimum and maximum spacing distances between the center of the holes, relative to the reference hole, are 83 mm and 83.3 mm, respectively, along the vertical and horizontal directions. In the perfect model (depicted in Fig. 2.1b), the spacing distance between holes is 83 mm in both the horizontal and vertical directions. To prepare the test model datasets, 14 holes are positioned by incrementally changing the spacing from 80 mm to 83 mm relative to the reference hole. This is done in only one of the axes: horizontal, vertical, or diagonal directions. The deviation in hole location across the three directions results in a total of 90 datasets.

The image datasets were prepared by exporting them as bitmap (BMP) extensions from the modeling software. To ensure consistency in pixel similarity, the software modeling workspace was maintained with constant settings. Subsequently, the exported images were loaded into Python and underwent appropriate image preprocessing. The original shape of each dataset was 897 by 1880 pixels, as illustrated in Fig. 2.2a. However, the images were subsequently cropped to 360 rows and 200 columns, as shown in Fig. 2.2b. This cropping was done to reduce the size of the datasets. Furthermore, in order to further decrease the dataset size, the BMP extension image files were converted to portable network graphics (PNG) extension files using Python. This conversion was performed with the aim of reducing the overall size of the datasets.

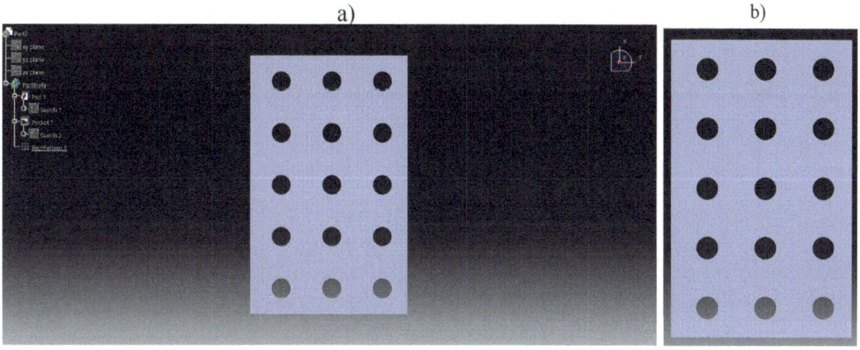

Fig. 2.2 The datasets sample: (**a**) before image preprocessing, (**b**) after coring the image

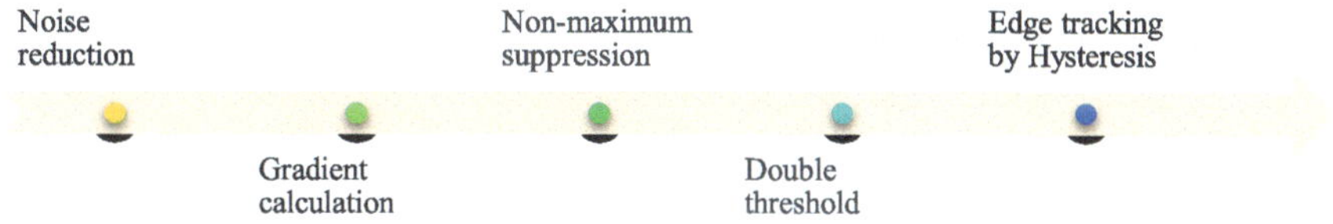

Fig. 2.3 General procedure of common edge detection method

Most edge detection algorithms typically involve five common steps, as depicted in Fig. 2.3: noise reduction, gradient calculation, non-maximum suppression, double thresholding, and edge tracking by hysteresis. It is important to note that edge detection processes are highly sensitive to image noise. To mitigate the impact of noise on the image, a common approach is to apply Gaussian blur, which smoothens the image and reduces the noise effect.

The general procedure followed in this chapter is illustrated in Fig. 2.4. The dataset models undergo a preprocessing operation, which involves converting the file types and cropping both the perfect model and the models to be tested.

Using machine vision, the edges of the perfect model and test model are compared, allowing any deviation in the hole location to be clearly identified. The programming part is designed to estimate the mean difference between the models if there is a deviation, and it will label the location of the difference using a bounding box. In cases where there is no mean difference, the mean difference will be zero, resulting in a dark image.

2.3 Results and Discussions

Edge detection plays a crucial role in detecting hole deviations. It encompasses various mathematical techniques that aim to identify edges or curves where there are significant changes in image brightness or discontinuities. Several approaches have been employed for successful edge detection, including the Canny algorithm, which

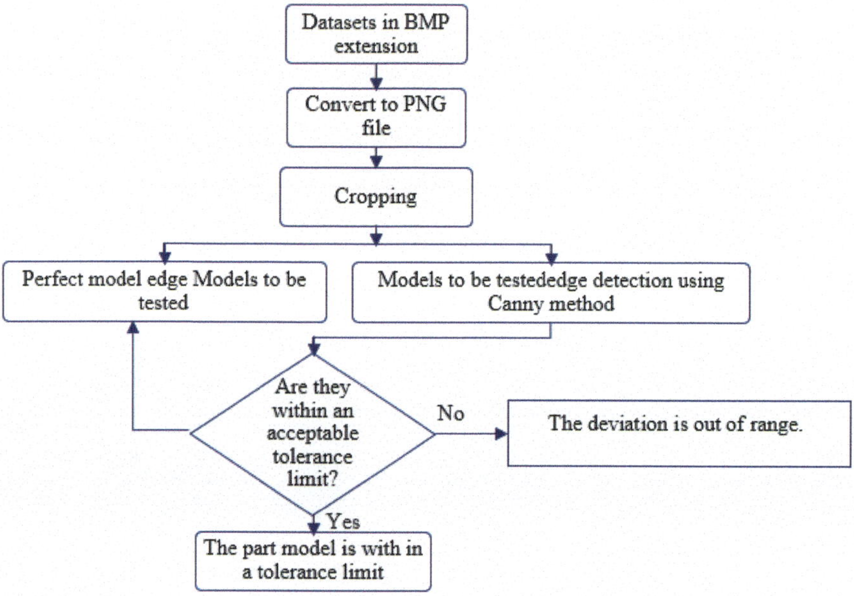

Fig. 2.4 General procedure for detecting deviation of location of hole

is a highly effective and complex multi-stage method compared to others. It is capable of identifying a wide range of edges. Another method is Laplacian detection, which performs second-order derivatives, making it sensitive to noise.

2.3.1 Edge Detection

The Canny edge detection method was used for this study, as depicted in Fig. 2.5a. The image was initially converted to grayscale, followed by the application of the Canny edge detection algorithm, as shown in Fig. 2.5b. This process reveals the edges of the models, enabling the analysis of differences. To improve detection efficiency and reduce dimensionality, the models were transformed into grayscale images. The implementation of this method was carried out using Python, utilizing the OpenCV library function.

The model without deviation is compared with the perfect model (depicted in Fig. 2.6a), and no differences are observed. The resulting output is a fully dark or blank image, as illustrated in Fig. 2.6b. This darkness indicates that there are no variations in the location of the hole.

In contrast, the model with a deviation from its original location exhibits clear white edge curves. Figure 2.7 displays a model with a 0.1 mm deviation, indicating a change in the spacing between holes from 80 mm to 80.1 mm in different axes.

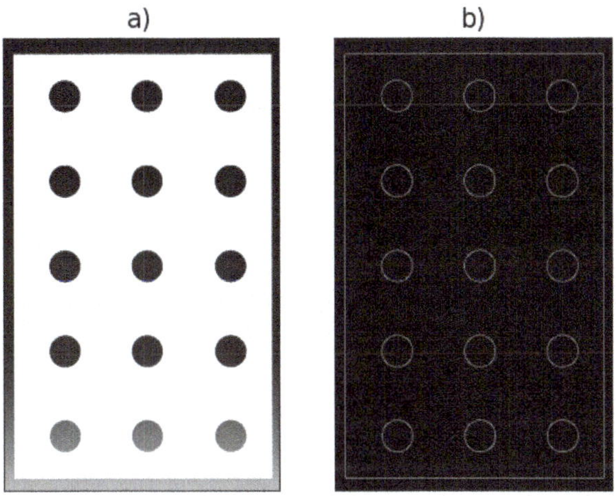

Fig. 2.5 Outputs: (**a**) gray and (**b**) Canny edge detection

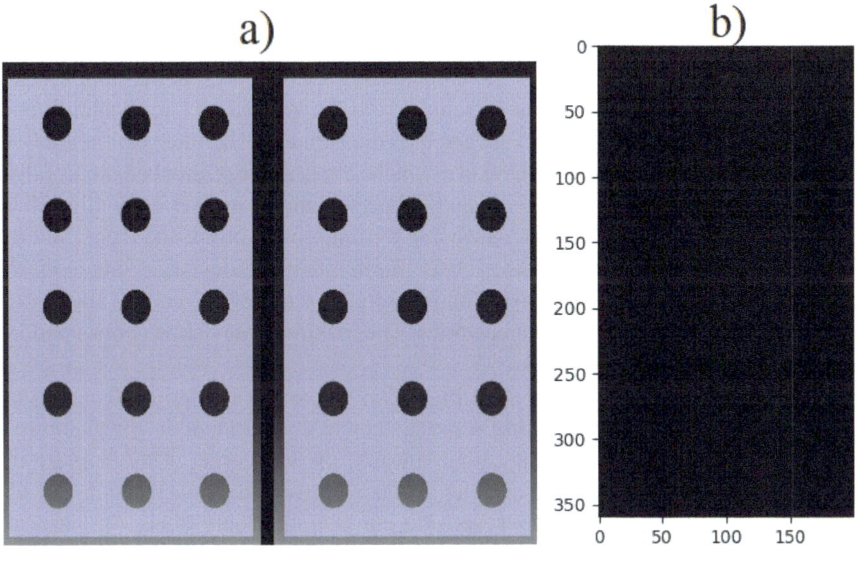

Fig. 2.6 Models with no deviation: (**a**) perfect and test model, (**b**) mean differences

While there is a hole in the top right corner, serving as a reference point for other holes, it is not visible in all axes because there is no deviation. The direction of the deviation may not be readily apparent due to the small magnitude of the deviation. However, as the deviation distance increases, the direction of the deviation becomes easily detectable to the naked eye.

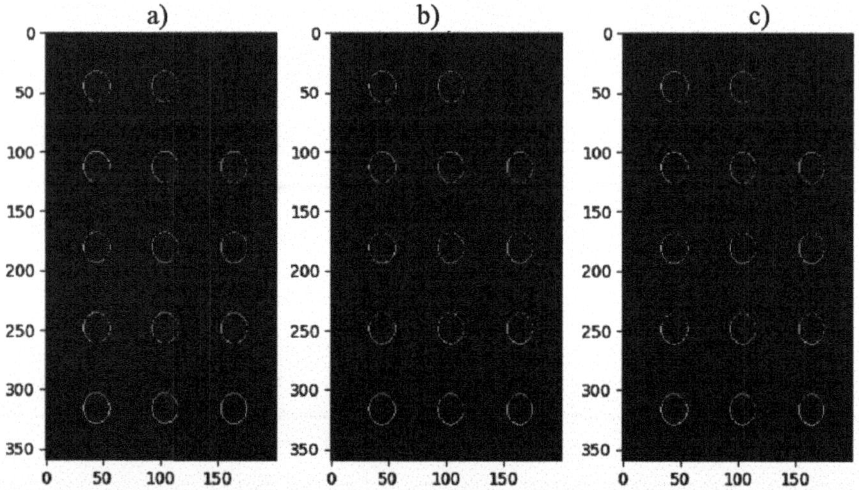

Fig. 2.7 Mean differences between perfect model and test model deviated by 0.1 mm in: (**a**) minor axis, (**b**) major axis, (**c**) both axes

Figure 2.8 illustrates the CAD model depicting the deviation of the hole's location when the spacing distance is 83 mm in both the vertical and horizontal directions. The amount of deviation is inversely proportional to the intersection area of the holes. In the image, the top right hole, which serves as the reference hole, is fully shaded indicating that the test model and the perfect model are perfectly aligned at that particular location. However, as we move from top to bottom and from right to left, the intersection area decreases. This implies that the deviation increases in these directions. In section line AA, the magnitude of deviation remains the same, but the shape of the intersection area varies. The maximum deviation is observed in the lower left hole.

The OpenCV result is depicted in Fig. 2.9a, where the amount of deviation is represented by the weight of the line. It is evident that the weight of the line increases when moving from the top right corner towards the left corner. This increase in weight is due to the variation in spacing distance between the hole's center and its intended position. Similarly, when considering deviation only along the vertical direction, the weight of the line also increases from the top corner to the bottom corner, as shown in Fig. 2.9b. In the third case, where the spacing distance is increased along a 45° angle by 0.1 mm increments, the weight of the line is significantly larger compared to the previous two cases. This is attributed to the existence of deviation in both the vertical and horizontal directions. Figure 2.9c demonstrates a similar pattern to the aforementioned cases.

The shape of the labeled white differences curves varies, as illustrated in Fig. 2.10. The results indicate that the presence of a gap or a contour with low line weight differs as we move from left to right. The direction of the gap varies and is directly correlated with the magnitude of the deviation. This correlation is also evident in Fig. 2.8, where the shaded area (intersection) corresponds to the dark area in Fig. 2.10.

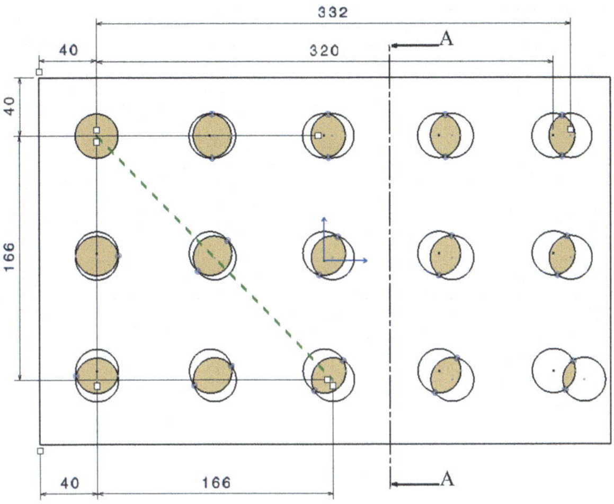

Fig. 2.8 Deviation between perfect model and test model which deviated 3 mm

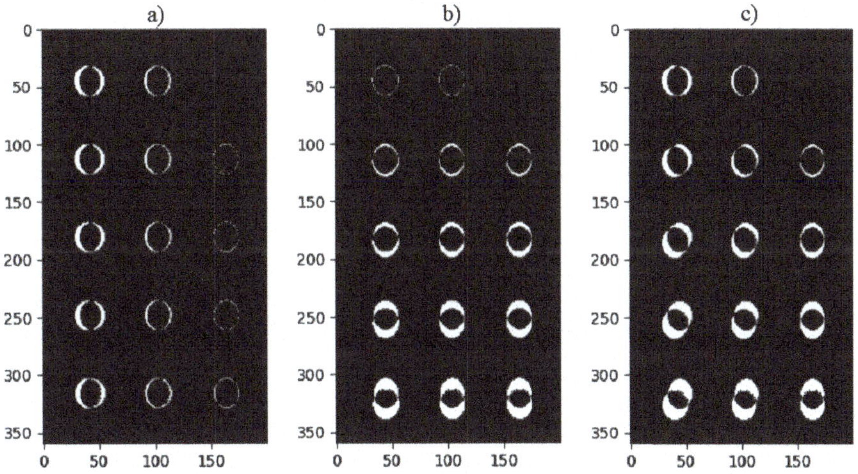

Fig. 2.9 Mean differences between perfect model and deviated model by 3 mm: (**a**) in horizontal directions, (**b**) in vertical directions, (**c**) in diagonal directions

2.3.2 Mean Square Error (MSE)

The mean square error (MSE) quantifies the quality of the predictor by evaluating the difference between the perfect model pixel ($P_{perfect}$) and the tested pixel model (P_{test}), as shown in Eq. (2.1).

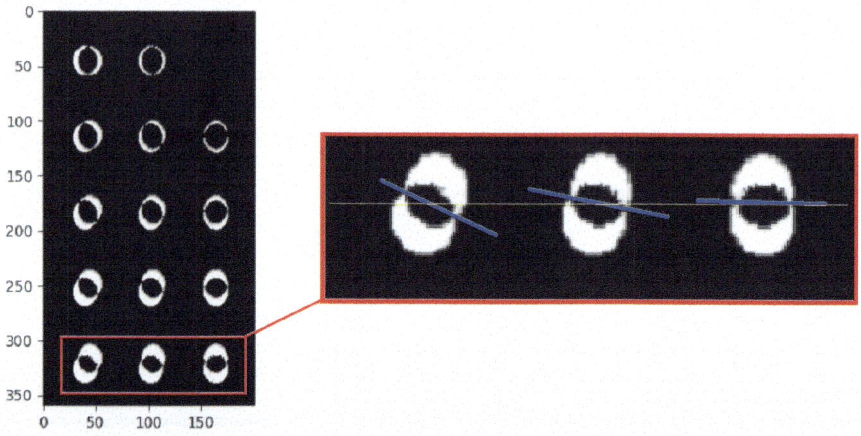

Fig. 2.10 Mean differences between perfect models

$$\mathrm{MSE} = \frac{1}{N}\sum_{i=1}^{N}\left(P_{\text{perfect}} - P_{\text{test}}\right)^2 \tag{2.1}$$

where N is the shape of the models which is the product of width and height pixels.

As depicted in Fig. 2.10, the deviation of the hole's center is indicated by thick weighted edges or curves. Following the same process, the remaining models with deviations ranging from 0 to 3 mm, incrementing by 0.1 mm, are shown in Fig. 2.11. For the model with a hole pattern spaced at 80.0 mm, the mean square error (MSE) is 0, indicating no deviation. However, as the deviation increases, the MSE shows a linear increment with some outliers.

The MSE results for deviations along the horizontal directions are illustrated in Fig. 2.11a, while deviations along the vertical directions exhibit different patterns, as shown in Fig. 2.11b, which are relatively unstable. The mean difference along the 45° direction shows a linear increment with deviations, accompanied by unexpected outliers, as depicted in Fig. 2.11c. Overall, there is a linear relationship between the spacing/deviation distance and the MSE in all cases. Notably, the MSE values indicate that the deviation along the 45° direction is much higher compared to deviations along the axes of the same magnitude. Additionally, the MSE appears larger for deviations along the vertical direction compared to the horizontal directions due to the differences in the total spacing distance along the last hole in each axis.

Furthermore, the differences between the holes of the perfect model and the test model have been labeled using bounding boxes for clear visualization, as depicted in Fig. 2.12.

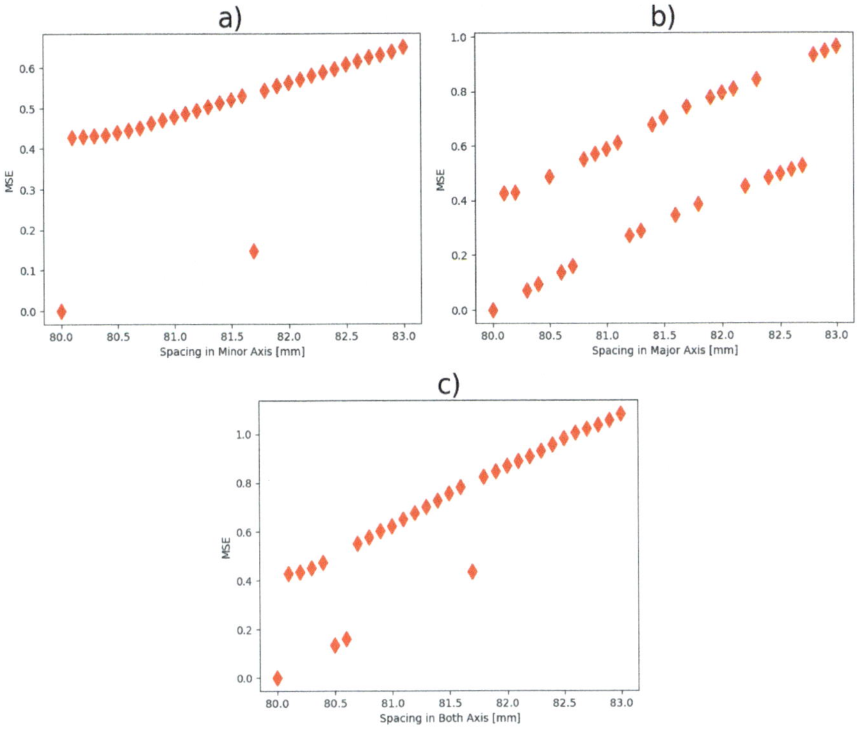

Fig. 2.11 MSE: (**a**) in horizontal directions, (**b**) in vertical directions, (**c**) in diagonal directions

2.4 Conclusion

Many industries have implemented machine learning applications to enhance the quality of their products. Integrating robots with machine vision to carry out repetitive and hazardous tasks is a common practice for quality improvement. Edge detection, a machine vision technique, finds applications in various fields. In this paper, the test model and perfect model were created using CAD software, with the perfect model having holes spaced at 83 mm in both the horizontal and vertical directions. The test datasets consisted of models with spacing distances ranging from 80 mm to 83 mm in increments of 0.1 mm, considering both vertical and horizontal directions, as well as both directions. The image processing stage utilized OpenCV and was implemented in Python for edge detection and analysis of mean differences. The results reveal the deviations between the perfect and test models and provide labels indicating the location of the deviations. Furthermore, the shape of the white contour lines can provide an estimate of the direction of deviation. It is important to note that directly implementing this approach in real-time situations may be challenging due to potential color (pixel) deviations between the models. However, with proper image processing and consideration of material consistency, the application can perform effectively.

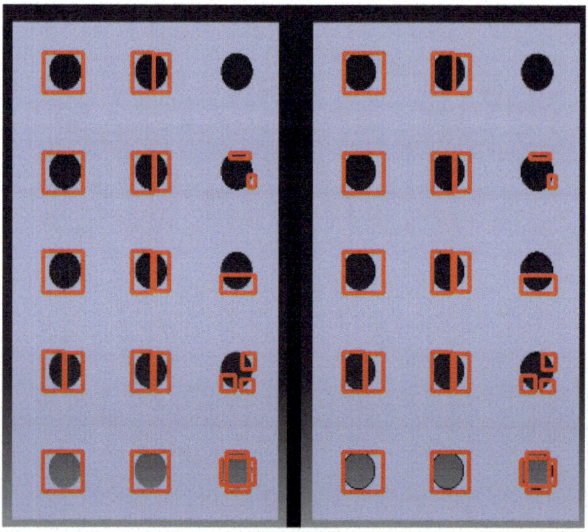

Fig. 2.12 Differences in labeling using bounding box

Conflict of Interest The authors declare no conflicts of interest.

References

1. Javaid, M., Haleem, A., Singh, R. P., Rab, S., & Suman, R. (2022). Exploring impact and features of machine vision for progressive industry 4.0 culture. *Sensors International, 3*, 100132.
2. Ambaye, G., Krishnan, K., & Boldsaikhan, E. (2023). *Detection of small screws using machine learning*. In 2023 International conference on information and communication technology for development for Africa (ICT4DA) (pp. 13–18). IEEE.
3. Goel, R., & Gupta, P. (2020). *Robotics and industry 4.0. A roadmap to industry 4.0: Smart production, sharp business and sustainable development* (pp. 157–169). Springer.
4. Zhang, S., Fan, Y., Huang, Y., Yang, X., Zhang, M., Luo, J., Deng, G., Zhou, Q., Song, H., & Wang, Y. (2023). Improvement of the surface condition of laser-drilled holes via a dual-wavelength double-pulse train. *Optics Laser Technology, 157*, 108681.
5. Aamir, M., Giasin, K., Tolouei-Rad, M., & Vafadar, A. (2020). A review: Drilling performance and hole quality of aluminium alloys for aerospace applications. *Journal of Materials Research and Technology, 9*(6), 12484–12500.
6. Król, P. M., Podziewski, P., & Hanincová, L. (2022). Deviation of coordinates of small-dimension holes in fiberboard. *Wood Material Science & Engineering*, 1–6.
7. Bagheri, B., Abdollahzadeh, A., Abbasi, M., & Kokabi, A. H. (2021). Effect of vibration on machining and mechanical properties of az91 alloy during fsp: Modeling and experiments. *International Journal of Material Forming, 14*, 623–640.
8. Khanna, N., Agrawal, C., Gupta, M. K., & Song, Q. (2020). Tool wear and hole quality evaluation in cryogenic drilling of inconel 718 superalloy. *Tribology International, 143*, 106084.
9. Polini, W., & Corrado, A. (2019). A general model to estimate hole location deviation in drilling: The contribution of three error sources. *The International Journal of Advanced Manufacturing Technology, 102*(1), 545–557.

10. Shen, N., Guo, Z., Li, J., Tong, L., & Zhu, K. (2018). A practical method of improving hole position accuracy in the robotic drilling process. *The International Journal of Advanced Manufacturing Technology, 96*, 2973–2987.
11. Wang, B., Pei, Y.-C., Xin, Q.-Y., Wang, Z.-Q., Wang, L.-L., & Luo, M.-Y. (2023). A measurement method for screw rod bending deflection by laser edge detection. *Measurement, 206*, 112272.
12. Luo, S., Hou, J., Zheng, B., Zhong, X., & Liu, P. (2022). *Research on edge detection algorithm of work piece defect in machine vision detection system*. In 2022 IEEE 6th information technology and mechatronics engineering conference (ITOEC) (Vol. 6, pp. 1231–1235). IEEE.
13. Basavaraju, H., Aradhya, V. M., Pavithra, M., Guru, D., & Bhateja, V. (2021). Arbitrary oriented multilingual text detection and segmentation using level set and gaussian mixture model. *Evolutionary Intelligence, 14*, 881–894.
14. He, M., Feng, L., & Fan, J. (2019). A method for real-time monitoring of running trains using φ-otdr and the improved canny. *Optik, 184*, 356–363.
15. Xu, J., Zhang, J., Zhang, K., Liu, T., Wang, D., & Wang, X. (2020). An apf-aco algorithm for automatic defect detection on vehicle paint. *Multimedia Tools and Applications, 79*(35), 25315–25333.
16. Kieselbach, K. K., Nöthen, M., & Heuer, H. (2019). Development of a visual inspection system and the corresponding algorithm for the detection and subsequent classification of paint defects on car bodies in the automotive industry. *Journal of Coatings Technology and Research, 16*(4), 1033–1042.
17. Akhtar, S., Tandiya, A., Moussa, M., & Tarry, C. (2019). An efficient automotive paint defect detection system. *Advances in Science, Technology and Engineering Systems Journal, 4*(3), 171–182.
18. Sizyakin, R., Cornelis, B., Meeus, L., Dubois, H., Martens, M., Voronin, V., et al. (2020). Crack detection in paintings using convolutional neural networks. *IEEE Access, 8*, 74535–74552.
19. Moffatt, A., Platt, E., Mondragon, B., Kwok, A., Uryeu, D., & Bhandari, S. (2020). *Obstacle detection and avoidance system for small uavs using a lidar*. In 2020 International Conference on Unmanned Aircraft Systems (ICUAS) (pp. 633–640). IEEE.
20. Ramalingam, B., Manuel, V.-H., Elara, M. R., Vengadesh, A., Lakshmanan, A. K., Ilyas, M., & James, T. J. Y. (2019). Visual inspection of the aircraft surface using a teleoperated reconfigurable climbing robot and enhanced deep learning technique. *International Journal of Aerospace Engineering, 2019*.
21. Xing, Y., Lv, C., Wang, H., Cao, D., & Velenis, E. (2019). Dynamic integration and online evaluation of vision-based lane detection algorithms. *IET Intelligent Transport Systems, 13*(1), 55–62.
22. Lei, B., Wang, N., Xu, P., & Song, G. (2018). New crack detection method for bridge inspection using uav incorporating image processing. *Journal of Aerospace Engineering, 31*(5), 04018058.
23. Ren, Z., Fang, F., Yan, N., & Wu. (2022). State of the art in defect detection based on machine vision. *International Journal of Precision Engineering and Manufacturing-Green Technology, 9*(2), 661–691.
24. Luo, Q., Fang, X., Liu, L., Yang, C., & Sun, Y. (2020). Automated visual defect detection for flat steel surface: A survey. *IEEE Transactions on Instrumentation and Measurement, 69*(3), 626–644.
25. He, Y., Song, K., Meng, Q., & Yan, Y. (2019). An end-to-end steel surface defect detection approach via fusing multiple hierarchical features. *IEEE Transactions on Instrumentation and Measurement, 69*(4), 1493–1504.
26. Tabernik, D., Šela, S., Skvarč, J., & Skočaj, D. (2020). Segmentation-based deep-learning approach for surface-defect detection. *Journal of Intelligent Manufacturing, 31*(3), 759–776.
27. Tan, X. J., Leow, W. Z., & Cheor, W. L. (2022). A simple, precise, and high-speed die edge detection framework based on improved k-mean and landscape analysis for the semiconductor industry. *Arabian Journal for Science and Engineering, 47*(2), 1643–1655.

28. Younas, M., Nawaz, Q., Hamid, I., Gilani, S. M. M., & Iqbal, M. J. (2022). A novel approach of ceramic tile crack detection using morphological operations. *Mehran University Research Journal of Engineering & Technology, 41*(2), 146–154.
29. Bivalkar, M., Agarwal, S., & Singh, D. (2022). Development of an efficient approach for detection and measurement of crack length in ceramic tile manufacturing using millimeter-wave imaging. *NDT & E International, 129*, 102656.
30. Zhao, Y., Zhao, Y., Liu, R., Yan, J., Zhang, Y., Peng, Y., Liang, Y., & Huang, Y. (2022). *Machine-vision based method for edge and surface crack detection in ceramic tile.* In MEMAT 2022, 2nd international conference on mechanical engineering, intelligent manufacturing and automation technology (pp. 1–4). VDE.
31. Devare, M. (2022). *Parallel image processing for liquid level detection.* In International conference on computing in engineering & technology (pp. 372–382). Springer.
32. Zhou, T., Zhang, K., Jiang, H., & Wang, G. (2022). *Stability analysis method of slow injection process based on edge detection theory*. In 2022 IEEE 6th information technology and mechatronics engineering conference (ITOEC) (Vol. 6, pp. 209–213).
33. Dong, X., & Li, Y. (2022). Online detection of turning tool wear based on machine vision. *Journal of Computing and Information Science in Engineering, 22*(5), 050903.
34. Zhang, S., Huang, H., Huang, D., Yao, L., Wei, J., & Fan, Q. (2022). Subtask-learning based for robot self-assembly in flexible collaborative assembly in manufacturing. *The International Journal of Advanced Manufacturing Technology, 120*(9), 6807–6819.
35. Zhang, K., Zhu, Y., Wang, W., & Zhang, J. (2022). Measurement for blade edge based on machine vision. In *New approaches for multidimensional signal processing* (pp. 139–150). Springer.
36. Xu, R., Hao, R., & Huang, B. (2022). Efficient surface defect detection using self-supervised learning strategy and segmentation network. *Advanced Engineering Informatics, 52*, 101566.
37. Sureshkumar, S., Mathan, G., RI, P., & Govindarajan, M. (2022). Deep learning framework for component identification. *International Journal of Information Technology*, 1–9.
38. Du, B., Hao, Z., & Wei, X. (2021). *Roundness detection of end face for shaft workpiece based on canny-zernike sub pixel edge detection and improved hough transform.* In (ICEIEC) 2021 IEEE 11th international conference on electronics information and emergency communication (ICEIEC) (pp. 1–4). IEEE.
39. Nahl, M. F., Dewi, I. A., & Putra, K. R. (2023). Holistically nested edge detection in chair ergonomic detection. In *AIP conference proceedings* (Vol. 2772). AIP Publishing.
40. Gui, R., Sun, Q., Wu, W., Zhang, D., & Li, Q. (2023). Transfer learning for cross-scene 3D pavement crack detection based on enhanced deep edge features. *Engineering Applications of Artificial Intelligence, 123*, 106452.
41. Vicens-Miquel, M., Medrano, F. A., Tissot, P., Kamangir, H., & Starek, M. (2022). Deep learning automatic detection of the wet/dry shoreline at fish pass, Texas. In *IGARSS 2022–2022 IEEE international geoscience and remote sensing symposium* (pp. 1876–1879). IEEE.
42. Ambaye, G. A., & Lemu, H. G. (2021). Dynamic analysis of spur gear with backlash using Adams. *Materials Today: Proceedings, 38*, 2959–2967.
43. Ambaye, G. A., & Lemu, H. G. (2021). Numerical stress analysis and fatigue life prediction of the classical external Geneva mechanism. In *International workshop of advanced manufacturing and automation* (pp. 176–186). Springer.
44. Tang, P., Guo, Y., Li, H., Wei, Z., Zheng, G., & Pu, J. (2021). Image dataset creation and networks improvement method based on cad model and edge operator for object detection in the manufacturing industry. *Machine Vision and Applications, 32*(5), 1–18.
45. Ambaye, G., Boldsaikhan, E., & Krishnan, K. (2023). Robot arm damage detection using vibration data and deep learning. *Neural Computing and Applications*, 1–13.

Chapter 3
Tractive Performances of Traction Devices of Two-Wheeled Tractors in Dry Field Conditions: A Review

Samuel Getachew Chekol, Smegnew Moges Mintesinot, Berihun Abebaw Mebratie, and Addisu Alamirew Meku

3.1 Introduction

3.1.1 Two-Wheeled Tractors

Two-wheeled tractors (2WTs) are multipurpose hand tractors designed primarily for land preparation. They are suitable for seedbed preparations in small farms and in hill farming [1]. Additionally, they can carry out other farm operations like pumping, planting, spraying, and transporting [2, 3]. They are designed in a way that the prime mover is in the direction of travel and the control for field operation is performed by the operator walking behind it. The operator walks behind the machine, holding the two control handles in his hands. For this reason, they are also called walk-behind tractors [4]. Different researchers call them by several names such as two-wheeled tractors [5], hand tractors [6], walking tractors [1], power tillers, garden tractors [7], and single axle tractors [8].

2WTs are used all over the world. Demand for these machines, which typically have less than 15Hp, have rapidly raised throughout the world because of their lower cost compared to four-wheeled tractors; their greater maneuverability, ability to be fully utilized in systems with small landholdings, their potential for off-farm use and economical compatibility to small and medium land holdings [9]. In Europe, they are used for garden and small area agricultural activities, while in Asia, they are the main sources of farm power [10]. Rice farming in South and South Eastern countries of Asia is powered by 2WTs and the demand for it is rapidly increasing. Whereas in Africa, even though 2WTs are 42% more economical than bullock-based plowing [11], the rate of adaptation of these machines is very slow. Researches

S. G. Chekol (✉) · S. M. Mintesinot · B. A. Mebratie · A. A. Meku
Faculty of Mechanical and Industrial Engineering, Bahir Dar Institute of Technology, Bahir Dar University, Bahir Dar, Ethiopia

S. W. Fanta et al. (eds.), *Sustainable Development Research in Manufacturing, Process Engineering, Green Infrastructure, and Water Resources*, Green Energy and Technology, https://doi.org/10.1007/978-3-031-77339-6_3

have shown that the current problem of mechanization in most Sub-Saharan African countries is similar to what was there in South and South East Asia such as Bangladesh, Thailand, Vietnam, and Sri Lanka, a few decades ago and it has been recommended that to rapidly mechanize the agricultural sectors 2WTs have to be used [12–15].

Despite the economic advantages of 2WTs, they have also some unfavorable conditions in operating them. The operators of 2WTs are subjected to temperature, noise, high vibration, and other adverse environmental conditions at work. These expose the operators to discomfort and pain. Especially, in time of doing primary tillage operations, the operators are subjected to severe vibration, which results in high fatigue. Dewangan et al. [10] found that while Rota-tilling, the amount of power that was absorbed by the operators' hands range from 253 mW to 422 mW with a frequency ranging from 6.3 Hz to 80 Hz, which is very high. These vibrations are transmitted to the hand, shoulder, and bone of the operator and cause "hand-arm vibration syndrome." This can lead to a serious problem of musculoskeletal and physical disorders [16].

2WTs have a great potential for farm applications since they are designed in a way the prime mover, i.e., the engine, can be accessed by implements and tools. Similar to the four-wheeled tractors, they have a drawbar and power take-off shaft. By connecting these parts to implements, primary and secondary tillage are performed. In addition to land preparation, they are also applied in plantations. In order to perform planting, the planting implement is connected to the drawbar of the 2WT by using a linkage. Whereas to do threshing, shelling, and milling, the output pulley of the engine is connected to the driven pulley by using V-belts as shown in Fig. 3.1.

2WTs basically have simple construction which can be broken down into four general operational systems. The power system is an internal combustion engine that can generate 3 to 15HP power by using petrol or diesel fuels. The engine has a flywheel and an out pulley, from which power is taken and transmitted to the proceeding system. One of the key factors for the versatile utilization of 2WTs is the orientation of the engine and its output pulley, as power can be taken for any use. Nowadays, to ensure sustainable power source, to reduce emissions to the atmosphere, and eliminate the exhaustion of fumes while working in greenhouses and plastic film–shielded houses, the prime mover of 2WTs is being transformed into electric motors. Electric 2WTs have a more simplified construction and fewer parts than the conventional 2WTs [18, 19].

The transmission system is found proceeding to the power system. It transmits power from the output pulley of the engine into the gearbox which finally drives the wheels. The third main system of 2WTs is the control system which enables the operator an overall authority while using it. This includes all the levers, gas pedals, steering clutches, and gear-selecting levers. The other basic system is the traction device. It directly gets engaged with the surface of the field to provide sufficient pull power during field preparation operations. It is the final drive component that determines the tractive field performance of the whole system.

(a) (b)

(c) (d)

Fig. 3.1 Two-wheeled tractors being used for (**a**) direct seeding (**b**) transport (**c**) shelling maize, and (**d**) pumping water [5, 17]

3.1.2 Traction Devices in 2WTs

Traction devices of tractors are the final and the most important devices where the power in the axles is brought into application on the ground. Gasparetto et al. [20], stated that *"The criterion for a good traction device is that, it should be able to deliver power closer to that supplied to its axle."* The selection of traction devices depends on the type of terrain, the physical and mechanical conditions of the soil and the kind of farm operation. Based on their construction, tractive devices can be divided into three. These are pneumatic tires (or rubber wheels), cage wheels, and tracks.

Pneumatic tires have so long been used on heavy machines and urban transportation vehicles. They are also the oldest tractive devices that are being used in 2WTs. Their lugs are designed to be circumferentially opposed as shown in Fig. 3.2. There are two types of pneumatic tire construction; Radial and Biased type. The biased ply tires are extinct since they have less tractive performance. Radial ply tires instead respond sensitively to wheel-loading conditions, inflation pressure variation, and make better interaction with the soil, which increases their tractive performance [21]. In addition to the tire section parameters (lug pattern, lug dimensions, lug width, lug height, lug spacing, etc.), the tractive performances of tires are highly dependent on the inflation pressure, effective diameter, and wheel loads [22, 23].

Fig. 3.2 Tractive devices in 2WTs (**a**) pneumatic tire [27] (**b**) cage wheel [50] (**c**) rubber track [27]

The second type of traction device is cage wheel. Cage wheels have been most widely used in power tillers as final drive elements in wet soil conditions. They are favorably used in the preparation of swampy fields for paddy and wheat production. Their primary purpose in such fields is mixing the soil properly and breaking down large clods of soil. But they are not suitable for use in asphalt bitumen roads since they are completely made of steel [24]. They have round rings and circumferentially welded lugs. The ability of cage wheels to provide high traction comes from their lug-to-soil contact area and depends on the configurations, positions, and orientations of the lugs connected to the ring rod throughout its circumference relative to each other. There are different patterns for the lugs of steel cage wheels; for instance, open flat lugs, opposed, staggered echelon, and perfect chevron types. Staggered echelon-type cage wheels are similar in lug arrangement patterns with circumferential opposed lugged pneumatic tires and are interchangeably named in the case of cage wheels [25]. The impacts of lug configurations on the tractive performance of the 2WTs are analyzed by changing the lug spacing, lug angle, lug pattern, lug opening, and circumferential angle.

Tracks can be made of a series of interlocking steel linkages (steel tracks) or a continuous loop of nylon-reinforced rubber belt (rubber track). Rubber tracks are simpler in construction and more suitable for road driving than steel tracks. For agricultural operations application, rubber tracks are mostly preferred since they have better tractive efficiency and less compaction of soil [26]. Though it is not strange to use rubber tracks in medium and large power tractors, they come to be used in 2WTs later than both cage wheels and tires. They are similar in their lug patterns with cage wheels and tires, but as they have more components, they are more complicated. In addition to drive wheels, they have front idlers, track tightening devices, couplings, springs, and frames as shown in Fig. 3.2c. Their tractive performances can be studied by changing the length of belt in contact with the ground, the width of the track belt, wheel load, and the number of wheel stations.

A number of studies indicate that the power that was generated by the engine is lost due to many reasons. One of the main reasons is the type of tractive device that is installed on the 2WTs. Because of this loss, the field performance of 2WTs highly drops and leads to low field coverage. As a result, the time required to cover a field increases and similarly does the fuel consumption. The goal of this review paper is to identify the tractive devices and mechanisms that can reduce power loss and develop high draft forces in 2WTs. The review will also focus on the principles of traction and draft forces, the different types of tractive devices, the factors that affect power loss in 2WTs, and the performance of different tractive devices and mechanisms. The findings of this review will be used to identify the most promising tractive device and mechanisms for reducing power loss and developing high draft forces in 2WTs.

3.2 Tractive Performance of Traction Devices

Traction is the capability of the tractor to grip the surface of the field in order to draw or pull the implement attached to it. It is a general attribute that can be explained and analyzed using several parameters, for instance, traction efficiency, drawbar pull, and motion resistance [1, 27, 28]. Nowadays, there are three main approaches to study the tractive performances of tractive devices. The first approach uses direct measurements to determine some parameters, such as using a cone penetrometer to measure the soil strength and the physical properties of soil such as cone index, bulk density, moisture content, and soil texture; and to use load cells, dynamometers, and torque meters to measure the drawbar force (net tractive force) and torque in the axles of the tractors [29]. The second approach is an analytical and empirical method of study based on the Coulomb properties of soil and internal angle of friction as a measure of shear strength [28]. Finite element analysis and discrete element analysis are the newly applied methods to study traction performances of tractive devices of tractors [30]. In this review, most of the references were done by the first approach.

3.2.1 Pneumatic Tires

A number of evaluations and analyses are conducted on the tractive performance of pneumatic tires under different field conditions. What makes most of the findings similar is that there is a high loss of power on operation: out of the total power of the 2WTs having pneumatic tires, only a quarter is gained in the drawbar on average. The evaluation of a 6.71 kW 2WT was conducted by Narang et al. [31] for evaluation of drawbar bar force and power on a tar track of 500 m length at an engine speed of 1500 rpm, and the results were recorded at a 15%-wheel slip. Its weight was 485 kg and the 6.00–12 pneumatic tires were inflated at a pressure of 147 kPa. The maximum drawbar pull and the maximum drawbar power were recorded at second low gear and third low gears, to be 2110 N and 2.042 kW, respectively. The fuel consumption at the maximum drawbar power was 0.969 lh^{-1}.

Narang et al. [1] evaluated the tractive performance of a 2WT of 8.95 kW engine power which weighs 485 kg and was wheeled with a 6.00–12 pneumatic tire inflated at a pressure of 147 kPa. The test was conducted on a tilled land (soil texture: sand 12.6%, silt 32.7%, and clay 54.7%) with a cone index of 239 kPa. The results of the study indicated that, at 25%-wheel slip and engine speed of 1500 rpm, the values of drawbar pull and power were 748.4 N and 0.348 kW, respectively. The fuel consumption at the maximum drawbar power was 0.582lh^{-1}.

An 8.1 kW 2WT having 6.00–12 pneumatic tires was tested on a tilled land (soil texture: sand 12.6%, silt 32.7%, and clay 54.7%) for its tractive performance by Sahay et al. [32]. The draught was measured at two different forward speeds corresponding to the slow (0.5 ± 0.03 ms^{-1}) and fast (1.3 ± 0.1 ms^{-1}) walking speeds of an operator walking behind the 2WT. A single axle trolley (weight: 360 kg) with a carrying capacity of 1 ton was attached behind it through specially fabricated brackets to hold an octagonal ring transducer (ORT) between the power tiller and trolley in a horizontal position while in operation. The drawbar pulls at 20%-wheel slip was 1077 N and 1.273 kW drawbar power was recorded. The results indicated that the draft developed by the power tiller at 20% slip was 1077 N and drawbar power available was 0.538 kW and 1.273 kW at the speeds of 0.49 and 1.18 ms^{-}, respectively.

Bisrat et al. [6] performed the evaluation of two 2WTs with engine power 11.185 kW and 8.95 kW on 6.00–12 pneumatic tires with inflation pressures 150 kPa which weighed 505 kg and 493 kg, respectively. The test was conducted at Melkasa Agricultural Research Center, Adama, Ethiopia, on a silty loam soil with 21.6% moisture content and having a penetrometer resistance of 460 kPa before the test. The results of the performance evaluation indicate that the maximum drawbar pull was 2524.9 N at first low gear and the maximum drawbar power was 1.6156 kW at third low gear, at a slip of 13%, and 1500 rpm engine speed for the 11.185 kW 2WTs. For the same engine speed, the 8.95 kW 2WTs gave a maximum drawbar pull of 2268.81 N on first low gear, a maximum power of 1.587 kW on third low gear, and at 10.2%-wheel slip.

3.2.2 Cage Wheels

Cage wheels are popular tractive devices of 2WTs in Asia, where field conditions in tillage are mostly swampy. Many researches indicate that the traction performances of cage wheels are highly dependent on their design parameters such as circumferential angle, lug angle, lug spacing, lug dimensions, and lug patterns [33–37]. Even there are a few who indicate the best combinations of design parameter values at which the cage wheels deliver superior performances in swamps [38, 39]. Yet, in the literature survey of this paper, a few researches are found that have been done in dryland conditions. Cage wheels perform better when there is more wheel slip than what is recommended for pneumatic wheels operation on dryland. Better pull forces and power are recorded at high slippage in wetlands and have high specific fuel consumption than pneumatic tires do in dryland [40].

Narang et al. [1], experimented on the tractive performance of an open flat lugged steel cage wheel. It was recorded that, at second low gear and third low gear the drawbar pull was 913.07 N and 953.15 N, respectively. In the same gear selection, the drawbar power was 0.39 kW and 0.54 kW with fuel consumptions of 0.683lh^{-1} and 0.746lh^{-1}. When compared with the tractive performances of the pneumatic wheels with and without a 40 kg wheel ballast at 1500 rpm engine speed, it gave 18% and 22.8% more pull at second low gear and third low gear, the same way 26.7% and 35.6% more power, and 14% and 22% more fuel consumption.

3.2.3 Rubber Tracks

Tractive performance of all devices is a strong function of the soil and wheel lug contact area. For both pneumatic tires and cage wheels, since they are annular in shape, to increase their contact area with the soil, necessarily their diameter has to be increased. But tracks are designed to provide larger contact area than comparably matching tire and cage wheel. As a result, tracks perform at least twice better than tires [41]. Studies on the comparison of the performances of tires and tracks when fitted to high-power tractors are plenty while a few researches were done in 2WTs with tracks. The tractive performances recorded by the application of rubber tracks in 2WTs have outperformed the results obtained by testing pneumatic tires in previous studies.

Rasool et al. [2] evaluated the traction performance of a track having a contact length of 0.8 m, track width of 0.1 m, and lug dimensions of 74 × 15 × 10 mm^3 (length × width × thickness) in a lateritic sandy clay loam soil at an average soil moisture content of 9%. The test was conducted by using a 7.5 kW three-phase electric motor in a soil bin system at 300 kPa, 500 kPa, 800 kPa, and 1000 kPa cone index. The result of the track system indicated that the maximum tractive efficiency was found to be in the range of 77–83% corresponding to a range of travel reduction ratio from 0.12 to 0.445. In comparison, at 250 kPa soil cone index condition, the track developed 2260 N pull force while the pneumatic tire developed 748.4 N pull force which is one-third of the pull developed by the track system [1].

Rasool et al. [27] tested a 9.7 kW 2WT by replacing the pneumatic tires with a rubber track wheel whose weight became 550 kg to evaluate its tractive performance in four different soil conditions. The test was also repeated on the same soil and operation conditions for a 6.00–12 pneumatic tire inflated at a pressure of 147 kPa with wheel ballast of 120 kg to remove the effect of mass variation with the track system and without wheel ballast which was 430 kg. The results indicated that in soft soil condition with an average cone index of 250 kPa at 15%- and 25%-wheel slippage, the rubber track, the tire with ballast and the one without ballast gave a drawbar force of 1509 N, 781 N, 669 N, and 1957 N, 1002 N, 892 N, respectively. In the same way, the tractive efficiency became 0.5115, 0.387, and 0.429.

3.3 Discussion

In all traction devices, the amount of power obtained in the drawbar is very less when compared with the power output of the prime mover in the test systems. There could be several factors hindering the traction performances. For instance, the soil's physical properties, the soil's mechanical properties, the type of terrain, the efficiency of the transmission systems, the type of the tractive device, the soil and wheel interactions, and others. This review paper focuses only on the relationships between the tractive devices and their effect on the performance of 2WTs. Table 3.1 presents the pull developed by different tractive devices at different field conditions. For uniformity, ease of analysis, and comparison of results, the data considered in this paper were those which were recorded at engine speed of 1500 rpm and pneumatic tires of size 6.00–12 inches.

3.3.1 Maximum Pull Developed By Tractive Devices

The table indicates that the highest pull was recorded when a 2WT with 6.17 kW power was tested on a tar road. All other tests were conducted on tillage field soils. Even though it was the lowest-powered 2WT, the highest traction was obtained when it was operated on tar roads. Pneumatic tires and rubber tracks are both made up of the same material (rubber). The increment in the soil strength resulted in rapid increment of pull force [27]. The pull developed by rubber tracks at cone indices 250 kPa, 500 kPa, and 1000 kPa at 25%-wheel slippage was 1975 N, 2495 N, and 2952 N, respectively. The trends of increment of pull force pneumatic tires provided are similar with that of the rubber track. At cone indices 250 kPa, 500 kPa, and 1000 kPa, at 25%-wheel slippage, was 1002 N, 1624 N and 2083 N, respectively. What pneumatic tires and rubber tracks got in common is, the rubber material they

Table 3.1 Tractive performance of tractive devices in different field conditions

Author	Terrain	Power (kW)	Traction device	Max. pull (N)	Max. power (kW)	Wheel slip (%)
Narang et al. [31]	Tar road	6.17	Pneumatic tire without wheel ballast	2110	2.042	15
			Pneumatic tire with 40 kg wheel ballast	2401	2.543	15
Narang et al. [1]	Tilled land, CI = 239 kPa	8.95	Pneumatic tire without wheel ballast	748.4	0.348	25
			Pneumatic tire with 40 kg wheel ballast	888	0.487	25
			Cage wheel (170 × 745 mm)	953	0.54	25
Sahay et al. [32]	Tilled land, CI = 42 ± 42 kPa	8.1	Pneumatic tire, pulling 1281.2 kg payload	113.3 ± 10	1.273	20
Rasool et al. [2]	Lateritic sandy clay loam soil, CI =250kPa[a] and 350kPa[b]	7.5	Rubber track 0.1 × 0.8 (length × width)	2260[a]	NR	15
				2260[b]	NR	15
Rasool et al. [27]	Sandy clay loam soil, CI = 250 kPa[a], 500kPa[b] and 1000kPa[c]	9.7	Rubber track 0.15 × 0.78 (length × width)	1509[a]	NR	15
				2179[b]	NR	15
				2665[c]	NR	15
			Pneumatic tire with 120 kg wheel ballast	781[a]	NR	15
				1289[b]	NR	15
				1640[c]	NR	15
			Pneumatic tire without wheel ballast	669[a]	NR	15
				1068[b]	NR	15
				1314[c]	NR	15

NR Not reported

are made up of. For both devices, as the cone index of the soil increases, the pull force also increases.

However, in tillage field conditions, the maximum pull was obtained by rubber tracks. Rasool et al. [2] tested rubber tracks in a soil bin system with a cone index of 250 kPa and compared the results with Narang et al. [1] findings. The results of the former test showed that the maximum pull developed was 2260 N while the latter, in which pneumatic tires were tested, was 748.4 N. This is about three times greater. In fact, the results were obtained in a soil bin system, not in the field; Rasool et al. [27] tested 2WTs by installing rubber tracks for 2WTs in the field. The performance results were strongly in agreement with the soil bin system performance results. Both experiments indicated that the performance of rubber tracks was better than pneumatic tires. The tractive performance of wheeled 2WTs cannot match that of comparable tracked 2WTs [42]. This comes in alignment with the studies made on heavy machineries that compared the performance of tracks and pneumatic tires [41–45].

Narang et al. [1] conducted a comparative experiment on pneumatic tires and cage wheels' tractive performance. The pneumatic tires were tested in two conditions: with a 40 kg ballast and without ballast. The one with ballast developed a maximum draft of 888 N while the latter gave a 748.4 N draft. The results of the test which was done by replacing the pneumatic wheels by steel cage wheels showed a much better draft of 953 N in the same soil condition. The power gained was also the highest value amongst other results of the test. In general, the results of this test indicate that cage wheels have better performances and potential than pneumatic tires.

Rasool et al. [2] indicated that rubber tracks developed better draft than pneumatic tires. Narang et al. [1] also indicated that steel cage wheels developed better draft force than pneumatic tires. Researches that compare performance of rubber tracks with cage wheels were not found in this review. However, there are some parameters that make the studies made to compare rubber tracks with pneumatic wheels and steel cage wheels with pneumatic wheels similar. Both were conducted in similar cone index of soils, which is 239 kPa and 250 kPa. In addition, in the track test system, the total weight was 250 kg while the cage wheel testing weighed 485 kg. As compared with the cage wheel, the rubber track is disadvantaged in weight. Taking these two considerations as a common ground, the tractive performance of the cage wheel and the rubber track can be compared. The cage wheels gave a draft of 953 N while the rubber track developed 2260 N. In this consideration, the rubber track draft force was more than the cage wheel draft force. Nonetheless, the field tests of Rasool et al. [27] do not align with these results, because they tested rubber tracks that totally weighed 550 kg and the draft developed was only 1509 N at the same cone index of soil.

In general, from the analyzed and discussed data, the maximum pull was developed by rubber tracks. However, researches have to be done which compare the tractive performances of rubber tracks and cage wheels.

3.3.2 *Effects of Weight on Tractive Performance*

The addition of wheel load (or wheel ballast) boosts the drawbar pull and power of a tractor. However, excessive wheel loading can result in high fuel consumption. A number of studies have shown that the addition of loads increases the pull force in the drawbar [1, 2, 21, 27, 31]. Narang et al. [31] tested a 6.71 kW 2WT and found that, by mounting a 40 kg wheel ballast on tires, the maximum drawbar pull and drawbar power were raised from 2110 N and 2.042 kW to 2401 N and 2.543 kW, respectively, and also fuel consumption did from 0.969lh^{-1} to 1.091lh^{-1}. This is a 14% and 25% increase in pull and power, respectively. But, the addition of wheel load has also slightly increased the fuel consumption by 12.6%. The study of Narang et al. [1] also indicates that the addition of a 40 kg wheel load raised the drawbar pull, power, and fuel consumption by 27.4%, 55.2%, and 28.3%, respectively. Rasool et al. [27] compared the performance of pneumatic tires by 120 kg of weight difference. The pull developed by the tire without ballast was 16% lesser. Similarly, Battaito et al. [21] conducted field tests on agricultural clay (C) Vertic Cambisol on the traction performance of mechanical front wheel drive 65 kW MFWD tractor with and without the addition of 9.8kN wheel loads. At a wheel slip of 25%, the drawbar pull raised from 28kN to 35kN.

In ballasts, the addition of weight increases the normal force that keeps the 2WTs with the soil. Unlike that, Sahay et al. [32] tested an 8.1 kW 2wt on tilled land by varying the weight of the trolley attachment. The attachments weighed 834 kg and 1113 kg during the test. In a similar way to the previously discussed results, the pull developed by the heavier attachment was 46% more than that of the lighter. This kind of weight addition improves the shear forces developed by the soil–tire contact instead of the increment of normal forces per unit contact area as wheel loads do.

In both ways of weight addition, significant increment in draft force has been gained. The increment of weight increases the pull. Weight addition can be the simplest mechanism to increase the tractive performance of 2WTs.

3.3.3 Effects of Design Parameters on Tractive Performance

According to Wong et al. [42], the traction gained from a tractive device is dependent on two main forces. The first one is the cohesion that is created between the device coming in contact with the terrain (which is a function of the contact area) and the second one is the friction of the terrain (which is a function of the normal load and internal shearing resistance of the terrain). Cohesion increases as contact area increases. The pneumatic tires which were referred in this review paper were all the same in design and that is 6.00–12 tire, which most manufacturers provide. Unlike tires and cage wheels, tracks have longer and flat contact area with the ground that enables them to have better grip [42].

Studies conducted on high-power tractor tires indicated, the increment of the contact area resulted in a significant improvement in the traction performances. Taylor et al. [23] conducted tests on 12.4–11 pneumatic tires of effective rim diameter 24 inches, 36 inches, and 42 inches. Except for the diameter, the tires were similar in all other parameters. As the results indicated, the increase in the diameter has improved the performance of the tire. Frank M [45]. test results agree that the larger the diameter, the better the tractive performance. However, these results were not merely achieved by the increment of diameter. As the diameter increases, the weight increases and the amount of pressure needed to reach proper working tire inflation also increases. For these reasons, even though a significant amount of pull was gained by the increase in wheel diameter, the inseparable effects of the wheel load increment and inflation pressure increment do not have to be ignored. Since pneumatic tires are available only in a standard design, contact area of the tires is predetermined and cannot be changed. The other design parameters of tires are lug angle and lug length. Studies indicate that the effects of these parameters have negligible influence on the tractive performance of pneumatic tires [22, 46].

The contact area does not only affect the tractive performance of devices, it also affects the compaction of the soil. For the same weight of 2WTs, the amount of static load per unit area determines the degree of compactness. Tractive devices having small contact area, i.e., pneumatic tires and cage wheels apply high pressure on the soil. Increasing the weight of the tractor increases the soil compaction. This leads to the deterioration of soil structure, and physical and physico-mechanical properties of the subsoil and topsoil layers [47]. In these regards, the soil permeability to water and air deteriorates, and as a result, the crop yield of the field reduces. Since rubber tracks have more surface area, the specific pressure exerted on the soil is lower compared to pneumatic tires and even lower compared to human foot [41, 48]. They have lower soil compaction than pneumatic tires and this reduces the negative effects on the soil compaction. From these points of view, rubber tracks outperform pneumatic tires [26]. In addition, tests conducted to evaluate the amount of rut formation, increment of soil cone index, and dry bulk density indicate that steel cage wheels produce very little amount of compaction than pneumatic tires do [49].

The tractive performance gained from cage wheels is a strong function of the design parameters [33]. There are several researches that have been done on the performance of cage wheels. However, unlike tires and tracks, in swampy field conditions, the measure of tractive performance of cage wheels is the development of sufficient lift forces on their lugs for floatation and pull forces for going forward. The reason for this is the sinkage of the 2WTs in the swamp. For this reason, the primary focus of the researches done so far has been limited to the agitation and floatation capacities of cage wheels in swampy field conditions.

Since cage wheels can be easily designed and produced, depending on the amount of traction needed and the type of field under work, they can be easily constructed to fit specific working conditions. But, in very firm terrain conditions for instance in tar roads, they perform very badly. In contrast, rubber tracks and tires smoothly roll over the terrain and give better operation comfort.

3.4 Conclusion

Several researches have been conducted so far to evaluate the tractive performances of pneumatic tires, cage wheels, and rubber tracks. This review mainly considered and discussed the comparison of draft forces developed by these devices. Accordingly, the following conclusions were drawn.

1. Rubber tracks develop the highest draft force than pneumatic tires and cage wheels at the same field conditions and comparatively matching design parameters in 2WTs.

2. Cage wheels have great potential to develop draft force better than pneumatic tires in dry field operations.
3. Pneumatic tires develop the least draft forces amongst other devices in off-road dry field operations.
4. The addition of weight on 2WTs significantly increases the draft force obtained in the drawbar.

3.5 Future Studies

Future studies have to be made on the potential of rubber tracks and cage wheels to improve field capacity. Optimization of tractive device design parameters, the tractive performances, degree of rut formation, and soil compaction of cage wheels in dry field conditions need to be studied.

References

1. Narang, S., & Varshney, A. C. (2006). Draftability of a 8.95 kW walking tractor on tilled land. *Journal of Terramechanics, 43*(4), 395–409.
2. Rasool, S., & Raheman, H. (2016). Suitability of rubber track as traction device for power tillers. *Journal of Terramechanics, 66*, 41–47.
3. Sam, B., & Kathirvel, K. (2006). Vibration characteristics of walking and riding type power tillers. *Biosystems Engineering, 95*(4), 517–528.
4. Prasanna Kumar, G. V., & Raheman, H. (2011). Development of a walk-behind type hand tractor powered vegetable transplanter for paper pot seedlings. *Biosystems Engineering, 110*(2), 189–197.
5. Baudron, F., Sims, B., Justice, S., Kahan, D. G., Rose, R., Mkomwa, S., & Gérard, B. (2015). Re-examining appropriate mechanization in Eastern and Southern Africa: two-wheel tractors, conservation agriculture, and private sector involvement. *Food Security, 7*, 889–904.
6. Bisrat, G., & Laike, K. (2017). Performance of single axle tractors in the semi-arid central part of Ethiopia. *Ethiopian Journal of Agricultural Sciences, 27*(1), 37–53.
7. Dharmawardene, M. W. N. (2006). Trends in farm mechanization by sugarcane small land holders in Sri Lanka. *Sugar Tech, 8*(1), 16–22.
8. Boshoff, W. H., & Innes, D. (1961). The performance of single axle tractors on peasant coffee farms in Buganda. *East African Agricultural and Forestry Journal, 27*(2), 77–84.
9. Diao, X, Silver, J., & Takeshima, H. (2016). International Food Policy Research Institute. African Transformation Report 2016: Transforming Africa's Agriculture.
10. Dewangana, K. N., & Tewari, V. K. (2008). Characteristics of vibration transmission in the hand–arm system and subjective response during field operation of a hand tractor. *Biosystems Engineering, 100*, 535–546.
11. Gupta, J. P. & Kumar, S. (2001). Status of power tiller use in Bihar – A case study in Nalanda district. 32: 19–22
12. Sheahan, M., & Barrett, C. B. (2017). Ten striking facts about agricultural input use in Sub-Saharan Africa. *Food Policy, 67*, 12–25.
13. International Development Enterprise report. (2012). *Study into the commercialization of selected agricultural machines in Bangladish.*
14. UNIDO report. (2008). Agricultural mechanization in Africa. Time for action.

15. Baker, D. (2008). *Agricultural mechanization in sub-Saharan Africa: Time for a new look*. FAO.
16. Waersted, M., & Westgaard, R. H. (1991). *Working hours as a risk factor in the devel-opment of musculoskeletal complaints*.
17. Ahmed, B., Bekele, A., & Hailu, G. (2018). On farm evaluation and demonstration of sifang power tiller for tillage and water lifting in Arsi zone. *Journal of Biology, Agriculture and Healthcare, 8*(7), 28–32.
18. Bodria, L., & Fiala, M. (1995). Design and testing of an electric-powered walking tractor. *Journal of Agricultural Engineering Research, 60*, 57–62.
19. Amongo, R. M. C., Quilloy, E. P., Ranches, M. A. F., Larona, M. V. L., & Madlangbayan, M. (2020). Development of an electric hand tractor (e-Tractor) for agricultural operations. In *IOP conference series: earth and environmental science* (pp. 1–11). IOP.
20. Gasparetto, E., Febo, P., Pessina, D., & Rizzato, E. (1992). The rolling resistance of narrow steel wheels in italian paddy fields. *Journal of Agricultural Engineering Research, 52*, 91–100.
21. Battiato, A., & Diserens, E. (2013). Influence of tyre inflation pressure and wheel load on the traction performance of a 65 kW MFWD tractor on a cohesive soil. *Journal of Agricultural Science, 5*(8), 197–215.
22. Taylor, J. H. (1973). Lug angle effect on traction performance of pneumatic tractor tires. *Transactions of the ASAE*.
23. Taylor, J. H., Vanden Berg, G. E., & Reed, I. F. (1967). Effect of diameter on performance of powered tractor wheels. *Transactions of the ASAE*.
24. Vijay Krishna, M., & Choudary, R. B. (2015). Design and analysis of a novel cage wheel with hydraulically actuated links. *International Journal of Engineering Research & Technology (IJERT), 4*(7), 619–625.
25. Watyotha, C., & Salokhe, V. M. (2001). Tractive performance of cage wheels with opposing circumferential lugs. *Journal of Agricultural and Engineering Research, 79*(4), 389–398.
26. Molari, G., Bellentani, L., Guarnieri, A., Walker, M., & Sedoni, E. (2012). Performance of an agricultural tractor fitted with rubber tracks. *Biosystems Engineering, 111*, 57–63.
27. Rasool, S., & Raheman, H. (2018). Improving the tractive performance of walking tractors using rubber tracks. *Biosystems Engineering, 167*, 51–62.
28. Alex, K., Nigel, H., Peeyush, S., Madhav, D. G., Simon, C., & Jannatul, F. (2013). A review of the tractive performance of wheeled tractors and soil management in lowland intensive rice production. *Journal of Terramechanics, 50*, 45–62.
29. Kim, W.-S., Kim, Y.-J., Baek, S.-Y., Baek, S.-M., Kim, Y.-S., Choi, Y., Kim, Y.-K., & Choi, I.-S. (2020). Traction performance evaluation of a 78-kW-class agricultural tractor using cone index map in a Korean paddy field. *Journal of Terramechanics, 91*, 285–296.
30. El-Sayegh, Z., El-Gindy, M., Johansson, I., Öijer, F., & Improved tire-soil interaction model using FEA-SPH simulation. (2018). *Journal of Terramechanics, 78*, 53–62.
31. Narang, S., & Varshney, A. C. (1995). Evaluation of a 6.7kW power tiller for draft and drawbar power on tar road. *Journal of Terramechanics, 32*(2), 91–97.
32. Sahay, C. S., Thomas, E. V., & Satapathy, K. K. (2009). Performance evaluation of a novel power-tiller-operated oscillatory tillage implement for dry land tillage. *Biosystems Engineering, 102*, 385–391.
33. Wang, X. L., Yamazaki, M., & Tanaka, T. (1995). Characteristics of soil reactions of open lugged wheel under paddy conditions. *Journal of Terramechanics, 32*(3), 115–125.
34. Salokhe, V. M., Gee-Clough, D., & Manzoor, S. (1990). Studies on effect of surface coating on forces produced by cage wheel lugs in wet clay soil. *Journal of Terramechanics, 27*(2), 145–157.
35. Hermawan, W., Yamazaki, M., & Oida, A. (2000). Theoretical analysis of soil reaction on a lug of the movable lug cage wheel. *Journal of Terramechanics, 37*, 65–86.
36. Soekarno, S., & Salokhe, V. M. (2003). Soil reactions on the cage wheels with staggered echelons of half-width lugs and perfect chevron lugs in wet clay soil. *Agricultural Engineering International: The CIGR Journal of Scientific Research and Development, 3*(002), 1–25.

37. Salokhe, V. M., Manzoor, S., & Gee-Clough, D. (1990). Pull and lift forces acting on single cage wheel lugs. *Journal of Terramechanics, 27*(1), 25–39.
38. Hendriadi, A., & Salokhe, V. M. (2002). Improvement of a power tiller cage wheel for use in swampy peat soils. *Journal of Terramechanics, 39*, 55–70.
39. Watyotha, C., Gee-Clough, D., & Salokhe, V. M. (2001). Effect of circumferential angle, lug spacing and slip on lug wheel forces. *Journal of Terramechanics, 38*, 1–14.
40. Salokhe, V. M., Hong, T. T., & Gee-Clough, D. (1994). Effects of low to medium slip, lug spacing, and moisture content on lug forces. *Journal of Terramechanics, 31*(6), 385–394.
41. Culshaw, D. (1988). Rubber tracks for traction. *Journal of Terramechanics, 25*(1), 69–80.
42. Wong, J. Y., & Huang, W. (2006). "Wheels vs. tracks" – A fundamental evaluation from the traction perspective. *Journal of Terramechanics, 43*, 27–42.
43. Bashford, L. L., & Kocher, M. F. (1999). Belts vs tires, belts vs belts, tires vs tires. *American Society of Agricultural Engineers, 15*(3), 175–181.
44. J. H. Esch, L. L. Bashford, K. Von Bargen, R. E. Ekstrom Tractive performance comparisons between a rubber belt track and a four-wheel-drive tractor. American Society of Agricultural Engineers. 33(4), 1109–1115 (1990).
45. Zoz, F. M. (1997). *Belt and tire tractive performance*. Society of Automotive Engineers.
46. Ali, O. S., & McKyes, E. (1979). Effects on soil thrust of lug angle, length and soil consistency. *Transactions of the ASAE*, 1294–1304.
47. Okello, J. A., Dwyer, M. J., & Cottrell, F. B. (1994). The tractive performance of rubber tracks and a tractor driving wheel tyre as influenced by design parameters. *Journal of Agricultural Engineering Research, 59*, 33–43.
48. Mudarisov, S., Gainullin, I., Gabitov, I., Hasanov, E., & Farhutdinov, I. (2020). Soil compaction management: Reduce soil compaction using a chain-track tractor. *Journal of Terramechanics, 89*, 1–12.
49. Dickson, J. W., Campbell, D. J., & Henshall, J. K. (1983). An assessment of seedbed compaction by open, flat-lugged, steel tractor wheels. *Journal of Agricultural Engineering Research, 28*, 45–60.
50. Talaythong Factory Co., Ltd. https://www.talaythong.co.th/EN/products/po wer-tiller123CB.html. Last accessed 2023/05/19.

Chapter 4
Recent Advances of Additive Manufacturing for Aerospace Industries: Methods, Materials, Challenges, and Future Outlooks

Melkamu Tadesse Getachew, Menberu Zeleke Shiferaw, and Bekalu Sintayehu Ayele

4.1 Introduction

The fundamental principle of additive manufacturing (AM), or 3D printing, has been around for many years [1]. Due to this rapid prototyping technology's versatility, affordability, and proven benefits from a production standpoint, interest in it has skyrocketed [1–3]. The process of gradually adding materials layer by layer to create three-dimensional components is known as additive manufacturing [1, 2]. The traditional subtractive manufacturing methods, which include turning, milling, drilling, and other material removal procedures, are incompatible with the additive manufacturing techniques [2–4]. This special feature eliminates the need for expensive forms or tools like dies, punches, or casting molds and enables the direct manufacturing of complex or custom parts from the design. It also removes the need for several common processing steps. Without the constraints of conventional processing methods, complex objects that are accurate to their design can be created in a single step [3–5].

The benefits of additive manufacturing, which have been mentioned above and more, have drawn interest from researchers in a variety of fields, including aerospace, automotive, medical devices, and other fields [2, 6, 7]. In addition to greater shape complexity and design freedom, another advantage of additive printing is the material's point of view. Complicated designs may already be realized through the use of many technological materials in additive manufacturing, such as metals, polymers, ceramics, and composites [8, 9]. Various types of materials are used in additive manufacturing nowadays, depending on the desired outcome of the product

M. T. Getachew (✉) · M. Z. Shiferaw · B. S. Ayele
Faculty of Mechanical and Industrial Engineering, Bahir Dar institute of Technology, Bahir Dar university, Bahir Dar, Ethiopia

S. W. Fanta et al. (eds.), *Sustainable Development Research in Manufacturing, Process Engineering, Green Infrastructure, and Water Resources*, Green Energy and Technology, https://doi.org/10.1007/978-3-031-77339-6_4

that will be produced [10]. The type of material used varies depending on the application, such as automotive, aerospace, and other industrial industries [8, 10]. The potential of additive manufacturing technology to facilitate multi-material AM processes has garnered considerable interest in recent times, owing to its capacity to fabricate intricate structures with customized material properties [11, 12]. Multi-material additive manufacturing (AM) facilitates the creation of components with improved functionality and performance by integrating many materials into a single build. Nevertheless, this methodology presents several obstacles, such as compatibility problems and interface delamination [2, 13].

When adjacent materials in a multi-material structure have a weak link, interface delamination happens. This may result in a finished product with less structural integrity and worse performance. Stress concentrations at the material interfaces can cause delamination because of mismatches in the properties of the materials, such as coefficients of thermal expansion and mechanical behaviors [14, 15]. However, when materials utilized in a multi-material AM process require distinct processing parameters, compatibility problems can occur. The ideal process parameters that concurrently meet the needs of all materials can be challenging to obtain since different materials may have distinct melting or cooling rates. This may result in flaws that further impair the component's performance, such as voids, porosity, or uneven material qualities [14, 15].

Researchers have developed tessellation-based morphological advances in multi-material AM methods to solve these issues [12, 16]. By altering a single material's internal structure, a process known as tessellation can replicate the characteristics of several different materials. Through precise internal geometry design, such as cellular architectures, gradients, or lattice structures, the material can display various capabilities depending on the location. Tessellation-based methods in multi-material additive manufacturing have various advantages. First of all, it decreases material waste and simplifies the fabrication process by doing away with the need for various materials. Second, by utilizing a single material, problems with compatibility between various materials are avoided, improving structural integrity and material compatibility. Last but not least, tessellation-based morphological advancements make it possible to produce parts with complex internal architectures that have better mechanical qualities, like higher energy absorption capacities and better strength-to-weight ratios [17].

Various nesting solutions have also been proposed by researchers to address the difficulties faced by multi-material AM [14]. Nesting is the procedure of positioning several materials inside a construction volume to minimize interface delamination and maximize material usage. By arranging components nearby that have similar processing requirements, these techniques lessen the possibility of heat stress and delamination. To decrease stress concentrations, nesting strategies also take into account the mechanical compatibility of the materials, making sure that neighboring sections have suitable mechanical properties. Using nesting methods in multi-material additive manufacturing processes can significantly lower the complexity of fabricating different components in the aerospace sector. Nesting, for instance, can be used to optimize the positioning of various materials with particular qualities, such as high-temperature resistance or lightweight qualities, in the manufacturing of aviation engine components [11, 14].

In addition to streamlining the production process, this raises the components' overall dependability and performance. Thus, tessellation-based morphological innovations present a viable way to improve the performance of aerospace components in multi-material additive manufacturing processes and reduce fabrication complexity. Multi-material-like qualities can be obtained without the use of numerous materials by altering the internal structure of a single material. Furthermore, nesting solutions improve compatibility and minimize interface delamination by further optimizing material placement. These developments, which make it possible to produce intricate, high-performance components, have the potential to completely transform the aerospace sector [12, 14, 17]. The primary challenge facing the aerospace sector is material complexity, which must be balanced with production complexity to ensure that aircraft weight and strength are balanced. Many metallic components, including those for engines, interiors, and structures, can now be 3D printed thanks to 3D printing [18, 19].

Moreover, Aerospace engineering is a demanding field that requires the use of advanced materials and manufacturing techniques to produce lightweight, complex geometries, customized components, and rapid prototypes. These elements are essential for the development of high-performance, efficient, and reliable aircraft and spacecraft. Lightweight materials, such as carbon fiber composites and titanium (Ti) alloys, are used to reduce the weight of aircraft, which in turn improves fuel efficiency and performance. Complex geometries are often required for aerodynamic surfaces and structural components, and advanced manufacturing techniques, such as additive manufacturing, allow for the precise fabrication of these complex shapes. Customization is important for tailoring components to specific applications and optimizing performance. Rapid prototyping enables the rapid creation of physical prototypes from digital designs, which facilitates testing and design iteration. The importance of lightweight, complex geometries, customization, and rapid prototyping in aerospace engineering cannot be overstated, as they are critical for the development of cutting-edge aerospace technology. The goal of this chapte is to provide an overview of the new advances in additive manufacturing for the aerospace industry, including everything from the fundamentals of the process to finished products.

4.2 Additive Manufacturing Methods (Techniques) for Aerospace Components

The aerospace sector has experienced a transformation because of additive manufacturing, it offers new production alternatives that are more affordable, versatile, and quick than traditional manufacturing. Prototyping and production part fabrication are just two of the many uses for additive manufacturing in the aerospace sector. Aerospace applications use a variety of additive manufacturing techniques, such as stereolithography (SLA), direct metal laser sintering (DMLS), selective laser sintering (SLS), electron beam melting (EBM), and fused deposition modeling (FDM) [8, 9, 20, 21].

The advantages and disadvantages of each type of additive manufacturing are distinct, and engineers select the best technique based on the particular needs of the item they need to make. Due to its ability to save production costs, increase design flexibility, and expedite production, additive manufacturing is quickly taking the lead in the aerospace industry.

4.2.1 Powder Bed Fusion Techniques

Powder bed fusion (PBF) is a group of additive manufacturing (AM) techniques that utilize a high-energy heat source to selectively melt or sinter powder particles to form a solid object layer by layer. PBF techniques are particularly well-suited for aerospace applications due to their ability to produce complex geometries with high precision and dimensional accuracy [22, 23]. Some of the most common PBF techniques used in aerospace applications include:

Selective Laser Melting (SLM) Selective laser melting (SLM) uses a focused laser beam to melt and fuse metal powder particles. This technique is capable of producing high-strength, high-quality metal components with excellent surface finish. SLM has developed as a transformative additive manufacturing (AM) technique, revolutionizing the aerospace industry (Fig. 4.1).

The ability of SLM to produce high-precision, complex geometries with minimal material waste and high strength-to-weight ratios makes it ideal for fabricating lightweight, structurally efficient aerospace components.

SLM is a highly advantageous additive manufacturing method for aerospace applications. It enables the precise fabrication of complex geometries, surpassing the limitations of traditional manufacturing techniques. This capability empowers aerospace engineers to design and manufacture parts that optimize aerodynamic performance, enhance structural integrity, and reduce overall weight. SLM's material versatility further enhances its value, as it can process a wide range of materials,

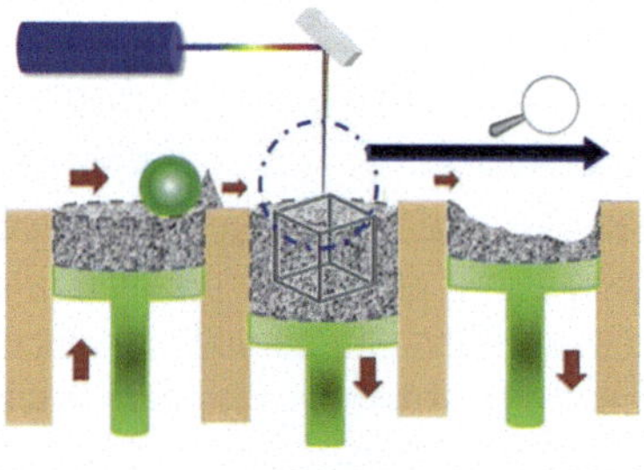

Fig. 4.1 Schematic diagram of selective laser melting [24]

including metals, polymers, and composites. This adaptability allows engineers to tailor materials to meet specific performance requirements, such as strength, weight, and temperature resistance. SLM's precision fabrication, complex geometry capabilities, and material flexibility facilitate the exploration of innovative designs, rapid prototyping, and fuel-efficient lightweight component manufacturing for the aerospace industry [25].

For aerospace component additive manufacturing, SLM offers several difficulties that must be resolved for optimum output. Support structures are often required during the build process to prevent warping and deformation, but removing these supports can be challenging and time-consuming. Ongoing research aims to develop self-supporting structures and dissolvable support materials to simplify post-processing [24, 25]. Residual stresses are another concern in SLM components, resulting from thermal gradients and the solidification process, which can impact performance and fatigue life. Stress relief techniques such as annealing are commonly employed to mitigate these residual stresses [24, 26]. Additionally, achieving the desired surface quality is crucial as SLM-produced components may have a rough surface finish, affecting aerodynamic performance and functionality. Surface finishing techniques like machining or polishing are necessary to achieve the required surface quality [24, 26]. By addressing these challenges through ongoing research and development, SLM can be further optimized for additive manufacturing of aerospace components, improving support structure removal, mitigating residual stresses, and enhancing surface quality.

SLM in additive manufacturing for aerospace components is being enhanced through various strategies and techniques. In situ monitoring and control systems are being developed to detect and mitigate potential defects during the SLM process, reducing the reliance on extensive post-processing [24, 25]. Laser peening, as a post-processing technique, utilizes a laser beam to induce compressive residual stresses on the component's surface, counteracting the tensile stresses caused by the SLM process [24–26]. Electropolishing, an electrochemical process, is employed to remove minute amounts of material from the surface, resulting in a smoother finish and improved surface integrity [24–26]. Additionally, hybrid AM processes, which combine SLM with other manufacturing techniques like machining or laser cutting, are utilized to address specific post-processing challenges. These hybrid approaches enable the removal of supports through laser cutting and achieve precise surface finishes through machining, further optimizing the SLM process [24–26]. Through the development of in-situ monitoring, laser peening, electropolishing, and hybrid AM processes, the aerospace industry continues to enhance the efficiency, reliability, and surface quality of SLM-produced components. SLM is continuously evolving, and advancements in material science, process control, and post-processing techniques are further expanding its capabilities for aerospace applications. As these advancements mature, SLM is poised to play an increasingly significant role in the design, development, and manufacturing of cutting-edge aerospace components.

Electron Beam Melting (EBM) Electron beam melting (EBM) utilizes a high-energy electron beam to melt and fuse metal powder particles. This technique is known for its ability to produce very dense and high-strength metal components, particularly for refractory metals like titanium and nickel alloys (Fig. 4.2).

Electron beam melting (EBM) is a high-precision additive manufacturing (AM) technique that utilizes a focused electron beam to selectively melt and fuse metal powder particles layer by layer to create complex three-dimensional objects. Unlike laser-based AM techniques, EBM utilizes a vacuum chamber, enabling the processing of reactive metals, such as titanium and nickel alloys, which are prone to oxidation in the presence of oxygen [27].

EBM has witnessed exceptional advancements in additive manufacturing for aerospace applications. One of its key strengths lies in its ability to produce components with exceptional material properties, such as high strength, toughness, and fatigue resistance. These properties are of paramount importance for aerospace components that must endure extreme stresses and harsh environments. Moreover, EBM stands out for its capability to process reactive metals like titanium and nickel alloys. This feature makes it an ideal choice for fabricating aerospace components that demand high strength-to-weight ratios, high-temperature resistance, and corrosion resistance, all critical factors in aerospace engineering [27, 28].

In addition to its remarkable material properties, EBM excels in delivering fine feature detail and a smooth surface finish, reducing the need for extensive post-processing. This level of precision is especially crucial for manufacturing aerospace

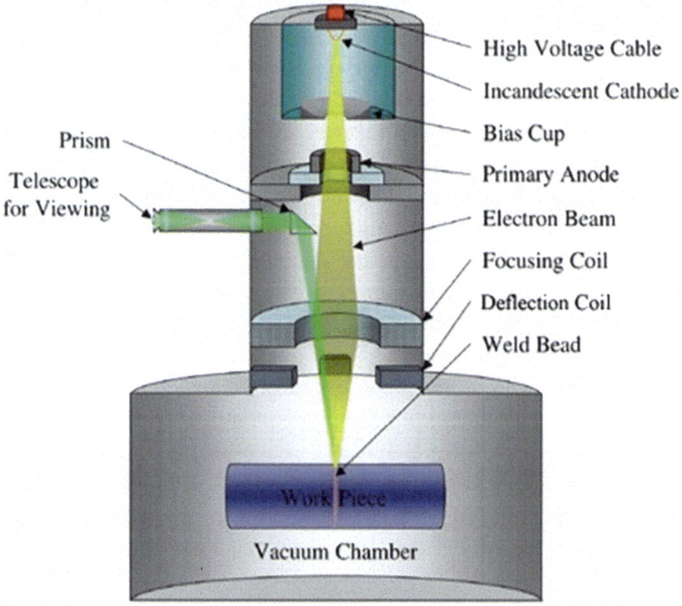

Fig. 4.2 Electron beam melting [26]

components with intricate geometries and critical surface requirements [27, 28]. Furthermore, EBM's layer-by-layer deposition process minimizes material waste, ensuring efficient utilization of valuable materials. This attribute holds particular significance in the aerospace industry, where weight optimization plays a crucial role in enhancing fuel efficiency and overall performance [27, 28].

Considering its exceptional material properties, the ability to process reactive metals, the precision in fine feature detail and surface finish, and the minimal material waste it generates, EBM has emerged as a prominent additive manufacturing technique for aerospace applications. Its unique capabilities address the stringent requirements of the aerospace industry, enabling the production of high-performance components that can withstand demanding operational conditions while optimizing weight and fuel efficiency. As research and development in EBM continue, it is anticipated that further advancements will be made, cementing its role as a key technology in aerospace additive manufacturing.

EBM, like other additive manufacturing techniques, faces certain challenges that need to be addressed for optimal component production in aerospace applications. One challenge involves the need for support structures during the build process to prevent warping and deformation. However, these support structures can be difficult to remove and may leave residual marks on the component surface. To overcome this, ongoing research focuses on developing self-supporting structures and dissolvable support materials, aiming to simplify the post-processing stage and eliminate the challenges associated with support structure removal [28, 29].

Residual stresses are another challenge encountered in EBM components due to thermal gradients and the solidification process. These stresses can have implications for component performance and fatigue life. To mitigate these effects, stress relief techniques such as annealing are commonly employed to release residual stresses and enhance component integrity. While EBM generally produces components with a smooth surface finish compared to other additive manufacturing techniques, certain applications may require even smoother surfaces. To address this, surface finishing techniques like machining or polishing can be applied to achieve the desired surface quality and meet specific aerospace requirements [28, 29].

Another challenge lies in the removal of EBM components from the build plate due to strong adhesion between the component and the plate. Advanced build plate designs and the use of release agents are being developed to facilitate easier component removal and reduce the risk of damage during the removal process. By addressing these challenges through ongoing research and development, the aerospace industry can further optimize the use of EBM in additive manufacturing. Overcoming issues related to support structures, residual stresses, surface roughness, and component removal from the build plate will contribute to the production of high-quality, reliable components using EBM technology, enhancing their suitability for aerospace applications [27, 30].

EBM has emerged as a leading AM technique for aerospace applications due to its ability to produce high-quality components from reactive metals with complex geometries and minimal material waste. As research continues to address post-processing challenges, EBM is expected to play an even greater role in the manufacturing of next-generation aerospace components [27, 30].

Direct Metal Laser Sintering (DMLS) Direct metal laser sintering (DMLS) is a variant of SLM that uses a focused laser beam to sinter metal powder particles rather than melting them completely. This technique is often used for processing materials that are prone to cracking or warping during melting, such as aluminum alloys.

The manufacturing of high-performance metal parts with intricate geometries has been made possible by DMLS, a 3D printing technology that has completely transformed the aerospace sector. When making aviation components that need to be extremely strong, durable, and resistant to high temperatures and pressures, DMLS is very helpful [2, 10, 31]. Heat exchangers, brackets, engine parts, and turbine blades are a few of the parts of aircraft that are frequently produced utilizing DMLS. Aerospace engineers may swiftly iterate and improve their ideas before going into production by using DMLS for prototyping and testing parts. All things considered, DMLS is an invaluable tool for the aerospace sector, facilitating the manufacture of superior, lightweight parts that are essential to the efficient and safe functioning of airplanes [5, 16] (Fig. 4.3).

Achieving dimensional accuracy and surface quality are post-processing problems in DMLS additive manufacturing for airplane components. Because of the nature of the process, DMLS may produce dimensional irregularities and uneven surface finishes. Specialized post-processing methods like chemical treatments, sanding, and polishing are used to solve these problems. By producing a consistent finish and getting rid of surface flaws, these methods aid in improving surface quality. Careful quality control procedures and testing techniques are used to guarantee that the printed components fulfill the necessary requirements in terms of dimensional correctness. DMLS can overcome these obstacles and create aircraft components with exceptional surface quality and dimensional precision by creating effective post-processing solutions [33, 34].

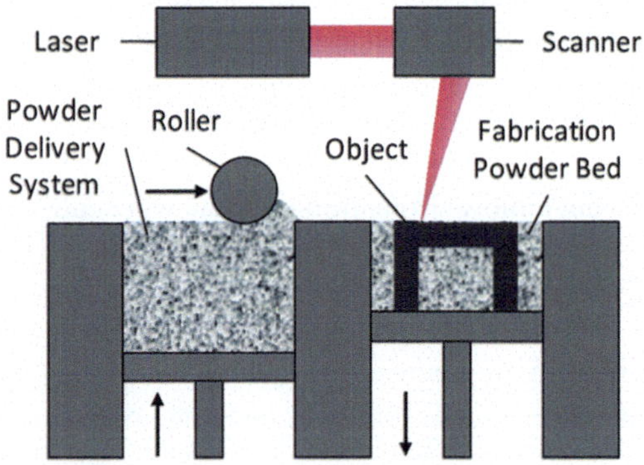

Fig. 4.3 Schematic diagram of DMLS [32]

4.2.2 Directed Energy Deposition (DED) Techniques

Directed energy deposition (DED) is a group of additive manufacturing (AM) techniques that utilize a concentrated heat source, such as a laser or electron beam, to melt and fuse material feedstock, typically in the form of powder or wire, onto a substrate layer by layer. DED techniques are particularly well-suited for aerospace applications due to their ability to produce large, complex geometries, repair damaged components, and process a wide range of materials [35–37]. Some of the most common DED techniques used in aerospace applications include:

Laser-Engineered Net Shaping (LENS) Laser-engineered net shaping (LENS) uses a focused laser beam to melt and fuse metal powder feedstock onto a substrate. This technique is known for its ability to produce high-strength, high-quality metal components with excellent surface finish. LENS is particularly well-suited for aerospace applications due to its ability to produce high-precision, high-strength, and complex geometries [32] (Fig. 4.4).

Laser-engineered net shaping (LENS), also known as direct metal laser sintering (DMLS), is a revolutionary additive manufacturing (AM) technique that has transformed the aerospace industry. Unlike traditional subtractive manufacturing methods that remove material to create a desired shape, LENS builds objects layer by layer from a digital model, enabling the fabrication of complex geometries and customized components with unprecedented precision and efficiency [30, 38].

LENS offers several advantages for aerospace applications. Firstly, it provides design freedom, enabling engineers to create complex geometries that are challenging or impossible to achieve using conventional manufacturing methods. This design freedom facilitates the development of lightweight, high-performance

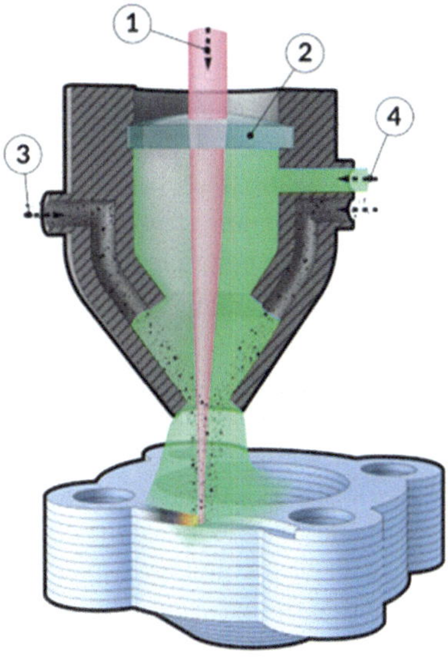

Fig. 4.4 Schematic of LENS (A laser beam [1] created by a laser generator is focused through a lens [2] onto the workpiece. Metal powder [3] or metal wire is introduced at the focus of the laser beam, where the metal powder and workpiece melt and thus build) [32]

components that optimize aerodynamic performance and structural integrity. Secondly, LENS ensures material efficiency by utilizing only the necessary material for building the desired structure, minimizing waste, and maximizing material utilization. This efficiency is particularly valuable in the aerospace industry, where lightweight materials are crucial for fuel efficiency. Additionally, LENS enables customization, allowing the fabrication of tailored components that meet specific application and performance requirements, which is essential in aerospace where stringent standards must be met. Moreover, LENS excels in high-precision and complex geometry fabrication, producing components with superior precision and the ability to fabricate intricate geometries that are difficult to achieve through traditional methods. Lastly, LENS offers a wide range of material versatility, capable of processing metals, polymers, and composites, catering to the diverse material requirements of aerospace applications [38].

LENS, like other additive manufacturing (AM) techniques, faces certain challenges that need to be addressed. One such challenge is the use of support structures to prevent warping and deformation during the build process, which can be difficult to remove and may leave marks on the component surface. Residual stresses are another concern in LENS components due to thermal gradients and the solidification process, potentially impacting performance and fatigue life. Although LENS produces components with a relatively smooth surface finish compared to other AM techniques, certain applications may require even smoother surfaces [39, 40]. By addressing these challenges, LENS can further optimize the production of aerospace components by improving support structure removal, mitigating residual stresses, and meeting specific surface quality requirements.

LENS technology in additive manufacturing for aerospace components is continuously advancing to overcome various challenges. Ongoing research aims to develop self-supporting structures, eliminating the need for additional supports and simplifying post-processing, while reducing residual marks. Advanced stress relief techniques, including annealing and high-temperature exposure, are being explored to mitigate residual stresses and enhance component performance. Surface finishing methods such as machining, polishing, and electropolishing are employed to achieve the desired surface quality required for specific aerospace applications. Additionally, the development of real-time monitoring and control systems enables in-situ detection and mitigation of potential defects during the LENS process, minimizing the reliance on extensive post-processing and improving overall component quality. By addressing these challenges through the advancement of self-supporting structures, stress relief techniques, surface finishing methods, and in-situ monitoring and control, LENS technology continues to enhance the efficiency, reliability, and surface quality of aerospace components [39, 40].

The combination of LENS's unique advantages with advancements in post-processing techniques is propelling this AM technology forward, making it an increasingly viable and impactful tool for manufacturing high-performance, complex aerospace components. As LENS continues to evolve, it is poised to play a pivotal role in shaping the future of aerospace design and manufacturing.

Electron Beam Additive Manufacturing (EBAM) Electron beam additive manufacturing (EBAM) uses a high-energy electron beam to melt and fuse metal powder feedstock onto a substrate. This technique is particularly well-suited for processing refractory metals like titanium and nickel alloys, which are difficult to melt using lasers. EBAM is revolutionizing the aerospace industry by advancing the production of various components. One area of advancement is the fabrication of complex geometries, allowing for the creation of intricate and lightweight aerospace components that were previously difficult or impossible to manufacture. EBAM also enables the production of large-scale components, such as fuselage sections and engine parts, in a single piece, eliminating the need for assembly and reducing weight while maintaining structural integrity [41, 42].

Furthermore, EBAM offers benefits in material efficiency, as it allows for the precise deposition of materials, minimizing waste and optimizing the use of expensive aerospace-grade alloys. The technology also enables the integration of multiple materials and functionalities within a single component, enhancing performance and reducing assembly complexity. Additionally, EBAM's ability to generate fully dense and near-net shape components reduces the reliance on traditional subtractive manufacturing processes, saving time and costs. With its advancements in complex geometry fabrication, material efficiency, integration of multiple materials, and reduced reliance on subtractive manufacturing, EBAM is driving significant progress in the aerospace industry, revolutionizing the way aerospace components are designed and produced [41, 43] (Fig. 4.5).

EBAM offers numerous advantages for the aerospace industry. Firstly, it enables the production of complex and lightweight components with intricate geometries that were previously challenging to manufacture. This capability allows for the design and fabrication of optimized structures that enhance performance and reduce weight, leading to improved fuel efficiency and overall aircraft performance. Additionally, EBAM enables the production of large-scale components in a single piece, eliminating the need for assembly and reducing potential points of failure. This results in stronger and more structurally sound components. Furthermore,

Fig. 4.5 Electron beam additive manufacturing [44]

EBAM offers material efficiency by allowing precise deposition of materials, reducing waste, and optimizing the use of costly aerospace-grade alloys. This efficiency is crucial in the aerospace industry, where lightweight materials and cost-effectiveness are paramount [42, 45].

While EBAM offers numerous advantages, it also presents post-processing challenges in the aerospace industry. One major challenge is the removal of support structures. The complex geometries and intricate designs of EBAM-produced components often require support structures during the printing process to prevent warping and distortion. However, removing these supports can be time-consuming and labor-intensive. To address this challenge, ongoing research focuses on developing self-supporting structures and dissolvable support materials, which would simplify the post-processing steps and reduce the need for manual support removal [41, 45].

Another post-processing challenge is surface roughness. EBAM-produced components may have a rough surface finish, which can impact aerodynamic performance and component functionality. Surface finishing techniques, such as machining, polishing, and post-processing treatments like shot peening or grinding, are commonly employed to achieve the desired surface quality. Additionally, advancements in in-situ monitoring and control systems are being developed to detect and mitigate potential defects during the EBAM process, reducing the need for extensive post-processing and improving component quality [43].

The ongoing research and development efforts in EBAM are focused on addressing these post-processing challenges. By advancing self-supporting structures, improving surface finishing techniques, and enhancing in-situ monitoring and control systems, EBAM is poised to overcome these challenges and further optimize the production of aerospace components [41, 43].

Wire Arc Additive Manufacturing (WAAM) Wire arc additive manufacturing (WAAM) utilizes a gas metal arc welding (GMAW) process to deposit metal wire feedstock onto a substrate. This technique is known for its ability to produce large-scale components with high deposition rates. WAAM has had a significant impact on the aerospace industry. WAAM enables the production of large-scale and complex components with reduced lead times and costs compared to traditional manufacturing methods. This technology allows for the deposition of metal layers using an electric arc between a wire electrode and the substrate, resulting in the gradual buildup of the desired component. The ability to fabricate near-net shape parts and repair worn or damaged components has made WAAM particularly valuable in the aerospace sector. Moreover, WAAM offers design flexibility, allowing for the creation of complex geometries and the incorporation of internal features, such as cooling channels, which can enhance the performance and efficiency of aerospace components. Furthermore, WAAM supports the use of a wide range of materials, including aluminum, titanium, and high-temperature alloys, catering to the diverse requirements of aerospace applications. With its advantages in cost-effectiveness, design flexibility, and material versatility, WAAM has become a promising technology in the aerospace industry, driving innovation and transforming the way components are manufactured [29, 46, 47] (Fig. 4.6).

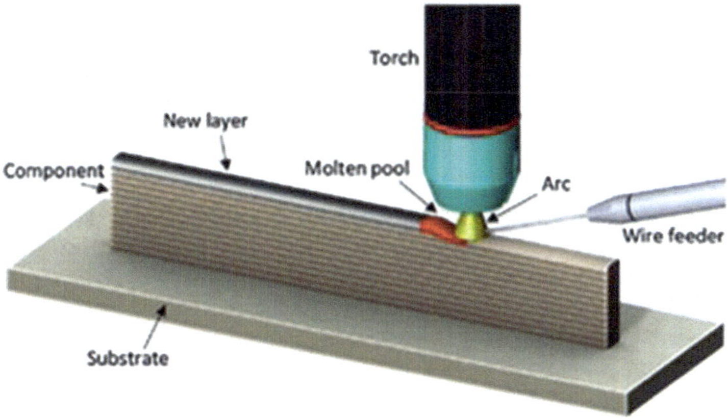

Fig. 4.6 Wire arc additive manufacturing [48]

Post-processing challenges in WAAM for aerospace applications include the removal of support structures and the need for surface finishing. Support structures are often required during the deposition process to prevent deformation, but their removal can be time-consuming and labor-intensive. Additionally, WAAM-produced components may exhibit rough surface finishes, which can negatively impact aerodynamic performance and component functionality [46, 48].

Upcoming solutions are being developed to address these challenges in WAAM for aerospace applications. Research and development efforts are focused on advancing self-supporting structures and dissolvable support materials, which would simplify the post-processing steps and reduce the need for manual support removal. Moreover, surface finishing techniques such as machining, polishing, and post-processing treatments are being optimized to improve the surface quality of WAAM-produced components. These advancements will enable the production of aerospace components with smoother surfaces, meeting the stringent requirements of aerodynamic performance and component functionality. By addressing these post-processing challenges, WAAM is poised to become even more efficient and reliable in the aerospace industry, offering cost-effective manufacturing solutions with improved surface quality [46, 48].

4.2.3 Material Extrusion Techniques

Material extrusion (ME) is a group of additive manufacturing (AM) techniques that utilize a heated nozzle to extrude a thermoplastic filament or paste layer by layer to form a solid object. ME techniques are particularly well-suited for aerospace applications due to their ability to produce complex geometries, customized components, and a wide range of thermoplastic materials. Some of the most common ME techniques used in aerospace applications include:

Fused Deposition Modeling (FDM) Fused deposition modeling (FDM) is the most widely used ME technique, utilizing a heated nozzle to extrude a thermoplastic filament, typically made from polymers like ABS, PLA, or nylon. FDM is known for its simplicity, affordability, and ability to produce high-resolution components. Advancements in FDM for aerospace applications have been significant [49]. FDM technology has evolved to include high-temperature materials capable of meeting the stringent requirements of aerospace components. This advancement allows for the production of parts with improved heat resistance and mechanical properties, expanding the range of applications for FDM in aerospace. Furthermore, advancements in FDM machines and software have improved the level of precision and accuracy, enabling the fabrication of complex geometries with greater detail and intricacy [49, 50] (Fig. 4.7).

Post-processing challenges in FDM for aerospace applications primarily revolve around achieving the desired surface finish and dimensional accuracy. FDM-produced parts often have visible layer lines and surface roughness, which can impact aerodynamic performance and aesthetics. Additionally, achieving tight tolerances can be challenging due to the inherent limitations of the layer-by-layer deposition process [49, 52].

To address these challenges, various solutions are being explored in FDM for aerospace applications. Post-processing techniques such as sanding, polishing, and chemical treatments can be applied to improve the surface finish of FDM parts. Additionally, advanced FDM systems and software algorithms are being developed to enhance dimensional accuracy and minimize the impact of visible layer lines. Furthermore, the use of support materials that can be easily removed or dissolved, reducing the need for manual support removal, is being explored to simplify the post-processing steps. These solutions aim to improve the overall quality and performance of FDM-produced components in aerospace applications [49, 52].

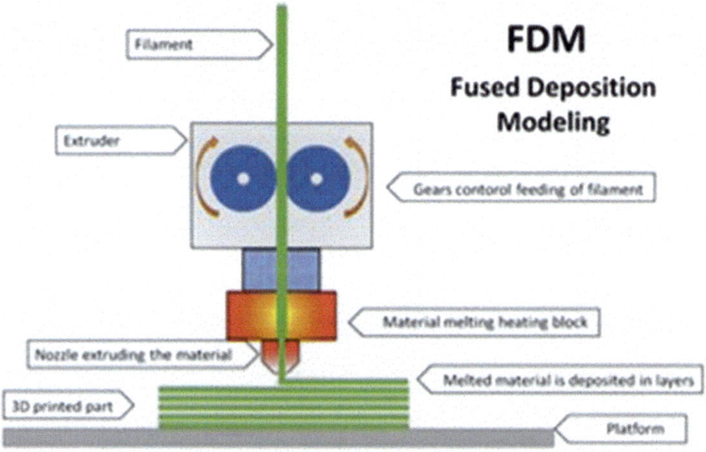

Fig. 4.7 Fused deposition modeling [51]

Big Area Additive Manufacturing (BAAM) BAAM is an emerging additive manufacturing technology that has the potential to revolutionize the fabrication of large-scale aerospace components. Unlike traditional AM techniques, which are limited to producing small to medium-sized parts, BAAM can produce components with dimensions of up to 100 m or more. This capability opens up new possibilities for the design and manufacturing of aerospace structures, such as aircraft fuselages, wings, and rocket bodies. BAAM works by extruding a continuous stream of molten material, typically plastic or metal, onto a substrate layer by layer. The material is then cooled and solidified, forming the desired shape. The process is controlled by a computer-aided design (CAD) system, which allows for precise control of the deposition pattern and material properties [49, 52] (Fig. 4.8).

BAAM offers several advantages for the aerospace industry. Firstly, it enables the production of large-scale components that surpass the limitations of traditional additive manufacturing (AM) techniques. This capability is particularly crucial for fabricating aerospace structures like aircraft fuselages and wings. Secondly, BAAM significantly reduces lead times in aerospace component fabrication by eliminating the need for traditional tooling or molds, which are time-consuming and expensive to manufacture. The absence of these processes allows for faster production cycles, enhancing overall efficiency. Finally, BAAM proves to be a cost-effective alternative to conventional manufacturing methods such as machining and welding. It achieves this by minimizing material waste and reducing labor costs associated with these traditional techniques. The cost efficiency of BAAM has a substantial impact on the overall production cost of aerospace components, making it an attractive option for manufacturers seeking optimized operations [53, 54].

BAAM brings several advancements to the aerospace industry. Firstly, researchers have developed new materials specifically tailored for BAAM, offering improved properties such as strength, stiffness, and heat resistance. These advanced materials are well-suited for aerospace applications, ensuring the production of components with superior performance. Secondly, BAAM systems have seen significant

Fig. 4.8 Big area additive manufacturing

progress in terms of deposition rates. They can now deposit material at much higher rates compared to traditional AM systems. This capability enables the fabrication of larger aerospace components in a shorter timeframe, enhancing production efficiency. Lastly, BAAM systems are equipped with advanced sensors and control systems, allowing for precise process control. This ensures the consistent production of high-quality components with desired properties, enhancing reliability and reducing variability. These advancements in material properties, deposition rates, and process control contribute to BAAM's growing prominence in the aerospace industry, providing manufacturers with enhanced capabilities for efficient and reliable production [53, 54].

As BAAM technology continues to mature, it is expected to play an increasingly important role in the fabrication of aerospace components. BAAM has the potential to revolutionize the aerospace industry by enabling the production of lighter, stronger, and more cost-effective aerospace structures.

4.2.4 Comparison of Additive Manufacturing Methods with Their Advantages and Limitations in Aerospace Applications

In aerospace applications, a comparison of different 3D printing methods reveals distinct advantages and limitations. PBF, such as SLM and EBM, offers high accuracy and excellent material properties, making it suitable for complex geometries and critical components. However, it can be time-consuming and costly. Material extrusion, including FDM, provides faster production rates and affordability, but sacrifices precision and material strength. Directed energy deposition (DED) allows for large-scale manufacturing and repair, but may have limitations in terms of surface finish and dimensional accuracy. Each method presents unique benefits and constraints, highlighting the importance of selecting the appropriate 3D printing technique based on specific aerospace requirements and component characteristics.

In general, SLM and EBM are the most widely used 3D printing methods in aerospace applications because they can produce high-quality components with complex geometries. However, these methods can be slow and expensive. FDM is a more affordable option, but it is limited in terms of precision and the range of materials that can be used. Material jetting (MJ) and robocasting (RC) can produce high-precision components, but they are limited in terms of the range of materials that can be used. WAAM can produce large-scale components, but it is limited in terms of precision and can have high residual stresses. The choice of 3D printing method for a particular aerospace application will depend on the specific requirements of the application. For example, if a high-precision component with a complex geometry is required, then SLM or EBM is a good choice. If a large-scale component is required, then WAAM is a good choice. If a simple, affordable component is required, then FDM is a good choice.

Table 4.1 Comparison of different additive manufacturing methods

Method	Advantages	Limitations
Selective laser melting (SLM)	High precision, can produce complex geometries, can use a wide range of materials	Can be slow, expensive, can be difficult to remove supports
Electron beam melting (EBM)	Can produce large, complex geometries, can repair damaged components, can process a wide range of materials	Can be slow, can be expensive, can have high residual stresses
Direct metal laser sintering (DMLS)	Simple, affordable, can produce high-resolution components	Limited range of materials, can be slow
Fused deposition modeling (FDM)	Affordable, versatile, can use a wide range of materials	Limited precision, can be slow
Material jetting (MJ)	High precision, can produce multi-material components	A limited range of materials can be expensive
Robocasting (RC)	Can produce high-strength, high-temperature-resistant components	Limited precision, can be slow
Wire arc additive manufacturing (WAAM)	Can produce large-scale components, can reduce lead times, can be cost-effective	Limited precision, can have high residual stresses

Here is a summary of the pros and cons of each technique [51] (Table 4.1):

4.3 Materials for Aerospace Additive Manufacturing

The variety of materials used in additive manufacturing is constantly growing, encompassing a wide range of industries like consumer products, automotive, aerospace, and healthcare. These include thermoplastics, metals, biocompatible resins, and sophisticated composites. Additive manufacturing (AM) of aerospace components enables the fabrication of intricate, lightweight, and high-performance components. Hence, the choice of material for AM depends on the specific application and the desired properties of the component. Common metals used in AM for aerospace applications include polymers, composites, ceramics, metals, and alloys such as titanium alloys, nickel-based superalloys, aluminum alloys, etc.

4.3.1 Polymers

Polymers are materials composed of repeating chains of molecules. Polylactic acid (PLA) and acrylonitrile-butadiene-styrene (ABS) are the most widely utilized polymers for 3D printing. These polymers are used in many industries, including the electronics, automotive, aerospace, and healthcare sectors [55]. Aerospace commonly uses synthetic thermoplastics like nylon and acrylonitrile-butadiene-styrene (ABS) to manufacture air ducts, seatbacks, wall panels, and other interior

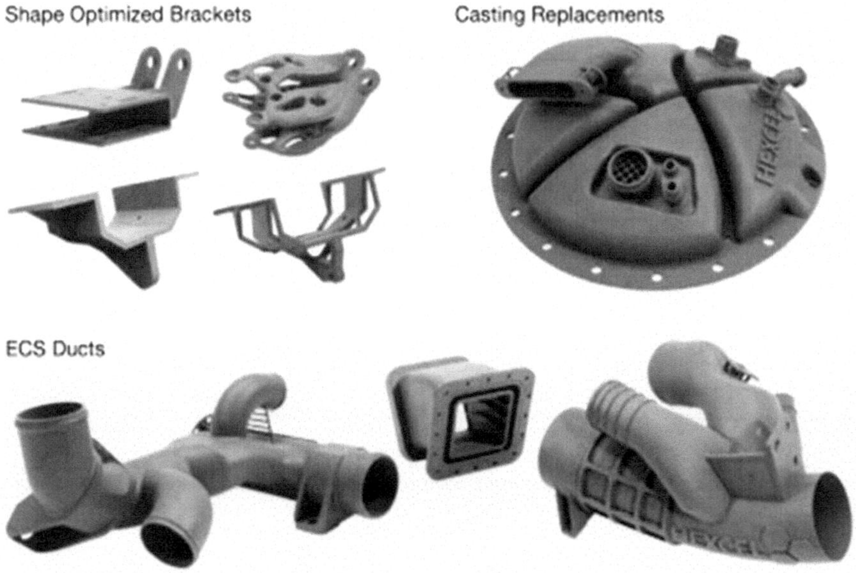

Fig. 4.9 Polymer-based 3D-printed aerospace components [56]

components. Polymers are generally excellent for use in aircraft applications since they are strong and lightweight. The various qualities exhibited by the polymer materials employed make them an adaptable option for a broad range of applications. Their versatility in reaching varying degrees of stiffness, density, and tensile strength makes them a noteworthy advantage that enables customized solutions to satisfy project-specific requirements. Their melting values are often lower than those of metals, which makes AM easier. However, polymers cannot, be utilized in heavy load-bearing applications where metal is frequently preferred due to their weakness in comparison to metals [56] (Fig. 4.9).

4.3.2 Ceramics

Ceramic materials offer capabilities including resistance to heat, brittleness, corrosion, and hardness. Their outstanding heat resistance, which qualifies them for high-temperature applications like those in the automotive and aerospace sectors, is one of their main advantages. They are also very effective electrical insulators. Ceramics guarantee long-lasting and robust parts because of their exceptional hardness and wear resistance. Their well-established qualities enable them to meet the constantly rising needs of numerous industry areas, such as aerospace and automotive. Their brittle nature restricts their use in applications where there is a risk of impact [57].

The different ceramics feedstock types utilized in additive manufacturing technologies are slurry-based, powder-based, and bulk solid-based. Liquid or semi-liquid systems are used in slurry-based ceramic additive manufacturing, where the feedstock is fine ceramic particles distributed. Conversely, loose ceramic particles enclosed within powder beds serve as the feedstock for powder-based ceramic additive manufacturing. Last but not least, bulk solid-based ceramic additive manufacturing uses a variety of ceramic materials as feedstock, including lead zirconate titanate (PZT), silicon carbide, glass-ceramic composites, tape-cast alumina, and zirconia green sheets. This process is inspired by the manufacturing of laminated objects [56].

The microstructure of the ceramic materials utilized is characterized by well-defined crystalline grain patterns within amorphous matrices. The constituents of printed things, ceramic powders, are made up of tiny particles with crystalline phases. These powders are layer by layer selectively fused during the AM process to form a network of closely spaced grains. The performance of the material can be precisely tuned by adjusting the microstructure through variables like dwell time and sintering temperature. This enables the engineering of ceramics with specialized properties for a range of applications, from biocompatible medical implants to high-temperature aerospace components [56].

Ceramic matrix composites (CMCs) are composites that consist of a ceramic matrix reinforced with fibers, such as carbon or silicon carbide. They are incredibly strong and lightweight, and they can withstand very high temperatures. CMCs are being developed for a variety of aerospace applications, including rocket engines, heat shields, and hypersonic vehicles. Additive manufacturing processes like binder jetting and SLA are used to create ceramic parts with high dimensional accuracy, that were traditionally challenging to produce, making them suitable for applications in aerospace engines, heat shields, and thermal protection systems.

However, because of their high melting temperatures, brittleness, and consequently low tolerance to processing faults, advanced ceramics have proven to be much harder to AM than metals or polymers. It requires further study and advancement to speed up the use of ceramic AM. Combining the development of AM machines specifically made to process ceramic materials with the optimization of material formulations and processing routes has proven to be a more effective way to address the processing challenges [57].

4.3.3 Metal and Alloys

The metal 3D printing process has garnered significant interest from a range of industries, including the automotive, aerospace, manufacturing, and medical sectors, owing to its numerous advantages. Their remarkable mechanical strength is one of their main features, which makes them perfect for applications like the aerospace sectors requiring durable and load-bearing elements. Metals are appropriate for high-temperature applications because of their superior heat resistance and thermal conductivity. They can be heavy, though, thus their usage in applications where weight is a factor is limited. Among these metals include steel, aluminum alloys, titanium alloys, nickel-based alloys, and cobalt-based alloys. Metallic objects created with additive manufacturing processes have finer microstructures and are denser overall than those generated with conventional methods [56].

Titanium Alloys Titanium (Ti) alloys are becoming increasingly popular in the aerospace and automotive industries due to their exceptional mechanical and chemical properties. Ti alloys are perfect for high-temperature and high-strength applications, including the blades and casings of engines and steam turbines. Additive manufacturing (AM) of titanium alloys offers several advantages over traditional methods because they are more suited to cutting tools than other materials. One additive manufacturing (AM) technology that can be used to fabricate big metallic components is DED, which has a high deposition rate. Several titanium-based alloys, including Ti6Al4V, Ti8Al1Er, TC11, TC21, and Ti5553, have garnered attention recently. These alloys have special qualities that make them attractive options for a range of aerospace-related applications. They are employed in aircraft structures, engine components, and landing gear [58] as shown in Fig. 4.10.

Ti64 alloy, or Ti6Al4V alloy, is a well-known $\alpha + \beta$ titanium alloy that accounts for 50% of the world's titanium supply. Ti64 is a highly sought-after material in many industries, including the aerospace, marine, energy, and automotive sectors. Its high strength, resilience to fatigue, and corrosion resistance combine with its low density. However, because of its high chemical reactivity, limited heat conductivity, and vulnerability to strain hardening, producing Ti6Al4V alloys can be difficult. In the past, conventional production processes including casting, shaping, and metal

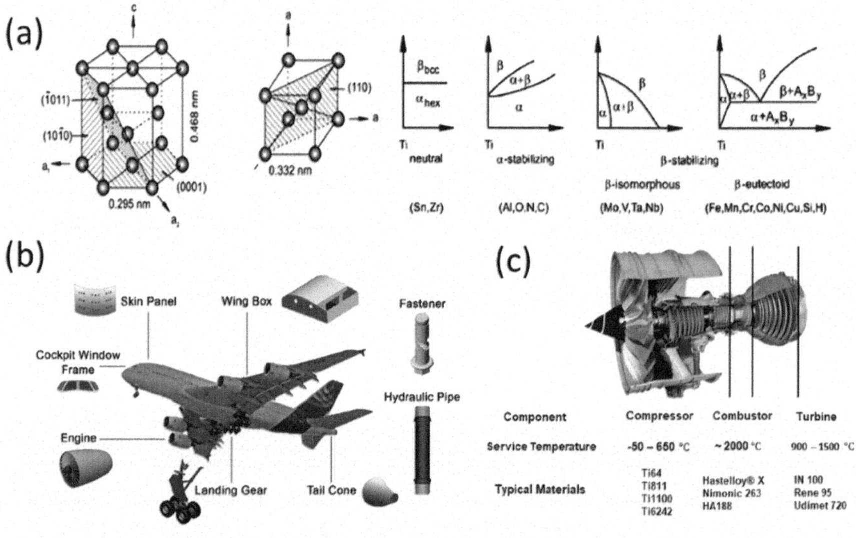

Fig. 4.10 (**a**) The titanium phase diagram categories with varying alloying (the α (hcp) and β (bcc) structures), (**b**) typical uses of titanium alloys in aircraft fuselage, and (**c**) typical gas turbine engine [58]

removal were employed; however, these processes frequently led to material waste, higher manufacturing costs, and longer lead times. Additive manufacturing techniques are utilized as a more effective substitute, providing the benefit of creating intricate structures while resolving the aforementioned issues. Adopting AM processes allows for the achievement of higher densities of Ti alloys as compared to conventional production techniques [56, 58].

In China, the alloy TC11 (Ti-6.5Al-3.5Mo-1.5Zr-0.3Si) is widely used in aerospace applications, particularly for compressor discs, due to its ability to withstand high strains and stress during loading. A study focused on the tensile strength of the TC11 alloy created by laser additive manufacturing, linking its strength to the configuration of columnar grains and the texture aligned along the deposition axis [59]. In comparison to other Ti-based alloys, Ti5553 alloys, which are made of Ti, Al, vanadium (V), molybdenum (Mo), and chromium (Cr), have substantially better mechanical qualities. Ti5553 alloys are extremely suited for applications where high-stress resistance components are needed. These alloys are appropriate for usage in crucial components such as the fuselage, wings, landing gear, and other elements of airplanes and helicopters that are subjected to significant mechanical loading. Using several Ti alloys, 3D-printed aircraft components are displayed in Fig. 4.11.

Aluminum Alloy Aluminum alloys are lightweight, corrosion-resistant, and relatively inexpensive, making them a good choice for a variety of aerospace applications. Aluminum (Al) alloys are appealing for additive manufacturing (AM) due to their low cost and excellent thermal conductivity, which allows for a simpler and quicker production process and reduces thermal stresses when compared to other metallic materials. Two popular alloys made of aluminum are AlSi10 and AlSi12. By adding scandium to Al alloys above the eutectic point, materials that are in high demand in aerospace applications can now be developed. An Al alloy that had scandium added showed a maximum tensile strength of 530 MPa. Because of its ductile

(a) (b)

Fig. 4.11 3D-printed parts made of titanium alloy. (**a**) Ti-5Al–2.5 Sn fuel pump impeller for the space shuttle, (**b**) Ti-6–4 and Ti-6242 compressor spool for a 747 engine [59]

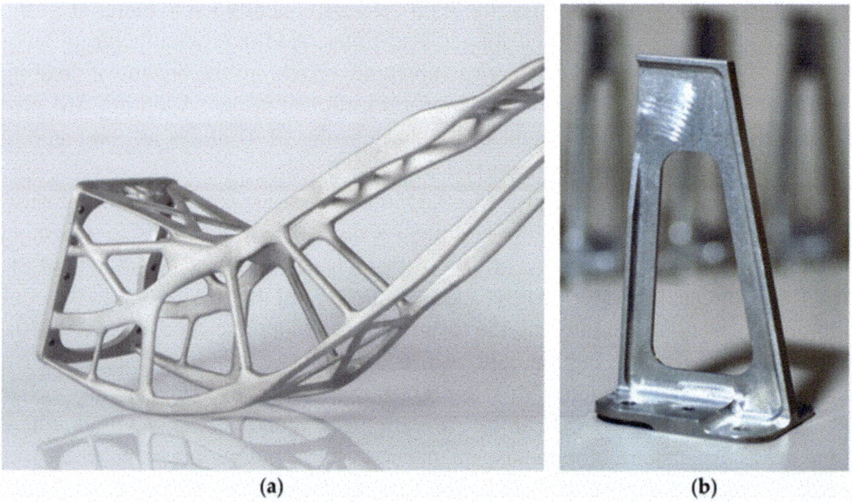

Fig. 4.12 AlSi10Mg antenna bracket manufactured by L-PBF (a) and brackets mounted on Boeing 787 (b) [60]

nature, the material also showed a high degree of flexibility, with a 20% reduction in cross-sectional area and a 14% elongation. Al-Li alloys 2060 and 2055 surpassed typical aerospace aluminum alloys like 2024–T3 and 7075–T6 in terms of corrosion performance, strength/toughness relationship, and thermal stability. Consequently, these alloys could be used as alternative components for the construction of the fuselage, lower wings, and upper wings. Moreover, Al alloys have been used in combustion engine AM lattice-structured cooling systems. The housing of the AlSi10Mg microturbine generator is made by Cobra Aero and nTopology. In contrast to traditional production, which requires six parts for each system, the manufacturers demonstrated a 50% decrease in waste material and the ability to construct the system with a single component [60] (Fig. 4.12).

Magnesium Alloys Magnesium (Mg) alloys are used in the aircraft industry primarily to replace heavier Al alloys because of their lower weight, reduced density, and promising strength characteristics. For example, Mg alloy AZ31 outperforms the Al-based alloy 2017A in cycle fatigue resistance below 150 MPa. Furthermore, because of their comparable density and increased rigidity, magnesium alloys are regarded as plastic substitutes. Additionally, there are several ways to improve the mechanical properties of magnesium alloys, including surface layer modification, grain size modifications, and composite conversion. Despite their advantages, magnesium alloys are highly susceptible to oxidation and degradation, mostly because of their hexagonal crystal structure. Studies revealed several processing-related problems as a result of this, including the possibility of an ignite when processing loose materials like chips or powders at high temperatures. The aircraft industry has used fewer magnesium alloys as a result of this characteristic [61].

4.3.4 Carbon Fiber-Reinforced Polymers (CFRPs)

Composites are also being used in AM to create components with tailored properties. Composites are made by combining two or more materials with different properties, such as metals and polymers, to create a material with improved properties. Polymer composites, particularly carbon fiber-reinforced polymers (CFRPs), offer a combination of high strength, lightweight, and stiffness. They are incredibly strong and lightweight, making them ideal for use in aircraft structures, such as wings, fuselages, and tailplanes [62]. With a weight savings of up to 1.5 tons when compared to the most sophisticated aluminum alloys, the A380 is the first aircraft in history to include a carbon fiber-reinforced plastic (CFRP) composite central wing box. CFRP is used for the rear pressure bulkhead and the upper deck floor beams. Various technologies, including automated fiber placement and resin film infusion, were tested for this final component [63] (Fig. 4.13).

In addition to the above materials, several new advanced materials are being developed for aerospace applications. These include glass fiber-reinforced polymers (GFRPs), nanocomposites that incorporate nanoparticles into the matrix material, and shape memory alloys are materials that can remember their original shape and return to it after being deformed.

Additive manufacturing has expanded the possibilities for producing composite materials with tailored properties. Techniques like fused filament fabrication (FFF) and continuous fiber reinforcement enable the fabrication of composite parts with high strength-to-weight ratios. By combining different materials, such as carbon fiber or glass fiber with polymer matrices, it is possible to create lightweight, strong,

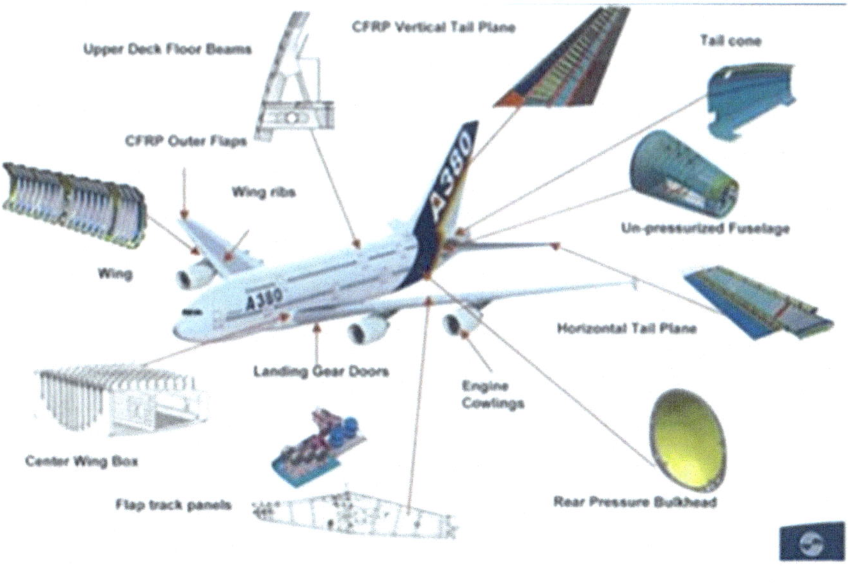

Fig. 4.13 Major monolithic CFRP and thermoplastics applications [63]

Table 4.2 Summary of materials used for additive manufacturing in the aerospace industries [1]

Material	Aerospace application	Advantages	Disadvantages
Polymers	Prototyping and concept development Interior components and cabin parts Non-structural parts and ducting systems Lightweight drone components and unmanned aerial vehicles (UAVs).	Low cost and ease of processing Lightweight, suitable for interior components Excellent design flexibility and complexity Good impact resistance and vibration-damping	Limited mechanical strength and load-bearing capacity Limited thermal stability and chemical resistance Limited dimensional accuracy and potential for warping or distortion during printing Limited recyclability and environmental concerns
Metals and alloys	Engine components and turbines Structural components and airframes High-temperature applications and aerospace structures Propulsion systems and rocket components Satellite components and communication systems	High strength-to-weight ratio Excellent mechanical properties High thermal conductivity Good wear resistance and fatigue resistance High resistance to extreme temperatures and harsh environments	High material and processing costs Limited design complexity and intricate features Limited availability of AM-grade materials for high-performance applications Potential for microstructural defects in printed parts Post-processing may be required to achieve the desired mechanical properties
Ceramics	Turbine blades and combustion systems Propulsion systems and rocket nozzles Ceramic matrix composites for aerospace structures Satellite components and communication systems	High-temperature resistance and thermal stability Excellent mechanical properties Low-density, lightweight Good chemical inertness and resistance to corrosion	Limited design complexity High processing temperatures are required for sintering Cost and availability of specialized ceramic powders Challenging to achieve dense and void-free prints due to high processing temperatures

and customizable components for aerospace applications. Additive manufacturing of advanced materials in aerospace engineering is a rapidly evolving field. Ongoing research focuses on further improving the process capabilities, expanding the range of printable materials, and optimizing the properties of additively manufactured components. These advancements have the potential to enhance the performance, efficiency, and sustainability of aerospace systems (Table 4.2).

4.4 Aerospace Components Manufactured Using Additive Manufacturing

Additive manufacturing technologies have opened up new avenues for the aerospace industry by making it possible to produce complex shapes with remarkable strength and accuracy. Some of the aircraft components that are commonly made with additive manufacturing technology are heat exchangers, ducting systems, brackets, interior parts, turbine blades, and engine parts. These components can be produced, tested, and prototyped with the help of additive manufacturing techniques including DMLS, SLS, and FDM. Aircraft performance, safety, and efficiency can all be increased by aerospace engineers using these technologies to produce creative designs [16, 64].

4.4.1 Engine Components

Additive manufacturing techniques have drastically changed the way that aviation engine components are produced. One of the additive manufacturing methods, DMLS, is widely used in the production of aircraft engine components [65]. Engine parts made with DMLS include heat exchangers, combustion liners, fuel nozzles, heat exchangers, and turbine blades [5, 66]. These components can be printed using high-temperature alloys or ceramic materials to withstand the extreme conditions within jet engines. DMLS is the appropriate production process for these applications since these components need to be extremely strong, long-lasting, and resistant to high pressures and temperatures. Innovative designs that raise the effectiveness and performance of aviation engines can be produced because of DMLS's ability to produce complicated geometries with exceptional strength and accuracy.

Through the utilization of additive manufacturing technologies such as design for additive manufacturing (DFAM), the aerospace sector can manufacture superior engine components that are essential for the secure and effective functioning of aircraft [1]. The use of an established DFAM allowed for the reduction of complexity from 127 to 4 welds in an impactful hot-fire test of an even larger additive manufacturing component that will be airlifted on NASA's RS-25 Pogo Z-Baffle (Space Launch System) (Fig. 4.14).

4.4.2 Interior Components

Additive manufacturing techniques have also been employed to build interior aircraft components. These techniques have several benefits, one of which is their ability to create complex, durable, and light parts. Interior components including panels, brackets, and air ducts are commonly made using FDM [67–69]. Not only

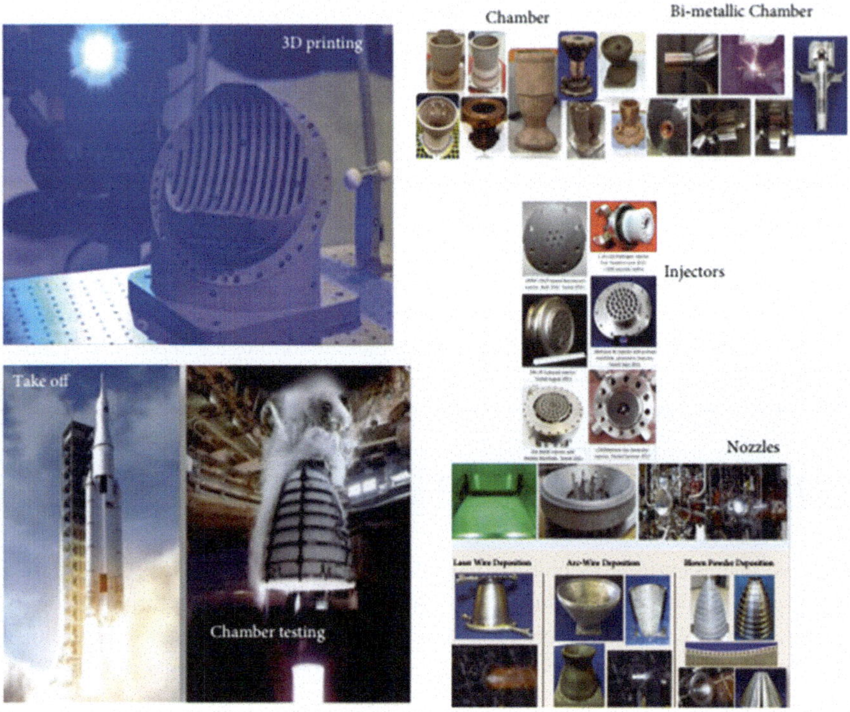

Fig. 4.14 3D-printed aircraft engine parts [1]

Fig. 4.15 3D-printed aircraft interior components [65]

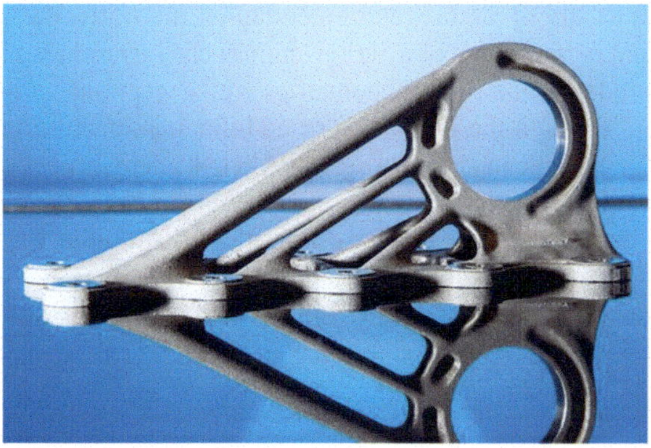

can FDM be used to create intricate geometries that are difficult or impossible to create with traditional manufacturing methods, but it is also particularly useful for creating low-stress components, which need less strength than engine components. Furthermore, FDM allows for the manufacture of personalized interior parts, giving aircraft designers more creative freedom [1, 2, 30]. Through the application of additive manufacturing technologies such as FDM, the aerospace industry can create

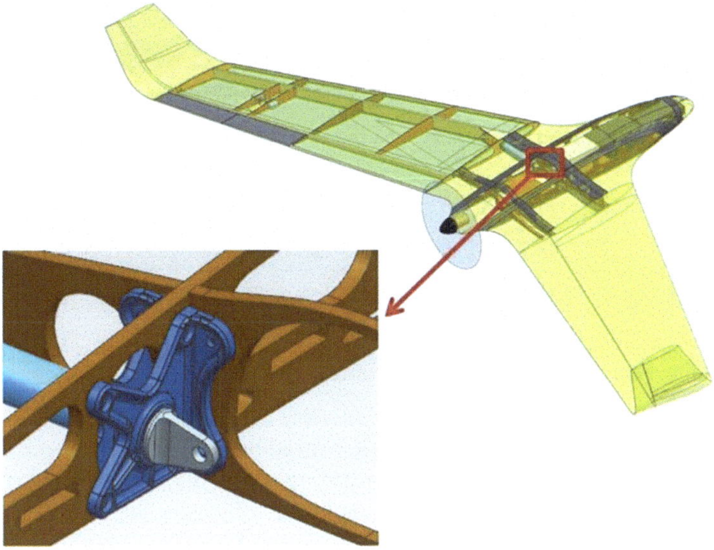

Fig. 4.16 Aircraft fitting structure design to be printed [65]

innovative interior designs that improve passenger comfort and safety while reducing weight and expenses (Fig. 4.15).

4.4.3 Aircraft Structures

Aircraft structures have also been produced using additive manufacturing methods, which have several advantages including the capacity to produce intricate, robust, and lightweight parts. The 3D printing process known as selective laser sintering (SLS) is frequently used to create parts for airplanes, including fuselage, wings, and brackets. SLS is perfect for aerospace applications because it allows parts to be produced with high strength-to-weight ratios and resilience to high temperatures and pressures [1, 2]. Furthermore, SLS enables the highly accurate and precise construction of complicated geometries, facilitating the creation of creative designs that enhance the effectiveness and performance of aircraft structures [1, 2, 56]. The aerospace sector may produce aircraft structures that are more durable, lightweight, and effective than those made using conventional manufacturing techniques by using SLS. Among the numerous parts produced by additive manufacturing in the aerospace sector are the following: rocket brackets, RF feed antenna, turbo-generator casing, satellite antenna, and airplane flap lever [1, 2, 30] (Fig. 4.16).

Fig. 4.17 3D-printed rocket nozzle [61]

4.4.4 Converging-Diverging Rocket Nozzles

Converging-diverging rocket nozzles made with additive manufacturing are a prominent use of this technology in the aerospace sector [70]. These nozzles may be efficiently and precisely designed and manufactured by using additive manufacturing techniques including DMLS, FDM, SLA, and SLS [61]. The optimization of the nozzle design, taking thrust performance and cooling into account, is made possible by the use of CAD software. The aerospace industry's manufacturing process is being revolutionized by additive manufacturing, which enables the rapid demonstration and verification of innovative ideas and concepts. Comprehensive testing at high Reynolds numbers is necessary to address issues about surface roughness and nozzle strength [1, 70]. Converging-diverging rocket nozzles made with additive manufacturing show how this technology can improve the effectiveness and performance of aerospace parts (Fig. 4.17).

4.4.5 Anti-icing Systems

Additive manufacturing served a significant role in the fabrication of aircraft anti-icing systems, which have revolutionized applications in aviation. In addition to greater performance and safety in inclement weather, these systems also offer increased energy efficiency and precise monitoring. An instance of an anti-icing system facilitated by additive manufacturing is the smart de-icing system. By integrating sensors and actuators into the aircraft's structure, this system enables accurate control and real-time monitoring of the de-icing procedure. The smart de-icing technology ensures effective ice removal while minimizing energy usage by sensing ice formation and delivering tailored heat or de-icing fluids to crucial regions [71, 72].

The thermal anti-ice system also makes use of additive manufacturing methods to create parts with intricate geometries and well-thought-out designs. Converging-diverging rocket nozzles with accurate cooling and thrust performance can be manufactured using SLA and FDM [72, 72]. These elements continuously heat and ventilate the area; thus, they efficiently prevent ice development. The aerospace sector has effectively used these additive manufacturing-enabled anti-icing devices, especially in commercial aircraft. The Boeing 737–300, 400, and 500, for example, use a hot bleed air anti-ice system [72]. To prevent ice formation, this mechanism warms the leading edge of the wing using hot air from the engine compressor's stages. Effective anti-icing is ensured by distributing and directing the hot bleed air to appropriate regions via the aircraft's pneumatic system [72, 72].

Thus, the creation of sophisticated anti-icing systems in airplanes, which provide increased safety, accurate monitoring, and energy efficiency, has been made possible by additive manufacturing. Additive manufacturing is revolutionizing aerospace applications by optimizing designs and fabricating complicated components, which ensures optimal performance under adverse weather conditions.

4.4.6 Lattice Heat Sinks

Additive manufacturing has drastically changed the design and construction of heat sinks, particularly lattice heat sinks. With the use of additive manufacturing, these heat sinks can produce complex lattice structures that increase performance and thermal management [73, 74]. For example, researchers examined the variance of chip temperature and pressure drop for various designs in a study examining heat sinks made using additive manufacturing. The results showed how additive manufacturing can increase thermal performance, highlighting the significance of particular design elements for improved cooling [14, 73, 74].

Furthermore, complicated lattice structures that are difficult to produce with conventional methods can now be produced by using DMLS [73, 75]. The enhanced surface area and improved airflow of these lattice heat sinks allow for better heat dissipation. With the use of this technique, complex lattice designs with optimal thermal characteristics may be created, guaranteeing effective cooling and heat transfer in a variety of applications [73, 74, 76]. As a result, additive manufacturing of lattice-designed heat sinks offers superior heat dissipation, increased airflow, and optimized cooling efficiency for a variety of industries and applications. Optimal performance in heat sink design is ensured by additive manufacturing, which is also revolutionizing the field of thermal management.

4.5 Challenges and Limitations of Additive Manufacturing in Aerospace Application

The aerospace sector has seen a transformation because of additive manufacturing technologies, which have made it possible to produce intricately designed, lightweight components of superior quality [33, 77]. Nonetheless, there are still several restrictions and difficulties in using additive manufacturing in the aircraft industry.

Finding premium materials that satisfy the exacting safety and performance requirements set forth by the aerospace sector is a significant obstacle [78, 79]. The requirement for dependable and consistent manufacturing procedures that may yield products with excellent accuracy and repeatability presents another difficulty. Furthermore, some businesses may find it prohibitive to purchase the tools and materials needed for additive manufacturing [66, 69, 80].

The aerospace sector is investing in additive manufacturing technologies despite these barriers and restrictions, and it is working to discover ways around them to capitalize on the advantages additive manufacturing has to offer. The aerospace sector is not without its constraints when it comes to additive manufacturing. The size of the pieces that can be created is one of the main restrictions [5, 6]. The majority of additive manufacturing equipment on the market today has a low build volume, which can limit the size of parts that can be manufactured. High-volume production runs may be limited by the slower production speed in comparison to conventional manufacturing techniques [28, 52, 68]. Additionally, elements like porosity and surface finish, which can affect the end product's strength and longevity, might have an impact on the quality of the parts produced. The production of specific kinds of aerospace components utilizing additive manufacturing technology may be challenging due to these constraints. These restrictions are being addressed by continued research and development, though, and additive manufacturing is anticipated to become more significant in the aerospace sector moving forward.

4.6 Future of Additive Manufacturing in the Aerospace Industry

The aerospace industry has already benefited much from additive manufacturing technologies, and this business looks forward to further developments in this field. Innovative aircraft components and structures have already been produced as a result of the great accuracy and precision with which complicated geometries and lightweight structures may be created [3, 13, 81]. It is anticipated that additive manufacturing will become even more significant in the aerospace sector in the future.

Engine component manufacture is one industry where additive manufacturing is anticipated to have a big influence. Engines that are more powerful and efficient will be possible thanks to the production of parts with complex internal geometries and high strength-to-weight ratios [5, 7, 82]. Performance will rise, pollution will be decreased, and fuel efficiency will increase as a result. Aircraft interior component manufacture is another area where additive manufacturing is anticipated to be very important. Improved comfort and safety in aircraft interiors will be made possible by the production of lightweight, durable, and customizable parts [2, 68, 83]. Both enhanced safety and better passenger experiences could result from this.

Spacecraft and satellite manufacturing is another industry where additive manufacturing is anticipated to have a big impact. More economical and efficient space vehicles will be able to be built thanks to the strong and lightweight structures made possible by additive manufacturing [2, 51, 68, 76, 83]. Furthermore, the capacity to create intricate geometries will facilitate the development of increasingly sophisticated and advanced satellites. However, to fully utilize additive manufacturing in the aerospace sector, several obstacles still need to be overcome. The ability to produce huge parts at a fair cost, consistent and dependable manufacturing methods, and the requirement for high-quality materials are some of these problems.

As a result, additive manufacturing technologies have already had a big impact on the aerospace sector, and their future in this field looks bright. More sophisticated and effective airplane structures and components will be made possible by the ability to manufacture lightweight, robust, and complicated parts. Furthermore, the capacity to produce unique and personalized designs would result in enhanced passenger experiences and higher levels of safety. The difficulties surrounding additive manufacturing will be resolved by continued research and development, and it is anticipated that this technology will become more and more significant in the aerospace sector in the future.

4.7 Conclusion

The review paper *Recent Advances of Additive Manufacturing for Aerospace Industries: Methods, Materials, Challenges, and Future Outlooks* provided a comprehensive overview of the materials used in aerospace applications concerning 3D printing technologies. The paper examined various additive manufacturing techniques and their specific applications within the aerospace industry. It also highlighted the post-processing challenges associated with each technique and presented innovative solutions to address these challenges.

The review paper emphasized the importance of material selection in aerospace additive manufacturing, considering factors such as strength, weight, temperature resistance, and performance characteristics. It explored the wide range of materials, including metals, polymers, and composites, that can be processed using 3D printing technologies, offering insights into their suitability for aerospace applications. Furthermore, the paper delved into different 3D printing techniques, such as SLM, wire arc additive manufacturing (WAAM), and FDM, providing an in-depth analysis of their capabilities, advantages, and limitations. It addressed the post-processing challenges associated with each technique, such as support structure removal, surface finishing, and achieving dimensional accuracy, and presented promising solutions and advancements in these areas.

Overall, the review paper offers a comprehensive and insightful analysis of additive manufacturing in the aerospace industry, covering materials, techniques, post-processing challenges, and future outlooks. It provides valuable information for researchers, engineers, and industry professionals, helping drive innovation and advancements in additive manufacturing for aerospace applications.

References

1. Balaji, D., Ranga, J., Bhuvaneswari, V., Arulmurugan, B., Rajeshkumar, L., Manimohan, M. P., et al. (2022). Additive manufacturing for aerospace from inception to certification. *Journal of Nanomaterials, 2022*(6), 1–18.
2. Bacciaglia, A., Ceruti, A., & Liverani, A. (2022). Towards large parts manufacturing in additive technologies for aerospace and automotive applications. *Procedia Computer Science, 200*, 1113–1124.
3. Rosso, S., Uriati, F., Grigolato, L., Meneghello, R., Concheri, G., & Savio, G. (2021). An optimization workflow in design for additive manufacturing. *Applied Sciences, 11*, 2572.
4. Sathish, K., Kumar, S. S., Magal, R. T., Selvaraj, V., Narasimharaj, V., Karthikeyan, R., et al. (2022). A comparative study on subtractive manufacturing and additive manufacturing. *Advances in Materials Science and Engineering*. https://doi.org/10.1155/2022/6892641
5. DebRoy, T., Wei, H. L., Zuback, J. S., Mukherjee, T., Elmer, J. W., Milewski, J. O., et al. (2018). Additive manufacturing of metallic components—Process, structure and properties. *Progress in Materials Science, 92*, 112–224.
6. Watson, J. K., & Taminger, K. M. B. (2018). A decision-support model for selecting additive manufacturing versus subtractive manufacturing based on energy consumption. *Journal of Cleaner Production, 176*, 1316–1322. [Internet]. Available from: https://doi.org/10.1016/j.jclepro.2015.12.009
7. Moon, S. K., Tan, Y. E., Hwang, J., & Yoon, Y. J. (2014). Application of 3D printing technology for designing light-weight unmanned aerial vehicle wing structures. *International Journal of Precision Engineering and Manufacturing-Green Technology, 1*, 223–228.
8. Collan, M., & Michelsen, K. E. (2020). *Technical, economic and societal effects of manufacturing 4.0: Automation, adaption and manufacturing in finland and beyond*. Palgrave Macmillan.
9. Thompson, M. K., Moroni, G., Vaneker, T., Fadel, G., Campbell, R. I., Gibson, I., et al. (2016). Design for additive manufacturing: Trends, opportunities, considerations, and constraints. *CIRP Annals, 65*, 737–760. [Internet]. Available from: https://doi.org/10.1016/j.cirp.2016.05.004
10. Bourell, D., Kruth, J. P., Leu, M., Levy, G., Rosen, D., Beese, A. M., et al. (2017). Materials for additive manufacturing. *CIRP Annals, 66*, 659–681. [Internet]. Available from: https://doi.org/10.1016/j.cirp.2017.05.009
11. Vishnu Prashant Reddy, K., Meera Mirzana, I., & Koti Reddy, A. (2018). Application of additive manufacturing technology to an aerospace component for better trade-off's. *Materials Today: Proceedings, 5*, 3895–3902. [Internet]. Available from: https://doi.org/10.1016/j.matpr.2017.11.644
12. Bhat, C., Kumar, A., Lin, S.-C., & Jeng, J. Y. Design of tessellation based load transfer mechanisms in additively manufactured lattice structures to obtain hybrid responses. [Cited 2023 July 24]. Available from: https://papers.ssrn.com/abstract=4459403
13. Karkun, M. S., & Dharmalingam, S. (2022). 3D printing technology in aerospace industry—A review. *International Journal of Aviation Aeronautics and Aerospace, 9*. https://doi.org/10.15394/ijaaa.2022.1708
14. Bhat, C., Kumar, A., Lin, S. C., & Jeng, J. Y. (2023). Design, fabrication, and properties evaluation of novel nested lattice structures. *Additive Manufacturing, 68*, 103510.
15. Diaz, A. (2019). Surface texture characterization and optimization of metal additive manufacturing-produced components for aerospace applications [Internet]. *Additive Manufacturing for the Aerospace Industry*. Available from: https://doi.org/10.1016/B978-0-12-814062-8.00018-2
16. Mohanavel, V., Ashraff Ali, K. S., Ranganathan, K., Allen Jeffrey, J., Ravikumar, M. M., & Rajkumar, S. (2021). The roles and applications of additive manufacturing in the aerospace and automobile sector. *Materials Today: Proceedings, 47*, 405–409.
17. Bhat, C., Kumar, A., Lin, S. C., & Jeng, J. Y. (2022). A novel bioinspired architectured materials with interlocking designs based on tessellation. *Additive Manufacturing, 58*, 103052.

18. Mehmood, Y., Aslam, J., Ullah, N., Chowdhury, M. S., Techato, K., & Alzaed, A. N. (2021). Adaptive robust trajectory tracking control of multiple quad-rotor UAVs with parametric uncertainties and disturbances. *Sensors, 21*, 2401.
19. Øùøó ÁÒØ, Óø ËØÑÊÓ, Ëùô ØØ, Òø ÔÖØÑÔÐÑ. ??del a?d Beha?i???Ba?ed ??b??ic G?alkee?e?
20. DMLS Leads to More Sustainably Manufactured Metal Parts | Engineering.com [Internet]. [cited 2023 May 15]. Available from: https://www.engineering.com/story/dmls-leads-to-more-sustainably-manufactured-metal-parts
21. O'Brien, M. (2018, March). Existing standards as the framework to qualify additive manufacturing of metals. In *IEEE aerospace conference proceedings* (pp. 1–10). IEEE.
22. Goodridge, R., & Ziegelmeier, S. (2017). Powder bed fusion of polymers. In *Laser additive manufacturing: Materials, design, technologies, and applications* (pp. 181–204). Woodhead Publishing.
23. Mehrpouya, M., Tuma, D., Vaneker, T., Afrasiabi, M., Bambach, M., & Gibson, I. (2022). Multimaterial powder bed fusion techniques. *Rapid Prototyping Journal, 28*, 1–19.
24. Spears, T. G., & Gold, S. A. (2016). In-process sensing in selective laser melting (SLM) additive manufacturing. *Integrating Materials and Manufacturing Innovation, 5*, 16–40.
25. Sofia, D., Chirone, R., Lettieri, P., Barletta, D., & Poletto, M. (2018). Selective laser sintering of ceramic powders with bimodal particle size distribution. *Chemical Engineering Research and Design, 136*, 536–547. [Internet]. Available from: https://doi.org/10.1016/j.cherd.2018.06.008
26. Gruber, H., Henriksson, M., Hryha, E., & Nyborg, L. (2019). Effect of powder recycling in electron beam melting on the surface chemistry of alloy 718 powder. *Metallurgical and Materials Transactions A, Physical Metallurgy and Materials Science, 50*, 4410–4422.
27. Sonkamble, V., & Phafat, N. (2023). A current review on electron beam assisted additive manufacturing technology: Recent trends and advances in materials design. *Discover Mechanical Engineering, 2*, 1–17. [Internet]. [Cited 2023 Nov 15]. Available from: https://link.springer.com/article/10.1007/s44245-022-00008-x
28. Sandell, V., Nilsson, J., Hansson, T., Åkerfeldt, P., & Antti, M. L. (2022). Effect of chemical post-processing on surfaces and sub-surface defects in electron beam melted Ti-6Al-4V. *Materials Characterization, 193*, 112281.
29. Shiyas, K. A., & Ramanujam, R. (2021). A review on post processing techniques of additively manufactured metal parts for improving the material properties. *Materials Today Proceedings, 46*, 1429–1436.
30. Alami, A. H., Ghani Olabi, A., Alashkar, A., Alasad, S., Aljaghoub, H., Rezk, H., et al. (2023). Additive manufacturing in the aerospace and automotive industries: Recent trends and role in achieving sustainable development goals. *Ain Shams Engineering Journal, 14*, 102516. [Internet]. [Cited 2023 Nov 15]. Available from: https://linkinghub.elsevier.com/retrieve/pii/S2090447923004057
31. Fiaz, H. S., Settle, C. R., & Hoshino, K. (2016). Metal additive manufacturing for microelectromechanical systems: Titanium alloy (Ti-6Al-4V)-based nanopositioning flexure fabricated by electron beam melting. *Sensors and Actuators A: Physical, 249*, 284–293. [Internet]. Available from: https://doi.org/10.1016/j.sna.2016.08.029
32. Guimarães, R. P. M., Minkowitz, L., Arneitz, S., Sommitsch, C., Giedenbacher, J., Müller, M., et al. (2022). Powder bed fusion processes: Main classes of alloys, current status, and technological trends. In *Advances in Metal Additive Manufacturing* (pp. 1–104). Elsevier.
33. Mishra, P., Sood, S., Pandit, M., & Khanna, P. (2021). Additive manufacturing: Post processing methods and challenges. *Advanced Engineering Forum, 39*, 21–42.
34. Malakizadi, A., Mallipeddi, D., Dadbakhsh, S., M'Saoubi, R., & Krajnik, P. (2022). Post-processing of additively manufactured metallic alloys—A review. *International Journal of Machine Tools and Manufacture, 179*, 103908.
35. (PDF) Principles of directed energy deposition for aerospace applications [Internet]. [Cited 2023 Nov 15]. Available from: https://www.researchgate.net/publication/349008743_Principles_of_Directed_Energy_Deposition_for_Aerospace_Applications

36. Friel, R. J. (2015). Power ultrasonics for additive manufacturing and consolidating of materials. In *Power ultrasonics: Applications of high-intensity ultrasound* (pp. 313–335). Woodhead Publishing.
37. Svetlizky, D., Das, M., Zheng, B., Vyatskikh, A. L., Bose, S., Bandyopadhyay, A., et al. (2021). Directed energy deposition (DED) additive manufacturing: Physical characteristics, defects, challenges and applications. *Materials Today, 49*, 271–295.
38. Izadi, M., Farzaneh, A., Mohammed, M., Gibson, I., & Rolfe, B. (2020). A review of laser engineered net shaping (LENS) build and process parameters of metallic parts. *Rapid Prototyping Journal, 26*, 1059–1078.
39. Laser engineered net shaping – Digital manufacturing products, support, and services [Internet]. [Cited 2023 Nov 15]. Available from: https://www.padtinc.com/digital-manufacturing/glossary/laser-engineered-net-shaping/
40. (PDF) Laser engineered net shaping process in development of medical implants: An overview [Internet]. [Cited 2023 Nov 15]. Available from: https://www.researchgate.net/publication/320298709_Laser_Engineered_Net_Shaping_Process_in_Development_of_Medical_Implants_An_Overview
41. All about electron beam additive manufacturing (EBAM) 3D printing [Internet]. [Cited 2023 Nov 15]. Available from: https://www.thomasnet.com/articles/custom-manufacturing-fabricating/electron-beam-additive-manufacturing-ebam-3d-printing/
42. Osipovich, K., Kalashnikov, K., Chumaevskii, A., Gurianov, D., Kalashnikova, T., Vorontsov, A., et al. (2023). Wire-feed electron beam additive manufacturing: A review. *Metals, 13*, 279. [Internet]. [Cited 2023 Nov 15]. Available from: https://www.mdpi.com/2075-4701/13/2/279/htm
43. Gong, X., Anderson, T., & Chou, K. (2014). Review on powder-based electron beam additive manufacturing technology. *Manufacturing Review, 1*, 2. [Internet]. [Cited 2023 Nov 15]. Available from: https://mfr.edp-open.org/articles/mfreview/full_html/2014/01/mfreview140001/mfreview140001.html
44. Metal Additive Manufacturing with EBAM® Technology | Sciaky [Internet]. [Cited 2023 Nov 15]. Available from: https://www.sciaky.com/additive-manufacturing/electron-beam-additive-manufacturing-technology
45. Progress in additive manufacturing | Special issue progress in electron beam additive manufacturing [Internet]. [Cited 2023 Nov 15]. Available from: https://www.springer.com/journal/40964/updates/21348140
46. Rodrigues, T. A., Duarte, V., Miranda, R. M., Santos, T. G., & Oliveira, J. P. (2019). Current status and perspectives on wire and arc additive manufacturing (WAAM). *Materials (Basel).*, 12, 1121.
47. Chintala, A., Tejaswi Kumar, M., Sathishkumar, M., Arivazhagan, N., & Manikandan, M. (2021). Technology development for producing Inconel 625 in aerospace application using wire arc additive manufacturing process. *Journal of Materials Engineering and Performance, 30*, 5333–5341. [Internet]. Available from: https://doi.org/10.1007/s11665-021-05781-6
48. Jin, W., Zhang, C., Jin, S., Tian, Y., Wellmann, D., & Liu, W. (2020). Wire arc additive manufacturing of stainless steels: A review. *Applied Sciences, 10*, 1563. [Internet]. 2020 [Cited 2023 Nov 15]. Available from: https://www.mdpi.com/2076-3417/10/5/1563/htm
49. Anitha, R., Arunachalam, S., & Radhakrishnan, P. (2001). Critical parameters influencing the quality of prototypes in fused deposition modelling. *Journal of Materials Processing Technology, 118*, 385–388.
50. Rathee, S., Srivastava, M., Maheshwari, S., & Siddiquee, A. N. (2017). Effect of varying spatial orientations on build time requirements for FDM process: A case study. *Defence Technology, 13*, 92–100.
51. Przekop, R. E., Gabriel, E., Pakuła, D., & Sztorch, B. (2023). Liquid for fused deposition modeling technique (L-FDM)—A revolution in application chemicals to 3D printing technology: Color and elements. *Applied Sciences, 13*, 7393 [Internet]. [Cited 2023 Nov 15]. Available from: https://www.mdpi.com/2076-3417/13/13/7393/htm

52. Parulski, C., Jennotte, O., Lechanteur, A., & Evrard, B. (2021). Challenges of fused deposition modeling 3D printing in pharmaceutical applications: Where are we now? *Advanced Drug Delivery Reviews, 175*, 113810.
53. Hill C, Bedsole R, Rowe K, Duty C, Ajinjeru C, Kunc V, et al. 2018 *Big area additive manufacturing (BAAM) materials development and reinforcement with advanced composites*. [Cited 2023 Nov 15]. Available from: https://www.osti.gov/servlets/purl/1434289/
54. Roschli, A., Gaul, K. T., Boulger, A. M., Post, B. K., Chesser, P. C., Love, L. J., et al. (2019). Designing for big area additive manufacturing. *Additive Manufacturing, 25*, 275–285.
55. Shiferaw, M. Z., & Gebremedhen, H. S. (2022). Recycled polymer for FDM 3D printing filament material: Circular economy for sustainability of additive manufacturing [Internet]. *Advances of science and technology. ICAST 2021* (Lecture notes of the Institute for Computer Sciences, Social Informatics and Telecommunications Engineering). Springer. Available from: https://doi.org/10.1007/978-3-030-93712-6_17
56. Hai, A., Ghani, A., Alashkar, A., Alasad, S., Aljaghoub, H., Rezk, H., et al. (2023). Additive manufacturing in the aerospace and automotive industries: Recent trends and role in achieving sustainable development goals. *Ain Shams Engineering Journal, 14*, 102516. [Internet]. Available from: https://doi.org/10.1016/j.asej.2023.102516
57. Lakhdar, Y., Tuck, C., Binner, J., Terry, A., & Goodridge, R. (2021). Additive manufacturing of advanced ceramic materials. *Progress in Materials Science, 116*, 100736. [Internet]. Available from: https://doi.org/10.1016/j.pmatsci.2020.100736
58. Liu, Z., He, B., Lyu, T., & Zou, Y. (2021). A review on additive manufacturing of titanium alloys for aerospace applications: Directed energy deposition and beyond Ti-6Al-4V. *JOM, 73*, 1804–1818. [Internet]. Available from: https://doi.org/10.1007/s11837-021-04670-6
59. Huang, R., Riddle, M., Graziano, D., Warren, J., Das, S., Nimbalkar, S., et al. (2016). Energy and emissions saving potential of additive manufacturing: The case of lightweight aircraft components. *Journal of Cleaner Production, 135*, 1559–1570.
60. Montanari, R., Palombi, A., Richetta, M., & Varone, A. (2023). Additive manufacturing of aluminum alloys for aeronautic applications: Advantages and problems. *Metals (Basel), 13*, 716.
61. Cieśla, M., Junak, G., & Marek, A. (2016). Fatigue characteristics of selected light metal alloys. *Archives of Metallurgy and Materials, 61*, 271–274.
62. Chen, Y., Zhang, J., Li, Z., Zhang, H., Chen, J., Yang, W., et al. (2023). Manufacturing technology of lightweight fiber-reinforced composite structures in aerospace: Current situation and toward intellectualization. *Aerospace, 10*, 206.
63. Maria, M. (2013). Advanced composite materials of the future in aerospace industry. *INCAS Bulletin, 5*, 139–150.
64. Uriondo, A., Esperon-Miguez, M., & Perinpanayagam, S. (2015). The present and future of additive manufacturing in the aerospace sector: A review of important aspects. *Proceedings of the Institution of Mechanical Engineers Part G Journal of Aerospace Engineering, 229*, 2132–2147.
65. Getachew, M. T., Shiferaw, M. Z., & Ayele, B. S. (2023). The current state of the art and advancements, challenges, and future of additive manufacturing in aerospace applications. Rajesh Kannan A, editor. *Advances in Materials Science and Engineering, 2023*, 1–13. [Internet]. Available from: https://www.hindawi.com/journals/amse/2023/8817006/
66. Alami, A. H., Ghani Olabi, A., Alashkar, A., Alasad, S., Aljaghoub, H., Rezk, H., et al. (2023). Additive manufacturing in the aerospace and automotive industries: Recent trends and role in achieving sustainable development goals. *Ain Shams Engineering Journal, 14*, 102516. [Internet]. [Cited 2023 Nov 15]. Available from: https://drive.google.com/file/d/1XvOV2NbpJsAWP1pj9LwrZjMyJX31w1HS/view?usp=drive_web&usp=embed_facebook
67. Čretnik, A., & Fekonja, A. (2022). The use of selective laser melting in mandibular retrognathia correction. *Metals (Basel), 12*, 1544.
68. Yusuf, S. M., Cutler, S., & Gao, N. (2019). Review: The impact of metal additive manufacturing on the aerospace industry. *Metals (Basel), 9*, 1286.

69. Borgue, O., Panarotto, M., & Isaksson, O. (2019). Modular product design for additive manufacturing of satellite components: Maximising product value using genetic algorithms. *Concurrent Engineering: Research and Applications, 27*, 331–346.
70. Chen, M., Baby, R., Dillard, S., Lee, Y. T., & Ekkad, S. (2022). Design and test a converging and de Laval nozzle using additive manufacturing. *Frontiers in Aerospace Engineering, 1*. https://doi.org/10.3389/fpace.2022.951987
71. Anti-icing systems in airplane engines | Turbomachinery blog [Internet]. [Cited 2023 July 25]. Available from: https://blog.softinway.com/anti-icing-systems-in-airplanes-boeing-737-300-400-500/
72. Cortés, A., Sánchez Romate, X. F., Jiménez-Suárez, A., Campo, M., Prolongo, M. G., Ureña, A., et al. (2021). 3D printed anti-icing and de-icing system based on CNT/GNP doped epoxy composites with self-curing and structural health monitoring capabilities. *Smart Materials and Structures, 30*, 025016.
73. Yaple, J., Bhate, D., Azeredo, B., & Phelan, P. (2021). Design and additive manufacturing of copper heat sinks for Microelectronics cooling. Master's Thesis, ARIZONA STATE UNIVERSITY.
74. Krishna, A., & Ram, G. (2019). Design and optimization of additively manufactured heat sink and heat spreader for microelectronics cooling, Master's Thesis, Department of Mechanical Engineering, The University of Texas at Arlington.
75. Mieloszyk, J., Tarnowski, A., Kowalik, M., Perz, R., & Rzadkowski, W. (2019). Preliminary design of 3D printed fittings for UAV. *Aircraft Engineering and Aerospace Technology, 91*, 756–760.
76. Nyamekye, P., Rahimpour Golroudbary, S., Piili, H., Luukka, P., & Kraslawski, A. (2023). Impact of additive manufacturing on titanium supply chain: Case of titanium alloys in automotive and aerospace industries. *Advances in Industrial and Manufacturing Engineering, 6*, 100112.
77. Ford, S., & Despeisse, M. (2016). Additive manufacturing and sustainability: An exploratory study of the advantages and challenges. *Journal of Cleaner Production, 137*, 1573–1587. [Internet]. Available from: https://doi.org/10.1016/j.jclepro.2016.04.150.
78. Rouf, S., Malik, A., Singh, N., Raina, A., Naveed, N., Siddiqui, M. I. H., et al. (2022). Additive manufacturing technologies: Industrial and medical applications. *Sustainable Operations and Computers, 3*, 258–274.
79. Rahman, M. M., Bin, R. S., Sultana, J., & Rahman, M. Z. (2023). Additive manufacturing techniques, their challenges, and additively manufactured composites for advanced engineering applications. In *Reference module in materials science and materials engineering*. Elsevier.
80. Shimoyama, K., & Komiya, A. (2022). Multi-objective Bayesian topology optimization of a lattice-structured heat sink in natural convection. *Structural and Multidisciplinary Optimization, 65*, 1.
81. Rosso, S., Savio, G., Uriati, F., Meneghello, R., & Concheri, G. (2019). Optimization approaches in design for additive manufacturing. In *Proceedings of the international conference on engineering design, ICED* (pp. 809–818). Cambridge University Press.
82. Zhang, C., Wang, S., Li, J., Zhu, Y., Peng, T., & Yang, H. (2020). Additive manufacturing of products with functional fluid channels: A review. *Additive Manufacturing, 36*, 101490. [Internet]. Available from: https://doi.org/10.1016/j.addma.2020.101490
83. Seifi, M., Salem, A., Beuth, J., Harrysson, O., & Lewandowski, J. J. (2016). Overview of materials qualification needs for metal additive manufacturing. *JOM, 68*, 747–764.

Chapter 5
Rotating Machinery Fault Diagnosis Using One-Dimensional Convolutional Neural Network

Fasikaw Kibrete, Dereje Engida Woldemichael, and Hailu Shimels Gebremedhen

5.1 Introduction

Rotating machinery is of utmost importance in various diverse industries, accounting for more than 90% of industrial machinery [1]. Since critical components in rotating machinery, such as gearbox, bearing, engine, motor, fan, etc., usually operate under changeable and complex environments, many types of faults are prone to occur. These faults not only negatively impact the performance of the equipment but also lead to substantial financial losses and jeopardize personal safety

F. Kibrete (✉)
Department of Mechanical Engineering, University of Gondar, Gondar, Ethiopia

Artificial Intelligence and Robotic Center of Excellence, Addis Ababa Science and Technology University, Addis Ababa, Ethiopia

Department of Mechanical Engineering, College of Engineering, Addis Ababa Science and Technology University, Addis Ababa, Ethiopia
e-mail: fasikaw.kibrete@uog.edu.et

D. E. Woldemichael
Artificial Intelligence and Robotic Center of Excellence, Addis Ababa Science and Technology University, Addis Ababa, Ethiopia

Department of Mechanical Engineering, College of Engineering, Addis Ababa Science and Technology University, Addis Ababa, Ethiopia

H. S. Gebremedhen
Artificial Intelligence and Robotic Center of Excellence, Addis Ababa Science and Technology University, Addis Ababa, Ethiopia

Department of Mechanical Engineering, College of Engineering, Addis Ababa Science and Technology University, Addis Ababa, Ethiopia

University, 1ICB UMR 6303 CNRS – University Bourgogne Franche-Comté, UTBM, 90010 Belfort, France, Belfort, France

S. W. Fanta et al. (eds.), *Sustainable Development Research in Manufacturing, Process Engineering, Green Infrastructure, and Water Resources*, Green Energy and Technology, https://doi.org/10.1007/978-3-031-77339-6_5

[2, 3]. Thus, it is indispensable to perform accurate fault diagnosis to uphold the dependable functioning of industrial systems and minimize maintenance costs.

Over the years, significant advancements have been achieved in intelligent methods for fault recognition in rotating machinery. Traditional approaches in fault diagnosis typically involve three stages: data acquisition, fault feature extraction, and fault classification [4]. In the data acquisition stage, valuable condition monitoring data or information is collected and stored from different sources of the target rotating machinery by sensor and data acquisition systems [5]. Subsequently, fault-sensitive features from the collected sensor data are extracted and selected using signal-processing techniques [6]. These features are then entered into traditional machine learning models for fault classification [7, 8]. However, these methods heavily rely on the quality of the extracted features, necessitating a solid understanding of signal-processing techniques as well as extensive expertise in fault diagnosis. Additionally, the feature extraction and classification steps are performed separately, resulting in a time-consuming and laborious process [9]. Moreover, traditional machine learning models often have shallow architectures, limiting their effectiveness in fault diagnosis tasks.

In contrast to shallow learning models, deep learning methods have the ability to automatically extract fault features and uncover hidden information from raw sensor data. Consequently, deep learning models have gained significant attention in fault diagnosis research. Among these models, convolutional neural networks (CNNs) have emerged as promising models for fault diagnosis, as they possess the capability to directly learn fault-specific features from original input data through convolutional operations [10]. CNNs possess beneficial attributes including sparse connectivity, weight sharing, and downsampling. These characteristics result in a reduction of training parameters, acceleration of the training process, introduction of translation invariance, and prevention of overfitting risks during training. Moreover, CNN models are relatively straightforward to train and fine-tune. In the field of fault diagnosis, researchers have primarily explored the use of two-dimensional (2D) CNNs [11–14]. However, the traditional 2D-CNN approach involves transforming one-dimensional (1D) signals into images using various conversion techniques. This transformation process is time-consuming, increases the workload, and necessitates specialized hardware for training. Furthermore, the transformation inevitably results in information loss, significantly affecting diagnostic performance. To overcome these limitations, 1D-CNNs have been introduced in this research field. The network architecture of 1D-CNN is simpler and more compact than that of a 2D-CNN, and it can effectively train and classify with small sample sizes. Researchers have explored the application of 1D-CNN models for fault diagnosis in various rotating components, including rolling bearings [4, 15], automobile engines [16], and gearboxes [17]. However, most existing studies have utilized 1D-CNN models with only small kernels, which can pose challenges during training.

In this study, a fault diagnosis method for rotating machinery is proposed, utilizing a 1D-CNN model. The proposed model incorporates wide kernels in the initial convolutional layer for enhanced noise immunity, while subsequent layers employ small kernels for robust feature representation. Batch normalization algorithm is incorporated after convolution layers to expedite the training process and improve the model's noise interference adaptability. Dropout regularization is employed in the dense layer to mitigate overfitting during network training. The proposed method is validated using roller-bearing vibration datasets and compared with other existing methods.

5.2 Convolutional Neural Network

A convolutional neural network (CNN) is a neural network designed for capturing localized features from the input data in a sequential and hierarchical manner to perform classification tasks [18, 19]. The typical configuration of a CNN is illustrated in Fig. 5.1.

The convolutional layer consists of multiple learnable filters, also known as kernels. Each kernel performs convolutions on local regions of the input signal to produce new feature maps [21]. In simpler terms, the convolutional kernel computes the dot product between the input of the convolutional layer x_j^{l-1}, the filter weight matrix w_{ij}^l, and a bias matrix b_i^l. The result of the convolution is then added to produce the output feature map x_i^l. During training, the weights and biases are adjusted using the backpropagation algorithm to find the optimal parameters. Mathematically, the convolution operation can be represented as follows:

$$x_i^l = f\left(\sum_{j=1}^{n} x_j^{l-1} * w_{ij}^l + b_i^l\right) \tag{5.1}$$

where n is the number of input feature maps; the notation $*$ denotes the convolution operator; and $f(.)$ denotes the activation function.

The activation function introduces nonlinearity to the model, allowing it to capture intricate patterns and make the extracted fault features more distinguishable. Rectified linear unit (ReLU) has gained popularity due to its effectiveness in

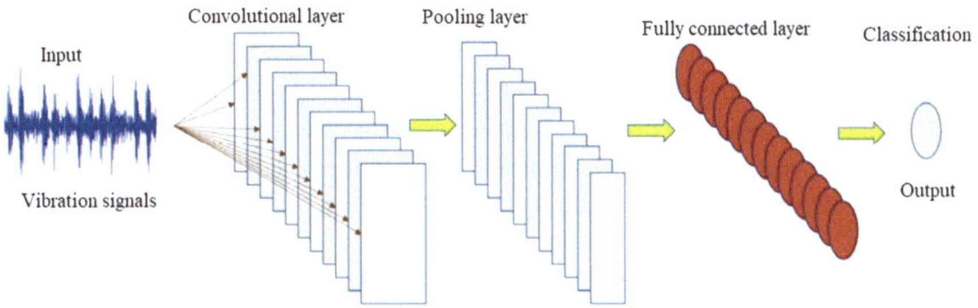

Fig. 5.1 Basic architecture of CNN [20]

hurrying training and alleviating the vanishing gradient [22]. The ReLU activation function can be defined as follows:

$$a_i^l = \mathrm{ReLU}\left(x_i^l\right) = \max\left\{0, x_i^l\right\} \tag{5.2}$$

In this formula, a_i^l is the output x_i^l of the convolution layer activated by ReLu.

After the convolutional and activation operations, the pooling layer performs downsampling for diminishing the extent of the extracted features, the number of network parameters, and remove redundant information, thereby reducing computational costs [23]. Downsampling also helps in preventing overfitting and improving the generalization ability of the CNN. The most widely employed pooling method is max-pooling, which can be described as follows:

$$p_i^l(j) = \max_{(j-1)W+1 \le i \le jW} \left\{a_i^l\right\} \tag{5.3}$$

where a_i^l is the input of the ith feature map of lth layer, W is the maximum value within a window of size, and j represents the step size of the filter.

After the last pooling layer, the output is flattened, transforming the extracted features into 1D feature vectors while preserving all the information. These vectors are then input into the fully connected or dense layer [24]. The dense layer can be mathematically represented as follows:

$$y_l = f\left(w_l * x_l + b_l\right) \tag{5.4}$$

where $f(.)$ represents the nonlinear transfer functions, x_l is the input of layer l, w_l is the weight matrix and b_l is the bias matrix.

The result of the final dense layer is typically fed into the classification function in the output layer. For multi-classification tasks, the softmax activation function is commonly used to distinguish between different classes [25]. Mathematically,

$$\mathrm{Softmax}\left(Z_j\right) = \frac{e^{Z_j}}{\sum_{k=1}^{n} e^{Z_k}} \tag{5.5}$$

where e^{Z_j} denotes the result of the jth neuron, z is the logit of the jth neuron, and n is the number of classes.

To address the internal covariate shift problem that arises when training a CNN with multiple layers and numerous parameters, batch normalization (BN) is often employed. BN addresses this issue by normalizing and scaling the layer inputs, reducing the shift in internal covariance [26]. BN also accelerates the network training and enhances the steadiness of the CNN. The BN process can be mathematically formulated as follows:

$$\hat{x}^{l-1} = \frac{x^{l-1} - \mu_B}{\sqrt{\sigma_B^2 + \varepsilon}} \tag{5.6}$$

$$x^l = \gamma^l \hat{x}^{l-1} + \beta^l \tag{5.7}$$

where $\mu_B = E[x^{l-1}]$, $\sigma_B = \mathrm{Var}[x^{l-1}]$. ε is added to the denominator in Eq. (5.6) for mathematical steadiness. γ^l and β^l are parameters that the BN layer learns to scale and shift the normalized inputs, respectively.

5.3 The Proposed Method

5.3.1 Network Architecture

In this study, a 1D-CNN model is proposed as an improved method for fault diagnosis in rotating machinery. The model's architecture consists of an input layer, three Conv1D layers, three max-pooling layers, a dense layer, and a softmax layer (Fig. 5.2). The Conv1D layers and max-pooling layers work together in an alternating manner to extract relevant features hierarchically. The output features from the final max-pooling layer are then transformed into a 1D feature vector and passed to the dense layer. To ensure accurate identification and classification of health conditions, a softmax classifier is utilized as the final output.

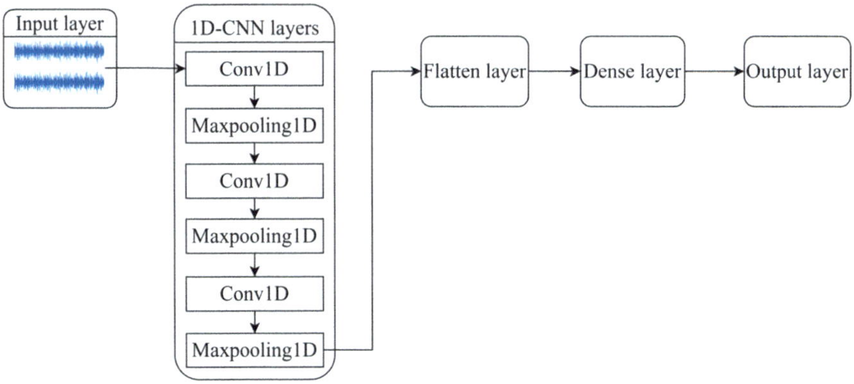

Fig. 5.2 Architecture of the proposed 1D-CNN model

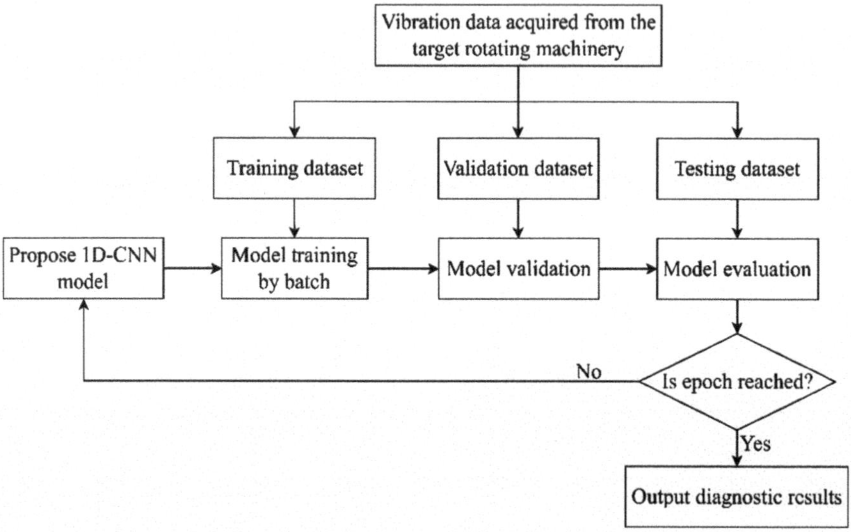

Fig. 5.3 Flowchart of the fault diagnosis process

5.3.2 *Fault Diagnosis Process*

The fault diagnosis process using the designed 1D-CNN model is depicted in Fig. 5.3. Initially, the vibration data from condition monitoring is randomly split into three sets: training set, validation set, and testing set. The training set is utilized to train the 1D-CNN, while the validation set is employed for parameter optimization. The testing set evaluates the diagnostic performance of the trained model. Notably, the diagnosis process automatically extracts pertinent features from the input data without requiring human intervention.

To avert overfitting and enhance the diagnosis performance, a dropout regularization technique is implemented in the dense layer of the 1D-CNN. This technique randomly drops out a fraction of the neurons with a specified dropout rate during training, making better the diagnostic performance and lessening the risk of overfitting. Moreover, the categorical cross-entropy is selected as the loss function to estimate the error between the actual distribution and the predicted distribution. It exhibits fast convergence and possesses global optimization capabilities [27]. Mathematically, it can be expressed as:

$$\text{Loss}(p,q) = -\sum p(x)\log q(x) \tag{5.8}$$

where $p(x)$ represents the actual distribution, and $q(x)$ represents the predicted distribution.

Many optimization algorithms are presented in the literature to reduce the loss values. In this study, the Adam optimizer is applied to diminish the proposed

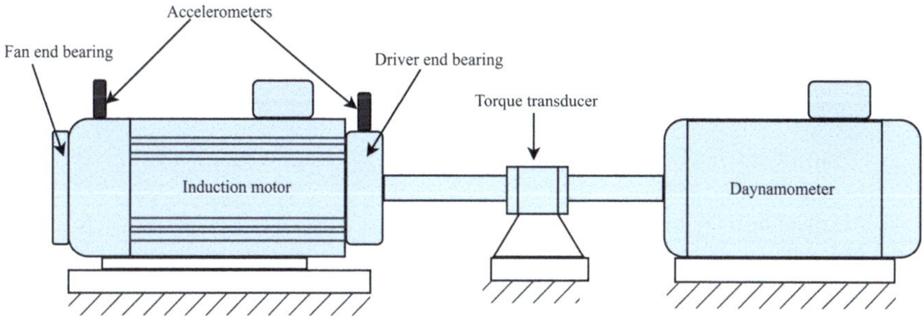

Fig. 5.4 Schematic of CWRU bearing test rig

1D-CNN loss during training because it is straightforward, memory-saving, and computationally effective [21].

5.4 Experimental Results and Discussion

5.4.1 Data Description

To validate the proposed method, the researchers utilized rolling bearing vibration data sourced from the case western reserve university (CWRU) Bearing Data Center [28]. The representation of the experimental test rig is displayed in Fig. 5.4.

The acceleration data from the drive end bearing under loads of 0, 1, 2, and 3 hp are considered as simulation data. The dataset consists of various health states, including healthy states, and three fault conditions: outer race, inner race, and ball faults. Each fault type has three different fault depths, specifically 0.007, 0.014, and 0.021 in., resulting in a total of nine distinct fault types. To organize the data, four datasets, namely A, B, C, and D, were created, corresponding to the loads of 0, 1, 2, and 3 hp, respectively. Each dataset comprises 10 health states, consisting of nine fault conditions and one healthy condition. For each health state, there are 1000 samples available. Each sample consists of 2048 sampling points, resulting in input signals with dimensions of 2048 × 1.

5.4.2 Setting Parameters of 1D-CNN

Previous studies have used 1D-CNN models with kernel sizes of 3 × 1 or 5 × 1 for fault diagnosis tasks [29]. However, smaller kernel sizes at the convolutional layer can be influenced by high-frequency background noise in practical applications. Additionally, constructing a model with only small kernel sizes can lead to deep feature extraction, which may result in training difficulties and overfitting [18]. Therefore, for the 2048 × 1 input vibration signal, a wide kernel of size 64 × 1 with

Table 5.1 Model parameters

No.	Layer type	Kernel size	Stride	Kernel number	Activation function	Padding
1	Input layer	2048 × 1	–	–	–	–
2	Conv1D	64 × 1	32	16	ReLU	'Same'
3	Maxpooling1D	2 × 1	2	16	–	No
4	Conv1D	3 × 1	1	32	ReLU	'Same'
5	Maxpooling1D	2 × 1	2	32	–	No
6	Conv1D	3 × 1	1	64	ReLU	'Same'
7	Maxpooling1D	2 × 1	2	64	–	No
8	Flatten	–	–	–	–	–
9	Dense	100	1	1	ReLU	–
10	Softmax	10	1	1	Softmax	–

Table 5.2 Other hyperparameters of 1D-CNN

Dropout rate	Learning rate	Batch size	Epochs	Loss function
0.2	0.001	256	200	Cross-entropy

a stride of 16 is employed in the first Conv1D layer. The subsequent two Conv1D layers utilize multi-layer small convolutional kernels of 3 × 1 with a stride of 2. The same padding strategy is applied to the Conv1D kernels. The ReLU is employed after each convolution layer to increase the model's nonlinearity. Max-pooling with a size of 2 × 1 is adopted in the pooling layers. The flattened layer converts the features from the third pooling layer into 1D feature vectors, which are then processed by a fully connected layer with ReLU activation to capture correlations among these features. The classification of different health states is performed using a softmax classifier. To enhance network training speed, batch normalization is applied after each convolutional layer, while dropout is implemented in the dense layer to mitigate overfitting. The parameters of the 1D-CNN model are detailed in Table 5.1.

The hyperparameters utilized for model training, including the dropout rate, learning rate, batch size, epochs, and the chosen loss function are listed in Table 5.2.

5.4.3 Diagnosis Results and Discussion

The proposed 1D-CNN model is executed using the TensorFlow library and Python version 3.8 on a computer with Intel(R) Core(TM) i5-7200U CPU@2.5 GHz and 8 GB of memory. In the experiments, the 1D-CNN model is evaluated under single load conditions, and the test accuracy is used to assess its diagnosis performance. To mitigate the effect of randomness, 10 repeated trials are conducted for each dataset. The results reveal that the method completes high accuracy, with 100% diagnosis accuracy for each dataset, as presented in Table 5.3.

Table 5.3 Accuracy results for each load condition

Dataset	Load condition	Average testing accuracy (%)
A	0 hp	100
B	1 hp	100
C	2 hp	100
D	3 hp	100

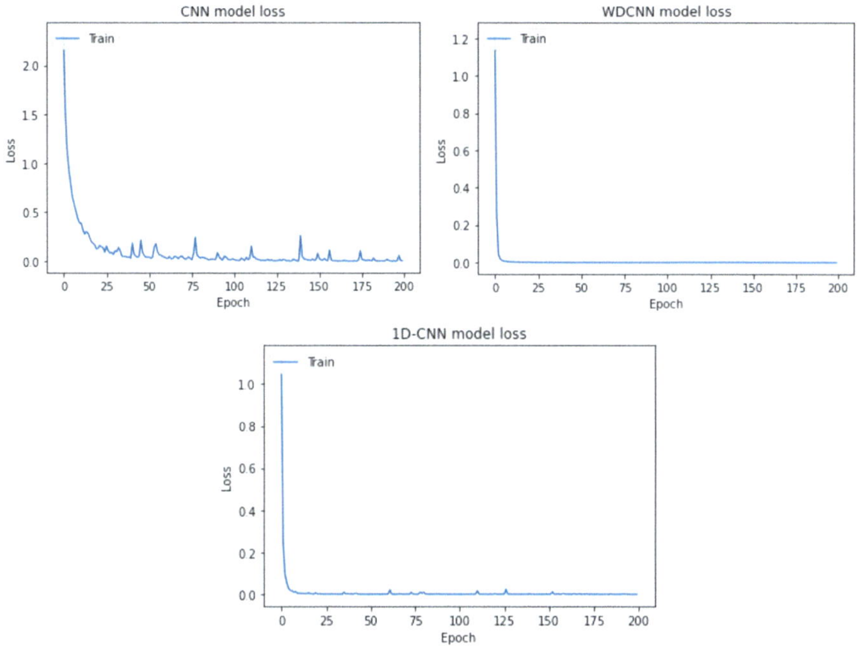

Fig. 5.5 Training loss curves per 200 epochs of the three methods

The classification performance of the model is further assessed under different loading conditions. The findings prove that the diagnosis method completes excellent performance, with a testing accuracy of 100% across all loading conditions. These findings highlight the method's effectiveness in diagnosing faults, even when faced with different loading scenarios. The high accuracy attained further confirms the reliability and practicality of the proposed approach for fault diagnosis in rotating machinery.

Furthermore, the diagnostic performance of the 1D-CNN model is compared with two state-of-the-art models such as CNN [21] and deep convolutional neural networks with wide first-layer kernels (WDCNN) [30], which further verifies its superior generalization performance. The training loss curves for CNN, WDCNN, and 1D-CNN are illustrated in Fig. 5.5. The training loss curves demonstrate the convergence and effectiveness of the proposed 1D-CNN model during the training process.

Table 5.4 Test accuracy of the three methods

References	Method	Average testing accuracy (%)
Zhang et al. [21]	CNN	99.6%
Li et al. [30]	WDCNN	100%
This work	1D-CNN	100%

The average test accuracy of the CNN, WDCNN, and 1D-CNN is displayed in Table 5.4. The test accuracy metric is employed to assess the effectiveness of each method in accurately classifying the health states.

The experimental findings validate that the 1D-CNN attains an average diagnosis accuracy of 100%. In comparison, CNN achieves an accuracy of 99.6% and WDCNN achieves 100%. This comparison highlights the superior diagnostic accuracy of the 1D-CNN compared to CNN. Furthermore, compared to WDCNN, the 1D-CNN model exhibits competitive performance despite its simpler network architecture. Thus, the developed method outperforms those state-of-the-art methods. The reasons for the superior performance of the method can be attributed to the following factors: (1) the utilization of wide kernels in the initial convolutional layer for extracting short-time features and mitigating high-frequency noise; (2) the use of multi-layer small convolutional kernels in subsequent layers to capture a robust representation of the input data, thereby improving diagnostic performance; (3) the incorporation of batch normalization after each convolutional layer for faster training and the application of dropout in the dense layer to address overfitting; and (4) the adoption of the Adam optimizer during backpropagation to continuously update the learning parameters and minimize the loss function. The combination of these factors collectively contributes to the exceptional performance of the proposed 1D-CNN model in diagnosing faults in rotating machinery.

5.5 Conclusion

This study presents an end-to-end fault diagnosis method based on a 1D-CNN model for rotating machinery. The method is assessed using the bearing vibration dataset and compared with two other deep learning methods, showcasing its competitive performance and superior efficiency. The study highlights the potential of 1D-CNN models in intelligent fault diagnosis, providing an automated and effective solution to enhance fault diagnosis in various machinery applications. Future work can explore the extension of this approach to different machinery types and investigate techniques to further improve the model's performance and robustness.

Funding This work was supported by the Artificial Intelligence and Robotics Center of Excellence, Addis Ababa Science and Technology University, under internal research Grant No. IGP012/2023.

Data Availability The data that support the findings of this study are available at the CWRU Bearing Data Center website https://engineering.case.edu/bearingdatacenter.

References

1. Han, T., Zhang, L., Yin, Z., & Tan, A. C. C. (2021). Rolling bearing fault diagnosis with combined convolutional neural networks and support vector machine. *Measurement, 177*, 109022. https://doi.org/10.1016/j.measurement.2021.109022
2. Chen, X., Wang, S., Qiao, B., & Chen, Q. (2018, June 1). Basic research on machinery fault diagnostics: Past, present, and future trends. *Frontiers of Mechanical Engineering, 13*(2), 264–291. https://doi.org/10.1007/s11465-018-0472-3
3. Guo, J., He, Q., Zhen, D., & Gu, F. (2022, April). Intelligent fault detection for rotating machinery using cyclic morphological modulation spectrum and hierarchical Teager permutation entropy. *IEEE Transactions on Industrial Informatics*. https://doi.org/10.1109/TII.2022.3185293
4. Peng, D., Liu, Z., Wang, H., Qin, Y., & Jia, L. (2019). A novel deeper one-dimensional CNN with residual learning for fault diagnosis of wheelset bearings in high-speed trains. *IEEE Access, 7*, 10278–12093. https://doi.org/10.1109/ACCESS.2018.2888842
5. Kibrete, F., & Woldemichael, D. E. (2023). Applications of artificial intelligence for fault diagnosis of rotating machines: A review. In *Artificial intelligence and digitalization for sustainable development* (Lecture notes of the Institute for Computer Sciences, Social-Informatics and Telecommunications Engineering, LNICST) (Vol. 455, pp. 41–62). https://doi.org/10.1007/978-3-031-28725-1_4
6. Yan, J., Liu, T., Ye, X., Jing, Q., & Dai, Y. (2021, August). Rotating machinery fault diagnosis based on a novel lightweight convolutional neural network. *PLoS One, 16*(8), 2021. https://doi.org/10.1371/journal.pone.0256287
7. Gundewar, S. K., & Kane, P. V. (2021). Condition monitoring and fault diagnosis of induction motor. *Journal of Vibration Engineering and Technologies, 9*(4), 643–674. https://doi.org/10.1007/s42417-020-00253-y
8. Liang, C., Chen, C., Liu, Y., & Jia, X. (2021). A novel intelligent fault diagnosis method for rolling bearings based on compressed sensing and stacked multi-granularity convolution denoising auto-encoder. *IEEE Access, 9*, 154777–154787. https://doi.org/10.1109/ACCESS.2021.3129061
9. Chen, J., Yin, X., Li, C., Yang, H., & Hong, L. (2022). One-dimensional convolutional neural network based bearing fault diagnosis. *Open Access Library Journal, 09*(04), 1–11. https://doi.org/10.4236/oalib.1108644
10. Yamashita, R., Nishio, M., Do, R. K. G., & Togashi, K. (2018). Convolutional neural networks: An overview and application in radiology. *Insights Into Imaging, 9*(4), 611–629. https://doi.org/10.1007/s13244-018-0639-9
11. Li, M., Wei, Q., Wang, H., & Zhang, X. (2019). Research on fault diagnosis of time-domain vibration signal based on convolutional neural networks. *Systems Science and Control Engineering, 7*(3), 73–81. https://doi.org/10.1080/21642583.2019.1661311
12. Guo, X., Chen, L., & Shen, C. (2016). Hierarchical adaptive deep convolution neural network and its application to bearing fault diagnosis. *Measurement, 93*, 490–502. https://doi.org/10.1016/j.measurement.2016.07.054
13. Wen, L., Li, X., Gao, L., & Zhang, Y. (2018). A new convolutional neural network-based data-driven fault diagnosis method. *IEEE Transactions on Industrial Electronics, 65*(7), 5990–5998. https://doi.org/10.1109/TIE.2017.2774777
14. Guo, S., Yang, T., Gao, W., & Zhang, C. (2018). A novel fault diagnosis method for rotating machinery based on a convolutional neural network. *Sensors (Switzerland), 18*(5), 1429. https://doi.org/10.3390/s18051429
15. Chen, C. C., Liu, Z., Yang, G., Wu, C. C., & Ye, Q. (2021). An improved fault diagnosis using 1D-convolutional neural network model. *Electronics, 10*(1), 1–19. https://doi.org/10.3390/electronics10010059
16. Du, C., et al. (2022). Research on fault diagnosis of automobile engines based on the deep learning 1D-CNN method. *Engineering Research Express, 4*(1), 015003. https://doi.org/10.1088/2631-8695/ac4834

17. Qian, W., Li, S., Wang, J., An, Z., & Jiang, X. (2018). An intelligent fault diagnosis framework for raw vibration signals: Adaptive overlapping convolutional neural network. *Measurement Science and Technology, 29*(9). https://doi.org/10.1088/1361-6501/aad101
18. Sun, H., & Zhao, S. (2021). Fault diagnosis for bearing based on 1DCNN and LSTM. *Shock and Vibration, 2021*. https://doi.org/10.1155/2021/1221462
19. Shao, Y., Yuan, X., Zhang, C., Song, Y., & Xu, Q. (2020, June). A novel fault diagnosis algorithm for rolling bearings based on one-dimensional convolutional neural network and INPSO-SVM. *Applied Sciences, 10*(12), 4303. https://doi.org/10.3390/app10124303
20. Ciaburro, G. (2022). Machine fault detection methods based on machine learning algorithms: A review. *Mathematical Biosciences and Engineering, 19*(11), 11453–11490. https://doi.org/10.3934/mbe.2022534
21. Zhang, W., Peng, G., & Li, C. (2017). Rolling element bearings fault intelligent diagnosis based on convolutional neural networks using raw sensing signal. In *Smart innovation, systems and technologies* (Vol. 64, pp. 77–84). https://doi.org/10.1007/978-3-319-50212-0_10
22. Wu, H., & Zhao, J. (2018). Deep convolutional neural network model based chemical process fault diagnosis. *Computers and Chemical Engineering, 115*, 185–197. https://doi.org/10.1016/j.compchemeng.2018.04.009
23. Li, G., Deng, C., Wu, J., Xu, X., Shao, X., & Wang, Y. (2019). Sensor data-driven bearing fault diagnosis based on deep convolutional neural networks and S-transform. *Sensors (Basel), 19*(12). https://doi.org/10.3390/s19122750
24. Skowron, M., Orlowska-Kowalska, T., Wolkiewicz, M., & Kowalski, C. T. (2020). Convolutional neural network-based stator current data-driven incipient stator fault diagnosis of inverter-fed induction motor. *Energies, 13*(6). https://doi.org/10.3390/en13061475
25. Chen, H. Y., & Lee, C. H. (2020). Vibration signals analysis by explainable artificial intelligence (XAI) approach: Application on bearing faults diagnosis. *IEEE Access, 8*, 134246–134256. https://doi.org/10.1109/ACCESS.2020.3006491
26. Zhao, K., Jiang, H., Li, X., & Wang, R. (2021). Ensemble adaptive convolutional neural networks with parameter transfer for rotating machinery fault diagnosis. *International Journal of Machine Learning and Cybernetics, 12*(5), 1483–1499. https://doi.org/10.1007/s13042-020-01249-6
27. Wang, Q., Yang, C., Wan, H., Deng, D., & Nandi, A. K. (2021). Bearing fault diagnosis based on optimized variational mode decomposition and 1D convolutional neural networks. *Measurement Science and Technology, 32*(10). https://doi.org/10.1088/1361-6501/ac0034
28. Neupane, D., & Seok, J. (2020, February). Bearing fault detection and diagnosis using case western reserve university dataset with deep learning approaches: A review. *IEEE Access, 8*, 93155–93178. https://doi.org/10.1109/ACCESS.2020.2990528
29. Zhou, S., Xiao, M., Bartos, P., Filip, M., & Geng, G. (2020). Remaining useful life prediction and fault diagnosis of rolling bearings based on short-time Fourier transform and convolutional neural network. *Shock and Vibration, 2020*(8857307), 14pp. https://doi.org/10.1155/2020/8857307.
30. Zhang, W., Peng, G., Li, C., Chen, Y., & Zhang, Z. (2017). A new deep learning model for fault diagnosis with good anti-noise and domain adaptation ability on raw vibration signals. *Sensors (Switzerland), 17*(2). https://doi.org/10.3390/s17020425

Chapter 6
Development and Evaluation of Metering Mechanism for Sowing of Berseem (*Trifolium alexandrinum L.*) and Lucerne (Medicagosativa) Crops

Smegnew Moges Mintesinot and Rohinish Khurana

6.1 Introduction

Legume fodder crops play a crucial role in the agricultural system, which supplies nutritious and palatable fodders for livestock and improves subsequent crop production by enriching soil fertility. However, the lack of high-quality feed and fodder throughout the year and the cost of fodder production activities are constraints in dairy farming. Sixty percent of the price of total milk production goes to feed and fodder [1, 2]. Green fodder is a nearly 30% cheaper source of digestible nutrients than concentrates. On the other hand, due to the increasing cost of using chemical fertilizers in both monetary and environmental terms, the current situation requires the implementation of integrated nutrient management of legumes-cereal cropping systems to achieve a sustainable agroecosystem [3]. Berseem (*Trifolium alexandrinum L.*) and Lucerne (*Medicago sativa*) are the most acceptable legume fodder crops. Berseem and lucerne are annual, very succulent, palatable, and nutritious fodder crops that grow in the winter season. Numerous studies have shown that they can increase soil organic matter, improve soil structure, and accumulate nitrogen reserves in topsoil by fixing free nitrogen from the atmosphere [4–6].

Mechanization of fodder crops is necessary for timeliness and efficiency in the fodder production system and for the achievement of a higher level of quality and productivity for sustainable crop and livestock production [7, 8]. A well-practiced

S. M. Mintesinot (✉)
Faculty of Mechanical and Industrial Engineering, Bahir Dar Institute of Technology, Bahir Dar University, Bahir Dar, Ethiopia

R. Khurana
Department of Farm Machinery and Power Engineering, Punjab Agricultural University, Ludhiana, Punjab, India
e-mail: rkhurana@pau.edu

S. W. Fanta et al. (eds.), *Sustainable Development Research in Manufacturing, Process Engineering, Green Infrastructure, and Water Resources*, Green Energy and Technology, https://doi.org/10.1007/978-3-031-77339-6_6

sowing operation implies the accurate placing of seeds in rows at the desired spacing and depth, covering seeds with the soil, and ensuring a uniform germination environment for each seed [9]. The principal requirement for accurate seeding of small-seeded fodder crops like berseem and lucerne with a given metering device includes: the size and shape of seeds must be the same, the critical parameters of the metering device must be accurately made, the seeds must be provided with an adequate opportunity to enter into holes and the seeds must not be damaged as it affects its germination. Traditionally, berseem and lucerne have sown in broadcasting methods, and these methods have limitations such as uneven distribution of seeds, and uneven space between rows and plants, which don't allow the use of mechanical weeders, thereby increasing labor requirements [10].

Thus, there is a need to replace the conventional sowing method of broadcasting fodder seeds with mechanical seeding to stabilize plant growth and improve productivity. Seed drills reduce the cost of subsequent farming operations while maintaining the desired seed rate and uniform depth of placement, which improve crop yield as plants, get enough space to grow. A well-practiced sowing of fodder crops is needed for better crop stand, saving time, and efficient use of seeds, water, fuel, etc. The metering mechanism device desired to collect the seed from the hopper and pass it to the seed tube at a uniform seed rate with minimum seed damage. Seed-metering mechanisms are designed to drop single or multiple seeds in the furrow at an equal distance. The seed-metering mechanism device types are classified into horizontal plate, inclined plate, vertical plate, indented cell type, rotor type, etc.; it is developed based on the physical properties of seeds, the agronomical requirement of the crops, and the experimental results. The development of the metering mechanism includes the determination of the length and diameter of the roller; and the number and shape of the grooves in it [11–13]. The present study was conducted to develop seed-metering mechanisms for sowing berseem and lucerne seeds and carried out laboratory testing for its performance evaluation.

6.2 Materials and Methods

6.2.1 Development Consideration for Seed-Metering Mechanism of Berseem and Lucerne Crop

Maintaining the plant-to-plant and row-to-row spacing in fodder cultivation is necessary for better yield. Therefore, the metering device for berseem and lucerne seeds was designed to place the seeds at an equal distance in a row. To provide sufficient row spacing for the use of mechanical power weeders and to reduce labor requirements for weeding, the crops need to be planted in rows.

Seed Characteristics The seeds of berseem and lucerne were categorized as small-seeded and irregular in shape. The size of the seeds of berseem and lucerne were measured based on their three principal axes dimensions of length, width, and thickness dimensions using a microscope. The microscope included a control box, a hand switch, a data logger, and a monitor. Cell imaging software for life science was used to measure seed size. To acquire the image of the seed, it was placed horizontally and naturally on the glass slide and placed under the microscope; then, the image was focused sharply and clicked to get the stationary image. The cell imaging software was then used to measure major and intermediate dimensions, and finally, the seed was rotated by 90°, and the third minor dimensions were determined by repeating the same procedure. Table 6.1 shows the physical properties of berseem and lucerne seeds. The bulk density was measured by using a 100 mL graduate measuring cylinder with a prepared sample of the seeds [14].

Then, the bulk density of the seed samples was computed from the known volume of seed that was filled in the graduated cylinder and the mass of the seed weighed on an electronic balance. Whereas, the test weight defined as the weight of 1000 seeds was used for the design of the seed-metering mechanism to specify seed-to-seed distance and the amounts of seeds to be sown. A random sample of 1000 berseem and lucerne seeds was manually counted and weighed on an electronic balance with an accuracy of 0.01 g [15, 16].

6.2.2 Development of Seed-Metering Mechanism

Based on the physical properties (Table 6.1) and the model of the metering mechanism used by researchers for alfalfa and berseem crops, different metering devices, i.e., trapezoidal grooved fluted roller, asymmetric grooved fluted roller and indented cell roller metering device were developed [11, 14]. A metering device draws seeds from bulk and delivers them at the desired rates in the seed tubes for uniform spacing and depth in the soil. The device should not cause mechanical damage to seeds. The metering mechanism was developed to sow both seeds with considering their agronomical requirements of the crops and the physical properties of the seeds. The agronomical requirements of the crops are tabulated in Table 6.2. The volume of seeds in the flute for one revolution of the metering mechanism, *Vr* (m^3 rev^{-1}), was determined by the following relationship:

$$V_{\mathrm{r}} = \frac{\pi S_{\mathrm{r}} * R_{\mathrm{s}} * (1+\phi) * D_{\mathrm{g}}}{B_{\mathrm{d}} * g_r * 10^4} \tag{6.1}$$

Table 6.1 The physical characteristics of berseem and lucerne seeds

Crop type	Test weight (g)	Bulk density ($kg.m^{-3}$)	Size (mm) (Range)	
			Length	Width
Berseem	2.6	813.94	1.84–2.68	1.30–1.98
Lucerne	2.93	779.52	1.66–3.10	0.95–1.77

Table 6.2 The agronomic requirement of the selected fodder crops

Crops	Seed rate (Kg.ha^{-1})	Row-to-row spacing (cm)	References
Berseem	10 ± 0.4	40	[16]
Lucerne	12–15	30	[17]

$$V_r = n * l * A \tag{6.2}$$

where, S_r = Seed rate (kg.ha^{-1}); D_g = Diameter of the ground wheel, m; B_d = Bulk density of the seed, kg m^{-3}; R_s = Row-to-row spacing of the crop, m; g_r = Speed ratio, ground wheel to roller; $\emptyset$ = skid fraction; n = number of flutes of metering mechanism; A = Area of the flute; l = length of the flute.

By using Eqs. 6.1 and 6.2, the details of different metering devices have been discussed.

Fluted Roller From Eqs. 6.1 and 6.2, considering the available gear ratio 1.79, the amount of skid of the ground wheel was assumed to be 10%, and based on the physical properties of the seeds and agronomical data in Tables 6.1 and 6.2, the volume of seeds in the flute for one revolution of the metering mechanism were calculated as 303.45 and 320.80 mm^3 for berseem and lucerne crop, respectively. Two types of fluted rollers with the same roller diameter of 85 mm having different groove shapes (trapezoidal and asymmetric) were used in the study, shown in Fig. 6.1a, b, respectively. The trapezoidal-type fluted metering mechanism had seven flutes with a 7 mm active flute length of the roller. The asymmetric-type fluted roller had five grooves and a 6.5 mm active groove length of the roller.

Indented Cell-type Roller The seed-metering mechanism consisted of an 85 mm diameter with indentations provided at the periphery rotating about a horizontal axis. Eight holes were drilled at regular intervals around the circumference of the roller. To accommodate both seeds of berseem and lucerne, the holes in the roller were adjustable as per the size of the seeds. The hole diameter was kept more than 1.5 times the maximum length of seeds, as shown in Table 6.1; it was 5 and 4 mm in depth. The detailed scheme of this metering mechanism has been shown in Fig. 6.1c.

6.2.3 Laboratory Evaluation of Metering Mechanisms

As shown in Fig. 6.2, three metering mechanism was mounted with the shaft in the metering mechanism holding device and hopper. The metering mechanisms were rotated when the power was delivered from the speed drive of the experiment setup by using a chain and sprocket mechanism. The chain and sprockets were selected according to the gear ratio of the ground wheel to the metering device. A 14-tooth sprocket was used at the ground wheel to drive the 25-tooth sprocket to the

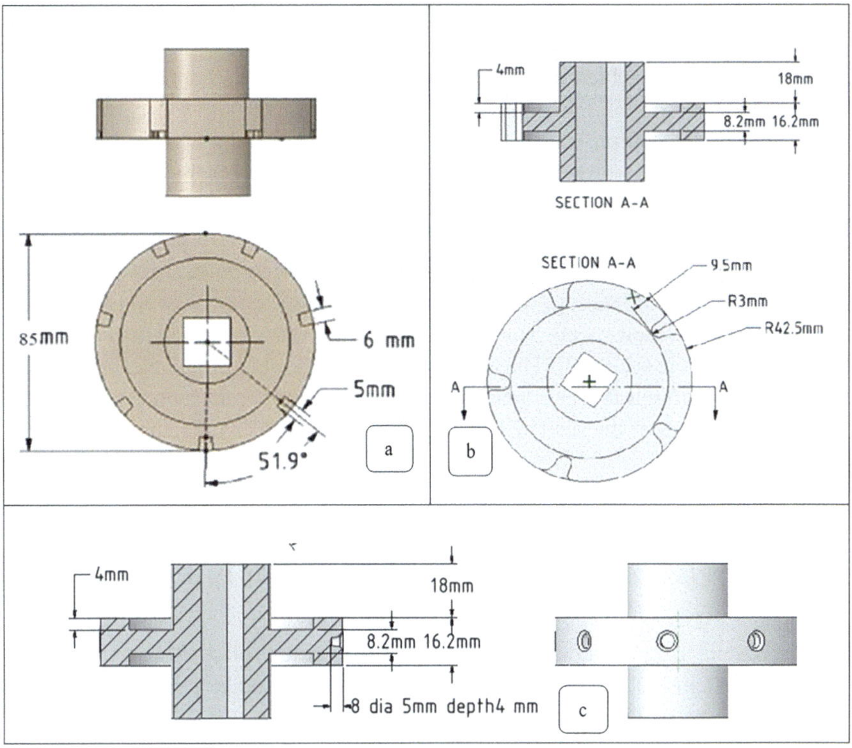

Fig. 6.1 The isometric view of used metering devices: (**a**) isometric view of trapezoidal fluted roller seed-metering device, (**b**) isometric view of asymmetric fluted roller metering device, and (**c**) isometric view of indented cell-type seed-metering device

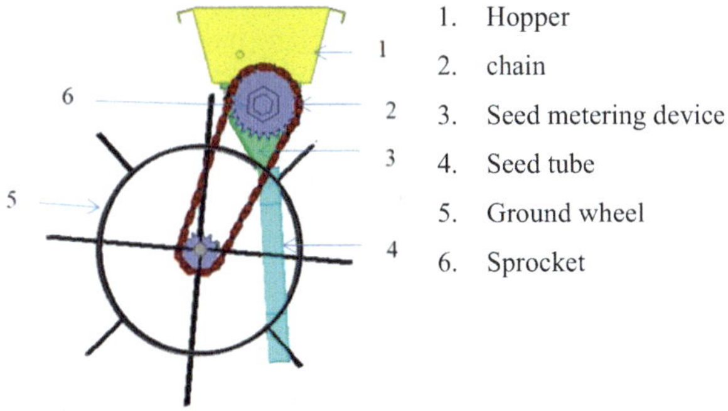

Fig. 6.2 A view of the metering mechanisms mounted on hopper and other main components

seed-metering shaft. The speed reduction ratio of the ground wheel and seed-metering device was 1.79:1. A single seed hopper was developed for the seed-metering mechanism for all three rows.

The seed hopper units had one drive for the metering mechanisms. To avoid frequent filling of seeds in the hopper, the capacity of the hopper was maintained at 5 kg of lucerne seed, which had a minimum bulk density. The hopper was provided with baffle plates to maintain a steady flow of the seeds.

Laboratory evaluation of seed-metering devices was carried out at the Farm Machinery Testing Centre of the Department of Farm Machinery and Power Engineering, Punjab Agricultural University, Ludhiana. The tests were conducted with an experimental setup consisting of a variable speed drive unit to change the speed of operation and a universal mounting frame to accommodate different sowing machines. During testing, the metering mechanism speed was adjusted to the same as the canvas belt speed. The canvas belt speed was developed to simulate the ground speed of the seed drill. The rotary power, which rotates the metering devices, was provided through a set of gearboxes, variable drives, and pulleys. Figure 6.3 illustrates the setup of the seed drill and testing rig used to evaluate different metering devices.

6.2.4 Selection of Independent and Dependent Parameters

Three seed-metering devices, i.e., trapezoidal-type fluted roller (M1), asymmetric-type fluted roller (M2), and indented cell type (M3); two fodder crops, i.e., berseem (C1) and lucerne (C2); and three forward speed levels, i.e., 1.5 (S1), 2.0 (S2), and 2.5 (S3) km.h^{-1} were taken as independent parameters for the lab experiments. The performance of the seed-metering devices was evaluated in terms of seed rate, seed-to-seed distance, variance coefficient of seeding uniformity, and seed damage. Each treatment was replicated thrice. The different variables and their levels have been presented in Table 6.3.

Seed Rate To determine the seed rate, polythene bags were kept under each seed tube and seeds were collected for 20 rounds of the ground wheel. Subsequently, the

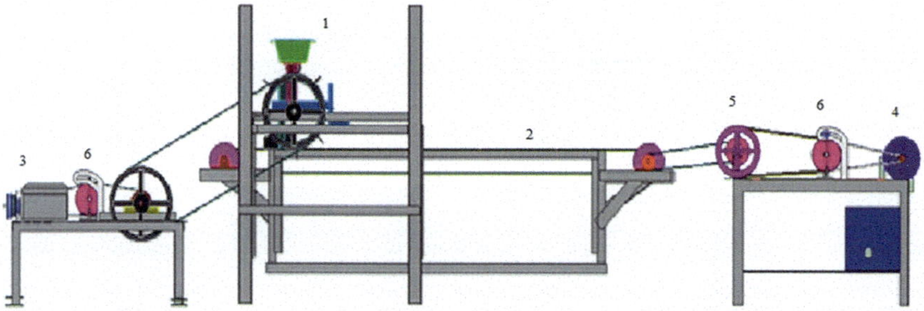

Fig. 6.3 A view of the setup used for the laboratory evaluation: (*1*) seed metering mounted on seed drill, (*2*) canvas belt, (*3*) gearbox for operated seed drill, (*4*) electric motor for transmission of canvas belt, (*5*) pully for canvas belt transmission, and (*6*) variable drive

Table 6.3 Independent and dependent variables selected for the study

Sr. no.	Independent parameters	Levels	Detail	Dependent parameters
1.	Crop	2	C1—Berseem (BL-10) C2—Lucerne (Composite LL-5)	Seed rate (kg.ha^{-1}) Seed-to-seed distance (mm)
2.	Speed	3	S1–1.5 km.h^{-1}	
			S2–2.0 km.h^{-1} S3–2.5 km.h^{-1}	Variance coefficient of seeding uniformity (%) Seed damage (%)
3.	Seed metering mechanism	4	M1—trapezoidal fluted roller M2—asymmetric fluted roller M3—indented cell roller M4—inclined plate type	

sample of the seeds was weighed at each speed level, crop type, and metering device. The seed rate (kg.ha^{-1}) was calculated from observed data using the following relationship [3, 11]:

$$S_r = \frac{S_w}{W_s * C_w} \tag{6.3}$$

where, S_r = Seed rate (kg. ha^{-1}); S_w =Weight of Seeds obtained by 20 revolutions of drive wheel; W_s = Width of seed drill, (m); C_w = Circumference of drive wheel, (m).

Seed-to-Seed Distance The distance between the seeds was measured using a wooden board that passes below the point of the seed drop. The board has a dimension of 2000 × 300 mm. To stick the seeds on the board, that fall from the seed tube, white grease was applied on the wooden board to reduce the effect of seed jumping. The distance between the seed (S_d) observation was recorded for 10 runs for each treatment, and its average was taken in mm. Finally, it was compared with its theoretical seed-to-seed spacing [18].

$$S_d = \frac{\Sigma S_a}{N} \tag{6.4}$$

where, S_d = the distance between the seeds; N = total number of observations.The theoretical seed-to-seed spacing was calculated from the following equations [14, 18, 19]:

$$S_t = \frac{T_{w*10}}{s_r * r_s} \tag{6.5}$$

where, S_t = Theoretical spacing between seeds, mm; T_w = Thousand grain weight, g; S_r = Seed rate (kg.ha^{-1}); R_s = Row-to-row spacing of the cropSimilarly, the variance coefficient of seeding uniformity was calculated to assess the seeding distribution performance of the seed-metering device based on the standard ISO-7256/2, 1984. Finally, the variance coefficient of seeding uniformity was calculated as follows [19]:

$$C_v = \frac{\text{SD}}{\overline{X}} \times 100\% \tag{6.6}$$

$$\text{SD} = \left\{ \sum_{i=1}^{n} \left[\left(X_i - \overline{X} \right)^2 \right] / (n-1) \right\}^{\frac{1}{2}} \tag{6.7}$$

$$\overline{X} = \frac{1}{n} \sum_{i=1}^{n} X_i \qquad (i = 1,2,3,\ldots n) \tag{6.8}$$

where, C_v was the variance coefficient of seeding uniformity in each row, %; SD was the standard deviation of seed spacing, mm; Xi was the spacing between seeds, mm; $\bar{X}$ was the average seed spacing for each row, mm and n was the total number of samples collected.

Seed Damage The seed damage was calculated based on the visible damage to metered seeds. Randomly selected 100 g seed weight of berseem and lucerne seeds was passed through each metering device. The treatment was replicated thrice. The seed damage (S_d) was calculated using the following relationship [14, 16, 18]:

$$S_d = \frac{W_{sd}}{W_s} * 100 \tag{6.9}$$

where, W_{sd} = observed weight of seed damaged; W_s = Weight of seeds selected through metering unit

6.2.5 Data Analysis

For conducting the performance evaluation of different metering devices, there were 24 treatments with three replications of each parameter. The data collected for all the dependent variables was subjected to analysis of variance using the SAS 9.3 software package at a 5% level of significance [20]. The mean of different treatments was compared by Turkey's method [21].

6.3 Results and Discussion

6.3.1 Effect of the Metering Device and Forward Speed on the Seed Rate

The effect of the type of metering devices on seed rate is shown in Fig. 6.4 and Table 6.4. For metering devices M1, M2, and M3, the seed rate of berseem crop ranged from 9.88–10.88, 13.44–14.10, and 9.05–9.70 kg.ha^{-1}, respectively, whereas for lucerne crop, the seed rate was 13.59–15.04, 15.69–17.07, and 11.38–11.88 kg. ha^{-1}, respectively at various levels of forward speed. Analysis of variance indicated that the seed rate was significantly affected by the type of metering devices.

The comparison of the mean seed rate of metering devices was conducted by Tukey's method (Table 6.4) at 5% level of significance. The mean seed rate was significantly lesser in M3 (10.53 kg.ha^{-1}) as compared to M1 (12.25 kg.ha^{-1}) and M2 (15.11 kg.ha^{-1}). The effect of the metering devices on the output of seed rate was found to be significant because the difference in configuration and shape of the groove of metering devices was different. As shown in Fig. 6.4, the seed rate decreased when the forward speed increased. The minimum seed rate of 9.05 and 11.38 kg.ha^{-1} was observed at a speed of 2.5 km.h^{-1}, and the maximum seed rate of 19.86 and 22.16 kg.ha^{-1} was observed at 1.5 km.h^{-1} for berseem and lucerne crops, respectively. Analysis of variance indicated that seed rate was significantly affected

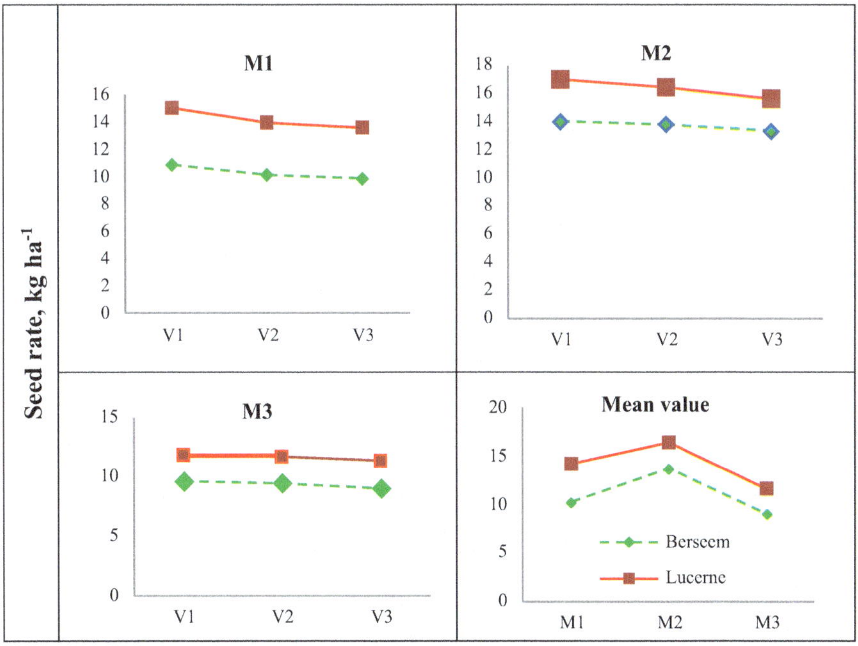

Fig. 6.4 Effect of metering device, crop type, and forward speed on seed rate

Table 6.4 Tukey's analysis for the effect of metering device (M), crop type (C), and forward speed (S) on different dependent parameters

Parameters		Seed rate (kg.ha^{-1})	Damaged seed (%)	Seed-to-seed distance (mm)
Metering device	M1	12.25^{b}	2.29^{c}	7.04^{c}
	M2	15.11^{a}	5.44^{a}	7.80^{b}
	M3	10.53^{c}	3.57^{b}	10.84^{a}
Crop type	C1	13.24^{b}	2.80^{b}	7.50^{b}
	C2	16.02^{a}	3.39^{a}	8.45^{a}
Speed	S1	15.08^{a}	1.98^{c}	7.35^{c}
	S2	14.65^{ab}	3.21^{b}	7.98^{b}
	S3	14.17^{b}	4.08^{a}	8.60^{a}

Note: Within the table, mean values with the same superscript letter are not significantly different ($p > 0.05$)

by forward speed. The comparison of mean seed rate within three different forward speed levels was done by Tukey's method at a 5% level of significance (Table 6.4), and it was found that there was a significant difference in seed rate between S1 and S3, but there was no significant difference between S1 and S2 and between S2 and S3. This may be because, at higher forward speed, the higher rotational velocity of the metering device resulted in additional seed disturbance, which led to reduced filling of seeds into the metering unit as the seed falls from the cells, thereby reducing seed rate. Malek et al [22] also stated that increasing the auger speed causes the auger flight to have a smaller opportunity to load the seeds from the seed chamber. This was because the auger speed become larger than the seed flow into the seed chamber. This trend was also consistent with that reported by Cortez et al [21] observed that the highest initial population of maize plant was found at the lowest sowing speed of 6 km.h^{-1} than 12 km.h^{-1}. Similar trends were observed during experiments on studies of the effect of forward speed on the number of seeds per meter length and the seed rate of different crops [21–24].

6.3.2 Effect of Metering Device and Forward Speed on Seed-to-Seed Distance

Figure 6.5 shows the effect of different combinations of metering devices, forward speeds, and crop types on average seed-to-seed distance. For berseem crop, seed-to-seed distance ranged from 6.57 to 7.12 mm, 6.96 to 7.71 mm, and 9.54 to 10.63 mm for metering device M1, M2, and M3, respectively; while for lucerne crop, seed-to-seed distance ranged from 6.49 to 7.90 mm, 7.57 to 8.81 mm, and 10.58 to 12.85 mm. The analysis of variance indicated that the average seed-to-seed distance was significantly affected by the type of metering devices. The comparison of mean seed-to-seed distance for metering devices was done by Tukey's method (Table 6.4) at a 5% level of significance. It was found that there was a significant difference in the

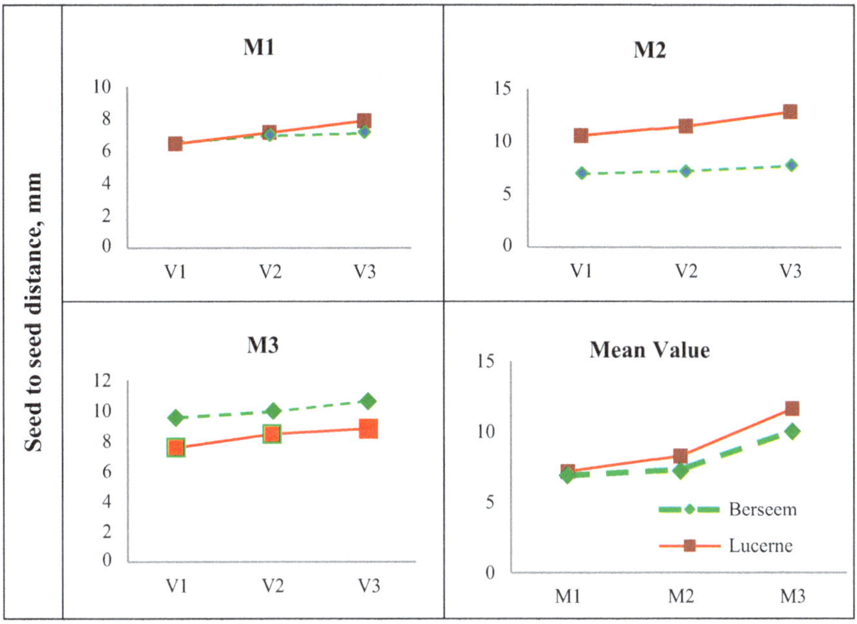

Fig. 6.5 Effect of metering device, crop type, and forward speed on seed-to-seed distance

seed-to-seed distance between M1, M2, and M3. However, the seed-to-seed distance was significantly lesser in M1 than in M2 and M3. The metering device had a significant effect because the shape of the groove of the metering devices was different. The minimum mean seed-to-seed distance was 7.35 mm at a forward speed of 1.5 km.h^{-1}, and the maximum was 8.6 mm at a forward speed of 2.5 km.h^{-1}. Analysis of variance indicated that the seed-to-seed distance increased significantly with an increase in forward speed. This was agreed with the findings [18, 20, 21, 23] as they observed the highest seed-to-seed spacing at the highest forward speed in their experiment due to a higher speed of the metering device might not have sufficient time for filling seeds from the seed hopper.

6.3.3 Effect of Metering Device and Forward Speed on Coefficient of Seed Distribution Variation

Figure 6.6 shows the effect of different combinations of metering devices and forward speeds for both crops on the average variance coefficient of seeding uniformity. For metering devices M1, M2, and M3, the coefficient of seed distribution variation was 4.39 ± 1.50, 9.15 ± 0.99, and 6.05 ± 3.61%, while for lucerne crop, it was 4.16 ± 2.24, 4.20 ± 2.20, and 6.81 ± 2.34%, respectively at various combinations of forward speeds of 1.5, 2, and 2.5 km.h^{-1}. It was observed that the mean

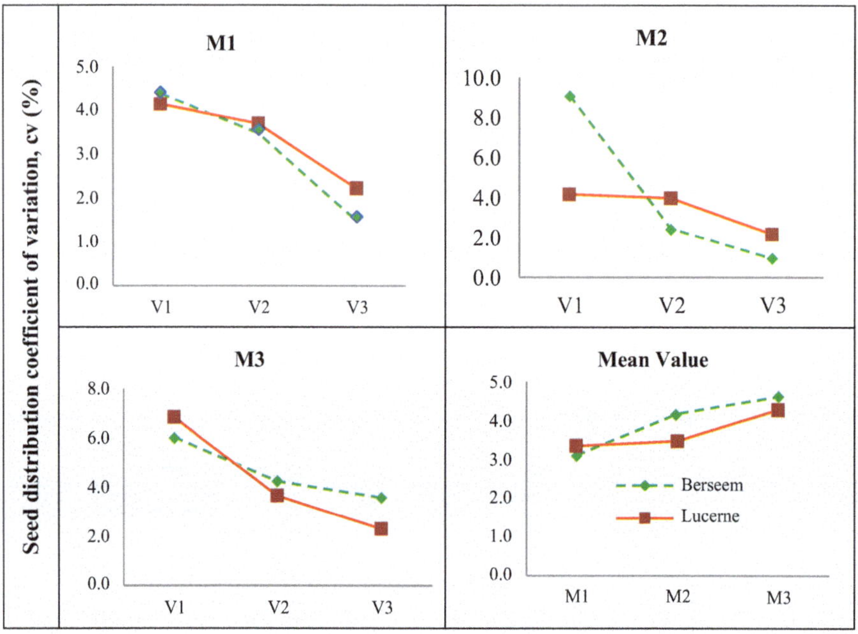

Fig. 6.6 Effect of metering device, crop type, and forward speed on the variance coefficient of seeding uniformity

coefficient of seed distribution variation with M1 (3.25%) was significantly lesser than M2 (3.84%) and M3 (4.47%). The values of the coefficient of variation (CV) were less than 5%, and it is considered very good and confirmed the acceptable limits [11]. It was found that the coefficient of seed distribution variation at 1.5 km.h^{-1} was 6.53% which was higher than attained at 2 km.h^{-1} (3.41%) and 3 km.h^{-1} (2.03%) for berseem crop, while for lucerne crop it was 5.08%, 3.80%, and 2.26%, respectively. Rajaiah et al [24] showed that the coefficient of variation (CV) of seeding distribution increased with a decrease in the seed rate of paddy using a fluted metering mechanism while the speed increased from 1.5 to 2.5 km.h^{-1}. For quantification of performance, the lower the CV value of precision better the type of metering device [21]. The CV value is in the acceptable range, as it is less than 10%.

6.3.4 *Effect of Metering Device and Forward Speed on Seed Damage*

The effect of the type of metering device on seed damage has been presented in Fig. 6.7. For metering devices M1, M2, and M3, seed damage percentage for berseem crop ranged 1.36–1.63, 4.96–5.66, and 2.18–4.21 while for lucerne crop the value of seed damage was 0.51–4.31, 3.92–7.23 and 1.55–5.94, respectively at

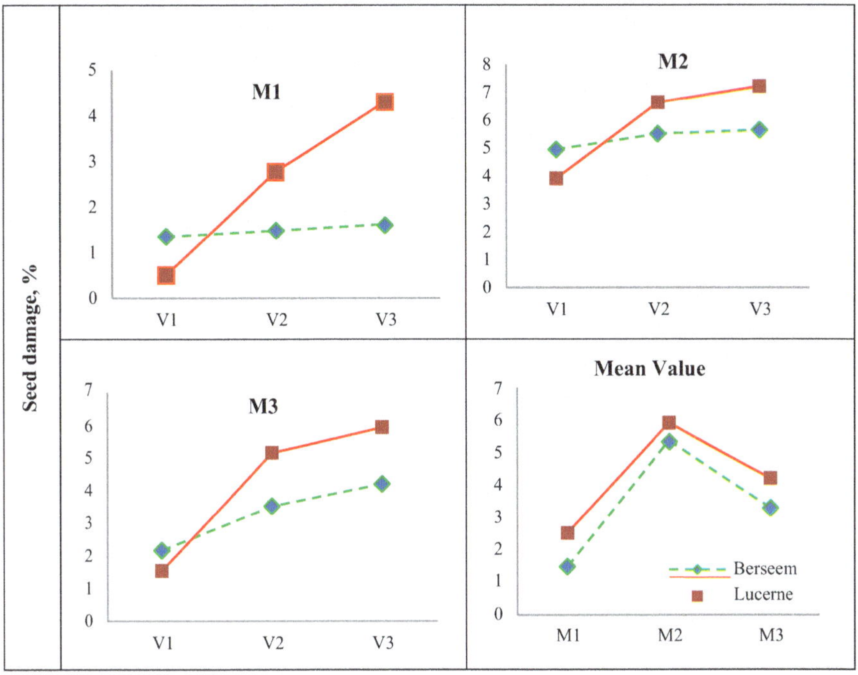

Fig. 6.7 Effect of metering device, crop type, and forward speed on seed damage

various combinations of forward speed. Analysis of variance indicated that seed damage was significantly affected by metering devices. The first-order interaction M × S was also found significant at 5% of the level of significance. The comparison of mean seed damage within the metering device was done by Tukey's method (Table 6.4) at a 5% level of significance. It was found that there was a significant difference in mean seed damage value as the minimum was M1 (1.08%) and the maximum was M2 (5.44%). The higher seed damage found for M2 could be because there was the tendency of multiple seeds to fill the seed cells and get exposed to rubbing against the baffle plate as it had a higher exposed area of grooves.

Analysis of variance indicated that the seed damage was significantly affected by crop type. There was a significant difference in damage for crops C1 and C2. Mean seed damage was significantly lesser for C1 (2.80%) as compared to C2 (3.39%). This could be because of the variation in the physical properties of both seeds. Forward speed had a significant direct effect on seed damage for both crops. The minimum seed damage was recorded for combination C2M1S1 and the maximum was recorded for combination C2M2S3. The first-order interaction of S x M and S x C was also found significant at 5% of the level of significance. The comparison of mean seed damage within three different forward speed levels was done by Tukey's method (Table 6.4) at a 5% level of significance. There was a significant difference in mean seed damage at S1 (1.98%), S2 (3.21%), and S3 (4.08%). This was in agreement with the findings of Ekka et al. and Zilpilwar et al. [18, 25] observed that the highest seed damage was found at the highest forward speed of their experiment.

6.3.5 Suitable Combination of Independent Parameters

From Table 6.4, for both berseem and lucerne crops, at a forward speed of S2 and S3, M1 resulted in a better seed rate as per the agronomical requirement of 10 ± 333 0.4 and 13.75 $kg.ha^{-1}$, respectively, than M2 and M3. At all levels of forward speed, M1 had lesser seed damage than M2 and M3 for both crops. Considering the theoretical seed spacing of 6.5 and 7.14 mm for berseem and lucerne crops, respectively, the seed-to-seed spacing was also better for M1 than M2 and M3 for both crops. Based on lab results and considering the desired seed rate and reduced seed damage as the main parameters of interest, a trapezoidal groove fluted roller metering device and forward speed of 2.0 $km.h^{-1}$ were selected as the suitable parameters.

At the selected independent parameters of the experiment, for berseem seeds, the seed rate, seed damage, seed-to-seed spacing, and seed distribution variation were 10.14 $kg.ha^{-1}$, 1.49%, 6.96 mm, and 4.3%, and for lucerne seeds, it was 13.97 kg. ha^{-1}, 2.78%, 7.17 mm, and 3.9%, respectively.

6.4 Conclusions

The seed rate was significantly affected by the type of metering device, i.e., M1, M2, and M3. In the trapezoidal fluted roller metering device (M1), a seed rate of 10.3 and 14.2 $kg.ha^{-1}$ was obtained for the berseem and lucerne crops, respectively; it is as per the agronomical requirements of the crops. The seed rate of 13.77 and 16.42 $kg.ha^{-1}$ was recorded for the berseem and lucerne crops, respectively, for asymmetric-type fluted roller metering device (M2). In all type of seed-metering mechanisms, with an increase in forward speed, the seed rate decreased significantly. The seed damage was significantly affected by the type of metering devices. The trapezoidal-type fluted roller metering mechanism resulted in minimum seed damage (1.08%), whereas the highest seed damage (5.44%) was observed in M2. The seed damage percentage was significantly lesser for berseem seed than for lucerne seed. The seed damage was the minimum at S1; it increased as the speed increased. The seed-to-seed spacing ranged between 6.49 and 7.90 mm; while using the trapezoidal metering mechanism, it was as per the theoretical seed-to-spacing. The seed distribution variation was significantly affected by the type of metering device; however, the value was found in the acceptable range, as it is less than 10%. There was a significant direct effect of forward speed on seed distribution variation. Considering the desired seed rate and reduced seed damage as the main parameters of interest, the trapezoidal groove fluted roller metering device and forward speed of 2.0 $km.h^{-1}$ were selected as suitable parameters for both crops.

Acknowledgments It is my proud privilege to express my devout gratitude and indebtedness to faculty and staff of Punjab Agricultural University, Ludhiana for helping me during present research work. I gratefully acknowledge the Indian Council of Cultural Relationship for providing me the opportunity to pursue Master's degree from Punjab Agricultural University, Ludhiana (India).

Conflict of Interest The authors declare no conflicts of interest regarding the publication of this paper.

Authors' Contribution Conceptualization and designing of the research work (SM and RK,); Execution of field/lab experiments and data collection (SM and RK); Analysis of data and interpretation (SM and RK,); Preparation of manuscript (SM).

Funding This research did not receive any specific grant from funding agencies in the public, commercial or not for profit sector.Data AvailabilityThe detail information was described in the body of the research paper.

References

1. Ranjan, A. K., Kumar, H., & Umrao, R. (2016). Efficacy of oat, berseem and lucerne under subabul (*Leucaena leucocephala*) and Poplar (*Populus deltoides*) Based Silvopastoral System. *The Bioscan, 11*(4), 2371–2374.
2. Halli, H. M., Rathore, S. S., Manjunatha, N., & Wasnik, V. K. (2018). Advances in agronomic management for ensuring fodder security in Semi Arid Zones of India – A review. *International Journal of Current Microbiology and Applied Sciences, 7*(2), 1912–1921. https://doi.org/10.20546/ijcmas.2018.702.230
3. Duchene, O., Vian, J. F., & Celette, F. (2017). Intercropping with legume for agroecological cropping systems: Complementarity and facilitation processes and the importance of soil microorganisms. A review. *Agriculture, Ecosystems and Environment, 240*, 148–161. https://doi.org/10.1016/j.agee.2017.02.019
4. Cuttle, S., Shepherd, M., & Goodlass, G. (2003). *A review of leguminous fertility-building crops, with particular refence to nitrogen fixation and utilisation* (p. 167). Department for Environment, Food and Rural Affairs (Defra) Project Report OF0316: The development of improved guidance on the use of fertility-building crops in organic farming Defra.
5. White, C., Holmes, H., & Morris, N. (2016). A review of the benefits, optimal crop management practices and knowledge gaps associated with different cover crop species. *AHDB Cereal Oilseed While, 90*, 93.
6. Sarwar, M., Ajmal Khan, M., Mahr-un-Nisa, & Touqir, N. A. (2005). Influence of berseem and lucerne silages on feed intake, nutrient digestibility and milk yield in lactating Nili buffaloes. *Asian-Australasian Journal of Animal Sciences, 18*(4), 475–478. https://doi.org/10.5713/ajas.2005.475
7. Kirkegaard, J. A., Peoples, M. B., Angus, J. F., & Unkovich, M. J. (2011). *Rainfed farming systems*. https://doi.org/10.1007/978-1-4020-9132-2
8. Devendra, C., & Thomas, D. (2002). Crop-animal interactions in mixed farming systems in Asia. *Agricultural Systems, 71*(1–2), 27–40. https://doi.org/10.1016/S0308-521X(01)00034-8
9. Karayel, D., & Özmerzi, A. (2008). Evaluation of three depth-control components on seed placement accuracy and emergence for a precision planter. *Applied Engineering in Agriculture, 24*(3), 271–276. https://doi.org/10.13031/2013.24494
10. Ah, S., & Thomas, U. (2020). Mechanization in fodder crop production – A review. *Forageresearch.in, 46*(1), 1–9.
11. Güler, I. E. (2005). Effects of flute diameter, fluted roll length, and speed on alfalfa seed flow. *Applied Engineering in Agriculture, 21*(1), 5–7.

12. Lv, H., Yu, J., & Fu, H. (2013). Simulation of the operation of a fertilizer spreader based on an outer groove wheel using a discrete element method. *Mathematical and Computer Modelling, 58*(3–4), 842–851. https://doi.org/10.1016/j.mcm.2012.12.017
13. Ahmad, E., Ghassemzadeh, H. R., Moghaddam, M., & Kim, K. U. (2008). Development of a precision seed drill for oilseed rape. *Turkish Journal of Agriculture, 32*(5), 451–458. https://doi.org/10.3906/tar-0707-52
14. Gautam, A., Khurana, R., Manes, G. S., & Dixit, A. K. (2016). Studies on engineering properties of pelleted carrot (*Daucus carota L.*) seeds. *International Journal of Bio-resource and Stress Management, 7*(5), 1044–1048. https://doi.org/10.23910/ijbsm/2016.7.5.1637a
15. Badgujar, C. M., Dhingra, H. S., Mannes, G. S., & Khurana, R. (2017). Development and evaluation of inclined plate metering mechanism for soaked okra (Abelmoschus esculentus) Seeds. *International Journal of Current Microbiology and Applied Sciences, 6*(12), 3948–3961. https://doi.org/10.20546/ijcmas.2017.612.455
16. Sahay, C. S., Pathak, P. K., & Singh, S. K. (2016). *Mechanization in fodder production* (42p). ICAR Indian Grassland and Fodder Research Institute.
17. Ahmed, S., Singh, S., & Roy, A. K. (2017, November). *Improved forage crop varieties of IGFRI.* ICAR-Indian Grassland and Fodder Research Institute.
18. Ekka, U., Singh, N., Bharti, N., Sahoo, P. K., & Singh, M. K. (2019). Design of seed metering system for jute seeds. *Agricultural Engineering International: CIGR Journal, 21*(3), 88–93.
19. Lei, X., et al. (2021). Seed motion characteristics and seeding performance of a centralised seed metering system for rapeseed investigated by DEM simulation and bench testing. *Biosystems Engineering, 203*, 22–33. https://doi.org/10.1016/j.biosystemseng.2020.12.017
20. Gautam, P. V., Kushwaha, H. L., Kumar, A., Khura, T. K., & Sarkar, S. K. (2023). Microcontroller-based low-cost seed metering module retrofit on cultivator. *Indian Journal of Engineering and Materials Science, 30*(1), 180–188. https://doi.org/10.56042/ijems.v1i1.45765
21. Cortez, J. W., Anghinoni, M., & Arcoverde, S. N. S. (2020). Seed metering mechanisms and tractor-seeder forward speed on corn agronomic components. *Engenharia Agrícola, 40*(1), 61–68. https://doi.org/10.1590/1809-4430-ENG.AGRIC.V40N1P61-68/2020
22. Maleki, M. R., Jafari, J. F., Raufat, M. H., Mouazen, A. M., & De Baerdemaeker, J. (2006). Evaluation of seed distribution uniformity of a multi-flight auger as a grain drill metering device. *Biosystems Engineering, 94*(4), 535–543. https://doi.org/10.1016/j.biosystemseng.2006.04.003
23. Ormond, A. T. S., Furlani, C. E. A., Oliveira, M. F., Noronha, R. H. F., Tavares, T. O., & Menezes, P. C. (2018). Maize sowing speeds and seed-metering mechanisms. *The Journal of Agricultural Science, 10*(9), 468. https://doi.org/10.5539/jas.v10n9p468
24. Rajaiah, P. (2020). Design and development of precision planter for paddy direct seeding. *Journal of Agricultural Engineering, 57*(4), 1–200. https://doi.org/10.52151/jae2020574.1723
25. Zilpilwar, S., Yadav, R., Gajjar, P., & Jakasania, R. (2021). Development and performance evaluation of metering mechanism for planting garlic cloves. *Agricultural Research, 10*(2), 274–284. https://doi.org/10.1007/s40003-020-00492-1

Chapter 7
Effects of Crop Parameters and Feed Rates on Performances of Axial Rice Thresher

Solomon Tekeste Hailemesikel, Mersha Alebachew, Mulugeta Admasu Delele, Addisu Negash Ali, Solomon Workneh Fanta, and Geta Kidanemariam

7.1 Introduction

Rice, *Oryza sativa* L., is the staple food of more than half the world's population. It is the main source of daily calorie intake for people in Asia and some parts of Africa. It is the most rapidly growing source of food in Africa. The rice paddy production in Africa increased from 24.4 million MT in 2008 to 33.2 million MT in 2018 [1]. Even though rice production is increasing rapidly in Africa, rice import (paddy, broken, milled, and husked rice) also increased from 8.9 million MT in 2008 to 16.8 million MT in 2018 [1].

Rice cultivation in Ethiopia had begun in the early 1970s. Even though rice is a newly introduced crop in Ethiopia, the production is expanding rapidly. The area under rice grew from 24,434 ha in 2008 [2] to 63,361 ha in 2018 [3]. The potential rice production area in Ethiopia is estimated to be about 30 million hectares [4]. According to this report, Ethiopia has the highest potential for rice production in Africa. Even though the area under cultivation is increasing rapidly, the country is still a net importer of rice. According to the Food and Agriculture Organization

S. T. Hailemesikel (✉) · A. N. Ali · G. Kidanemariam
Faculty of Mechanical and Industrial Engineering, Bahir Dar Institute of Technology, Bahir Dar University, Bahir Dar, Ethiopia

M. Alebachew
Ethiopian Institute of Agricultural Research, Fogera National Rice Research and Training Center, Woreta, Ethiopia

M. A. Delele · S. W. Fanta
Faculty of Chemical and Food Engineering, Bahir Dar Institute of Technology, Bahir Dar University, Bahir Dar, Ethiopia

S. W. Fanta et al. (eds.), *Sustainable Development Research in Manufacturing, Process Engineering, Green Infrastructure, and Water Resources*, Green Energy and Technology, https://doi.org/10.1007/978-3-031-77339-6_7

(FAO) report, in 2018, the government of Ethiopia imported 294,000 metric tons of rice (paddy, broken, milled, and husked rice). In monetary terms, this costs about 202.5 million USD [1].

In Ethiopia, rice production increases every year through the expansion of cultivation area and yield (ton/ha), introduction of improved varieties, and little emphasis is given to postharvest handling practices. According to [5], the current rice postharvest handling practices in Africa cause around 15–25% physical loss and, because of poor quality, an additional financial loss of 20–30% would occur at the market. In Ethiopian conditions, with the existing traditional harvesting, threshing, and storage methods, postharvest losses of rice are estimated to constitute as high as 30–35% [6].

Moreover, local rice has received lesser prices than the imported one in the domestic market. This is because the local rice contains a high percentage of broken rice, more than 50%, due to the poor postharvest handling practices [7]. Threshing operation can be mentioned as one of the major factors that influence postharvest quality and quantity losses of rice in Ethiopia [8].

In Ethiopia, threshing operation is commonly practiced in a traditional way, by trampling with a bunch of oxen and/or by beating the crop with a stick. Then, the cleaning operation is performed by tossing the threshed material in the air to remove the light elements from the grain.

This method of threshing and cleaning is usually tedious and time consuming. In addition, expose the grain for contamination with inert matter such as soil, small stones, weed seed, straw, and immature and unfilled grains. These extraneous materials will degrade the final grain quality and market value of rice. In this method of threshing, losses occurred due to some rice grains remaining unthreshed and gone with the straw, some grains scattering around, some grains sticking to the mud floor and not being recovered, and cracking and breakage due to the hoof pressure of trampling animals. Therefore, in this system, quantitative and qualitative postharvest losses are often very high. Therefore, introducing and promoting appropriate threshing technologies was found to be an important element to minimize losses and maintain the quality of the paddy rice, which has a direct implication during the processing.

The performances of grain threshers are affected by machine, crop, and operating parameters. These include threshing cylinder peripheral speed, threshing mechanism, threshing cylinder length and diameter, clearance between threshing cylinder and concave, feed rate (FR), grain and straw moisture content (MC), and grain-straw ratio [9]. Numerous previous works have been conducted in light of these factors [10–17]. Even though many previous research has been conducted in this area, threshing performances vary with factors including crop cultivar under study. In this research, the performance of locally fabricated axial flow rice threshers was examined for three local cultivars.

7.2 Materials and Method

7.2.1 Description of the Test Thresher

An axial flow mechanical thresher was manufactured at local public enterprise called Amhara Metal Industry and Machine Technology Development Enterprise, using the conventional design process. The thresher had a spike tooth-type threshing drum with a cleaning mechanism. Figure 7.1. depicts the main parts of the thresher, including the feeding table, the threshing unit, the cleaning sieve, and the blower. Furthermore, the detailed specification of the thresher is shown in Table 7.1.

(a) (b)

Fig. 7.1 Axial rice thresher machine (**a**) during the testing of the machine, (**b**) assembling the spike tooth

Table 7.1 Specification of the thresher

Parameters	Description
Power source (hp)	14 hp diesel engine
Drum diameter (mm)	300
Drum length (mm)	880
Number of threshing bars and spikes	10 bars and100 spikes
Spike spacing (SS), diameter (D), and length (L) (mm)	SS = 70, D = 10 mm and L = 80
Number of straw thrower blade	5
Concave clearance (CC) (mm) and length (CL (mm)	20
Concave mesh size (L(mm) × W(mm))	18 × 28
Cleaning mechanism	Oscillating sieve with blowing fan
Oscillating stroke length (mm)	20
Blowing fan	Number of blades = 4

7.2.2 Rice Cultivars Used in This Study

The experiment was conducted in North-Western Ethiopia, Amhara region, Fogera district, in the compound of Fogera National Rice Research and Training Center. Three commonly grown rice varieties, '*Shaga*,' '*Ediget*,' and '*Fogera-1*,' which are widely farmed in the study area, were used for this experiment. '*Shaga*' and '*Ediget*' are locally improved rain-fed lowland rice varieties whereas '*Fogera-1*' is an upland variety. '*Shaga*' is brown in color whereas '*Ediget*' and '*Fogera-1*' are white.

Some important parameters of the rice cultivars were measured before the test was conducted. These include grain moisture content, straw moisture content, grain-straw ratio, straw length, and length of the panicle. From the material that is to be threshed, three samples were randomly taken of approximately 1 kg each. The grains and straw are separated by hand from each sample and are kept paired. After weighing, the samples are oven-dried at 105 ± 1 °C for 24 h and then reweighed. The moisture content on wet basis, MC (%), for the paddy and the straw is determined using Eq. (7.1).

$$\%MC = \frac{W_w - D_w}{W_w} * 100 \tag{7.1}$$

where

- %MC = moisture content wet basis
- W_w = weight of the sample before oven-dry (wet weight)
- D_w = weight of the sample after oven-dry (dry weight)

Then grain-straw ratio was determined using Eq. (7.2). Grain-straw ratio (R) is given as follows:

$$R = \frac{W_g}{W_s} \tag{7.2}$$

where:

- W_g—is the weight of dry grain, g
- W_s—is the weight of dry straw, g

Length of the panicle and the straw was measured using measuring tape for 10 samples and the mean value is presented.

7.2.3 Experimental Design

This experiment was conducted to determine the effects of feed rate and moisture content on the threshing performance of three different local paddy rice cultivars. In doing so, a factorial experiment with three factors, feed rate (FR), moisture content (MC), and paddy rice variety (VAR), was designed. Feed rate and variety were investigated at three experimental levels and moisture content was examined at two levels. The levels of feed rate are FR-1 = 2000 kg/h, FR-2 = 3000 kg/h and FR-3 = 4000 kg/h. The three cultivars used in the study include '*Ediget*' (VAR-1), '*Shaga*' (VAR-2), and '*Fogera-1*' (VAR-3). The levels of moisture content are MC-1 = 12.4% and MC-2 = 17.3% wet bases. For three replications a total of 3*2*3*3 = 54 tests were conducted.

7.2.4 Performance Evaluation of the Thresher

The performance of the thresher was evaluated according to the test procedure stated in [18]. Before the start of the test, the thresher and the engine were subjected to an idle running-in period of 2 h. During the running-in period, the speed of threshing drum and blowing fan were adjusted as per the requirement of the test, and fuel consumption was measured with no load. Then preliminary testing was done to select the working range for feed rate and the rpm of threshing drum.

The threshing experiment is conducted for three cultivars with three feed rates. Each trial was conducted for 10 min and replicated three times. After each test, the threshing area was cleaned and then prepared for the next trial. The speed of the threshing drum was adjusted to 750 rpm, by using the throttle lever of the engine. The drum speed (RPM) was measured using a Cole-Parmer (model 4060) digital tachometer with an accuracy of ±1 RPM or 0.05%.

During each test trial, samples were collected from the main grain outlet, straw thrower outlet, chaff outlet, and scattered grain based on the procedures as stated in [18] to analyze for losses, purity, and grain quality.

The weight of threshed paddy per unit time, W_{tg}, from the threshed grain outlet was measured using a weighing scale with a capacity of 60 kg and accuracy of 0.01 kg model (ZMISSIL F2-60). The unthreshed paddy and the straws were collected and weighed using a laboratory scale-sensitive balance with a capacity of 6000 g and accuracy of 0.1 g model (EK6100I) separately. The thresher performance was evaluated based on actual threshing capacity ,C_a(kg/h), threshing efficiency, T_e(%), cleaning efficiency, C_e(%), separation loss, S_l(%), scattering losses, SC_l(%), unthreshed grain ,UT_g(%), and grain damage, G_d(%). These performance parameters were computed using Eqs. (7.3), (7.4), (7.5), (7.6), (7.7), (7.8), (7.9), and (7.10)

$$C_{\mathrm{a}}\left(\frac{\mathrm{kg}}{\mathrm{h}}\right)=\frac{W_{tg}}{T_{\mathrm{o}}} \tag{7.3}$$

$$T_{\mathrm{e}}(\%)=100-UT_{\mathrm{g}}(\%) \tag{7.4}$$

$$T_{\mathrm{r}}(\%)=\frac{W_{tg}}{W_{\mathrm{t}}} \tag{7.5}$$

$$C_{\mathrm{e}}(\%),=\frac{W_{\mathrm{c}}}{W_{\mathrm{u}}} \tag{7.6}$$

$$UT_{\mathrm{g}}(\%)=\frac{W_{ug}}{W_{\mathrm{t}}}*100 \tag{7.7}$$

$$S_{\mathrm{L}}(\%)=\frac{W_{sl}}{W_{\mathrm{t}}}*100 \tag{7.8}$$

$$SC_{\mathrm{l}}(\%)=\frac{W_{scl}}{W_{\mathrm{t}}}*100 \tag{7.9}$$

where:

- T_r—is threshing recovery (%)
- W_{tg}—is the weight of threshed grain from threshed grain outlet, kg
- T_{o}—is the time of operation, h
- W_{u}—is the weight of uncleaned grain, g
- W_{c}—is the weight of cleaned grain, g
- W_{ug}—is the weight of unthreshed grain (kg)
- W_{t}—is the weight of total threshed grain (kg)
- W_{sl}—is the weight of clean grain collected from straw and chaff outlet
- W_{scl}—is the weight of scattered grain

Three samples of 100 g each were taken from threshed grains for analysis. Separate those grains that were broken crushed or dehulled (partially or fully) and weigh. Percentages of damaged grains, G_{d}(%), were computed using Eq. (7.10).

$$G_{\mathrm{d}}(\%)=\frac{W_{\mathrm{dg}}}{100\,\mathrm{gram\,sample}}*100 \tag{7.10}$$

where W_{dg}—is the weight of damaged grains in the sample (g).

7.2.5 Fuel Consumption and Specific Energy Consumed

Fuel consumed for each threshing operation was measured using a graduated cylinder with accuracy ±1 mL, by filling the engine fuel tank completely at the start and finish of each threshing operation. Energy consumed for threshing a unit kg of paddy was calculated using Eqs. (7.11) and (7.12). Engine power (EP) consumed for threshing operation is estimated using Eq. (7.11) as given by [19]

$$EP = FC * \frac{1}{3600} \rho_f * LCV * 427 * \eta_m * \eta_{th} * \frac{1}{75} * \frac{1}{1.36} \text{kW} \tag{7.11}$$

where:

- FC = fuel consumption (L/h)
- ρ_f = density of diesel fuel (kg/L)
- LCV = Lower calorific value of diesel fuel (k.cal/kg)
- E = energy consumed (kJ/kg)
- η_m = mechanical efficiency of diesel engine
- η_{th} = thermal efficiency of diesel engine
- 427 = mechanical equivalent of heat = 427 kg. m/kcal

The following constants were used in determining the specific energy consumption of the machine.

- Density of diesel fuel = 0.85 (kg/m^3) [20]
- Calorific value of diesel fuel (C_a) = 11 kWh per liter (kJ/kg) = 11,000 k.cal/kg [20]
- Mechanical efficiency of diesel engine, η_m, was assumed to be 80% for diesel engine [20]
- Thermal efficiency of diesel engine, η_{th}, was assumed to be 35% for diesel engine [20]

Specific energy (SE) is given by Eq. (7.12).

$$SE\left(\frac{kWh}{\text{kg}}\right) = \frac{EP(\text{kW})}{C_a\left(\frac{\text{kg}}{\text{h}}\right)} \tag{7.12}$$

7.2.6 Data Analysis

Analysis of variance (ANOVA) with factorial design, for three feed rates, two moisture content, and three varieties combination, was done using JMP 17 statistical package software (JMP statistical Discovery LLC, 2022). Treatment means were considered statistically different at $P < 0.05$ Pearson's correlation coefficient.

7.3 Result and Discussion

7.3.1 *Crop Parameters*

Table 7.2. displays a number of important crop factors for the varieties tested that are believed to have an impact on the threshing conditions. These variables include grain-straw ratio (*R*), grain-straw moisture content (MC-1 and MC-2), straw moisture content (SMC-1 and SMC-2), and straw length. Two different moisture contents were used in the experiment. In the lower moisture experiment, paddy rice and straw's moisture content are indicated by MC-1 and SMC-1, whereas in the higher moisture experiment, these parameters are indicated by MC-2 and SMC-2.

As seen in Table 7.2, there was a minor variability in grain moisture content between varieties. These differences, which are less than 0.5%, were disregarded and grand mean moisture contents of 12.4% and 17.3% were used as MC-1 and MC-2 in order to simplify data analysis and result reporting. The result shows that there is a significant variation among varieties in grain-straw ratio.

7.3.2 *Threshing Performances*

Threshing Capacity Analysis of variance as shown in Table 7.3 indicates that the main effect, variety, feed rate (FR), moisture content (MC), and interaction effect (FR * MC) all significantly affect the threshing capacity of the tested thresher at 0.0001 (0.01%) level of significance. As shown in Fig. 7.2, threshing capacity increases with increasing feed rate and decreases with increasing moisture content. Additionally, there is a substantial difference in the threshing capacity of the thresher with different varieties. The highest threshing capacity of the thresher is reported while threshing '*Fogera-1*' (VAR-3) which has the highest grain-straw ratio; then comes '*Shaga*' (VAR-2) and '*Ediget*' (VAR-3) consecutively. In general, the tested thresher's threshing capacity typically rises as the grain-straw ratio does.

Threshing Efficiency The observed mean threshing efficiency is presented in Table 7.4. The result showed that at lower moisture content (MC-1 = 12.4%), threshing efficiency of 100% was recorded for all varieties at all feed rate levels, since no unthreshed grain is observed at lower moisture content. This would be due to the fact that the paddy was overdried and easily detached from the straw during the

Table 7.2 Mean values and standard deviations of crop parameters

Varieties	MC-1 (%)	SMC-1 (%)	MC-2 (%)	SMC-2 (%)	*R*	Straw length (cm)
Ediget	12.3 ± 0.3	11.8 ± 1.3	17.5 ± 0.26	40.8 ± 3.2	0.56 ± 0.03	89.3 ± 8.42
Shaga	12.4 ± 0.26	11.4 ± 2.26	17.3 ± 0.17	37.7 ± 3.04	0.63 ± 0.03	107.3 ± 5.6
Fogera-1	12.5 ± 0.26	11.2 ± 1.51	17.1 ± 0.2	40.9 ± 2.35	0.72 ± 0.02	102.3 ± 4.5

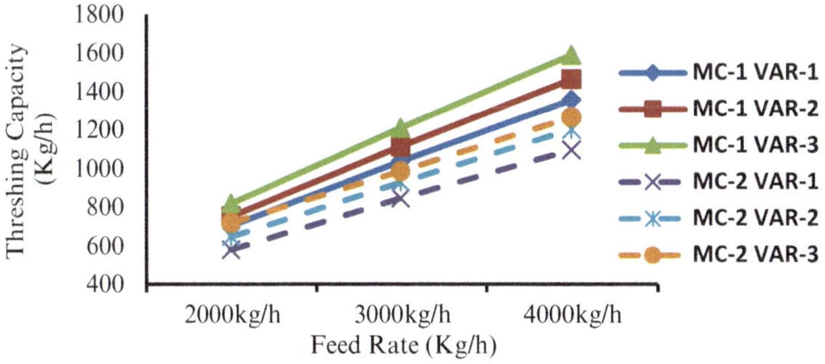

Fig. 7.2 Mean threshing capacity for different feed rates, moisture content, and varieties

Table 7.3 Analysis of variance for threshing performance parameters

	Sources of variation (independent variables)			
	Variety	FR	MC	FR*MC
Response variables	Prob > *F*	Prob > *F*	Prob > *F*	Prob > *F*
Threshing capacity (kg/h)	<0.0001	<0.0001	<0.0001	<0.0001
Threshing efficiency (%)	<0.0001	0.0001	<0.0001	0.0001
Cleaning efficiency (%)	<0.0001	<0.0001	<0.0001	<0.0001
Separation loss (%)	<0.0001	<0.0001	<0.0001	<0.0001
Scattering loss (%)	<0.0001	<0.0001	0.8998	0.7750
Grain damage (%)	<0.0001	<0.0001	<0.0001	0.3992
Engine power (kW)	<0.0001	<0.0001	<0.0001	<0.0001
Specific energy (kWh/kg)	<0.0001	<0.0001	<0.0001	0.0003

Table 7.4 Mean cleaning and threshing efficiency

	Variety-1		Variety-2		Variety-3	
	Ce(%)	Te(%)	Ce(%)	Te(%)	Ce(%)	Te(%)
FR-1*MC-1	92.5[e]	100[d]	92[f]	100[b]	91.9%[f]	100[c]
FR-2*MC-1	90.9[d]	100[d]	90.1[d]	100[b]	89.9%[e]	100[c]
FR-3*MC-1	89.9[c]	100[d]	89.1[f]	100[b]	88.8%[c]	100[c]
FR-1*MC-2	90.7[d]	99.9[b]	89.8[d]	99.8[f]	89.3%[d]	99.9[b]
FR-2*MC-2	88.3[b]	99.9[b]	87.1[d]	99.6[c]	86.6%[b]	99.6[a]
FR-3*MC-2	86.3[a]	99.8[a]	85.2[c]	99.4[a]	84.8%[a]	99.6[a]

[a–f]Means within the same column with the same letter superscript are not significantly different at $p < 0.05$

threshing operation. At higher moisture content (MC-2 = 17.3%) a small decrement in threshing efficiency is recorded with increasing moisture content and feed rate, and this decrement is statistically significant at 0.05 (5%) level for *Shaga* variety at all feed rate levels. In addition, relatively lesser threshing efficiency is recorded for the *Shaga* variety, this would be because *Shaga* is relatively harder to thresh than the other two varieties.

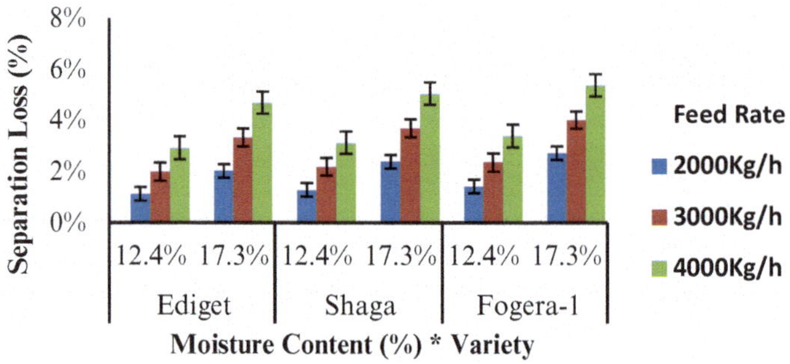

Fig. 7.3 Mean separation loss for different feed rates, moisture content, and varieties

Cleaning Efficiency Table 7.4. shows the observed cleaning efficiency. The result demonstrated that cleaning efficiency declines as moisture and feed rate rise [10–12] all found similar findings. This is due to the fact that at higher moisture contents (MC-2), as shown in Table 7.4, the moisture content of straw is significantly higher than that of grain. As a result, using the air from the blowing fan to separate the grain from the straw is challenging. This increases the pressure on the oscillating sieve while the separation is taking place. Similar to this, at greater feed rates, more material will be separated at once; as a result, the air from the blowing fan may not have sufficient time to carry out the separation operation effectively. Moreover, at a higher feed rate, the oscillating sieve is overloaded with grain and straw this makes the cleaning process difficult. Additionally, at all levels of the interaction between feed rate and moisture content, significant variation is observed among the varieties. It has been observed that *Ediget*, the variety with the smallest grain-straw ratio, threshes with a higher cleaning efficiency, while *Fogera-1*, the variety with the highest grain-straw ratio, threshes with a lower cleaning efficiency.

Separation Loss For the three cultivars under investigation, Fig. 7.3. illustrates the impact of feed rate and moisture content. With an increase in feed rate or moisture content, separation loss rises. Similar results were reported by [10, 12]. Separation loss varies significantly among varieties as well. The variety with the larger grain-straw ratio, *Fogera-1*, has the highest separation loss, whereas the variety with the lower grain-straw ratio, *Ediget*, has the smallest separation loss.

Scattering Loss No significant ($P < 0.05$) variation in scattered grain is recorded as the moisture content varies. As the feed rate increases, scattering loss increases significantly ($P < 0.05$). This might be due to as the feed rate increases, more grains will inter in the threshing drum so that the scattering loss increases accordingly.

Visible Grain Damage Figure 7.4. shows the result of grain breakage and hulled grain percentage. The result showed that visible grain damage decreases with the increase in moisture content. Similar results were reported by [13, 16, 17]. It was

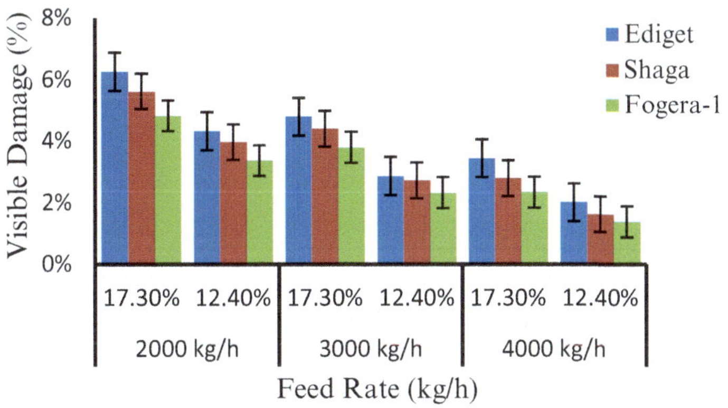

Fig. 7.4 Mean visible damage (%) for different varieties, feed rate, and moisture content

also observed that visible grain damage decreases with the increased feed rate. This might be because at a higher feed rate, the grain might receive a smaller number of impacts by the biting action of the threshing cylinder spikes. The other reason might be that, as suggested by [9], increasing the feed rate might provide more cushioning for the grain.

Engine Power and Specific Energy Requirement Figure 7.5 shows the mean engine power (EP) consumption under the test condition. The results demonstrated that when feed rate and moisture content increase, so does engine power consumption. This is due to the fact that as the feed rate increases, more material will need to be threshed in the threshing chamber, which increases the load on the threshing drum and increases the amount of fuel consumed [11, 21–23], all found similar findings.

Similarly, as the moisture content increases the load on the threshing drum increases and more fuel will be consumed. This may be because grain becomes more difficult to thresh at greater moisture levels. As a result, the highest engine power consumption of 5.67 kW was recorded at a feed rate of 4000 kg/h and moisture content of 17.3% wet basis.

The results of specific energy consumption (W/kg) is depicted in Fig. 7.6. The result shows that specific energy consumption also increases with the increase in moisture content as that of engine power consumption. This would be because, as mentioned earlier, at higher moisture content more power is required to detach the grain from the straw. Contrary to engine power consumption, specific energy consumption decreases ad the material feed rate increases. This is because at a higher feed rate, threshing capacity is also higher and specific energy consumption is the quotient of total energy consumption to that of threshing capacity. Hence, as the threshing capacity increases, specific energy consumption decreases.

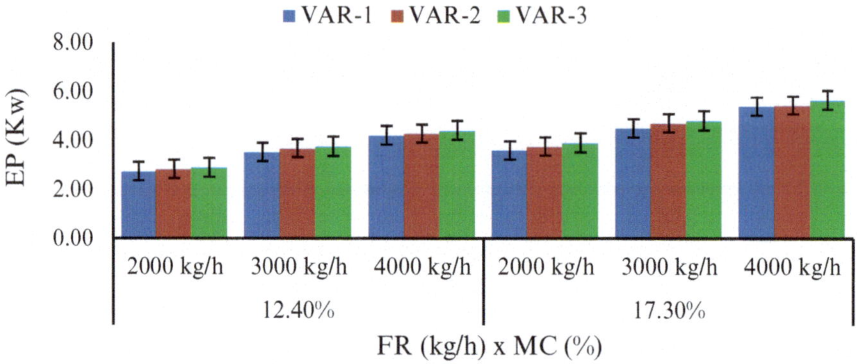

Fig. 7.5 Mean engine power consumption (kW) for different FR, MC, and VAR

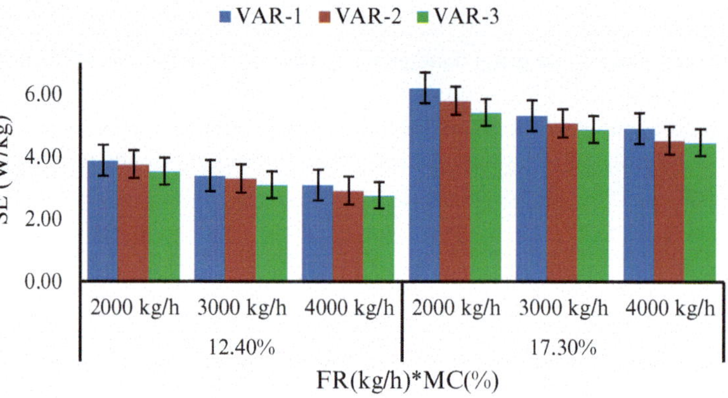

Fig. 7.6 Mean specific energy consumption (W/kg) for different FR, MC, and VAR

7.4 Conclusion

From the present study, the following conclusions can be made. Threshing efficiency significantly increased with the decrease in moisture content and 100% efficiency is recorded at 12.4% moisture content. Threshing capacity increases the decrease in moisture content and increase in feed rate. A threshing capacity of 1358 kg/h was recorded at a lower MC of 12.4% and a higher FR of 4000 kg/h. Cleaning efficiency decreases with MC and FR. A higher cleaning efficiency of 92.5% was recorded at lower MC of 12.4% and FR of 2000 kg/h. Separation loss increases with MC and FR. A higher separation loss of 5.4% was recorded at a higher MC of 17.3% and FR of 4000 kg/h. Visible grain damage of 5.58% is recorded at lower MC and FR. A significant variation in specific energy requirement with moisture content and decreases with feed rate. Hence, from the perspective of minimizing energy consumption working at a higher feed rate is recommended.

Acknowledgments The authors would like to express their gratitude to Amhara Metal Industry and Machine Technology Development Enterprise for providing the test thresher, as well as Fogera National Rice Research and Training Center for hosting the experiment and providing test material.

References

1. Food and Agriculture Organization of the United Nations. (2021). *FAOSTAT statistical database*. FAO. http://www.fao.org/faostat/en/#data/QC. Accessed September 17, 2022
2. CSA. (2013). *The Federal Democratic Republic of Ethiopia Centeral Statistical Agency Agricultural Sample Survey 2007/8 (2000 E.C.): Report on area and production for major crops (Private Peasant Holdings, Meher Season)* 2013 (July 2012).
3. CSA. (2018). *The Federal Democratic Republic of Ethiopia Centeral Statistical Agency Agricultural Sample Survey 2017/18 (2010 E.C.): Report on Area and Production for Major Crops (Private Peasant Holdings, Meher Season)*. Statistical Bulletin No. 584 I: 57.
4. Ministry of Agriculture and Rural Development (MoARD). (2010). *National rice development strategy of Ethiopia*. MoARD.
5. Africa Rice Center (AfricaRice). (2011). *Boosting Africa's rice sector: A research for development strategy 2011–2020*. Cotonou Benin ii. Cotonou, Benin. http://www.africaricecenter.org/publications/StrategicPlan/AfricaRice_Strategic_Plan_2011-2020.pdf
6. AgroBIG Team. (2016). *Rice value chain analysis*. AgroBIG Team.
7. Ayalew, Z. (2006). *Performance evaluation of three rice mills and determination of their impact on physico-chemical properties of some local rice varieties*. MSc Thesis. Alemaya University.
8. Alliance for a Green Revolution in Africa (AGRA). (2013). *Establishing the status of postharvest losses and storage for major staple crops in eleven African countries (Phase I)*. AGRA.
9. Srivastava, A. K., Goering, C. E., Rohrbach, R. P., & Buckmaster, D. R. (1993). *Engineering principles of agricultural machines* (2nd ed.). American Society of Agricultural and Biological Engineers.
10. Simonyan, J. K., & Yiljep, D. Y. (2008). Investigating grain separation and cleaning efficiency distribution of a conventional stationary rasp-bar sorghum thresher. *Agricultural Engineering International: CIGR EJournal, X*, 1–13.
11. Amponsah, S. K., Addo, A., Dzisi, K., Moreira, J., & Ndindeng, S. A. (2017). Comparative evaluation of mechanised and manual threshing options for Amankwatia and AGRA rice varieties in Ghana. *Journal of Agricultural Engineering, 48*(4), 181–189.
12. Olaye, A. R., Moreira, J., Hounhouigan, J., & Amponsah, S. K. (2016). Effect of threshing drum speed and crop weight on paddy grain quality in axial-flow thresher (ASI). *Journal Multidisciplinary Engineering Science and Technology, 3*, 3716–3721.
13. Weerasooriya, G. V., Bandara, M. H., & Rambanda, M. (2011). Performance evaluation of four wheel tractor driven high capacity combined paddy thresher. *Tropical Agricultural Research, 22*(3), 273–281.
14. Manes, G. S., Dixit, A., Singh, A., Singh, M., & Singh, B. P. (2015). Comparative performance evaluation of axial flow and tangential axial flow threshing system for basmati rice (Oryza sativa). *Agricultural Research, 4*, 303–308.
15. Waghmode, A., & Patel, R. (2019). Performance evaluation of flow through paddy thresher. *International Journal of Agricultural Engineering, 12*(1), 18–24.
16. Shittu, S. K., & Ezeh, N. M. (2021). Performance evaluation of a Rice (Oryza sativa) thresher. *FUOYE Journal of Engineering and Technology, 6*(3), 5–9.
17. Alizadeh, M. R., & Khodabakhshipour, M. (2010). Effect of threshing drum speed and crop moisture content on the paddy grain damage in axial-flow thresher. *Cercetări Agronomice în Moldova, 4*(144), 5–11.

18. Smith, D. W., Sims, B. G., & O'Neill, D. H. (1994). *Testing and evaluation of agricultural machinery and equipment: Principles and practices*. FAO.
19. Hunt, D. (2008). *Farm power and machinery management*. Waveland Press.
20. Heywood, J. B. (1988). *Internal combustion engine fundamentals*. Massachusetts Institutes of Technology, McGraw-Hill Press.
21. Abo-El-Naga, M. H. M., El-Gendy, H. A., & Mosa, E. H. (2015). Evaluation of locally threshing machine performance for threshing lentil crop. *Misr Journal of Agricultural Engineering, 22*, 376–388.
22. Emara, Z. M. I. (2006). Modification of the threshing drum of a stationary thresher to suit separating flax crop. *Misr Journal of Agricultural Engineering, 23*, 324–345.
23. El Saiedy, E. A., Gamea, G. R., & El Ashker, A. M. (2014). Development of the threshing machine for improving its performance under egyptian conditions. *Misr Journal of Agricultural Engineering, 31*(1), 65–90.

Chapter 8
Enhancing Mass Customization Using a Digital Twin: Review and Model Development

Mehari Kumilachew Bezuneh, Assefa Asmare Tsegaw, Bereket Haile Woldegiorgis, Teshome Mulatie Bogale, Matthias Brossog, and Jörg Franke

8.1 Introduction

At first glance, the word "mass customization"(MC) appears to combine two opposing concepts, namely mass production, and customization, both of which receive a lot of attention and are very much a reality today [1]. Customizing products and services to fit the specific needs of individual customers while retaining near-mass production efficiency is called mass customization. MC is generally accepted to be deferring the work of delaying a product for the customer to the most recent possible point in the supply network, called the customer order decoupling point (CODP) [2]. It marks the location where the customer's input is required to define the final model and appearance of the product according to the customer's preferences. The degree of customization depends on the point at which the customer is involved [3]. This customization strategy needs high integration between the customer and the manufacturer with the help of digital technologies [4].

Over the last decade, digital manufacturing has been viewed as a potential combination of technologies for addressing the need for customization, and faster market reaction [5]. Nowadays, one of the most important concepts associated with the Industry 4.0 wave is the digital twin (DT), which provides virtual representations of systems throughout their product lifecycle [6, 7]. Despite the fact that DT is becoming used in manufacturing processes, the power of DT on CODPs-based MC strategies is not well discussed. As a result, the main objective of this research was to investigate and develop a model of how DT technology can enhance MC by

M. K. Bezuneh (✉) · A. A. Tsegaw · B. H. Woldegiorgis · T. M. Bogale
Bahir Dar Institute of Technology, Bahir Dar University, Bahir Dar, Ethiopia

M. Brossog · J. Franke
Friedrich-Alexander-Universität Erlangen-Nürnberg, Erlangen, Germany

S. W. Fanta et al. (eds.), *Sustainable Development Research in Manufacturing, Process Engineering, Green Infrastructure, and Water Resources*, Green Energy and Technology, https://doi.org/10.1007/978-3-031-77339-6_8

integrating customers, and manufacturers using real-time data through the analysis of relevant literature on both CODP-based MC and DT technologies.

This research is structured as follows: Section 8.2 deals with the methodology of the research; Section 8.3 reviews related works on CODP-based MC strategies; Section 8.4 reviews related works on DT; Section 8.5 deals with the application of DT on CODPs-based MC strategies; and Sect. 8.6 contains discussion, conclusions, and recommendations.

8.2 Methodology

To address the objective of this research, the following research approach was conducted. First, a literature search was conducted on "mass customization"and "customer order decoupling point"sequentially to understand the requirements and data integration between customers and manufacturers at each value-added process of the product. Additionally, a literature search on the "digital twin"focused on its concept and application in the design, manufacturing, and assembly stages of a product was reviewed. Based on these findings, an analysis was conducted and a model was developed that integrates the customer and the manufacturer to enhance CODP-based mass customization. The research methodology and procedure are shown below (see Fig. 8.1).

Fig. 8.1 Research methodology and procedure

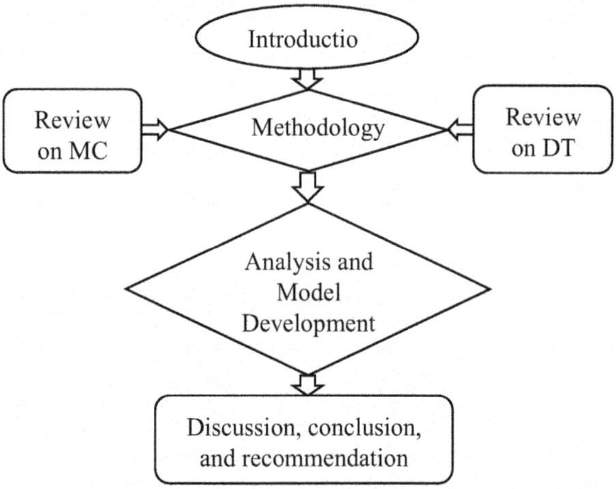

8.3 Related Literature on Mass Customization

Several scholars have offered definitions of MC. Stanley Davis described MC as treating a large number of customers simultaneously as in mass markets while addressing their individual needs as in customized markets [8]. Pine emphasized the primary objective of MC as delivering goods and services with sufficient variety and customization to meet the unique preferences of nearly everyone [9]. MC involves producing goods and services with near-mass production efficiency to meet the individual needs of customers [10]. MC is manufacturing goods for a sizable market while catering to the specific requirements of each customer [11]. The degree of customization depends on how early or far the consumer is integrated into the manufacturing process. The customer order decoupling point (CODP) is the point in the manufacturing cycle where the customer is involved or input begins.

The earlier the customer is involved in the production cycle, the higher the degree of customization [13] (see Fig. 8.2).

According to J. C. Wortmann [14], there are four basic CODP configurations, namely, engineer-to-order (ETO), make-to-order (MTO), assemble-to-order (ATO), and make-to-stock (MTS). As stated by Bogner et al. and Ivanov et al. [15, 16], each CODP has an impact on each of the value-added processes in the product life cycle, which is more clearly stated as follows:

Engineer (Design)-To-Order (ETO or DTO): A new product development or high adaptation is taking place.

Make-To-Order (MTO): The product development and design processes are completed before a customer order is received. Only the manufacturing process can be customized.

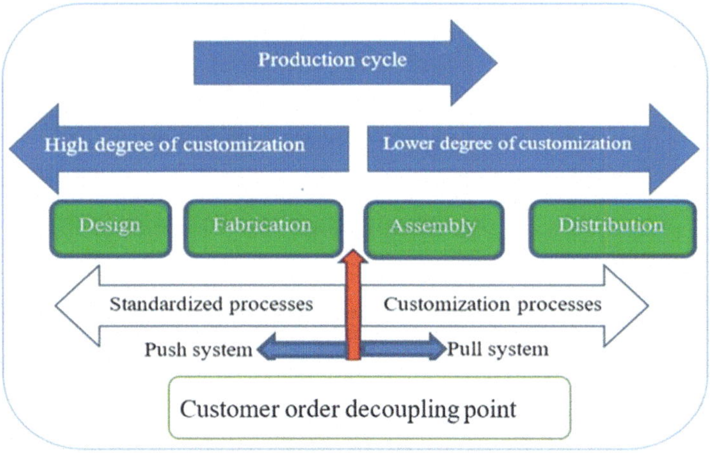

Fig. 8.2 Degree of customization from a customer involvement point of view [12]

Assemble-To-Order (ATO): Variety of components are manufactured as forecasted in advance. The customer selects components that are available in stock, and the assembly will be done after the customer orders.

Make-To-Stock (MTS): All the design, fabrication, assembly, and finishing processes are forecast-driven [15, 16]. (see Fig. 8.3).

According to [17], a high level of customization requires early customer involvement in the production cycle, and vice versa. Customer-centric product development is an approach that maximizes customer involvement by placing customer truths at the forefront of new development. This ensures that every new feature and functionality can be directly linked to addressing real customer problems [18]. Customer-driven design and manufacture actively consider individual customer requirements during the design, manufacturing, and delivery of the products [19, 20]. Customization, which starts in the design stage, is fundamental to maximizing the MC strategy [21, 22].

Industry 4.0 allows manufacturers to respond to the trend of increasing customization since customers attribute a higher added value to personalized products [23]. It also enables direct customer input in design, which enables companies to produce customized products [24]. Cheng Zhang [25] introduced a data-driven smart customization framework to facilitate smart customization across various aspects, including product design, manufacturing, pricing, marketing, and service. The findings highlight that MC is influenced by data from multiple domains, such as customer, factory, and service. So that the decision-making process can be made smarter.

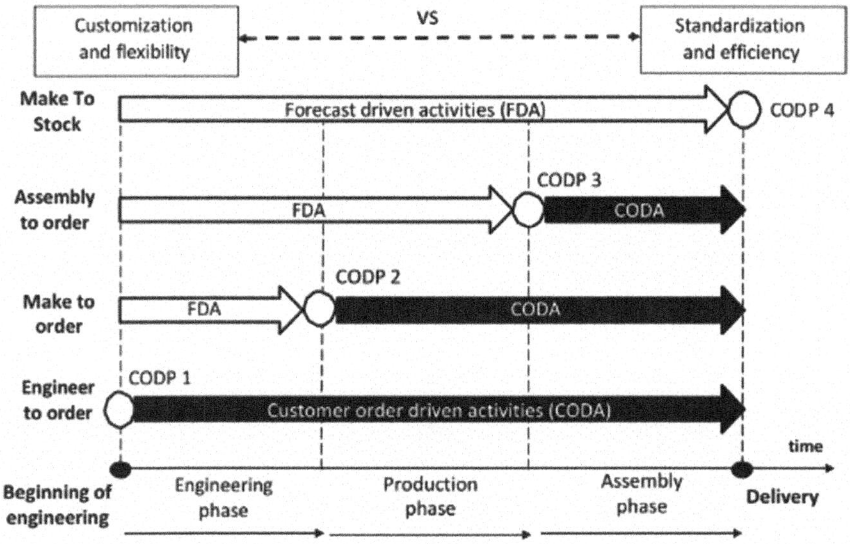

Fig. 8.3 Configuration of a manufacturing system based on CODP positions [14]

Personalized product development process comprising the physical, cyber, and user experience (UX) models is proposed by [26]. This approach emphasizes increased user involvement in the product development stage, leading to the creation of personalized products with enhanced efficiency.

In the context of MC, user experience, co-creation, and product change are recognized as key characteristics of the product development process. However, according to [27], there can be a discrepancy between customers' perceptions of designing their products and the actual level of their involvement. The manufacturer designs a product platform with diverse modules, customers select modules, place orders, make payments, and the company finally manufactures the product.

From this review, it is clear that the level of MC and customer satisfaction, specifically at the CODP, depends on the degree of customer involvement. However, the interaction between customers and manufacturers often lacks real-time communication. This gap hinders the ability to gather up-to-date customer requirements, real-time resource availability, process data, and other information crucial for customization.

8.4 Review on Digital Twin

Digital manufacturing is a new concept in the manufacturing industry that promotes collaboration throughout the product life cycle [28]. With the advent of Industry 4.0 and its digitalized technologies, manufacturers can adopt a comprehensive and interconnected approach to digitizing their operations [29]. DT has gained significant attention in recent times as a crucial strategy for integrating virtual elements with physical components in the digital manufacturing process and a key enabler of digital transformation in manufacturing, despite being considered for over a decade [30, 31].

8.4.1 *Digital Twin Concepts*

The first definition of DT was formulated by NASA, as a multi-physics, multi-scale simulation that replicates the life of a physical system using the best available models, sensor updates, and historical data [32]. A DT is a virtual representation of a physical system that undergoes constant updates with information regarding its performance, maintenance, and health status throughout its entire life cycle [33, 34]. It possesses the ability to continuously collect and accumulate data and knowledge pertaining to various stages, including design, production, quality inspection, etc. [35] (see Fig. 8.4).

According to [36, 37], DT dimensions are physical entities, virtual equipment, DT data, services, and connection models. Physical entities represent the physical counterpart of the DT, including its physical attributes, product, physical system,

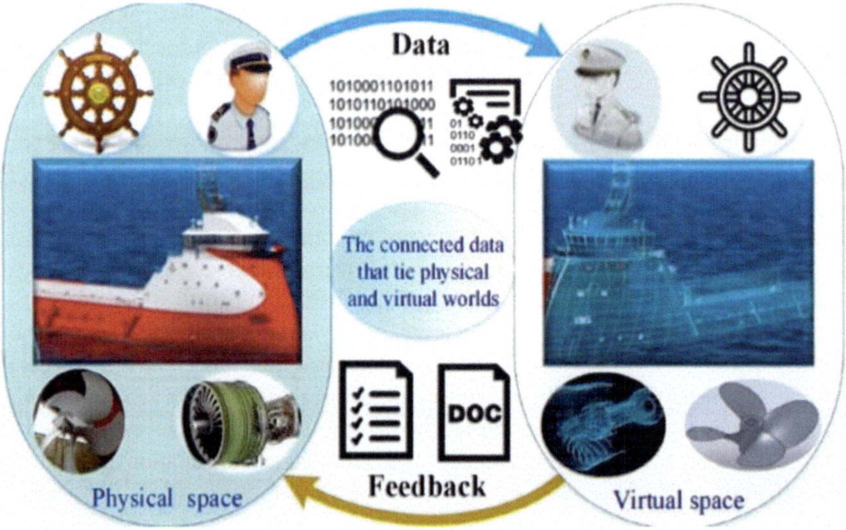

Fig. 8.4 General DT model for a product [35]

actions, processes, and behavior. The virtual models are representations of the physical world incorporating digital simulations, algorithms, three-dimensional, behavioral, environmental, rules, etc. The service model offers the operations needed to run both virtual and physical equipment. The connection models are the bidirectional connections of the physical, virtual, data, and service models to exchange real-time data. The DT data deals with multi-scale, dimension, source, and heterogeneous data obtained from physical and virtual models and service systems (see Fig. 8.5).

The DT structural model (DTSM), which combines existing DT definitions, models, and theories to identify future applications using well-known models such as RAMI4.0 and product, process, and resource (PRR) model is introduced by [38] (See Fig. 8.6).

As stated by [36], the five-dimensional model of a DT requires a range of technologies to support its implementation. In the physical model, technologies such as sensing, measuring, material processing, and others play crucial roles. Simulation and modeling, visualization, model evaluation, and related technologies are essential for the modeling dimension. DT connections rely on technologies like the internet, interfaces, communication, interaction, and security. The DT service dimension involves technologies like knowledge management, software development, platforms, algorithms, and more. Lastly, data storage, collection, transmission, processing, fusion, visualization, and related technologies are vital for managing DT data effectively [36].

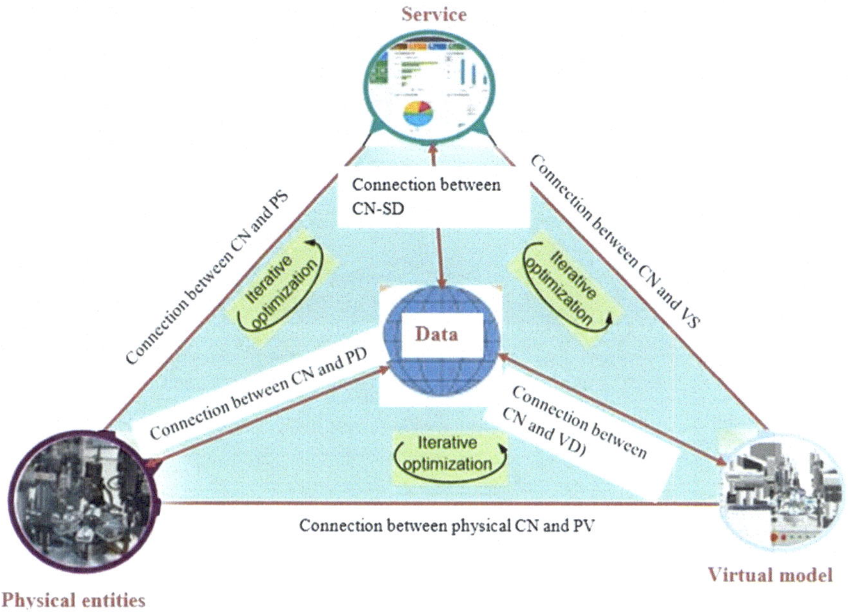

Fig. 8.5 Dimensions of digital twin model adapted from [36, 37]

8.4.2 *Application of Digital Twin in the Manufacturing Industry*

DT is used to develop a new design, redesign, and assess the latent customer requirements and manufacturing flow [39]. According to the design specifications, the raw materials or components are processed and assembled into products. The manufacturing execution process needs to be monitored and managed. Therefore, the features, performance, parameters, and process conditions of production factors (e.g., human, machine, material, environment) are collected in real time and recorded to monitor the production process [40]. After the product is manufactured, the assembly process of product development forms a complete product [41]. A DT approach for assembly process optimization is a useful mechanism for complex product assembly, because of its ability to provide a precise simulation result and avoid a lot of trial and error during the assembly process [42]. The following table shows the application of DT to product design, manufacturing, and assembly processes (see Table 8.1).

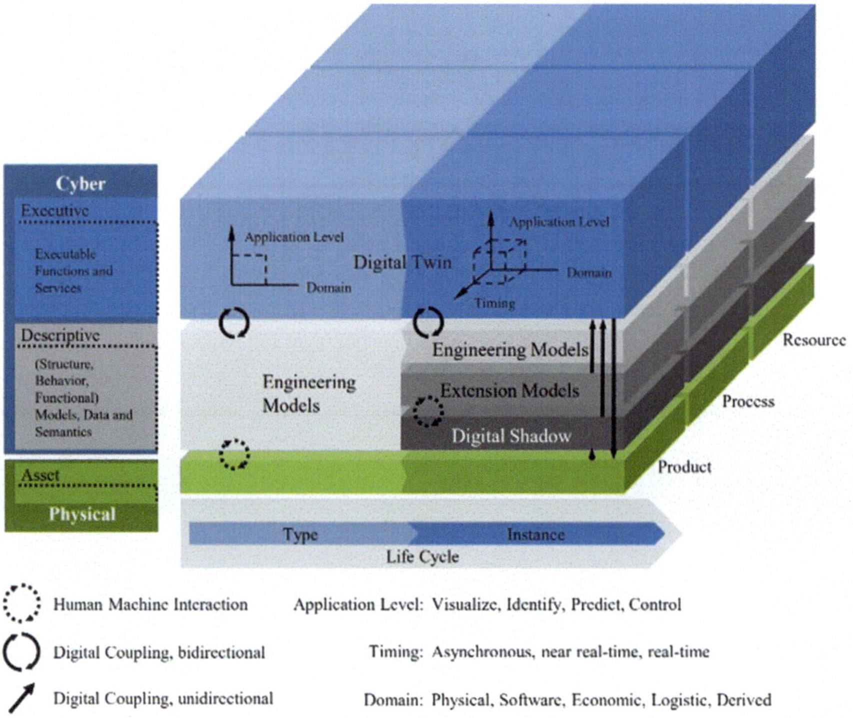

Fig. 8.6 The proposed structure model of the digital twin consists of a life cycle axis and a PRR axis as well as several layers [38]

8.5 Application of the Digital Twin in Mass Customization

Digital twin technology plays a vital role in addressing MC by bridging the gap between the physical and digital worlds. It provides real-time monitoring, remote control, prediction, scheduling, and maintenance functionalities, offers empirical evidence, and enhances responsiveness [64, 65]. DT technologies enable manufacturers to enhance their situational awareness, improve operational resilience, and increase flexibility, particularly in the context of MC using truthful information flow [39]. According to [66], DT significantly expands customization capabilities from the following three main aspects. Firstly, it enables manufacturers to have a deeper understanding of their customers. Secondly, with the help of the DT, customers can better participate in customization activities in virtual spaces. Thirdly, the digital twin renders manufacturing systems more flexible and predictive, thus reducing risks caused by uncertainties. Digital Twin enables customers to design and validate their products, even with limited domain knowledge. This empowers customers to make modifications or upgrade components to meet their needs through collaboration with suppliers [66].

Table 8.1 Application of DT on product design, manufacturing, and assembly

Design phase		
DT dimensions	Application	Ref.
Physical entities designers, tools, product design, software, etc.	Used to get real-time data of product design and parameters, design resources, and process data to create a virtual model of product design.	[40, 43]
DT data Physical, virtual model, service, historical data.	To collect, transmit, fuse, and visualize product design data, resource data, and process data from physical entities, DT services, and virtual models.	[39, 44]
DT virtual model virtual models of designers, design tools, product design, software, and other resources.	To simulate, evaluate, monitor prediction, and verify failures of product design like product performance, geometrical variations, product functions, structures, and manufacturability, and product properties like mechanical, thermal, and electrical, etc.	[45, 46]
DT connection to exchange real-time data between DT dimensions.	To transmit real-time data to physical (designer, manufacturer, and customer), design service, and design modeling.	[40]
	To incorporate customer demand and experience in the design process to develop a mutual understanding.	[47, 48]
DT services provide third-party users with value-adding services.	Used to design, model, and evaluate, optimization and integration, diagnosis, predictive, fault-tolerance management algorithms, and dynamic updating techniques.	[49, 50]
Manufacturing phase		
Physical entities machine tools, operators, software, and others	Used to get real-time data on a product, production resources and processes, customers' requirements, and suppliers to create a virtual model.	[51]
DT data product, process, resource, service, historical knowledge, supplier, etc.	To collect, transmit, and visualize real-time data of products, resources, customers' and suppliers' data to and from physical entities, DT services, and virtual models.	[52]
DT virtual model virtual models of products, production tools, product design, production software, operators, customers, suppliers, etc.	To simulate, evaluate, monitor, verify, and predict failures of the production process. To evaluate and optimize production processes, resources, cycle time, inventory cost, etc. To test, evaluate, validate, and make corrections to the required performance of products.	[53, 54]
DT connection to connect and distribute real-time data of the manufacturing process to DT dimensions.	To create collaboration between product developers, manufacturers, vendors, and customers to handle a variety of configurations, provide better customization alternatives, and ensure reliability.	[41]
	Increases human-machine collaboration and interaction and allows workers to interact with machines safely.	[42]
DT services system Knowledge, application, resource, manufacturing process, service.	To develop a plan, simulate, model, evaluate, optimize, and integrate diagnostic and predictive faults in the product and production processes.	[36, 55]

(continued)

Table 8.1 (continued)

Design phase		
DT dimensions	Application	Ref.
Assembly phase		
Physical entities assembly components, tools, operators, and other resources.	To get real-time data on assembly components, assembly resources and processes, customers' requirements, and suppliers to create a virtual model.	[56, 57]
DT data component, process, resource, model, service, knowledge, customer, etc.	Collect, transmit, and visualize assembly components, resources, and processes customer, supplier, etc. data to and from physical entities, DT services, and virtual models.	[56, 58]
DT virtual model Virtual models of assembly components, assembly tools, operators, customers, suppliers, and other resources.	To simulate, evaluate, and verify the overall assembly process and resources to create more realistic, accurate, and efficient product assembly.	[59, 60]
	If something is wrong with the quality, assembly error, or dimensional accuracy during the assembly, testing aids scientific decision-making based on the facts.	[61]
DT connection connect and distribute real-time data of the assembly process to all DT dimensions.	Transfer multi-source real-time data of the components, assembly tools, operators, resource data, customer data, and supplier data. Increases human-machine collaboration and interaction and allows workers to operate safely. Increase collaboration between assemblers, vendors, and customers.	[58, 62]
DT services system knowledge, application, resources, and process services of the assembly.	Used to develop a plan for the assembly process and resources and simulation, model and evaluate, optimize and integrate diagnosis, and predict faults on the assembly process plan, processes, and resources.	[63]

8.5.1 Digital Twin for DTO Customization

In DTO customization, all the customers' requirements are expected to be new and customized from the design phase to the manufacturing, assembly, and service phases. One good example of product customization at the DTO stage is an elevator. Due to modern buildings, elevators are highly customized industrial products [66]. If the elevator is customized during the DTO phase, the customer needs a unique elevator that fits their building. Therefore, detailed design of the elevator, production process, resource management, and assembly process are the basic requirements (see Table 8.2).

8.5.2 Digital Twin-Based MTO Customization

During the MTO stage, the customization starts from the production or manufacturing stage. Customer requirements are product features, manufacturing processes, and assembly processes. The customization of the machine element shaft is a good example of the MTO phase customization strategy [15] (see Table 8.2).

Table 8.2 Digital twin application on CODP-based customization

DTO customization (elevator)		
DT model	Function of DT dimensions	Application of DT on elevator customization
Physical entities	Represents product design, designers, customers, design tools, design software, and other resources	To create a virtual model so that customers and designers communicate with each other by using real-time data of elevator design, resources, and process data
Data model	Includes customer data, design data and model data, designer's data, resources data, historical data, etc	To collect data from the customer and designer and to transmit, fuse, and visualize elevator design data, resource data, and process data
Virtual model	Includes virtual models of customer requirements, product design, designers, design tools, and other resources	Model assembly of the new house and elevator to simulate, evaluate, monitor prediction, and verify the product performance, geometrical variations, product functions, structures, etc. of the elevator
Connection model	To connect and exchange real-time data between the customer and designer	To transmit real-time data of the house and elevator from the customer to the designer and real-time data of the elevator design and vice versa
Services model	Preparing requirements, specifications, design, model, mathematical analysis, process requirement, material requirement, etc	The customer side prepares the house and elevator requirements and the designer side prepares the elevator design, model, mathematical analysis, process, material requirement, simulation, etc
MTO customization (machine element shaft)		
DT model	Function of DT dimensions	Application of DT on shaft customization
Physical entities	Represents machine tools, operators, manufacturing software, and other resources	Used to get real-time virtual models and data of a shaft like dimension, material, production resources, machine tools, process types, etc
Data model	Includes customer requirement, product requirement data, process data, operators and resources, service, historical knowledge, supplier, etc. data	To collect and transmit data from customer to manufacturer and vice versa and visualize the current status of a shaft by using real-time data on a shaft, resources, customers and suppliers' data DT services, and virtual models, etc
Virtual model	Includes virtual models of customer requirements, product requirements, operators, machine tools, and other resources	Model a shaft and its assembly to simulate, evaluate, monitor, and verify the performance, geometrical variations, product functions, structures, and product properties
Connection model	To connect and exchange real-time data between the customer and manufacturer	To create collaboration between product developers, manufacturers, vendors, and customers to handle a variety of configurations, provide better customization alternatives, and ensure reliability

(continued)

Table 8.2 (continued)

DTO customization (elevator)		
DT model	Function of DT dimensions	Application of DT on elevator customization
Services model	Preparing product specifications, requirements, manufacturing process, resources, material etc	From the customer side preparing the requirements of a shaft. the manufacturer prepares all the resources and processes and other analyses
ATO customization (bicycle components)		
DT model	Function of DT dimensions	Application of DT on bicycle assembly
Physical entities	Represents customer requirements, suppliers, assembly components, assembly resources, and assembly processes	To get real-time virtual model data of bicycle components, like dimension, material, assembly machine tools, assembly sequences, customers' requirements, and bicycle components suppliers
Data model	Components suppliers and customers data assembly process, resource data, model, service data, knowledge data, etc	To collect and transmit data from customer to manufacturer and vice versa and visualize the status of assembly components, resources, processes, and assembly models
Virtual model	Virtual models of assembly components, assembly tools, operators, customers, suppliers, etc	To simulate, evaluate, and verify the assembly process and to create a more realistic, accurate, and efficient product assembly of a bicycle
Connection model	To connect and exchange real-time data between the customer, supplier, and manufacturer	To transmit real-time data and create collaboration between bicycle assemblers, suppliers, and customers on bicycle components, assembly processes, assembly simulations, etc
Services model	Preparing assembly specifications, component requirements, assembly process, resource requirements, material requirements, etc	Customers prepare the requirements or specifications of bicycle components to be assembled. assemblers and suppliers prepare bicycle components, machine tools, assembly sequences, and simulations of bicycle assemblies

8.5.3 *Digital Twin-Based ATO Customization*

During ATO customization, components are produced in advance, and the assembly of a product is implemented individually after the receipt of a customer order. Only the assembly process and component selection from the available components are the basic requirements of the customer. Selecting bicycle components to be assembled is one example of ATO customization strategies [13]. A customer can select a component of a bicycle like a frame, saddle, seat, wheels, etc. from the available components. Component selection and assembly using DT technologies allow the manufacturer and the customer to know the real-time status of the assembly process and resources, available components, assembly conditions, etc. (see Table 8.2).

The application of DT in all three stages of CODPs and the model which shows the integration between the customer, CODP points, and the application of DT is shown below (see Fig. 8.7).

8.6 Discussion, Conclusion, and Recommendation

8.6.1 Discussion

The main objective of this research was to explore the potential of DT technology in enhancing MC through the utilization of real-time data at each stage of the CODP and develop a framework that shows the integration between the customer and the manufacturer. To achieve the objectives, a literature review was conducted on CODP-based MC to understand the concepts and identify the gaps during the data exchange between the customer and the manufacturer. Our review indicates that a CODP-based MC strategy is a customer-centric approach that maximizes MC by placing customer needs at the core of new product development. By prioritizing customer truths, every new feature and functionality released can be directly linked to addressing real customer interest. The review also indicates that there is a real-time data exchange gap between the customer and the manufacturer to get what the customer needs currently. The second portion of the research focused on discussions on the concept, application, and how DT technology is used in the manufacturing industry during each stage of product development. The review shows that DT technology is used at each stage of product development by providing real-time data for all participants in the processes. Based on the findings of the two reviews, the impact of DT technologies on CODPs-based MC strategies has been surveyed, and a framework that shows the integration between customer and manufacturer using DT has been developed using example products. The study's findings indicate that DT technologies enable clear understanding and communication between customers and manufacturers at every stage of CODPs (ETO, MTO, and ATO) through real-time data exchange. This enhances customization, improves information flow, and ensures efficient fulfillment of customer requirements.

8.6.2 Conclusion and Recommendation

In conclusion, the study highlights the effectiveness of the CODP-based MC strategy in addressing customer requirements and maximizing customization. By integrating customers and manufacturers through real-time data exchange, this strategy enables a comprehensive understanding of customer needs at each stage of product customization. The utilization of DT technologies facilitates this integration and enhances communication and collaboration between both parties. Overall,

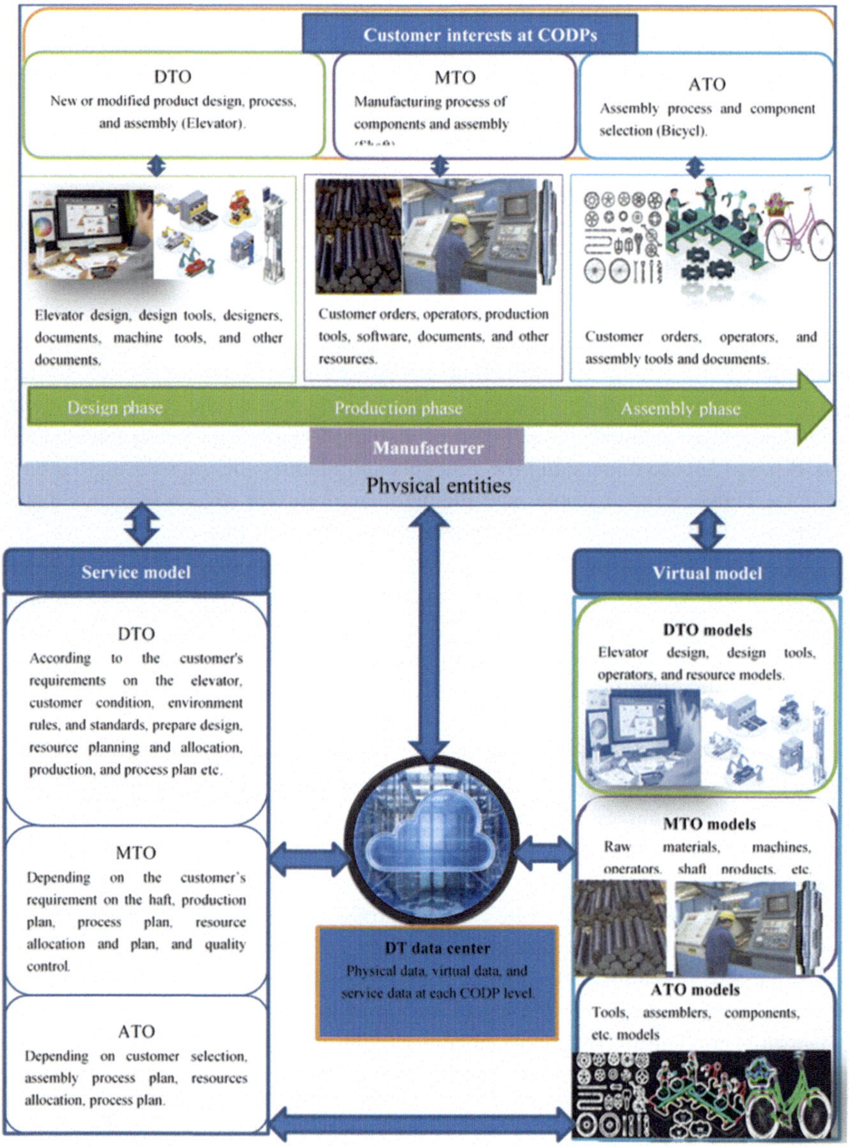

Fig. 8.7 A model of DT-based CODP customization

leveraging DT technologies in CODP-based MC contributes to improving customer integration, and communication between customer and manufacturer leading to effective and efficient customization and customer satisfaction.

As a recommendation, this research focused on the impact of DT technologies on CODP-based customization, specifically DTO, MTO, and ATO, based on the inte-gration of customers and manufacturers. However, there are other CODPs in the processes of MC, like maintenance and products on service. Therefore, further research is needed to maximize MC by using DT technologies and increasing the integration of customers during the processes.

References

1. Kumar, A., Gattoufi, S., & Reisman, A. (2007). Mass customization research: Trends, directions, diffusion intensity, and taxonomic frameworks. *International Journal of Flexible Manufacturing Systems, 19*, 637–665.
2. Cannas, V. G., Pero, M., Rossi, T., & Gosling, J. (2018). Integrate customer order decoupling point and mass customisation concepts: A literature review. In S. Hankammer, K. Nielsen, F. T. Piller, G. Schuh, & N. Wang (Eds.), *Customization 4.0* (pp. 495–517). Springer.
3. Rudberg, M., & Wikner, J. (2007). Mass customization in terms of the customer order decoupling point. *Production Planning & Control, 15*(4), 445–458.
4. Piller, F. T. (2002). Customer interaction and Digitizability — A structural approach to mass customization. In C. Rautenstrauch, R. Seelmann-Eggebert, & K. Turowski (Eds.), *Moving into mass customization: Information systems and management principles* (pp. 119–137). Springer.
5. Liu, M., Fang, S., Dong, H., & Xu, C. (2021). Review of digital twin about concepts, technologies, and industrial applications. *Journal of Manufacturing Systems, 58*, 346–361.
6. Negri, E., Fumagalli, L., & Macchi, M. (2017). A review of the roles of digital twin in CPS-based production systems. *Procedia Manufacturing, 11*, 939–948.
7. Tao, F., Sui, F., Liu, A., Qi, Q., Zhang, M., Song, B., Guo, Z., Lu, S., & Nee, A. (2018). Digital twindriven product design framework. *International Journal of Production Research, 57*, 1–19.
8. Tu, Q., Vondcrcmbsc, M. A., & Ragu-Nathan, T. S. (2001). Thc impact of timc-bascd manufacturing practices on mass customization and value to customer. *Journal of Operations Management, 19*, 201–217.
9. Hvam, L. (2006). Mass customisation in the electronics industry: Based on modular products and product configuration. *International Journal of Mass Customisation, 1*, 410–426.
10. Tseng, M. M., & Jiao, J. (2001). Mass customization, handbook of industrial engineering. Technology and operation. *Management.*
11. Piller, F. T., & Stotko, C. M. (2002). Mass customization: Four approaches to deliver customized products and services with mass production efficiency. In *IEEE international engineering management conference* (pp. 773–778). IEEE.
12. Spahi, S. (2020). Optimizing the level of customization for products. In *Mass customization systems*. University of Central Florida STARS. (pp 7).
13. Duray, R. (1997). *Mass customization configurations: An empirical investigation of manufacturing practices of customization*. The Ohio State University.
14. Wortmann, J. C. (1992). Production management systems for one-of-a-kind products. *Computers in Industry, 19*, 79–88.
15. Bogner, E., Löwen, U., & Franke, J. (2017). Systematic consideration of value chains with respect to the timing of individualization. *Procedia CIRP, 60*, 368–373.
16. Ivanov, D., Tsipoulanidis, A., & Schönberger, J. (2017). *Global supply chain and operations management*. Springer.
17. Jost, P.-J., & Süsser, T. (2020). Company-customer interaction in mass customization. *International Journal of Production Economics, 220*, 107454.
18. Customer-Centered New Product Development. https://uservoice.com/blog/customer-centered-new-product-development, last accessed 2023/05/19.

19. Bai, X., Huerta, O., Unver, E., Allen, J., & Clayton, J. E. (2021). A parametric product design framework for the development of mass customized head/face (eyewear) products. *Applied Sciences, 11*, 5382.
20. Jagdev, H., & Browne, J. (1998). The extended Enterprise—A context for manufacturing. *Production Planning & Control, 9*, 216–229.
21. Baranauskas, G., Raišienė, A. G., & Korsakienė, R. (2020). Mapping the scientific research on mass customization domain: A critical review and bibliometric analysis. *Journal of Risk and Financial Management, 13*, 220.
22. Du, G., Jiao, R. J., & Chen, M. (2014). Joint optimization of product family configuration and scaling design by Stackelberg game. *European Journal of Operational Research, 232*, 330–341.
23. Torn, I. A. R., & Vaneker, T. H. J. (2019). Mass personalization with industry 4.0 by SMEs: A concept for collaborative networks. *Procedia Manufacturing, 28*, 135–141.
24. Wang, Y., Ma, H.-S., Yang, J.-H., & Wang, K.-S. (2017). Industry 4.0: A way from mass customization to mass personalization production. *Advanced Manufacturing, 5*, 311–320.
25. Zhang, C., Chen, D., Tao, F., & Liu, A. (2019). *Data driven smart customization* (pp. 564–569).
26. Zheng, P., Yu, S., Wang, Y., Zhong, R. Y., & Xu, X. (2017). User-experience based product development for mass personalization: A case study. *Procedia CIRP, 63*, 2–7.
27. Koren, Y., Shpitalni, M., Gu, P., & Hu, S. J. (2015). *Product Design for Mass-Individualization. Procedia CIRP, 36*, 64–71.
28. Paritala, P., Manchikatla, S., & Yarlagadda, P. (2017). Digital manufacturing- applications past, current, and future trends. *Procedia Engineering, 174*, 982–991.
29. Freelancer, A. (n.d.). Industrial revolution 4.0 | important things you need to know.
30. Zhou, G., Zhang, C., Li, Z., Ding, K., & Wang, C. (2020). Knowledge-driven digital twin manufacturing cell towards intelligent manufacturing. *International Journal of Production Research, 58*, 1034–1051.
31. Boschert, S., & Rosen, R. (2016). Digital twin—The simulation aspect. In P. Hehenberger & D. Bradley (Eds.), *Mechatronic futures: Challenges and solutions for mechatronic systems and their designers* (pp. 59–74). Springer.
32. Glaessgen, E.H., Stargel, D.S. (n.d.). The digital twin paradigm for future NASA and US air force vehicles.
33. Madni, A., Madni, C., & Lucero, S. (2019). Leveraging digital twin Technology in Model-Based Systems Engineering. *Systems, 7*, 7.
34. Fuller, A., Fan, Z., & Day, C. (2020). Digital twin: Enabling technologies, challenges and open research. *IEEE Access, 99*, 1–1.
35. Saddik, A.E.: Digital twins: The convergence of multimedia technologies. Undefined. (2018).
36. Qi, Q., Tao, F., Hu, T., Anwer, N., Liu, A., Wei, Y., Wang, L., & Nee, A. Y. C. (2021). Enabling technologies and tools for digital twin. *Journal of Manufacturing Systems, 58*, 3–21.
37. Sjarov, M., Lechler, T., Fuchs, J., Brossog, M., Selmaier, A., Faltus, F., Donhauser, T., & Franke, J. (2020). The digital twin concept in industry–a review and systematization. In *2020 25th IEEE international conference on emerging technologies and factory automation (ETFA)* (pp. 1789–1796). IEEE.
38. Tobias Lechler; Jonathan Fuchs; Martin Sjarov; Matthias Brossog; Andreas Selmaier; Florian Faltus, et al. (2020). Introduction of a comprehensive structure model for the digital twin in manufacturing | 2020 25th IEEE international conference on emerging technologies and factory automation (ETFA) IEEE conference IEEE Xplore.
39. Lu, Y., Liu, C., Wang, K. I.-K., Huang, H., & Xu, X. (2020). Digital twin-driven smart manufacturing: Connotation, reference model, applications and research issues. *Robotics and Computer-Integrated Manufacturing, 61*, 101837.
40. Tao, F., Cheng, J., Qi, Q., Zhang, M., Zhang, H., & Sui, F. (2018). Digital twin-driven product design, manufacturing and service with big data. *International Journal of Advanced Manufacturing Technology, 94*, 3563–3576.

41. Redeker, M., Weskamp, J. N., Rössl, B., & Pethig, F. (2021). Towards a digital twin platform for industrie 4.0. In *2021 4th IEEE international conference on industrial cyber-physical systems (ICPS)* (pp. 39–46). IEEE.
42. Wang, T., Li, J., Kong, Z., Liu, X., Snoussi, H., & Lv, H. (2021). Digital twin improved via visual question answering for vision-language interactive mode in human–machine collaboration. *Journal of Manufacturing Systems, 58*, 261–269.
43. Tao, F., Qi, Q., Wang, L., & Nee, A. Y. C. (2019). Digital twins and cyber–physical systems toward smart manufacturing and industry 4.0: Correlation and comparison. *Engineering, 5*, 653–661.
44. Lutters, E. (2018). Pilot production environments driven by digital twins. *South African Journal of Industrial Engineering, 29*, 40–53.
45. Gusev, M. P., Nikolaev, S. M., Uzhinsky, I. K., Padalitsa, D. I., & Mozhenkov, E. R. (2018). Application of optimization in the early stages of product development, using a small UAV case study. In *Product lifecycle management to support industry 4.0: 15th IFIP WG 5.1 international conference, PLM 2018.*, Turin, Italy, July 2–4, 2018, Proceedings 15 (pp. 294–303). Springer.
46. Schleich, B., Anwer, N., Mathieu, L., & Wartzack, S. (2017). Shaping the digital twin for design and production engineering. *CIRP Annals, 66*, 141–144.
47. Li, X., Cao, J., Liu, Z., & Luo, X. (2020). Sustainable business model based on digital twin platform network: The inspiration from Haier's case study in China. *Sustainability, 12*, 936.
48. Liu, Q., Zhang, H., Leng, J., & Chen, X. (2019). Digital twin-driven rapid individualised designing of automated flow-shop manufacturing system. *International Journal of Production Research, 57*, 3903–3919.
49. Cai, Y., Starly, B., Cohen, P., & Lee, Y.-S. (2017). Sensor data and information fusion to construct digital-twins virtual machine tools for cyber-physical manufacturing. *Procedia Manufacturing, 10*, 1031–1042.
50. Tao, F., Zhang, H., Liu, A., & Nee, A. Y. (2018). Digital twin in industry: State-of-the-art. *IEEE Transactions on industrial informatics, 15*, 2405–2415.
51. Bécue, A., Maia, E., Feeken, L., Borchers, P., & Praça, I. (2020). A new concept of digital twin supporting optimization and resilience of factories of the future. *Applied Sciences, 10*, 4482.
52. Zhang, H., Liu, Q., Chen, X., Zhang, D., & Leng, J. (2017). A digital twin-based approach for designing and multi-objective optimization of hollow glass production line. *IEEE Access, 5*, 26901–26911.
53. Min, Q., Lu, Y., Liu, Z., Su, C., & Wang, B. (2019). Machine learning based digital twin framework for production optimization in petrochemical industry. *International Journal of Information Management, 49*, 502–519.
54. Liu, J., Zhou, H., Liu, X., Tian, G., Wu, M., Cao, L., & Wang, W. (2019). Dynamic evaluation method of machining process planning based on digital twin. *IEEE Access, 7*, 19312–19323.
55. Digital Twin Service towards Smart Manufacturing—ScienceDirect. https://www.sciencedirect.com/science/article/pii/S2212827118302580, last accessed 2022/09/08.
56. Ma, Y., Zhou, H., He, H., Jiao, G., & Wei, S. (2019). A digital twin-based approach for quality control and optimization of complex product assembly. In *2019 international conference on artificial intelligence and advanced manufacturing (AIAM)* (pp. 762–767). IEEE.
57. Zhang, D., Leng, J., Xie, M., Yan, H., & Liu, Q. (2022). Digital twin enabled optimal reconfiguration of the semi-automatic electronic assembly line with frequent changeovers. *Robotics and Computer-Integrated Manufacturing, 77*, 102343.
58. Sun, X., Bao, J., Li, J., Zhang, Y., Liu, S., & Zhou, B. (2020). A digital twin-driven approach for the assembly-commissioning of high precision products. *Robotics and Computer-Integrated Manufacturing, 61*, 101839.
59. Ma, X., Tao, F., Zhang, M., Wang, T., & Zuo, Y. (2019). Digital twin enhanced human-machine interaction in product lifecycle. *Procedia CIRP, 83*, 789–793.
60. Wang, Z. (2020). Digital twin technology. In *Industry 4.0 - impact on intelligent logistics and manufacturing*. IntechOpen.

61. Zhuang, C., Liu, J., & Xiong, H. (2018). Digital twin-based smart production management and control framework for the complex product assembly shop-floor. *International Journal of Advanced Manufacturing Technology, 96*, 1149–1163.
62. Jiang, Y., Li, M., Guo, D., Wu, W., Zhong, R. Y., & Huang, G. Q. (2022). Digital twin-enabled smart modular integrated construction system for on-site assembly. *Computers in Industry, 136*, 103594.
63. Wang, Y., Feng, J., Liu, J., Liu, X., & Wang, J. (2022). Digital twin-based design and operation of human-robot collaborative assembly. *IFAC-PapersOnLine, 55*, 295–300.
64. Aheleroff, S., Zhong, R. Y., & Xu, X. (2020). A digital twin reference for mass personalization in industry 4.0. *Procedia CIRP, 93*, 228–233.
65. Stavropoulos, P., & Mourtzis, D. (2022). Chapter 10 - digital twins in industry 4.0. In D. Mourtzis (Ed.), *Design and operation of production networks for mass personalization in the era of cloud technology* (pp. 277–316). Elsevier.
66. Wang, X., Wang, Y., Tao, F., & Liu, A. (2021). New paradigm of data-driven smart customisation through digital twin. *Journal of Manufacturing Systems, 58*, 270–280.

Chapter 9
Physical Properties of Berseem (*Trifolium alexandrinum*) and Lucerne (*Medicago sativa*) Seeds

Smegnew Moges Mintesinot, Rohinish Khurana, Samuel Getachew Chekol, and Mequanint Demeke Aynalem

9.1 Introduction

Lucerne (*Medicago sativa*) and berseem (*Trifolium alexandrinum*) are two of the most important legume crops used for livestock feed around the world [1, 2]. These crops play a significant role in agriculture by providing nutritious and palatable feeds for animals, while also enriching the soil fertility and improving subsequent crop production [1, 2]. Legume fodder crops are currently receiving more research attention due to their numerous advantages over other fodder crops [3]. As a result, there is a need for new technologies and systems to keep up with the increasing demand for fodder production. In India, berseem and lucerne cover two million hectare and one million hectare areas and give 85 tons of green forage per hectare and 60 to 130 tons of green fodder per hectare, respectively [4]. For all species of animals, highly nutritious feed can be obtained from lucerne fodder crops; it contains 16 to 25% of crude protein and 20 to 30% fiber content [4]. Berseem contains 20 to 24% of rough protein and 70% digestible dry matter content, which is a nutritious fodder crop for dairy cattle, particularly for milch animals; also, it enhances the soil organic carbon by 0.38 to 0.46%, nitrogen by 46%, and accessible phosphorus by 15 to 26 kg [5, 6].

S. M. Mintesinot (✉) · S. G. Chekol
Faculty of Mechanical and Industrial Engineering, Bahir Dar Institute of Technology, Bahir Dar University, Bahir Dar, Ethiopia

R. Khurana
Department of Farm Machinery and Power Engineering, Punjab Agricultural University, Ludhiana, Punjab, India
e-mail: rkhurana@pau.edu

M. D. Aynalem
Faculty of Chemical and Food Engineering, Bahir Dar Institute of Technology, Bahir Dar University, Bahir Dar, Ethiopia

S. W. Fanta et al. (eds.), *Sustainable Development Research in Manufacturing, Process Engineering, Green Infrastructure, and Water Resources*, Green Energy and Technology, https://doi.org/10.1007/978-3-031-77339-6_9

Increasing the production of feeds and fodder is crucial for enhancing milk production. Nonetheless, obtaining high-quality feed and fodder year-round through cost-effective techniques is a challenge. The cost of milk production is largely dependent on feed and fodder, which accounts for a significant 60% of the total expenses [7]. Green fodder is a 30% cheaper source of digestible nutrients than concentrates [8].

Farmers use a mechanical sowing method to ensure seeds' placement at the desired distance and depth. However, currently, the sowing of these crops is carried out with the hand broadcasting sowing method, where the seeds are spread randomly over an area. Mechanical seeding in rows results in the even distribution of plants in rows; this maximizes crop productivity and allows subsequent operations to take place easily and efficiently. The conventional broadcasting of fodder seeds needs to be replaced by mechanical seeding to stabilize plant growth and improve productivity. Mechanical sowing allows an even, fast, and complete germination with good sowing and a highly mechanized cultivation system [9]. Good plant emergence and early growth performance are factors that have a significant impact on crop yields. This is the reason why sowing for the correct establishment of the plant is crucial to the production of fodder crops. In addition, the use of seed drills reduces the cost of sowing seeds when sowing at the desired seed rate and uniformity, and the crop yield also improves as the plants have enough space to grow [10]. To design, develop, and adjust the sowing machine with small-seeded fodder crops, the determination of the engineering properties of the seeds is a pre-request. In addition to the seed drill, to develop suitable technologies such as a harvester, thresher, seed cleaner, seed grader, and sorter machines, the physical characteristics of the seeds are needed. Thus, the size (length, width, and thickness), shape (geometric mean diameter and sphericity), thousand-grain weight (TGW), bulk density, true density, angle of repose, and coefficient of static friction are the main influencing physical properties of the berseem and lucerne seeds. In recent years, for the design and development of mechanical seeding and processing machinery, different authors have studied the physical properties of several crops, including the engineering properties of tef [11], the physical properties of onion seeds [12], the physical properties of pelleted radish seeds [13], etc. The objective of this research was to study the physical properties of berseem and lucerne seeds. The findings of this study could be used as input data for numerous systems and technologies design and development for sowing, harvesting, and postharvest handling of these crops.

9.2 Materials and Methods

9.2.1 Sample Collection, Preparation, and Materials Used

According to the conditions recommended for Punjab state, India, the local varieties of berseem and lucerne seeds were selected [14]. The crop variety selected for berseem was *BL-10*, and for lucerne seed was *LL composite-5* for studying their physical properties. The seeds were manually cleaned to remove foreign matter, broken and immature seeds. The moisture content of the samples was determined according to association of official analytical chemists (AOAC)-approved vacuum oven method [15]. The value of the seed mean moisture content of the samples of berseem and lucerne was 13.64% and 9.18% on a dry basis, respectively. Since transportation, storage, and handling operations of the seeds are performed in this moisture range, all physical properties of berseem and lucerne seeds were characterized under this mean moisture content. A digital microscope, cell imaging software, and measuring ruler were used for measuring the size of seeds [15]. A graduated measuring cylinder and an electronic balance with 0.01 g precision were used to calculate the bulk density of the seeds; in addition to these materials, toluene (C_7H_8) was used to quantify the true density [15]. For measuring the angle of repose, a plastic cylinder with a 70 mm internal diameter and 270 mm height was used, and the coefficient of static friction testing apparatus and mild steel sheet metal were used to quantify the coefficient of static friction [15].

9.2.2 Determination of Size and Shape

As shown in Figs. 9.1 and 9.2, the three principal axis dimensions of length (major), width (intermediate), and thickness (minor) are used to describe the size of the seeds. Because the seed size of lucerne and berseem crops are considered as small-seeded, their sizes were studied under a microscope; then, the images were analyzed with cell imaging software to measure their seed size. To capture the image of the seed, it was placed horizontally on a glass slide and positioned under a microscope. The image was then focused to obtain a clear view and captured as a still image. The seeds were rotated by 90°, and the measurements of the minor third dimensions were taken using the same procedure [15]. This process was repeated for 50

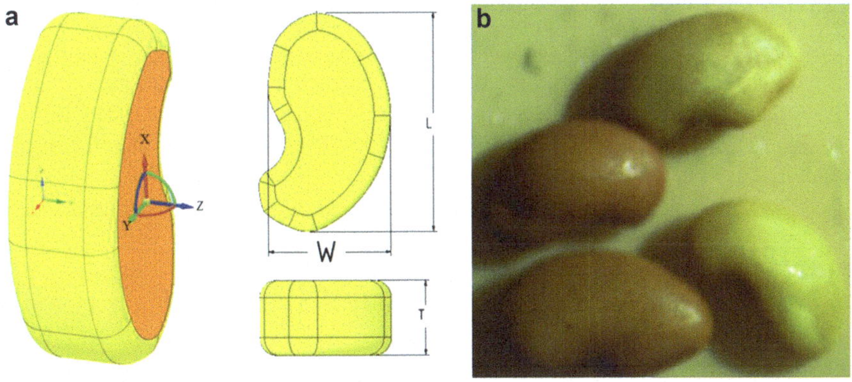

Fig. 9.1 Characteristic dimension of lucerne seed: *L*, length; *W*, width; *T*, thickness (**a**) three-dimensional geometry; (**b**) side view and front view; (3) microscopic view of lucerne seed

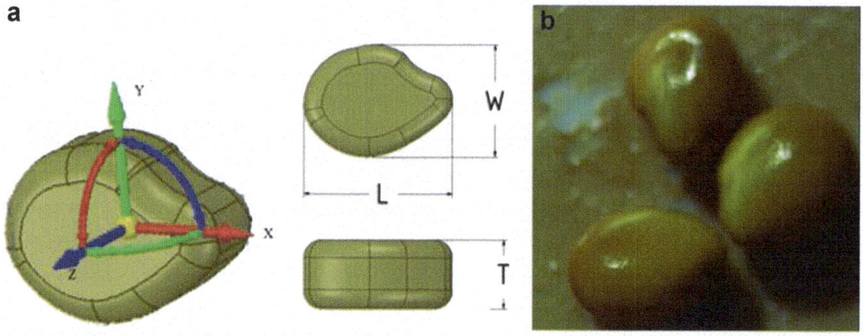

Fig. 9.2 Characteristic dimension of berseem seeds: *L*, length; *W*, width; *T*, thickness (**a**) three-dimensional geometry; (**b**) side view and front view; (3) microscopically view of berseem seeds

randomly selected berseem and lucerne seeds. The third minor dimensions of the seeds were measured and recorded using the same procedure. Additionally, the geometric mean diameter of the seeds (Dp) was calculated using the following equations [16]:

$$Dp = 3\surd LWT \tag{9.1}$$

where, L = length (mm), W = width (mm), T = thickness (mm).

Similarly, the sphericity of seeds was calculated from the following equations [15]:

$$SP = \frac{\sqrt[3]{LWT}}{\mathrm{L}} \tag{9.2}$$

9.2.3 Determination of Thousand-Grain Weight (TGW)

A random sample of 1000 numbers for each berseem and lucerne seeds was manually counted and weighed on an electronic balance with an accuracy of 0.01 g [8, 12, 14–16]. The experiment was repeated for different ten samples, and then the mean value of TGW was determined.

9.2.4 Determination of Bulk Density and True Density

The bulk density of berseem and lucerne seeds were determined according to Gautam et al [15] by using a 100 ml graduate measuring cylinder with a prepared sample of the seeds. No manual compaction of the seeds was carried out. Then, the bulk density of the seed samples was computed from the known volume of seed that was filled in the graduated cylinder, and the mass of the seed was weighed on an electronic balance [16]. The experiment was replicated ten times to determine the mass of seed samples and the volume occupied by the seed samples for calculating the average value of the bulk density.

The true density of the seed was computed using the toluene displacement method [15]. The toluene (C_7H_8) having 865 kg/m^{-3} was first poured into the graduated cylinder, then the known weight of the seed sample was immersed in toluene and the displaced volume of toluene in the measuring cylinder was noted. The experiment was replicated ten times to determine the weight of the seed and displaced volume by toluene. The true density of berseem and lucerne seeds is computed from the following relationships [16].

$$T_D = \frac{W_S}{V_d} \times 10^9 \tag{9.3}$$

where, T_d = true density (kgm^{-3}), W_s= weight of the sample (g), V_d = displaced volume (cm^3).

9.2.5 Determination of Angle of Repose and Coefficient of Static Friction

The angle of repose for both seeds was measured by placing a plastic cylinder with a diameter of 70 mm and a height of 270 mm vertically on a flat wooden surface. The cylinder was then filled with samples of the seeds under free-falling conditions. To ensure even distribution and minimize any wall effects, any excess grains surrounding the platform were removed. The diameter of the resulting pile and the height of the heap were measured using a measuring scale. This experiment was repeated five times to determine the average diameter of the base and the height of the cone. The angle of repose was calculated according to the following equation [20]:

$$\Phi_{r^1} = \tan^{-1}\left(\frac{2H_h}{D_p}\right) \tag{9.4}$$

where, Φ_r= angle of repose, H_h = height of heap, D_p = diameter of pile, mm.

The coefficient of static friction for both berseem and lucerne seeds was measured on a mild steel sheet surface. The testing apparatus used in the coefficient of static friction measurement consists of a 105 mm × 105 mm × 50 mm rectangular wooden box attached to a pan with string and frictionless pulley fitted on a frame and mild sheet surface [15]. Known amounts of seed were filled in seed containers and placed on the test mild steel surface. Then, weights were added to the loading pan until the containers began to slide. The experiment was replicated five times to measure the weight of samples of the seed and the total weights added to the loading pan. The coefficient of static friction was calculated using the formula below [17]:

$$\mu f = (\mathrm{W2} - \mathrm{W1}) / Ws \tag{9.5}$$

where, μ_f = coefficient of static friction, W_1 = total weights added just to start sliding when seed box container is empty, *g*, W_2 = total weights added just to start sliding when seed box container contains seed, *g*, and W_s= weight of sample seed, *g*.

9.3 Result and Discussion

This study was conducted on the physical properties of berseem and lucerne seeds that are crucial in the design and development of seed drills and other processing machines for these crops. The size, shape, thousand-grain weight, true and bulk density, angle of repose, and coefficient of static friction of berseem and lucerne crops were characterized. The result of the different physical properties of berseem and lucerne seeds is present under this section.

9.3.1 Size and Shape of Seeds

Table 9.1 shows the size of the seeds of berseem and lucerne crops based on three principal axes of the seeds, such as length, width, and thickness. Length, width, and thickness of berseem seeds ranged from 1.84 to 2.68 mm, 1.30 to 1.98 mm, and 0.97 to 1.53 mm with mean values of 2.30 mm, 1.63 mm, and 1.30 mm, respectively. For berseem seeds, length, width, and thickness were varied from 1.84 to 2.68 mm, 1.30 to 1.98 mm, and 0.97 to 1.53 mm with mean values of 2.30 mm, 1.63 mm, and

Table 9.1 Values of the size of berseem and lucerne seeds

Crop	Lengths (mm)			Width (mm)			Thickness (mm)			
type	Range	Mean	±SD	Range	Mean	±SD	Range	Mean	±SD	Reference
Beseem	1.84–2.68	2.30	0.18	1.30–1.98	1.63	0.17	0.97–1.53	1.30	1.39	This study
Lucerne	1.66–3.10	2.52	0.32	0.95–1.77	1.44	0.2	0.61–1.22	0.88	0.14	This study
Alfalfa	–	2.04	–	1.25	–	–	0.98	–	–	[23]
Clover	–	2.43	–	–	1.64	–	–	0.84	–	[22]
Quinoa	–	1.90	–	–	1.89	–	–	0.98	–	[21]

Table 9.2 Values of geometric mean diameter, sphericity, and roundness of berseem and lucerne seeds

Crop type	Geometric mean diameter (mm)			Sphericity (%)			Roundness (mm)			Reference
	Range	Mean	±SD	Range	Mean	±SD	Range	Mean	±SD	
Berseem	1.4–1.96	1.69	0.125	0.60–0.84	0.74	0.05	0.53–0.78	0.7	0.05	This study
Lucerne	1.21–1.8	1.47	0.136	0.46–0.76	0.59	0.07	0.31–0.63	0.42	0.48	This study
Alfalfa	1.14–1.5	–	–	0.61–0.70	–	–	–	–	–	[23]
Clover	–	1.75	–	–	0.72	–	–	–	–	[22]
Quinoa	1.4–1.6	–	–	–	0.78	–	–	–	–	[21]

1.30 mm, respectively. Although, for lucerne seeds, the range of lengths, width, and thickness varied from 1.67 to 3.10 mm, 0.95 to 1.77 mm, and 0.61 to 1.22 mm having mean values of 2.52 mm, 1.44 mm, and 1.88 mm, respectively. Similar studies show that the average thickness of berseem seeds were within the same range as quinoa seeds [21], but the length and width of berseem were observed in the same range as clover seeds [22]. Similarly, the size of lucerne seeds were observed within the same range as alfalfa seeds at dry basis moisture content of 7.98% [23]. However, the dimensions of both berseem and Lucerne seeds were higher than those of tef and amaranth seeds [21, 24] but lower than onion seeds [25].

The geometric mean diameter of berseem and lucerne seeds ranged from 1.40 mm to 1.96 mm and 1.21 mm to 1.80 mm, with mean values of 0.74 mm and 0.59 mm, respectively, as shown in Table.9.2. Depending on the size of the seeds, the size of the seed metering mechanism has been designed and developed by various researchers to fit the size of the seeds in any position without causing external injury of the seeds [26]. Thus, the size of the seed metering mechanism for berseem and lucerne seeds is designed to be adjusted from the maximum value of length and width of the seeds of lucerne to accommodate both seeds and for berseem adjusted accordingly.

The shape of berseem and lucerne seeds were studied and expressed in terms of roundness and sphericity, and the values are given in Table.9.2. The sphericity of berseem and lucerne seeds varied between 0.60 to 0.84% and 0.46 to 0.76%, with the mean value of 0.74 and 0.59%, respectively. The roundness of berseem and lucerne seeds varied between 0.53 to 0.78% and 0.31 to 0.63%, with a mean of 0.70 and 0.42%, respectively. For both berseem and lucerne seeds, the value of the coefficient of variation (CV) of roundness and sphericity was more than 10.00%. The value of the berseem seed's roundness and sphericity percentage was higher than lucerne seeds. The studies show that the average sphericity of berseem was higher than clover seeds [22] and lower than quinoa seeds [21]. Similarly, the sphericity of lucerne seeds was observed within the same range as alfalfa seeds at a moisture content of 7.98% (db.) [20]. When designing a common seed drill machine for both crops, the lower value of roundness and sphericity of lucerne seeds is needed to design the slope of the seed transfer unit and metering mechanism, which facilitates the filling and release of the seeds because it has lower tendency of rolling than berseem seeds.

9.3.2 *Thousand-Grain Weight (TGW) of Seeds*

The value of thousands of grain weight of berseem and lucerne seeds is given in Table 9.3. The mean thousands grain weights values obtained for lucerne seeds were 2.93 g at an average moisture content of 9.18% db.; this was higher than the value of alfalfa seeds [23], while for berseem seeds it was 2.60 g at a mean moisture content of 13.64% db.; it was lower than clover seeds [22]. However, these results were comparable to the ones reported for quinoa seeds [21], and exhibited higher values than tef and chia seeds [8, 15]. This parameter is useful for computing plant-to-plant spacing and seed drill development as it influences the flow of seed from the metering plate to the boot and the size of the hopper.

9.3.3 *Bulk Density and True Density of Seeds*

The observed value of bulk density and true density of berseem and lucerne seeds under this study is given in Table 9.3. The value of mean bulk density was observed at 813.94 and 779.52 kg.m^{-3} for berseem and lucerne seeds, respectively. The mean value of true density is 810.65 kg.m^{-3} and 827.34 kg.m^{-3}, respectively, for berseem and lucerne crops. The bulk density was highest for berseem seeds, but the true density was highest for a lucerne seed due to its higher value of thousands of grain weights. The reason for this is that berseem seeds possess greater spherical and roundness values, resulting in less empty space when gathered in large quantities, in comparison to lucerne seeds. As lucerne seeds, Togo et al [20] and Vilche et al [18] discovered that both alfalfa and quinoa seeds displayed a lower bulk density in comparison to their true density. Considering the bulk and true density of seeds is essential for achieving accurate seeding of these fodder crops. These small seeds can be

Table 9.3 Values of physical properties of berseem and lucerne seeds

Crop type	Thousand grains weight (g)	True density (kgm^{-3})	Bulk density (kgm^{-3})	Reference
Berseem	2.6	810.7	813.94	This study
Lucerne	2.93	827.34	779.52	This study
Alfalfa	1.79–2.58	1333–1483	830–1199	[23]
Clover	3.36	–	1377.84	[22]
Quinoa	2.53–3.11	928–1188	747–667	[21]

easily affected by wind drift, so it is crucial to keep the distance between the seed outlet and the opened seeding furrow as close as possible. This will help to determine the optimal distance between the tip of the delivery tubes and the seeding furrow, ensuring successful seeding.

9.3.4 Angle of Repose and Coefficient of Static Friction of Seeds

The angle of repose is a crucial physical characteristic of seeds that helps to determine the appropriate size and slope of seed transportation components in a seed drill. Table 9.4 provides data on the angle of repose for berseem and lucerne seeds. On average, the angle of repose for berseem seeds was 19.15°, while for lucerne seeds, it was 20.31°. The maximum angle of repose was 22.92° for lucerne seeds and 21.80° for berseem seeds. This is due to the sphericity value of berseem seeds being greater than lucerne seeds. The studies show that the average angle of repose of berseem and lucerne was lower than quinoa, clover, and alfalfa seeds [18–20]. Table 9.4 provides data on the coefficient of static friction for the crop under study. The average coefficient of static friction for berseem and lucerne seeds on mild steel was 0.35 and 0.59, respectively. The values of lucerne seeds were in the same range as those reported for alfalfa seeds [20], and higher than those of quinoa seeds [18]. Lucerne seeds had the highest coefficient static friction on a mild steel surface, indicating that they are rougher than berseem seeds and may have difficulty sliding over parts of the machine made of mild steel.

9.4 Conclusion

This study was conducted on the physical properties of berseem and lucerne seeds that are crucial in the design and development of seed drills and other processing machines. These crops are grown to produce nutritious and cost-effective green fodder to meet the ever-increasing demand for animal feed. The size, shape,

Table 9.4 Values of angle of repose and coefficient of static friction berseem and lucerne seeds

	Angle of repose (°)				Coefficient of static friction (%)				
Crop types	Range	Mean	±SD	CV	Range	Mean	±SD	CV	Reference
Berseem	16.16–21.8	19.15	1.87	9.80	0.29–0.44	0.35	0.05	13	This study
Lucerne	16.16–22.92	20.31	0.14	9.30	0.46–0.76	0.59	0.07	11	This study
Alfalfa	27.05–28.68	–	–	–	0.54–0.72	–	–	–	[23]
Clover	–	34.5	–	–	–	–	–	–	[22]
Quinoa	18–25	–	–	–	0.14–0.27	–	–	–	[21]

thousand-grain weights, bulk density, true density, coefficient of static friction, and angle of repose of berseem and lucerne seeds were characterized. For selected varieties of berseem and lucerne seeds, the mean values of their major, intermediate and thickness were 2.30 mm, 1.63 mm, 2.54 mm, and 1.44 mm, respectively. The geometric mean diameter for berseem was 1.69 mm, and for lucerne, it was 1.47 mm. In terms of shape, berseem seeds were found to be more rounded (0.70) and spherical (0.74) as compared to lucerne seeds, which had values of 0.42 and 0.59, respectively. The mean value of thousand-grain weight of lucerne seeds was higher than berseem seeds, which weighed 2.93 g and 2.60 g, respectively. The bulk density of berseem was higher than its true density, while the bulk density of lucerne was lowered than its true density. The coefficient of static friction on the surface of mild steel plate was higher (0.59) for lucerne seeds and lower (0.35) for berseem seeds. The range of angle of repose value was 16.17 to 22.92° for lucerne seeds and 16.16 to 21.80° for berseem seeds. Hence, this research data can serve as valuable input for designing and developing appropriate technologies for these fodder crops, like seed drills, harvesters, threshers, seed cleaners, seed graders, sorter machines, and postharvest handling technologies.

References

1. Mielmann, A. (2013). The utilisation of lucerne: A review. *British Food Journal, 115*(4), 590–600. https://doi.org/10.1108/00070701311317865
2. Balazadeh, M., Zamanian, M., Golzardi, F., & Torkashvand, A. M. (2021). Effects of limited irrigation on forage yield, nutritive value and water use efficiency of Persian clover (Trifolium Resupinatum) compared to Berseem clover (Trifolium Alexandrinum). *Communications in Soil Science and Plant Analysis, 52*(16), 1927–1942. https://doi.org/10.1080/00103624.2021.1900228
3. Enahoro, D., Herrero, M., & Johnson, N. (2019). *Promising options for improving livestock production and productivity in developing countries. ILRI Project Report*. ILRI.
4. Roy, A. K., Malaviya, D. R., & Kaushal, P. (2016). Genetic improvement of fodder legumes especially dual purpose pulses. *Indian Journal Of Genetics And Plant Breeding, 76*(4), 608–625. https://doi.org/10.5958/0975-6906.2016.00076.6
5. Pandey, K. C., & Roy, A. K. (2011). *Forage crops varieties*. IGFRI Jhansi.
6. Shukla, A. K., Behera, S. K., Pakhre, A., & Chaudhari, S. K. (2018). Micronutrients in soils, plants, animals and humans. *Indian Journal of Fertilisers, 14*(4), 30–54.
7. Singh, D. N., Bohra, J. S., Tyagi, V., Singh, T., Banjara, T. R., & Gupta, G. (2022). A review of India's fodder production status and opportunities. *Grass and Forage Science, 77*(1), 1–10. https://doi.org/10.1111/gfs.12561
8. Hossain, S. A., Sherasia, P. L., Phondba, B. T., Pathan, F. K., & Garg, M. R. (2017). Effect of feeding green fodder based diet in lactating buffaloes: Milk production, economics and methane emission. *Indian Journal of Dairy Science, 70*(6), 767–773.
9. Johansen, C., Haque, M. E., Bell, R. W., Thierfelder, C., & Esdaile, R. J. (2012). Conservation agriculture for small holder rainfed farming: Opportunities and constraints of new mechanized seeding systems. *Fields Crops Research, 132*, 18–32. https://doi.org/10.1016/j.fcr.2011.11.026
10. Ahmad, E., Ghassemzadeh, H. R., Moghaddam, M., & Kim, K. U. (2008). Development of a precision seed drill for oilseed rape. *Turkish Journal of Agriculture Forestry, 32*(5), 451–458. https://doi.org/10.3906/tar-0707-52

11. Zewdu, A. D., & Solomon, W. K. (2007). Moisture-dependent physical properties of tef seed. *Biosystems, 96*(1), 57–63. https://doi.org/10.1016/j.biosystemseng.2006.09.008
12. Yalçin, I., Çetin, M., & Ozarslan, C. (2009). Moisture-dependent physical properties of onion (Allium cepa L.) seed. *Asian Journal of Chemistry, 21*(2), 1341–1349.
13. Gautam, A. G., Khurana, R., Singh Manes, G., & Dixit, A. K. (2016). Studies on physical properties of pelleted onion (*Allium cepa L.*) seeds. *International Journal of Bio-resource and Stress Management, 7*(6), 1279–1283. https://doi.org/10.23910/ijbsm/2016.7.6.1693
14. Anonymous. (2018). *Package of practice for Rabi crops of Punjab* (pp. 91–105). Punjab Agricultural University.
15. Gautam, A. G., Khurana, R., Singh Manes, G., & Dixit, A. K. (2016). Studies on engineering properties of pelleted carrot (Daucus carota L.) seeds. *International Journal of Bio-resource Stress and Management, 75*, 1044–1048. https://doi.org/10.23910/ijbsm/2016.7.5.1637a
16. Wang, B., Li, D., Wang, L., Huang, Z., Zhang, L., Chen, X. D., & Wang, Z. (2007). M.: Effect of moisture content on the physical properties of fibered flaxseed. *International Journal of Food Engineering, 3*(5). https://doi.org/10.2202/1556-3758.1281
17. Patel, K. K., Naik, R. K., Patel, G., & Pandey, S. (2022). Studies on some physical and engineering properties of finger millet for designing thresher. *Biological Forum, 14*(4), 387–393.
18. Guiotto, E. N., Ixtaina, V. Y., Tomás, M. C., & Nolasco, S. M. (2011). Moisture-dependent physical properties of chia (Salvia Hispanica L.) seeds. *Transactions of the ASABE, 54*(2), 527–533. https://doi.org/10.13031/2013.36455
19. Aralbayev, N., Dikhanbayeva, F., Serikbayeva, A., Yusof, Y., & Chang, L. (2020). Physical education for youth – Macau. *Determination of Some Physical Properties of Dry Shubat, 2*(2), 39–46. https://doi.org/10.29013/vii-conf-usa-7-39-46
20. Bajpai, A., Kumar, Y., Singh, H., Prabhakar, P. K., & Meghwal, M. (2020). Effect of moisture content on the engineering properties of Jamun (Syzgium cuminii) seed. *Journal of Food Process Engineering, 43*(2), 1–8. https://doi.org/10.1111/jfpe.13325
21. Vilche, C., Gely, M., & Santalla, E. (2003). Physical properties of quinoa seeds. *Biosystems Engineering, 86*(1), 59–65. https://doi.org/10.1016/S1537-5110(03)00114-4
22. El-Sayed, A. B., Abd El-Wahab, M. K., El-Shal, H. M., & Abd Allah, W. E. (2023). Performance of cleaning unit for sesame and clover seeds based upon some physical and mechanical properties. *Agricultural Engineering International: CIGR Journal, 25*(2), 296–311.
23. Togo, J. M., Wang, D., Ma, W., & He, C. (2018). Effects of moisture content on selected physical and mechanical properties of alfalfa seeds. *Journal of Biology Agriculture and Healthcare, 8*(14), 8–18.
24. Abalone, R., Cassinera, A., Gastón, A., & Lara, M. A. (2004). Some physical properties of amaranth seeds. *Biosystems Engineering, 89*(1), 109–117. https://doi.org/10.1016/j.biosystemseng.2004.06.012
25. Abhayawick, L., Laguerre, J. C., Tauzin, V., & Duquenoy, A. (2002). Physical properties of three onion varieties as affected by the moisture content. *Journal of Food Engineering, 55*(3), 253–262. https://doi.org/10.1016/S0260-8774(02)00099-7
26. Petrů, M., Novák, O., Herák, D., Mašín, I., Lepšík, P., & Hrabě, P. (2014). Finite element method model of the mechanical behaviour of Jatropha curcas L. bulk seeds under compression loading: Study and 2D modelling of the damage to seeds. *Biosystems Engineering, 127*(1), 50–66. https://doi.org/10.1016/j.biosystemseng.2014.08.011

Chapter 10
Recent Developments in Biomechanics-Based Prediction of Musculoskeletal Disorders: A Review

Alehegn Melesse Semegn, Bereket Haile Woldegiorgis, and Zerihun Wondimu Lemessa

10.1 Introduction

During their workdays, employees are physically in contact with different types of tools, equipment, and materials; this physical interaction causes motions in various human body parts. Musculoskeletal disorders (MSDs) are caused when a condition affects the mobility of muscles, tendons, bones, joints, ligaments, and nerves in various regions of the human body. According to Donisi et al., 2022, work-related MSDs are collection of excruciating degenerative and inflammatory conditions that affect the ligaments, muscles, tendons, peripheral nerves, spinal disks, cartilage, as well as joints. It results from three main factors, namely, duration, intensity, and repetition, due to which employees have more risk and injury during activities. Prolonged on unfavorable work circumstances the risk is increased when the load is too heavy, repetitive, executed unsafely, and the work has already experienced back disorders [2]. In the US, MSDs account for 30% of lost days due to injuries and incur expenses of 45–54 billion USD [3].

A nonneutral posture causes overloading of the joint structures, which results in MSDs, which are a major problem for workers in labor-intensive industries [4, 5]. According to L. Donisi et al. [1], lifting is the most potential harmful action for work-related MSDs. The manual handling of materials contributes to the high prevalence of MSDs among airline baggage handlers [6]. Labor-demanding and repetitive tasks are common in the construction sector, workers are prone to get MSDs-related injuries while performing such a given activities [7]. The meat processing sector has gotten attention due to the high frequency of linked MSDs because of the manual nature of the labor force more exposed of the upper body and

A. M. Semegn (✉) · B. H. Woldegiorgis · Z. W. Lemessa
Faculty of Mechanical and Industrial Engineering, Bahir Dar Institute of Technology, Bahir Dar University, Bahir Dar, Ethiopia

S. W. Fanta et al. (eds.), *Sustainable Development Research in Manufacturing, Process Engineering, Green Infrastructure, and Water Resources*, Green Energy and Technology, https://doi.org/10.1007/978-3-031-77339-6_10

neck, and more movement and stress on the workers [8]. Applying biomechanical principles and models in industry has a major role in controlling MSDs of the lower back and upper extremity of workers [5, 9]. One of the advantages of biomechanical models explained by Ilaria Conforti and Ilaria Mileti, 2020, are kinematics factors related to the lower limb joints and trunk body segment that help differentiate between a safe and dangerous posture [11]. Biomechanical models are important to determine spinal load and muscle force on the lower back of the human body [11].

In risk assessments and injury prevention, recent advancements in biomechanics and ergonomics that use integrated developing wearable technology and prediction algorithms are crucial. Biomechanical-based risk assessment and evaluation in ergonomics through measurement of biomechanical parameters using such tools have a huge possibility for reducing and eliminating nonfatal occupational coincidences especially work-related MSDs [8, 12].

In this review, smart wearable devices that are used for data collection and prediction models (learning algorithms) used for the purpose of MSDs prevention are considered. Thus, the study aimed to assess the current state of art on biomechanical-based MSDs prediction and to map the application of sensor-based devices and other recent techniques with prediction of MSDs. To meet the objective, the study addressed the following research questions: (1) Do wearable devices and prediction models have significant advancement toward MSDs risk assessment? (2) Which application areas applied wearable smart devices to improve biomechanical risk assessment performance? (3) What prediction models are more usable in biomechanics-based MSDs prediction?

In this study, we reviewed smart wearable devices used for data extraction and prediction algorithms (mainly machine learning (ML)) for the purpose of estimation of MSDs. Then the research gaps for future work were identified, and finally, conclusions were drawn.

The integration of wearable sensors and prediction algorithms in biomechanics for the purpose of assessing and preventing MSDs are not yet comprehensively reviewed. Therefore, this review will help researchers to use it as an additional input to observe their integration and potential for future studies.

10.2 Methodology

A strategy was developed under the guidance of an experienced and well-known database library to search the articles. Studies in peer-reviewed journals related to the subject area were carefully identified. The search was made in databases of IEEE Explore, PubMed, and Google Scholars, mainly focusing on recent publications since 2012. Keywords used to identify studies on biomechanics-based prediction of MSDs included biomechanical, musculoskeletal disorders or ergonomics, smart wearable device or sensors, and prediction algorithms or machine learning. In addition, we looked through the retrieved papers' reference lists to find pertinent studies that we may incorporate into our review.

In this review, articles without full text, commentaries, conference abstracts, and editorials, were not considered. English language written studies were searched. Studies were then identified and extracted as being of prediction models applied, used of biomechanical models, and wearable devices. Studies that had a high or moderate relation with the current review study were included.

Critically read the studies based on the collected data to analyze and show recent advancement of biomechanical-based prediction important to prevent MSDs-related injuries and risks. The initial search identified and downloaded studies based on title, abstract, and keywords, then evaluated them based on their relevance as shown in Fig. 10.1 based on [13] standard.

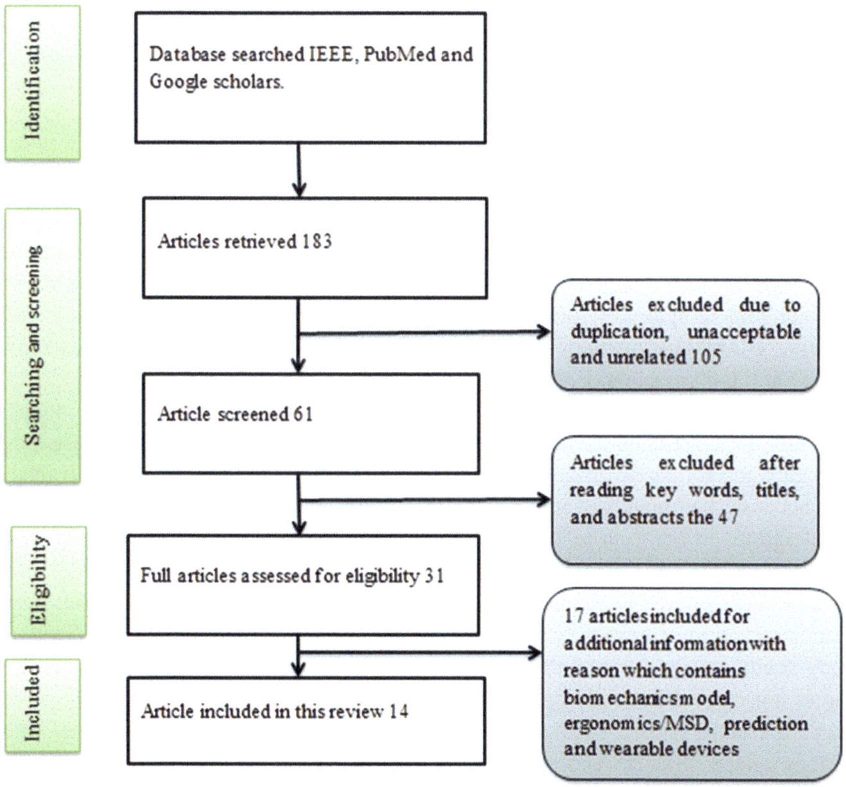

Fig. 10.1 Flowchart summarized the steps of the study search and selection processes

10.3 Results and Discussion

In this systematic review, using the aforementioned databases searched since 2012, about 183 studies were retrieved. Most of them included keywords without direct relation to the aim of this study. 105 articles were excluded because of duplication, inaccessible, and unrelated, 47 studies again were excluded after reading their titles, abstracts, and whole document. Finally, 14 articles in this review were included in the studies. The inclusion procedure was conducted in accordance with the flowchart as shown in Fig. 10.1 on the methodology part.

10.3.1 Biomechanical Models

The study of how living things move is known as biomechanics; occupational biomechanics is closely tied to ergonomics. The study of biomechanics focuses on how the human body moves in accordance with Newton's laws of motion to forecast work positions. According to Parida and Ray [14], biomechanics models classified static and dynamic (kinetics and kinematics) models depending on the methods and characteristics of the task to be performed.

Static Model: When the total of all net external forces acting on a body equals zero, the model is said to be static. The body is in a condition of static equilibrium, which is determined by applying Newton's third law of motion and accounting for the effects of gravity on posture. Multiple-link coplanar modeling, planar models of internal and nonparallel forces, National Institute for Occupational Safety and Health (NIOSH) lifting models, and three-dimensional strength modeling are examples of static biomechanical models.

Dynamic Models: More intricate by nature than static models. Static models typically underestimate the stresses associated with dynamic activities because they disregard the forces resulting from the load's and body segments' inertia. It takes into account the impacts of motion kinematics and kinetics, including velocity and acceleration, in addition to external forces acting on the body and posture. Two levels of complexity are introduced by the dynamic biomechanical model: the motion is considered to be characterized kinematically, which includes a body segment's direction of motion, velocity, and acceleration-deceleration profile. Various body segments accelerate and decelerate as a result of these changes, and inertial forces are created by applying Newton's second law. A variety of model types are used, including single-segment, multiple-segment (i.e., optimization-based and EMG-assisted models), and specific models for body links/segments (such as the model for the shoulders, the trunk, the push-pull low-back model, and the hands).

Other scholars Mihcin [15] also evaluate the difference between the anticipated and measured profiles, upper body motion in vitro is performed using a skeletal model of the flexion-extension movements of the elbow and shoulder. Estimating joint moments at the hips, knees, shoulders, and L5/S1 is done using a bottom-up inverse dynamics model. Biomechanical models in two and three dimensions are utilized to calculate the compressive force on the lower back, determine the strength needed for various tasks, and apply strain index and threshold limit value to treat musculoskeletal disorders affecting the distal upper extremities [9].

Using such models can be measured directly using technical smart tools and then it is possible to predict the risk level of the human body parts. Therefore, the workers can take action to reduce or prevent the risks and injuries that will happen in future when performing the activities.

10.3.2 Wearable Devices Advancement in Biomechanical-Based Risk Assessment

Wearable technologies are one of the smart devices that are emerging technology used for different applications. Thus smart devices are for wearing or embedding different body parts of the human body and can collect important data for the purpose of performance enhancement and risk assessment. In recent times, wearable devices have been more common instruments in biomechanical-based risk assessment to prevent MSDs. The main techniques that have been employed in human kinematics and kinetics measurement of body movement and activities are using camera systems and wearable sensors. Among the two, the latter one is the concern of this review because of wearable sensor applications ever increasing in biomechanical risk assessment techniques.

Most studies indicated that sensors are applied in athletics (sport), ergonomics, clinical purposes, and industrial applications. Smart wearable devices consist of different types of technology to identify and assess biomechanical-based, human-related injuries and risks during tasks activities and movements [16]. By employing data-driven decision-making from information gathered using such devices, health safety professionals, ergonomists, and human factor practitioners can finally help them to maximize the industries' performance [17].

The types of wearable sensors used, its purpose, where the body part is attached, number of sensors applied, and the application area of the study are stated clearly in Table 10.1. The most applicable smart wearable sensors as a single or a combined system used for various applications to collect MSDs-related information (about 80%) were inertial wearable units (IMUs) ((such as, accelerometer, gyroscopes, and magnetometers). The one reason to compare it to other sensors is its validity in real environment, and it can also provide real-time estimation at the least cost of money [18].

Different types of motion sensors and systems such as accelerometer, gyroscope, magneto-resistive sensors, flexible goniometer, electromagnetic tracking system (ETS), sensing fabric, force sensor, and sensors for electromyography (EMG) as a single or a combination of multiple types of sensors are applied for various applications to mitigate MSDs-related issues by collecting relevant information. Weijun

Table 10.1 Data extraction devices and techniques

Authors	Wearable devices used	Purpose	Attached body location	Number of sensors used	Application area
[1]	IMU	To get inertial signals	Sternum	1	Laboratory
[2]	Barometric altimeters, IMUS	Record signals.	Legs	4	The laboratory
[10]	IMUs	Measurements of kinematic parameters	Shank, foot thigh, lumbar	8	In lab setup
[20]	IMU	Acceleration data extract	Pelvis	1	In lab setup
[21]	IMUs	To predict segment trajectories	Thighs, shanks, pelvis, and feet	7	In lab setup
[23]	Kinect	Postural data	Based on the frontier concept	Not mentioned	In lab setup
[24]	Surface electromyography integrates	Use of internet of things (IoT) devices directly in the factory	Trunk	4	Fiat Chrysler automobiles (FCA) assembly line
[22]	EMG, IMUs	Estimation of daily shoulder load	Wheelchair frame, wheel, thorax, right upper arm, and forearm.	Not mentioned	In laboratory
[26]	Pressure sensors with camera	Predicts lower extremities and L5/S1 moments	Forefoot, mid-foot, rear-foot	99	In lab setup
[11]	Steadicam vest	To investigate biomechanical loading of the lumbar spine	Arm with wrench hand tools.	Not mentioned	In lab setup

Tao, Tao Liu [19] indicated that wearable sensors have a major application in gait analysis to improve human health performance in the working area. As stated on the result part, IMUs are more applicable sensors compared to others type of wearable sensors and it is clearly shown in table1.

Some application areas of IMUs are: inertial signals (linear acceleration and angular velocity) at the strum during repeated rhythmic lifting tasks [1, 20]; measure kinematic parameters of range of motion in the lower limb joints and lumbosacral, as well as the trunk's displacement in relation to the pelvic [10]; to forecast five phases of gait (mid-stance, pressing, loading response, swing, and terminal stance) as well as segment trajectories (angular velocity of the shank, thigh, and foot segments) [21]; estimate daily shoulder load to know the magnitude, frequency, and duration by attached on wheelchair frame, forearm, right upper arm, and thorax

[22]. Furthermore [8], also IMUs are applied to measure human activity in slaughterhouse worker wrists to capture information on their movements.

Other devices like Steadicam vest device are used to investigate biomechanical loading of the lumbar spine measured peak and mean muscle forces are another application [11]. Reflective safety jackets equipped with a triaxial accelerometer, triaxial magnetometer, and barometric altimeters to record continuous signals during work activities for the purpose of classifying safe lift and unsafe lift [2], and also [23] used as depth sensor camera (Kinect) based on the frontier concept, to assess workers' safety in a changing construction environment. In order to record the positions of the body segments, IMUs integrated with motion capture systems in two distinct scenarios. In each scenario, a video camera is positioned in front of, behind, and laterally to the worker [24]. Using an insole sensor, the foot was separated into three zones which means forefoot, mid-foot, and posterior foot, and the sensor position was retrieved for each zone. MCam2 cameras were utilized to anticipate L5/S1 and lower extremities moments [6]. Other than wearable sensors, Modified Nordic Musculoskeletal Questionnaire [6, 25] was used to gather data on symptoms of MSDs.

As shown in Table 10.1, different authors have explained data collection tools, particularly wearable devices, for various purposes. This includes a detailed discussion of the number of sensors used, the body parts where the devices are located, and the application areas of the studies, as mentioned above. Some of the authors, such as [1, 2, 10, 20, 21], applied different IMU sensors, attaching them to various body parts for a range of purposes. Most studies indicated that IMUs were more applied because they measure different factors, including speed, acceleration, direction, angular rate, specific force, magnetic fields surrounding the device, as well as IMUs' input from different sensors in order to accurately output movement [23]. Kinect was used to record workers' postures while they completed the task, and then an inverse dynamics approach was applied to calculate the force applied to the joints [23]. Steadicam vest was used to investigate biomechanical loading of the lumbar spine [11].

Due to exposure to different occupational hazards, workers who perform physically demanding activities may be at risk for sickness, injury, or other unfavorable health outcomes. Applications of sensor technology are helpful in the workplace to measure and track job exposures in an inconspicuous manner and to give employees access to data whenever they need it or real-time feedback. Tabel 10.1 shows us that it still needs improvement to get the required information in the workplace because the research output was not accurate in the controlled environment.

10.3.3 Prediction Algorithms in Biomechanics to Predict MSDs

One of the tremendous opportunities to advance science is the integration of real-time posture analysis through smart wearable technology and prediction models (e.g., machine learning algorithms) is a useful tool that helps to reduce MSDs. Moreover, it is now gaining popularity in ergonomic for risk assessment and determines the risk level of workers by means of data acquired with the help of wearable devices in recent years. Most applied algorithms are machine learning algorithms such as neural network (NN), decision tree (DT), supervised vector machine (SVM), logistic regression model (LRM), random forest (RF)), etc. The study showed that the output of wearable sensors fed to prediction algorithms to predict different human biomechanics to prevent MSDs were used to supervise machine learning algorithms, such as regression (linear regressions, support vector regression) and classification (logistic regression, decision tree, Naïve Bayes classifier, neural network), as summarized in table 10.2.

Recently different learning algorithms have been involved in different industrial applications. Modern machine learning techniques are beginning to supplement conventional statistical tools in order to fully utilize the potential of data retrieved from sensors and improve the efficiency of research output [27]. For example, predictive models are often used to forecast the injury and risk depending on the previous experience.

The science of achieving human intellect on a machine is known as artificial intelligence (AI). Over the last 70 years, AI has been extensively used in numerous fields. Machine learning (ML) is a branch of AI that deals with algorithms and statistical models that are capable of autonomous learning, identifying patterns in data, and producing precise predictions to enhance task performance in the real environment. ML is primarily classified into four categories based on the method used to learn the data: unsupervised learning, semi-supervised learning, reinforcement learning, and semi-supervised learning [28]. The rapid development of information and AI technology, particularly ML, has greatly influenced research, industry, and society. These algorithms are widely used in many industries as a result of technology advancements that have increased data collection volume and improved processing capability. Thus techniques from various algorithms, training data, and/or prior knowledge can be used to optimize performance [25]. Using ML techniques for risk assessment and injury prevention represents a new trend in ergonomics research [10, 29]. The collected data using smart wearable devices is then fed to learning algorithms of machine learning and artificial intelligence for the purpose of forecasting MSDs-related risks as well as injuries, forecast work-related MSDs on bus driver workers using Decision tree, Random forest, and Naïve Bayes, and measure the performance of their accuracy (100%), except Naïve Bayes (93.28%) which was less accurate [1, 25]. The data collected from wearable sensors was fed into a logistic regression model to classify the risk associated with lifting tasks, determining whether there was a risk or no risk. The model's performance, based on

Table 10.2 Prediction algorithms applied in biomechanical-based MSDs prediction

Authors	Prediction models used	Purpose	Key findings	Application area
[25]	Naïve Bayes, random forest, and decision tree	Forecasting MSDs	100% accuracy, decision tree and random forest and Naïve Bayes yields 93.28% accuracy.	Bus drivers
[1]	Logistic regression model	Classify risk	Accuracy, sensitivity, and specificity of 82.8%, 84.8%, and 80.9%, respectively.	Lifting tasks
[2]	Neural classifier	Assess the safety level	Achieved an accuracy of 95.6%.	Lifting loads
[10]	SVM	Measurements of kinematic parameters	An accuracy of 99.4% (specificity of 100%) measurements of all kinematic parameters as features.	Manual material handling tasks
[20]	Decision tree, LR, multiple perceptron, SVM, RF	Classify risk using the revised NIOSH lifting equation.	Decision tree accuracies greater than 90%.	Lifting tasks
[21]	Long-short term memory	Lower limb segment trajectories	All predicted trajectories with correlation coefficients greater than 98%.	Gait trajectory and gait phase
[23]	Random forest classifier and regression model	Identify the theoretical maximum attainable level of safety	Implement prediction model velocity as input features.	Construction workers
[8]	Different AI and ML algorithms	To assess the level of ergonomic risk suffered by workers	RF achieved 98% accuracy prediction of the rapid upper limb analysis (RULA) score on the test dataset.	
[33]	LSTM	Predict segment trajectories and five gait phases	All predicted trajectories were strongly correlated coefficients greater than 0.98.	Gait trajectory and gait phase

accuracy, sensitivity, and specificity, yielded results of 82.8%, 84.8%, and 80.9%, respectively [1, 25].

On the other hand, neural networks were applied to classify load lifting as safe, unsafe, or none, i.e., lifting of a light object, and achieved an accuracy of 95.6% [2]. Machine learning algorithms were also implemented for automation and real-time monitoring in construction workers by applying biomechanical model of inverse kinematics and dynamic for training data set which is stated by [7]. Decision tree, logistic regression, multiple perceptron, supervised vector machine, and RF were applied to classify risk according to the revised NIOSH lifting equation for a binary risk/no-risk classification and tree-based algorithms, having accuracies of about 90% [20].

In the meat processing business, artificial intelligence and machine learning techniques are integrated to perform task classification and ergonomic assessments [8]. In order to determine the proper and wrong postures, a support vector machine was used to estimate kinematics characteristics such as lower limb range of motion, lumbosacral joints, and trunk displacement with respect to the pelvis [10].

This article Mora, 2012, suggested use of artificial neural networks to capture real data using a glove device and apply it in a biomechanical model. A novel ML approach that incorporates long short-term memory (LSTM) model, SVM, time series prediction deep learning models, and planar pressure as an input is used to predict lower extremities during systematic lifting [26]. Based on biomechanical principles, this [30] predictive model implemented on a smartphone user depicts standing posture in the sagittal plane, joint torques, and support forces. Arterial stiffness was monitored using back propagation neural network, decision tree, and multivariate linear regression, by extracting 21 features [31]. Convolutional neural network and random forest regression-based regression models were applied to estimate fatigue levels in a continuous fashion [32].

As shown in table 10.1, 75% of studies performed in controlled laboratory experimental level were not applied in real working environment. Among different factors, some of them were temperature and light differences, external pressure sensitivity, number of sensors attached/embedded on the body part leading to discomfort, humidity difference, and position difference because it needs high expected accuracy [34].

10.4 Conclusion and Future Research Opportunities

10.4.1 Conclusion

Biomechanical models can be used to simulate the stresses and strains placed on the body during specific activities. It can be used to predict the risk of developing MSD by combining data from wearable devices with prediction algorithms. The use of smart wearable devices integrated with prediction models for different industrial applications which were worn or embedded sensors to provide important data. Then it is fed to different prediction models to mitigate MSDs-related issues by collecting relevant information.

This study's main finding shows that wearable sensors and prediction algorithms have dramatically increased as well as good application prospects to prevent MSD-related injuries. The use of smart wearable devices integrated with prediction models for different industrial applications to mitigate MSDs. The research output was almost limited in the laboratory which shows there are opportunities to implement it at the real environment. There is no evidence found in the literature about the number of sensors increased to affect the quality or accuracy of the collected information. The studies indicated that the application area applied when the task was repetitive motion, the load was heavy, and the work was performed under unsafe conditions.

Understanding of how biomechanical factors can affect the onset and progression of musculoskeletal disorders is provided by recent developments in biomechanics-based prediction of MSDs. This information can aid in the development of more potent MSD treatments and preventative measures by researchers and physicians. It can also assist companies in creating better working environments that lower the risk of MSDs.

10.4.2 Future Research Opportunities

According to research findings, there have been significant advancements and promising application prospects when using wearable sensors and prediction algorithms. Even now, minor technological issues still need to be fixed before this technology can be widely used in real-world human life in the future. Current commercial wearable devices have limitations in estimating tissue-level MSDs' performance inside the body. The constancy of sensor signals, reliability of algorithms for kinematics and kinetics in MSDs, making low-cost, appropriate-size wearable sensor systems, and best positioning of wearable sensors on the human body are some technical aspects that need to be addressed in future research.

The integration of wearable sensors and learning algorithms for MSDs prediction presents numerous research opportunities. Some of the potential future work will be early detection, prevention, and management of MSDs, ultimately improving the quality of life and increasing productivity.

References

1. Donisi, L., et al. (2022). *A logistic regression model for biomechanical risk classification in lifting tasks* (pp. 1–15).
2. Pistolesi, F., & Lazzerini, B. (2020). *Assessing the risk of low back pain and injury via inertial and barometric sensors* (Vol. 3203, pp. 1–10). IEEE Transactions on Industrial Informatics. https://doi.org/10.1109/TII.2020.2992984
3. Harari, Y., Bechar, A., & Riemer, R. (2019). Simulation-based optimization methodology for a manual material handling task design that. *IEEE Transactions Human-Machine Systems, PP*, 1–9. https://doi.org/10.1109/THMS.2019.2900294
4. Conforti, I., Del Prete, Z., & Palermo, E. (2019). *Assessing ergonomics and biomechanical risk in manual handling of loads through a wearable system* (pp. 388–393).
5. Tröster, M., et al. (2022). Biomechanical analysis of stoop and free-style squat lifting and lowering with a generic Back-support exoskeleton model. *Internationl Journal of Environmental Research and Public Health, 19*(15), 9040.

6. Tafazzol, A., Aref, S., Mardani, M., Haddad, O., & Parnianpour, M. (2016). Epidemiological and biomechanical evaluation of airline baggage handling. International Journal of Occupational Safety and Ergonomics, 22(2), 218–227. https://doi.org/10.1080/10803548.2015.1126457
7. Subedi, S., & Pradhananga, N. (2021). Automation in construction sensor-based computational approach to preventing back injuries in construction workers. *Automation in Construction, 131*(September), 103920. https://doi.org/10.1016/j.autcon.2021.103920
8. Villalobos, A., & Mac Cawley, A. (2022). Prediction of slaughterhouse workers ' RULA scores and knife edge using low-cost inertial measurement sensor units and machine learning algorithms. *Applied Ergonomics, 98*(October 2020), 103556. https://doi.org/10.1016/j.apergo.2021.103556
9. Taylor, P., Garg, A., & Kapellusch, J. M. (2009). Applications of biomechanics for prevention of work- related musculoskeletal disorders. *Ergonomics, 52*, 37–41. https://doi.org/10.1080/00140130802480794
10. Del Prete, Z., Palermo, E., Conforti, I., & Mileti, I. (2020). Measuring biomechanical risk in lifting load tasks machine-learning approach. *Sensors, 20*(6), 1557.
11. Weston, E. B., Alizadeh, M., Knapik, G. G., Wang, X., & Marras, W. S. (2018). Biomechanical evaluation of exoskeleton use on loading of the lumbar spine. *Applied Ergonomics, 68*, 101–108. https://doi.org/10.1016/j.apergo.2017.11.006
12. Ranavolo, A., et al. (2020). The sensor-based biomechanical risk assessment at the base of the need for revising of standards for human ergonomics. *Sensors, 20*(20), 5750.
13. Moher, D., Liberati, A., Tetzlaff, J., & Altman, D. G. (2009). Academia and clinic annals of internal medicine preferred reporting items for systematic reviews and meta-analyses. *Annals of Internal Medicine, 151*(4), 264–269.
14. Parida, R., & Ray, P. K. (2015). Biomechanical modelling of manual material handling tasks : A comprehensive review. *Procedia Manufacturing, 3*., Ahfe, 4598–4605. https://doi.org/10.1016/j.promfg.2015.07.539
15. Mihcin, S. (2019). Investigation of wearable motion capture system towards biomechanical modelling. *IEEE International Symposium Medical Measurements and Applications*, 1–5. https://doi.org/10.1109/MeMeA.2019.8802208
16. Mcdevitt, S., et al. (2022). Wearables for biomechanical performance optimization and risk assessment in industrial and sports applications. *Bioengineering (Basel), 9*(1), 33.
17. Van Hooren, B., Goudsmit, J., Restrepo, J., & Vos, S. (2019). Real-time feedback by wearables in running : Current approaches, challenges and suggestions for improvements. *Journal of Sports Sciences, 00*(00), 1–17. https://doi.org/10.1080/02640414.2019.1690960
18. Zhao, J. (2018). *A review of wearable IMU (inertial-measurement-unit) -based pose estimation and drift reduction technologies – based pose estimation and drift reduction technologies.*
19. Zheng, R., Feng, H., Tao, W., & Liu, T. (2012). Gait analysis using wearable. *Sensors*, 2255–2283. https://doi.org/10.3390/s120202255
20. Donisi, L., Cesarelli, G., Coccia, A., Panigazzi, M., Capodaglio, E. M., & Addio, G. D. (2021). Work-related risk assessment according to the revised NIOSH lifting equation : A preliminary study using a wearable inertial sensor and machine learning. *Sensors, 21*(8), 2593.
21. Binbin, S., Elena, M., & Gutierrez-Farewik. (2020). *Gait trajectory and gait phase prediction based on an LSTM network* (pp. 1–17). https://doi.org/10.3390/s20247127
22. Veeger, H. E. J. (2022). Classification of wheelchair related shoulder loading activities from wearable sensor data : A machine learning approach. *Sensors, 22*(19), 7404.
23. Subedi, S., & Pradhananga, N. (2021). Automation in construction sensor-based computational approach to preventing back injuries in construction workers. *Automation in Construction, 131*(February), 103920. https://doi.org/10.1016/j.autcon.2021.103920
24. Greco, A., Gerbino, S., Caputo, F., Lamberti, M., & Miraglia, N. (2019). *Integrated wearable devices for evaluating the biomechanical overload in manufacturing* (pp. 93–97).
25. Hanumegowda, P. K., & Gnanasekaran, S. (2022). *Prediction of work-related risk factors among bus drivers using machine learning* (Vol. 19).

26. Chae, S., Choi, A., Jung, H., Kim, T. H., Kim, K., & Mun, J. H. (2021). Machine learning model to estimate net joint moments during lifting task using wearable sensors : A preliminary study for design of exoskeleton control system. *Applied Sciences, 11*(24), 11735.
27. Halilaj, E., Rajagopal, A., Fiterau, M., Hicks, J. L., Hastie, T. J., & Delp, S. L. (2018). Machine learning in human movement biomechanics: Best practices, common pitfalls, and new opportunities. *Journal of Biomechanics, 81*, 1–11. https://doi.org/10.1016/j.jbiomech.2018.09.009
28. Li, X., Liu, X., Deng, X., & Fan, Y. (2022). *Interplay between artificial intelligence and biomechanics modeling in the cardiovascular disease prediction* (pp. 1–21).
29. Lee, S., Member, S., Liu, L., Radwin, R., & Member, S. (2021). Machine learning in manufacturing ergonomics : Recent advances, challenges, and opportunities. *IIEEE Robotics and Automation Letters, 6*(3), 5745–5752. https://doi.org/10.1109/LRA.2021.3084881
30. Gorce, P., Jacquier-bret, J., & Merbah, J. (2021). While standing with support ac an us cr t. *Ergonomics, 0*(0), 000. https://doi.org/10.1080/00140139.2021.1963845
31. Miao, F., Wang, X., Yin, L., & Li, Y. (2018). A wearable sensor for arterial stiffness monitoring based on machine learning algorithms. *IEEE Sensors Journal, 1748*. https://doi.org/10.1109/JSEN.2018.2880434
32. Jiang, Y., Hernandez, V., Venture, G., Kuli, D., & Chen, B. K. (2021). A data-driven approach to predict fatigue in exercise based on motion data from wearable sensors or force plate. *Sensors*, 1–16.
33. Su, B., & Gutierrez-Farewik, E. M. (2020). Gait trajectory and gait phase prediction based on an LSTM Network. *Sensors*, 1–17. https://doi.org/10.3390/s20247127
34. Gastel, A., Scataglini, S., Zelck, S., Denteneer, L., Vanden Bossche, H., & Verwulgen, S. 2021). *Advances in Simulation and Digital Human Modeling. Proceedings of the AHFE 2021 Virtual Conferences on Human Factors and Simulation, and Digital Human Modeling and Applied Optimization (LNNS 264).*

Chapter 11
Opportunities and Challenges of E-SCM in Developing Countries

Zerihun Wondimu Lemessa, Bereket Haile Woldegiorgis, and Alehegn Melesse Semegn

11.1 Introduction

The industrial world has undergone numerous revolutions, leading to advancements in production over time. The fourth industrial revolution follows mechanization, electrification-based mass manufacturing, and automation aided by the development of electronics and computerization [31]. Today, Industry 4.0 has emerged due to the robust assimilation of information and communication technologies to link the physical and virtual worlds [67]. This is a great possibility to incorporate businesses and their assets to improve resource utilization (Ron [51]). Human–machine interaction (HMI), artificial intelligence (AI), advanced additive manufacturing technologies, Big Data, and Cloud Computing are all incorporated into I4.0. Many researches revealed that supply chain has undergone significant transformations because of the necessity to incorporate emerging new technologies [33, 40, 62]. As technology has advanced, electronic supply chain management (E-SCM) has become an increasingly important topic in business [34]. Many literature studies were conducted to identify how the main elements of Industry 4.0 relate to supply chain management (SCM) [16, 38, 39, 50, 53, 54, 59]. Nevertheless, the proposed supply chain measures and approaches were developed based on industry standards established in developed sectors.

Without a doubt, implementing AI technologies in consumer provision shortens the time to market. However, it is still unclear whether the use of it in the manufacturing sector will make supply chains in developing countries' contexts more nimble [29]. To the best of our knowledge, currently, there are only a few studies investigating E-SCM from the perspective of developing countries implying that no

Z. W. Lemessa (✉) · B. H. Woldegiorgis · A. M. Semegn
Faculty of Mechanical and Industrial Engineering, Bahir Dar Institute of Technology, Bahir Dar University, Bahir Dar, Ethiopia

S. W. Fanta et al. (eds.), *Sustainable Development Research in Manufacturing, Process Engineering, Green Infrastructure, and Water Resources*, Green Energy and Technology, https://doi.org/10.1007/978-3-031-77339-6_11

effort has been made to conduct literature review studies of existing research on supply chain digitalization. Considering the numerous challenges faced by manufacturers in developing countries, such as resource scarcity, non-integrated and underdeveloped information and communication technology (ICT) infrastructure, cultural and organizational issues, a shortage of skilled labor, and labor-intensive supply chain operations [29], we intended to carry out organized literature review on the deployment of digital supply chain in developing countries. Consequently, the main goal of this paper is to explore the determinants of e-commerce adoption in developing countries, identify and analyze opportunities and challenges of e-supply chain in developing countries, and propose possibilities for future research. Three key research questions have been formulated based on the analysis:

Q1. What are the main determinants of e-commerce adoption in developing countries?
Q2. What are the barriers to successful e-supply chain deployment in developing countries?
Q3. What are the opportunities for implementing e-supply chain in developing countries?

The remaining work is organized as follows: Section 11.2 presents a comprehensive definition of E-SCM. In Sect. 11.3, we provide a detailed explanation of the research methodology employed in this study. Section 11.4 presents a concise summary of the research results, highlighting the key findings. Finally, in Sect. 11.5, we offer concluding remarks and discuss potential future research directions in the field of E-SCM.

11.2 Theoretical Background

Supply chain management refers to the systematic coordination and control of suppliers, manufacturers, warehouses, and stores to efficiently produce and distribute goods. It involves managing the flow of products in the appropriate quantity, to the correct location, and at the right time, with the aim of minimizing overall costs while meeting specified service-level standards [55]. Conventionally, these collaborations are separate and occur in the form of sequential stages [7]. Through increased international unpredictability and dynamic consumer needs, this discrete manner of SCM is not any more practical. Because, it lacks the quickness, translucence, adaptability, and reliable evidence required to compete in today's shared environment. By straightly linking dealers and customers and vice versa, digital technology can give solutions to a number of most persistent concerns in SCM by lowering complexity, cost, and boosting capacity and flexibility, leading to superior service levels [1].

Contrasting conventional ones, electronic supply chain (E-SC) greatly depends on systems to maintain actions carried out by internationally dispersed partners to purchase, create, stock, transfer, and sell a product. In a typical SC, organizational structures are constrained by geography/function, and are hesitant to segment information willingly. Traditional supply chains employ technology to collect, store, and show data, whereas E-SC uses technology to make strategic decisions. As a result, E-SC can be defined as a new interrelated intelligent system that expands supply chain applications outside a sole company's local/regional application in order to progress customer service and commercial success [57].

11.2.1 Introduction to E-SCM

The literature presents numerous definitions of E-SCM, reflecting the diverse perspectives and interpretations within the field. Lankford [35] defined E-SCM as a blend between the Internet and supply chain management. Pulevska-Ivanovska and Kaleshovska [49] provided a more extensive definition explaining the E-SCM as a network of independent partners who not only act as distributors of specific products and services within the supply chain but also play a vital role in stimulating demand and leading the synchronization of capabilities and resources across the entire supply chain. This collaboration aims to achieve high levels of operational efficiency and establish market leadership. Today, Internet technology has altered the concept of SCM and introduced new scopes to the operation, changing SCM into E-SCM [35].

11.2.2 e-Supply Chain Management and E-Commerce

E-SCM is the collaborative use of the internet to improve B2B and C2C processes and increase speed, agility, real-time control, and customer satisfaction [30]. The digital connectivity facilitated by new technologies in E-SCM has a significant impact on supply chain performance through enabling e-commerce, information sharing, and knowledge sharing. The first impact, e-commerce, pertains to the online buying and selling of products and services [30, 49]. So, today's supply chain and electronic marketplaces are dependent on one another. The second impact has to do with information sharing, namely how the network may be utilized to approach and convey information among supply chain participants. But, the network also makes it possible for supply chain participants to obtain data statements and modeling, which helps everyone involved in the process improve projection, design, and administration. Such collaborative preparation and supervisory process is the other way that the Internet has an impact on SCM [61].

11.3 Methodology

A systematic literature review was conducted according to Okoli [44] guidelines to address the research questions. Systematic literature reviews are clear, evidence-based activities [63]. It reveals the study's decisions, procedures, and results [11]. Our methodology is arranged in three parts, similar to that used by Ammirato et al. [3] and Grimaldi et al. [19] in their respective literature reviews: paper location, paper selection, and paper analysis.

11.3.1 Paper Location

The reviews' papers were found by a number of queries made in January 2023 but without a time limit to the database Scopus. We explored for the multiple mixtures of keywords including "digital supply chain," "e-supply chain," "smart supply," "digitalization," "developing countries," "E-commerce," and "E-market place." The keywords were chosen after conducting a preliminary literature review. Table 11.1 presents the whole list of the keywords used in the review. The search yielded 310 publications, which were reviewed for the study. A total of 57 research publications were found in several research inquiries; additionally, three articles could not be found using an Internet search. The remaining 250 papers were examined in the following step. A summary of the methodology used in this study is presented in Fig. 11.1.

11.3.2 Paper Selection

Following the removal of duplicates and unavailable articles, the remaining 250 articles were exposed to additional rejection criteria. Articles that were not published in English and did not include the e-supply chain or e-commerce in their title

Table 11.1 Keywords used for literature search

Query					No. results
e-supply chain	and	E-commerce			21
e-supply chain	and	Developing countries			52
Digital supply chain	and	Developing countries			28
Smart supply chain	and	Developing countries			34
Supply chain	and	Digitalization			155
e-supply chain	and	Developing countries	and	Challenges	9
e-supply chain	and	Developing countries	and	Benefits	3
Digitalization	and	Supply chain			8
Total					310

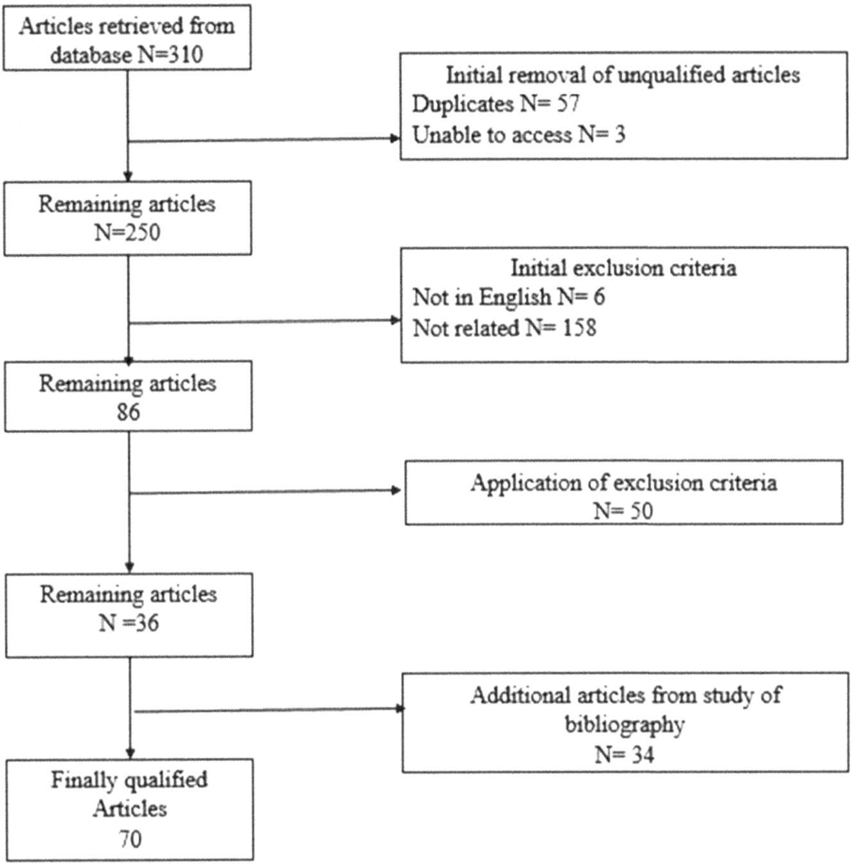

Fig. 11.1 The flowchart of literature search process

or abstract were excluded. Following a thorough analysis and examination of the remaining 86 publications, 50 articles were excluded. These extra exclusion criteria centered on the article's goal: an article was rejected if it did not mention e-supply chain management or e-commerce in emerging countries. Thus, 70 papers related to the subject were discovered.

To extend the literature, the references of all appropriate papers were correspondingly examined. The bibliographies generated a total of 34 articles, bringing the total number of publications used in this study to 70.

11.3.3 Paper Analysis

The investigation was carried out by thoroughly reading each whole paper and noting the opportunities and challenges of the e-supply chain in developing countries.

11.4 Results and Discussion

As a result of strictly analyzing all selected papers, determinants of e-commerce adoption in developing countries, barriers, and opportunities of E-SCM were identified. In the literature, the primary elements of electronic-commerce implementation in developing countries are extensively studied. Consequently, organizational e-readiness (awareness, resource, commitment, and governance) and external e-readiness (government e-readiness, market force e-readiness, supporting industries e-readiness, and curiosity) were recognized. The outcome of the study is presented in Table 11.2. The obstacles to electronic SC chains in developing countries are identified as cultural and technical challenges, trust, perceived risk, strategy, organizational, technological, and social influence. Whereas E-SCM benefits are found to be information perceptibility and distribution, in-time communication, order tracing, improved instantaneous forecasting decisions, faster delivery; strengthened partner relationships, better understanding of the customer, increased customer satisfaction, quicker market adaptation, reduced resources requirement, cost savings, increased profits, etc. These advantages contribute to optimizing supply chain operations and achieving improved performance outcomes.

Table 11.2 Identified elements of e-commerce implementation in developing countries

		Sources
Organizational E-readiness	Awareness	[18, 29, 37, 56, 14, 22, 42, 43, 64]
	Resources	[36, 60, 9, 66, 17, 36, 27]
	Commitment	[13, 66, 15, 4]
	Governance	[10, 47]
Eternal E-readiness	Government E-readiness	[69, 2, 66, 46, 64]
	Market force readiness	[70, 5, 23]
	Supporting industry readiness	[48, 64, 23, 45]
	Curiosity	[12, 21].

11.4.1 E-Commerce Implementation in Developing Countries

New technologies are booming, particularly in the service sector (IT, Telecommunications, Internet, etc.). Economic transactions have grown significantly easier and faster because of technical improvements, owing mostly to the rise of e-commerce. Regarding information technology and electronic-commerce, businesses in industrialized and emerging nations operate differently. Several authors have concluded that due to managerial, organizational, and environmental constraints, enterprises in developing countries generally face much higher implementation risk than those in developed nations [32]. However, there has not been any rigorous research into the reasons why certain businesses in developing nations use e-commerce while others do not. And there has not been much research done on the variables that explain for the variations in e-commerce use sophistication (i.e., the degree of institutionalization) among companies in developing countries. In general, e-commerce status indicator [41] can be categorized into six phases that are pertinent to the electronic-commerce certainties of developing nations as shown in Table 11.3.

Kabanda and Parker [26] conducted an empirical study in South Africa and found that institutionalization of digital trade is strongly prejudiced by external issues of the firm. The technical factors, information deficiency as well as communications technology setup, outdoor legitimate and supervisory differences, absence of digital literacy among stakeholders, language issues, no delivery system, and lack of trust in e-commerce are among the factors that obstruct the practicability of electronic-commerce in Ethiopia, according to Singh and Sahu [58]. Nonetheless, the measure of cross-border digital market of China broke fresh ground at 10.5 trillion yuan, and the commerce's saturation rate surpassed 30% to 33.2%. This demonstrates how much more quickly e-commerce institutionalization is occurring in

Table 11.3 E-commerce status indicator of developing countries

Phase	Status	Description	Implication
Phase 1	No e-commerce	Those organizations that have not considered e-commerce.	
Phase 2	Connected e-commerce	Organizations that have considered e-commerce but are only connected to the Internet via e-mail and do not have a website, or have a static website that publishes basic company information on the web but does not allow for interactivity.	These are organizations in the early stages of adoption or entry-level adoption
Phase 3	Static e-commerce		
Phase 4	Interactive e-commerce	Organizations that accept user queries, e-mail, and form entry; perform online selling and purchasing of products and services, including customer service; and integrate organizational systems with suppliers, customers, and other back office systems, allowing the majority of business transactions to be conducted electronically.	These organizations are said to be at the e-commerce institutionalization maturity stage.
Phase 5	Transactive E-commerce		
Phase 6	Integrated e-commerce		

China than in other developing countries [65]. The potential cause could be China's developed e-commerce infrastructure. Whereas organizations in developing countries, Africa in particular, tend not to reach the institutionalization stage of e-commerce [26].

Based on perceived characteristics and their underlying importance, various scholars have identified the primary elements of electronic-commerce implementation in developing nations. Additionally, the literature has extensively covered organizational and external e-readiness for successful e-commerce adoption in developing countries. E-readiness is defined as the readiness of individuals or organizations to participate in the development process [29].

11.4.1.1 Organizational E-Readiness

Awareness The study on technology transmission entirely believes that cognizance of an invention and its profits is a critical early step in influencing the resolution to implement or throw away the innovation [18, 29]. Thus, successful adoption of e-commerce depends on how an organization perceives, understands, and projects its advantages and threats. Once a company recognizes the latent of electronic-commerce and decides to capitalize in incomes to fetch in the technology, the awareness of additional e-commerce benefits, risks, and success and failure cases influences its decision to continue investing to expand beyond entry-level functionality [37, 56]. However, most of managers in emerging nations has no necessary knowledge to effectively use the Internet as a tool for information, marketing, and communication [14, 22, 42]. Both early implementation of electronic-commerce plus subsequent level-of-utilization decisions are likely to be influenced by their capacity to create knowledge and monitor ecological tendencies in standings of the implementation of e-commerce by competitors, suppliers, and business partners [22, 43, 64].

Resources The accessibility of resources influences an organization's ability to handle the problems and prospects of electronic-commerce [36]. Resources refers to a company's human, technological, and business resources. Organizations differ significantly in terms of the stock, uniqueness, longevity, specificity, heterogeneity, and non-imitability of their resources, and these differences may result in variances in the amount of e-commerce implementation [60]. This is supported by a number of studies that point to qualified employees, a high level of resources [9], the availability of slack funding to advance and preserve electronic-commerce systems [66], and a supportive managerial trend [17] as factors that either influence or define the sophistication of e-commerce at the time of initial adoption. Commerce in emerging nations sometimes move slower than their equivalents in industrialized nations regarding the use of information technology, including electronic-commerce, due to intense managerial limits on monetary, technical, and workers [36]. Additionally, the majority of companies in developing countries have very little experience conducting business online, managing digital transaction, and managing electronically mediated intra- and inter-company transactions [27]. Therefore, the degree to which they are prepared on a human, technological, cultural, and structural level may have a significant impact on whether e-commerce is adopted initially and then institutionalized.

Commitment Many electronic-commerce ventures are ruined to disaster or will not advance above the initial level without commitment [13]. Commitment is the support of important company figures, particularly the CEO, in promoting e-commerce. Lack of commitment (managerial and organizational) has been identified as a significant factor affecting information systems (IS) projects in various researches of information systems (IS) in developing nations [4, 66]. There is a significant managerial turnover rate in several developing countries, and new hires have a tendency to be skeptical of the efforts of their predecessors [15]. To adopt and institutionalize e-commerce, it is therefore essential to create and maintain the momentum of commitment.

Goverance The importance attributed to electronic-commerce implementation is determined by organization's governance model [10]. Governance is the strategic, tactical, and operational framework that defines how organizations are structured in order to set goals, distribute resources, and make choices. It has an impact on how effectively a company handles its e-commerce transition, business integration, and advancement past the first acceptance stage. Even though governance frameworks are crucial once an organization decides to implement electronic-commerce, they are woefully absent in companies in developing nations [47].

11.4.1.2 External E-Readiness

Government E-Readiness Government e-readiness is the institutional context in which business operate to foster e-commerce, which also affects company confidence and e-commerce goals [46]. Participation of government, particularly, in developing countries is a key element which could effect the implementation of innovations [69]. Government can provide supportive infrastructure, legal and regulatory frameworks, and e-commerce use directions. It's to encourage public and private sectors to use e-commerce [2]. A number of studies that studied the influence of governmental pressure on the adoption of e-commerce discovered that adopter enterprises felt more governmental pressure than non-adopter firms [66]. In comparison to developed countries, structural characteristics relating to institutional arrangements are less organized in developing countries [64]. More frequently than market forces, the collaboration between business and the government forms the competitive climate in developing nations. Decisions regarding the introduction of e-commerce might be influenced by the governments of developing nations because they frequently oversee technological progress. The use of e-commerce can be encouraged if a government demonstrates a strong commitment to it through its policy initiatives. Nonetheless, various businesses may have varying perspectives on the readiness of the government. This will undoubtedly have an impact on their conclusions to implement electronic-commerce, institutionalize it, and eventually progress further than basic implementations.

Market Forces Readiness E-commerce reacts to network externalities principles in that the drivers of implementation are more commonly found in the organization's social system than in the qualities of the innovation [5]. These networks include an organization's competitors, customers, suppliers, and other business partners. For example, the primary motivations for the implementation and succeeding stage of the use of electronic-commerce have been recognized as a burden for a business's stakeholders [70]. Companies that believe market dynamics are set for electronic-commerce are more likely to implement electronic-commerce or start on more complex e-commerce employment out of fright of competitive disadvantage, anticipated e-commerce achievement, or just "to keep up with the trend" [66]. Similar pressures are less likely for some developing-country businesses [23].

Supporting Industries Readiness The operation of electronic-commerce is dependent on a variety of supporting industries. The growth of such sectors provides benefits by providing efficient, quick, and often preferential access to e-commerce inputs [48]. Although e-commerce-supporting industries can cover almost every aspect of an economy, three of them are especially important in emerging nations. They are the accessibility and cost-effectiveness of IT industry services, banking sector institutionalization and development, and carrier and transportation facility penetration and reliability [64]. Most developing countries' IT industries are not established enough to play supply push roles, the monetary division is not mature enough to lever electronic transactions, and transportation infrastructure is inadequate [23]. Yet, various firms may evaluate these industries' e-commerce readiness differently, resulting in variances in e-commerce implementation and arrangement [45].

Curiosity Curiosity fuels our behaviors, according to current research, and our deficiency of understanding those aspects contributes greatly to our desire in seeking extra knowledge [12]. Curiosity refers to a state of human desire in which they seek the unknown and attempt to discover the hidden and unexplored things that interest them. According to Hill, Fombelle, and Sirianni [21], enthusiastically concerned clients, without a doubt, are highly motivated in their buying decisions compared to dispassionate ones. Drawing from this debate and previous data, we can conclude that e-commerce is a novel concept for consumers in developing countries, and curiosity likely plays a crucial role in its adoption.

11.4.2 Barriers to Successful E-SC Adopting in Developing Countries

There are similarly some of obstacles or issues that the organizations in developing countries are fronting in functioning electronic supply chains such as cultural and technical challenges, trust, perceived risk, and strategy.

Cultural and Technical Challenges E-SCM is all about effectively integrating the flow of information and material inside the exchange process across network partners. Integration among partners is a difficulty in itself. The Internet and information technologies and systems enable such integration; nevertheless, simply developing and adopting new software is not the answer or solution to achieving long-term competitive advantage. Although the use of Internet-based technology has the potential to reduce costs and increase efficiencies, this approach raises several cultural and technical challenges in developing countries. As a result, in order to reap the benefits of integration, enterprises must overcome cultural obstacles [8]. On the other hand, technical problems include the Internet unsecured character and the difficulty of application incorporation across business allies. Supply chain cooperation has been challenging to implement in reality due to an extreme dependency on technology. Lack of commitment to comprehend what information has to be exchanged and, more importantly, how business practices must remain adjusted when new technology is installed.

Trust Teamwork, on the other hand, is established on the efforts of trust and commitment, as well as cooperative standards, and it needs firms to overcome their natural repugnance to unveiling commercial secrets to other partners. Confidence in virtual buying relies on the guarantee that companies will uphold their promises and tasks, without engaging in deception toward the customer [28]. We feel that consumers in emerging nations have little income. They are apprehensive regarding virtual fee and they are hesitant in using digital market since they are not used to paying online. As a result, if consumers are skeptical of electronic transactions and virtual purchases, they might be hesitant engaging in electronic-commerce.

Perceived Risk According study, the risk in digital market is the greatest important hindrance stopping customers from engaging in e-commerce [20]. One of the main risks to an e-commerce business is stolen credit card information or unauthenticated transactions [24, 28]. Consumer frustration is the idea that using mobile payments could expose them to private data leakages or cash fatalities. Recent digital payment research has found a negative association between user behavior and perceived risk [28].

Strategy According to early adopters, technology adoption and transition into virtual integration require a well-planned approach. Wigand [68] has stressed the significance of good organizational fitting and configuration for information technology adoption. He has made it very evident that effective coordination between strategy and technologies rather than information technology is what adds value to a company. Before beginning the implementation, it is critical to identify the right partners, the suitable number of allies, and the skill and knowledge to be used. Therefore, investing time and money in the early stages of strategy formation is vital. Launching a defined e-supply chain roadmap and plan after analyzing SCM opportunities give the company the chance to navigate itself toward a successful adoption and boosted performance.

11.4.3 Benefits of Adopting E-SCM

E-SCM has proven its importance by improving operations, enlightening subcontracting, increasing profits, taming customer satisfaction, producing quality results, addressing competitive stresses, increasing globalization, increasing importance of e-commerce, and increasing profits despite the difficulties [25]. The concept of E-SCM empowers firms to thrive in the ever-changing global market by providing real-time information and fostering collaboration among trading partners. Key success factors of E-SCM include fast and cost-effective communication as well as effective collaboration with customers and suppliers. One of an organization's most vital resources is the ability to add value for the consumer [6].

11.5 Conclusion and Future Work

In this review, we identified and analyzed the main determinants of e-commerce adoption, the barriers to successful e-supply chain deployment, and the opportunities of implementing e-supply chains in developing countries. When discussing the main determinants of e-commerce implementation in emerging nations, we categorized them under two macro groups: organizational E-readiness and external E-readiness. Based on the review, organizational E-readiness elements are awareness, resources, commitment, and governance. Furthermore, external E-readiness factors include government E-readiness, market force readiness, supporting industry readiness, and curiosity. Focusing on the barriers that organizations in developing countries are facing in operating e-supply chains, our findings demonstrate that cultural and technical challenges, trust, perceived risk, strategy, organizational, technological, and social influence related were identified. Despite the difficulties, E-SCM has demonstrated its value by enhancing operations, enlightening subcontracting, increasing profits, improving customer pleasure, producing quality results, addressing competitive pressures, increasing globalization, increasing the importance of e-commerce, and increasing profits.

Several potential avenues for further research have been identified, including: Empirical studies on the impact of the Internet on various supply chain management

(SCM) processes, such as reverse logistics and demand management. These areas have received limited attention from researchers thus far and warrant further investigation. Increased emphasis on empirical studies exploring the implementation of the Internet in the product development process. This area presents opportunities to examine the benefits, challenges, and best practices associated with integrating Internet technologies into product development activities. How to make further implementation of e-commerce needs support from governments to prepare infrastructures so that it can make the diffusion of e-commerce effective. Investigating the benefits and barriers of B2B procurement, and to analyze how to strategically segment e-procurement strategies research should concentrate on the use of e-business within the supply chains of small and medium enterprises. An additional promising area for further research involves investigating the specific circumstances and conditions under which various Internet-enabled supply chain collaboration tools have a more significant impact on performance.

References

1. Agrawal, P., & Narain, R. (2018). Digital supply chain management: An overview. *IOP Conference Series: Materials Science and Engineering, 455*(1), 012074.
2. Alyoubi, A. A. (2015). E-commerce in developing countries and how to develop them during the introduction of modern systems. *Procedia Computer Science, 65*(Iccmit), 479–483. https://doi.org/10.1016/j.procs.2015.09.127
3. Ammirato, S., et al. (2020). A systematic literature review of revenue management in passenger transportation. *Measuring Business Excellence, 24*(2), 223–242.
4. Antoni, Darius. 2019. "E-supply chain management value concept for the palm oil industry."
5. Au, Y. A., & Kauffman, R. J. (2001). Should we wait? Network externalities and electronic billing adoption. *Proceedings of the Hawaii International Conference on System Sciences, 00*(2), 237.
6. Barlow, A. (2007). *E-supply chains: Understanding current and future opportunities and barriers* (Vol. 6, pp. 286–298).
7. Büyüközkan, G., & Göçer, F. (2018). Digital supply chain: Literature review and a proposed framework for future research. *Computers in Industry, 97*, 157–177. https://doi.org/10.1016/j.compind.2018.02.010
8. Chou, D. C., Tan, X., & Yen, D. C. (2004). Web technology and supply chain management. *Information Management and Computer Security, 12*(4), 338–349.
9. Chwelos, P., Benbasat, I., & Deter, A. S. (2014). E-business adoption by travel agencies: Prime candidates for mobile e-business. *Information Systems Research, 12*(3), 304–321.
10. Concordia, C. O., & Mbarika, V. W. A. (2008). *E-business in developing countries: How much does culture really matter?*" (August).
11. Cook, D. J., Mulrow, C. D., Brian Haynes, R., & Mcmaster, F. (1997). Systematic reviews: Synthesis of best evidence for clinical decisions. *Annals of Internal Medicine, 126*, 376–380. http://annals.org/
12. Dahabiyeh, L., Najjar, M. S., & Agrawal, D. (2021). When ignorance is bliss : The role of curiosity in online games adoption ☆. *Entertainment Computing, 37*(July 2020), 100398. https://doi.org/10.1016/j.entcom.2020.100398
13. Daniel, E. (2002, September). *An exploratory comparison of electronic commerce adoption in large and small enterprises.*

14. Fintz, E. C., & Julia, S. C. (2002). *Small businesses' acceptance and adoption of e-commerce in the western-cape province of South-Africa* (pp. 1–13).
15. Galliers, R. D., et al. (1998, November). *Information systems and culture: Applying' stages of growth' concepts to development administration* (p. 1102).
16. Ghobakhloo, M. (2018). The future of manufacturing industry: A strategic roadmap toward industry 4.0. *Journal of Manufacturing Technology Management, 29*(6), 910–936.
17. Goode, S., & Stevens, K. (2000). An analysis of the business characteristics of adopters and non-adopters of world wide web technology. *Information Technology and Management, 1*, 129–154.
18. Goyal, K., & Chauhan, A. (2022). Emerging trends of e-commerce perspective in developing countries. *Journal of Positive School Psychology, 6*(4), 7090–7096.
19. Grimaldi, M., Corvello, V., De Mauro, A., & Scarmozzino, E. (2017). A systematic literature review on intangible assets and open innovation. *Knowledge Management Research and Practice, 15*(1), 90–100.
20. Hew, J. J., et al. (2017). The age of Mobile social commerce: An artificial neural network analysis on its resistances. *Technological Forecasting and Social Change, 144*(October), 311–324.
21. Hill, K. M., Fombelle, P. W., & Sirianni, N. J. (2016). Shopping under the in Fl Uence of curiosity : How retailers use mystery to drive purchase motivation ☆. *Journal of Business Research, 69*(3), 1028–1034. https://doi.org/10.1016/j.jbusres.2015.08.015
22. House, M., & Montealegre, R. (2010). Implications of electronic commerce for managers in less-developed countries. *Information Technology for Development, 7*(March 2015), 37–41.
23. Humphrey, J., et al. (2003). The reality of e-commerce with developing countries prepared by. In *London school of economics and political science and institute of development studies*. University of Sussex. (March): 7, 8. http://www.sri.com/sites/default/files/brochures/dec-05.pdf
24. Işık, C. (2013). The importance of creating a competitive advantage and investing in information Technology for Modern Economies: An ARDL test approach from Turkey. *Journal of the Knowledge Economy, 4*(4), 387–405.
25. Ivanov, D., & Sokolov. (2019). The impact of digital technology and industry 4.0 on the ripple effect and supply chain risk analytics. *International Journal of Production Research, 57*, 829.
26. Kabanda, Salah, and Ziyaad Parker. 2011. "Barriers towards E-commerce institutionalization in South Africa."
27. Kabango, C. M., & Asa, A. R. (2015). Factors Influencing E-commerce development: Implications for the developing countries. *International Journal of Innovation and Economic Development, 1*(1), 59–66.
28. Kalinic, Z., Marinkovic, V., Molinillo, S., & Liébana-Cabanillas, F. (2019). A multi-analytical approach to peer-to-peer mobile payment acceptance prediction. *Journal of Retailing and Consumer Services, 49*(December 2018), 143–153. https://doi.org/10.1016/j.jretconser.2019.03.016
29. Khalifa, N., Elghany, M. A., & Elghany, M. A. (2021). Exploratory research on digitalization transformation practices within supply chain management context in developing countries specifically Egypt in the MENA region. *Cogent Business and Management, 8*(1). https://doi.org/10.1080/23311975.2021.1965459
30. Khan, S. A., Liang, Y., & Shahzad, S. (2014). Adoption of electronic supply chain management and E-commerce by small and medium enterprises and their performance: A survey of SMEs in Pakistan. *American Journal of Industrial and Business Management, 04*(09), 433–441.
31. Koumas, M., Dossou, P.-E., & Didier, J.-Y. (2021). Digital transformation of small and medium sized enterprises production manufacturing. *Journal of Software Engineering and Applications, 14*(12), 607–630.
32. Kshetri, N. (2007). Barriers to E-commerce and competitive business models in developing countries: A case study. *Electronic Commerce Research and Applications, 6*(4), 443–452.
33. Kunrath, Tobias Leonardo, Aline Dresch, and Douglas Rafael Veit. 2023. "Supply Chain Management and Industry 4. 0 : A Theoretical Approach." 20(1): 1–24.

34. Lancaster, S., Yen, D. C., & Cheng Yuan, K. (2006). E-supply chain management: An evaluation of current web initiatives. *Information Management and Computer Security, 14*(2), 167–184.
35. Lankford, W. M. (2004). Supply chain management and the internet. *Online Information Review, 28*(4), 301–305.
36. Lawrence, J. E., & Tar, U. A. (2011). Persistent barriers to E-commerce in developing countries. *Journal of Global Information Management, 19*(3), 30–44.
37. Lawrence, K. (2002). *Factors inhibiting the collaborative adoption of electronic commerce amang Australian SMEs* (pp. 178–180).
38. Li, B., & Li, Y. (2017). Internet of things drives supply chain innovation : A research framework. *The International Journal of Organizational Innovations, 9*(January 2017), 71–93.
39. Majeed, M. A. A., & Rupasinghe, T. D. (2017). Internet of things (IoT) embedded future supply chains for industry 4.0: An assessment from an ERP-based fashion apparel and footwear industry. *International Journal of Supply Chain Management, 6*(1), 25–40.
40. Martins, F. D. C., Simon, A. T., & De Campos, R. S. (2020). Supply chain 4.0 challenges. *Scielo Brazil, 27*(3), 1–22.
41. Molla, A., & Licker, P. S. (2005). Perceived E-readiness factors in e-commerce adoption: An empirical investigation in a developing country. *International Journal of Electronic Commerce, 10*(1), 83–110.
42. Moodley, S. (2003). E-commerce and export markets: Small furniture producers in South Africa *. *Conceptual Framework, 41*(3), 317–324.
43. Mirchandani, D. A., & Motwani, J. (2001). Understanding small business electronic commerce adoption: An emperical analysis. *Journal of Computer Information Systems, 41*(3), 70–73.
44. Okoli, C. (2015). A guide to conducting a standalone systematic literature review. *Communications of the Association for Information Systems, 37*(1), 879–910.
45. Oxley, J. E., & Yeung, B. (2001a). E-commerce readiness: Institutional environment and international competitiveness. *Journal of International Business Studies, 32*(4), 705–723.
46. Oxley, J. E., & Yeung, B. (2001b). E-commerce readiness : Environment competitiveness. *Journal of International Business Studies, 32*(4), 705–723.
47. Palacios, J. J. (2006). *The development of e-commerce in Mexico: A business-led passing boom or a step toward the emergence of a digital economy?*" (October 2014) (pp. 37–41).
48. Porter, M. E. (2000). The competitive advantage of nations."
49. Pulevska-Ivanovska, L., & Kaleshovska, N. (2013). Implementation of E-supply chain management. *TEM Journal Journal, 2*(4), 314–322. www.temjournal.com
50. Rodriguez, L., & Cunha, D. (2018). Impacts of big data analytics and absorptive capacity on sustainable supply chain innovation: A conceptual framework. *LogForum, 14*(2), 151–161. https://doi.org/10.17270/J.LOG.2018.267
51. Davies, R. (2015). Digitalisation for productivity and growth. *European Parliamentary Research Service (September).*
52. Ross, D. F. (2011). *Introduction to supply chain management technologies: Series on resource management*. Routledge.
53. Rubmann, M., et al. (2015). *Industry 4.0: World economic forum* (pp. 1–20). The Boston Consulting Group.
54. Roger, S., & Antonella, Z. (2017). Industry 4.0, global value chains and international business. *Multinational Business Review, 25*, 174–184. http://sro.sussex.ac.uk
55. Schmidt, M., Maier, J. T., & Härtel, L. (2020). Data based root cause analysis for improving logistic key performance indicators of a Company's internal supply chain. *Procedia CIRP, 86*, 276–281. https://doi.org/10.1016/j.procir.2020.01.023
56. Seok, K., et al. (2016). Critical failure factors that discourage of electronic commerce the growth. *International Journal of Electronic, 4*(2), 25–43.
57. Seyedghorban, Z., Tahernejad, H., Meriton, R., & Graham, G. (2020). Supply chain digitalization: Past, present and future. *Production Planning and Control, 31*(2–3), 96–114.

58. Singh, N., & Sahu, O. (2022). Feasibility assessment for E-commerce: A data collection from developing country (Ethiopia). *MethodsX*. https://doi.org/10.1016/j.mex.2022.101639
59. Szozda, N. (2020). *LogForum*. (2015).
60. Tabor, S. W. et al. (2000). *Electronic commerce adoption & success: A study of organizational factors and influences*.
61. Taghipour, A., Murat, S., & Huang, P. (2021). E-supply chain management: A review. *International Journal of e-Education, e-Business, e-Management and e-Learning, 11*(2), 51–61. https://doi.org/10.17706/ijeeee.2021.11.2.51-61
62. Tjahjono, B., Esplugues, C., Ares, C., & Pelaez, G. (2017). Ng engineering society international conference 2017, MESIC 201. *Procedia Manufacturing, 13*, 1175–1182. https://doi.org/10.1016/j.promfg.2017.09.191
63. Tranfield, D., Denyer, D., & Smart, P. (2003). Towards a methodology for developing evidence-informed management knowledge by means of systematic review* introduction: The need for an evidence-informed approach. *British Journal of Management, 14*, 207–222.
64. Travicaa, B. (2014, November). Diffusion of electronic commerce in developing countries: The Case of Costa Rica. *Journal of Global Information Technology*, 37–41.
65. Wang, H.-d., Zheng, C.-f., & Xu, X. (2022). An AMOS model for examining the factors influencing the development of china cross-border E-commerce comprehensive pilot areas. *Discrete Dynamics in Nature and Society, 2022*, 1–7.
66. Wang, S., & Cheung, W. (2004). E-business adoption by travel agencies: Prime candidates for mobile e-Business. *International Journal of Electronic Commerce, 8*(3), 43–63.
67. Weyer, S., Schmitt, M., Ohmer, M., & Gorecky, D. (2015). Towards industry 4.0 - standardization as the crucial challenge for highly modular, multi-vendor production systems. *IFAC-PapersOnLine, 28*(3), 579–584. https://doi.org/10.1016/j.ifacol.2015.06.143
68. Wigand, R. T. (2006). The information society: An electronic commerce: Definition, theory, and context. *The Information Society: AnInternational Journal*, 37–41. (January 2015).
69. Yusuf, S. (2022). How does COVID-19 Impact the Supply Chain Management of the E-Commerce Industry in Indonesia? *International Journal of Business and Technology Management, 4*(3), 70–88.
70. Zeinab, M., & El, G. (2005). *The impact of E-commerce on developed and developing countries* (pp. 193–202).

Chapter 12
Utilization, Nutritional, and Microbiological Quality of *Awaze*: Fermented Domestic Spicy Red Pepper Paste of North Western Part of Amhara, Ethiopia

Biadge Kefale Anteneh, Mulugeta Admasu, Solomon Workneh Fanta, and Solomon Abate

12.1 Introduction

Ethiopian *Awaze* is an ethnic food made with the fermentation of red pepper paste. Spicy red pepper paste (*Awaze*) or local name *Awaze* or *Dikuse*, a hot condiment made with more than 50 types of spices, is used as Ethiopian-style ketchup, served on the side to up the heat quotient and to prepare stews, enhancing the taste of food, seasoning, and for aiding the digestive system. There are many traditional red pepper paste in different parts of the world. Such traditional red pepper paste as *Awaze* [12] of Ethiopia, Gochujang of Korea [21], Doubanjiang of China [15, 16] are used as flavor and taste enhancers in many traditional dishes. Fermented foods contain microorganisms that convert original food into organic acids and other compounds beneficial for health such as, decrease antinutrients, increase vitamin, antioxidant activity, and antidiabetic activity [25].

B. K. Anteneh (✉)
Faculty of Chemical and Food Engineering, Bahir Dar Institute of Technology, University, Bahir Dar, Ethiopia

Ethiopian Institute of Agricultural Research, Holeta Agricultural Research Center, Food Science and Nutrition Research, Addis Ababa, Ethiopia

M. Admasu · S. W. Fanta
Faculty of Chemical and Food Engineering, Bahir Dar Institute of Technology, Bahir Dar University, Bahir Dar, Ethiopia

S. Abate
Ethiopian Institute of Agricultural Research, Head Quarter, Food Science and Nutrition Research, Addis Ababa, Ethiopia

S. W. Fanta et al. (eds.), *Sustainable Development Research in Manufacturing, Process Engineering, Green Infrastructure, and Water Resources*, Green Energy and Technology, https://doi.org/10.1007/978-3-031-77339-6_12

Spices such as red pepper, red onion, ginger, garlic, thyme, coriander, long pepper, cinnamon, *kemune*, cardamom, clove, fenugreek, nutmeg, basil, white cumin, black cumin, white cumin, rue, rosemary, and ground pepper are used for *Awaze* preparation at household level. Red pepper, red onion, ginger, and garlic are the main ingredients for *Awaze* preparation at household level. These spices are aromatic or fragrant products used to flavor foods or beverages [19]. Spices are used as flavoring agents and improve the taste and aroma of food [27]. Spices also have health benefits such as antioxidant, antibiotic, antiviral, anticoagulant, anticarcinogenic, and anti-inflammatory agents [17].

Red pepper is a cash crop in the Amhara region and is used as the main ingredient for *Awaze* preparation at household level. 68% of the produced red pepper in the Amhara region is sold in the market. The share of the region in the total production of red pepper in the country is 25%, which is next to the southern region. Bure, Dembecha, Fogera, Anedede, and Gozamine districts are among the potential districts for red pepper and utilize *Awaze* paste.

Previous studies on laboratory-made and commercial *Awaze* quality characteristics reported particularly on the microbial and some physicochemical qualities of *Awaze*, and the fermentation of *Awaze* for microbial quality [12]. The behavior of *E.coli O157:H7* during fermentation of *Awaze* at different temperatures was reported by [31]. Probiotic properties of lactic acid bacteria isolated from *Awaze* were investigated [30] and evaluated in vitro [7]. However, there is no study on the utilization and quality profiling of *Awaze*. Therefore, the main objective of this study was to assess and determine the utilization and quality characteristics of *Awaze* collected from different locations.

12.2 Materials and Methods

12.2.1 Study Site

The study was done in five selected districts of the Amhara region from March to August 2022. These districts were selected based on high production area for red pepper, garlic, and red onion and the wide utilization area of *Awaze*. The five districts were Andeded, Gozamin, Dembecha, Bure, and Fogera, as shown in Fig. 12.1.

12.2.2 Study Design and Sample Size

A cross-sectional study design was employed to conduct the study on the utilization of *Awaze* in the North West Amhara. The inclusion criteria included women respondents aged 20 to 60 years living in the household of interviewed women and men in rural and urban areas (spice merchants). For the sample size determination for large

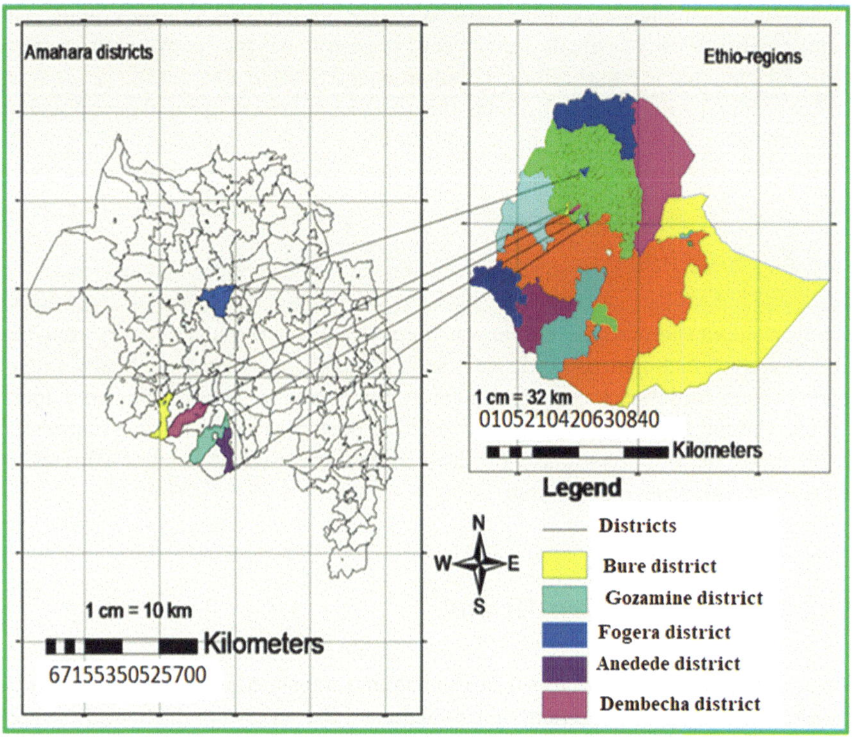

Fig. 12.1 Study area for traditional *Awaze* in Amhara region. (Drawn by Biadge Kefale)

Table 12.1 The total population size of the five districts (study districts)

Districts	Zone	Region	Population
1. Aneded	East Gojjam	Amahara	110,429
2.Gozamin	East Gojjam	Amahara	160,853
3.Dembecha	West Gojjam	Amahara	171,133
4. Bure	West Gojjam	Amahara	195,349
5. Fogera	South Gonder	Amahara	296,242
Total			934,006

Source: Ethiopian statistics service (web)

population, Cochran's formula (1963:75) (Table 12.1) was used because the study area population was large and the population number was affected by different factors from time to time such as birth, death, and migration.

Sample size calculation

$$n = \frac{Z^2 pq}{e^2} \tag{12.1}$$

where n is the sample size, Z^2 is the abscissa of the normal curve that cuts off an area α at the tails at 95%, *e* is the desired level of precision, *p* is the estimated proportion of an attribute that is present in the population, and *q* is 1-*p*.

According to the above formula, the total sample size from the population was 196.

12.2.3 Sampling

Simple random sampling technique was used. Upon arrival in each survey area, the surveyor consulted community leaders such as kebele managers and health extension workers to define the geographical confines which corresponded to a local village. Enumerators used respondents of 20 to 60 years of age for each household respondents and only one household level survey was used. A total of 40 respondents were used from each district.

12.2.4 Data Collection Method

The questionnaire was developed in English and translated to Amharic and back-translated to English. The enumerators conducted the surveys in the household and market. The questions were developed in Google form in smart mobile phones, filled by enumerators, converted into Excel spreadsheet, and exported to SPSS for data analysis.

12.2.5 Physicochemical Analysis

Home-made domestic *Awaze* samples collected from five districts in Amhara region were used for the study (Fig. 12.2).

12.2.5.1 Sample Collection

A total of 15 *Awaze* sample was collected from five districts from each 3 sample from different households were collected for this study. The sample before collection was stored in *sheklla pot* and for 4–6 months according to the respondent's response. The collected samples were stored in a glass bottle in a refrigerator until microbial and physicochemical analysis.

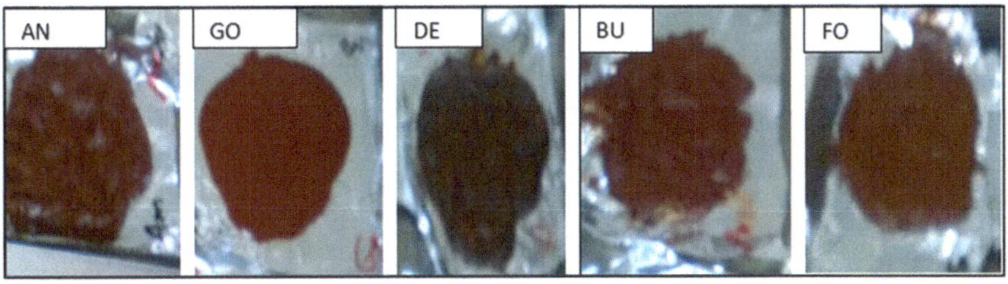

Fig. 12.2 *Awaze* samples collected from Aneded (AN), Gozamin (GO), Dembecha (DE), Bure (BU), and Fogera (FO) districts in Amhara region

12.2.5.2 Antioxidant Activity

Antioxidant activity was evaluated by measuring the radical scavenging effect of collected *Awaze* samples and the methanolic extracts toward the 2,2-diphenyl-1-picrylhydracyl (DPPH) as reported previously [4, 26]. Five mL of a 0.1 mM methanol solution of DPPH were added to 20, 40, 60, 80, and 100 μL of concentrations of methanol extracts of *Awaze* samples. The absorbance of the sample was recorded at 517 nm in a spectrophotometer (Shimadzu UV-vis spectrophotometer 1900). Free radical scavenging activity was expressed as inhibition percentage (%).

12.2.5.3 Color Measurement (L*, a*,b*)

The L*, a*, and b* color values of collected *Awaze* samples were measured by Hunter colorimeter instrument following the method [15, 16].

12.2.5.4 pH Value

The pH content of collected *Awaze* was obtained following the method [6]. 45 ml of distilled water was added to 5 g of sample and homogenized for one minute. The solution was kept at room temperature for one hour, and the pH of the supernatant was measured with a pH meter (Mettler Toledo, China).

12.2.5.5 Titratable Acidity

Titratable acidity (TA) of collected *Awaze* samples was determined titrimetrically according to the method [6]. For (5 g) sample, 100 ml of deionized water was added, then stirred gently, and allowed the mixture to stand for one hour. Pipette 20 ml into a 100 ml flask and add 0.5 ml phenolphthalein and titrate with 0.1 N NaOH until a pink color appears for 30 s.

12.2.5.6 Total Soluble Solids

Total soluble solids (TSS) of collected *Awaze* samples were obtained according to the association of official analytical chemists (AOAC) method [1]. 50 mL of distilled water was added to 10 g of the sample and subsequently, the mixture was filtered. One milliliter of aliquot was placed in a digital refractometer.

12.2.5.7 Macronutrients

Crude protein and ash content were quantified using the AOAC method [2]. Ash content of the collected samples was pre-weighted in a crucible and combusted in muffle furnace at 55 °C for 4 h. Then, after cooling the crucibles were weighted and the ash content was calculated as weight difference. Crude fat was determined using nuclear magnetic resonance spectroscopy (NMR) [10]. Crude fiber, Carbohydrate(CHO) and energy were quantified according to the method [29].

$$\text{Carbohydrate content}\left(\text{CHO}\right)=\left[100-\left(\text{MC}+\text{PC}+\text{Fat}+\text{fiber}+\text{ash}\right)\right] \tag{12.2}$$

$$\text{E}\left(\text{Kcal.}\right)=\text{CPx4}+\text{CHOx4}+\text{fatx9} \tag{12.3}$$

12.2.5.8 Micronutrients

The calcium, potassium, magnesium, iron, and zinc contents of the samples were analyzed following the AOAC method [2]. 0.5 g *Awaze* sample was digested by using a muffle furnace at 550 °C for 3 h until the residue ash became white. Then, the sample was digested by using concentrated HCl. The digested sample was diluted up to 100 ml for mineral analysis. Calcium, potassium, magnesium, iron, and zinc contents were determined using an atomic absorption spectrophotometer (Model: Agilent Series, America).

12.2.5.9 Microbial Profile

For the detection of microorganisms, standard methods of AOAC [2] were used. 25 g of *Awaze* sample was homogenized with 225 ml of sterile saline containing 0.1% peptone water and 0.85% NaCl. Serial dilution was performed and diluted samples (1 ml) were applied to the surface of the media using the poring technique. *Total bacterial count* was estimated by applying plate count agar and incubating at 30 °C for 48 h. Potato dextrose agar (PDA) was used to isolate *yeast and mold* and incubate at 25 °C for 72 h. *Lactic acid bacteria* was estimated by using de man, rogosa and sharpe agar (MRS) agar and incubated aerobically at 30 °C for 48 h.

Coliform was estimated by pouring violet red bile agar (VRBA) agar and incubating at 35 °C for 24 h. The count was calculated as a log colony forming unit per gram.

12.2.6 Data Analysis

Descriptive statics and inferential statics were used to summarize the collected data. Descriptive analysis is given as proportions (%). Bivariate analysis of *Awaze* consumption with independent variables was conducted using Pearson chi-square for independent categorical variables. Multivariable logistic regression was used to calculate the adjusted odds ratio (OR) and 95% confidence interval. One adjusted model was presented in this study. This model includes respondents' residence, age group, education, and occupation. The model-building strategy was similar to that outlined by [8]. The initial selection of variables was based on a priori thought of their availability in the data. For physicochemical and microbial analysis, ANOVA and Duncan multiple range test were employed by using SPSS software (IBM SPSS Version 24) to explore the significant difference between sample collection districts with a 95% confidence interval at $p < 0.05$.

12.3 Result and Discussion

12.3.1 Socio-demographic Characteristics of Households

Among the 200 sample surveys, 93% were females and 7% were male. Above three-fourths (84.5%) of the respondents were in the age groups of 25–50 years. Above two-thirds (76%) of the study participants were from a rural area. The majority (87%) of the respondents were married. With regard to the educational background, 56% of the respondents were at the primary level of education. Two-thirds (71%) of the household relyied on agricultural practices as their income source and 27% of the respondents were spice merchants who participated in the study. The sampled households who consumed the product were respondents living in rural (64%) and farmer respondents (60%) which are greater than urban (20%) and merchants (19%). Gender, age, and marital status had no significant effect on the consumption of *Awaze*, while respondents' residence, education, and occupation significantly ($p = 0.000$) affected the consumption of *Awaze* (Table 12.2).

Table 12.2 Socio-demographic characteristics of household and *Awaze* consumption of the study participants from five districts in Amhara region, Ethiopia, 2022 ($n = 200$)

Variable	Consumption of *Awaze*			
	Yes	No	*P*- value	Total
	n(row%)	n(row%)		n(col%)
Gender			0.147	
Male	10(5)	4(2)		14(7)
Female	126(63)	60(30)		186(93)
Age			0.059	
25–40 year	59(29.5)	32(16)		91(45.5)
41–50 year	52(26)	26(13)		78(39)
51–60 year	25(12.5)	6(3)		31(15.5)
Marital status			0.268	
Married	121(60.5)	52(26.5)		173(87.5)
Single	15(7.5)	11(5.5)		26(13)
Education			0.00	
None	1(0.5)	1(0.5)		2(1)
Primary	88(44)	24(12)		112(56)
Secondary	45(22.5)	32(16)		77(33.5)
Higher	2(1)	7(3.5)		9(4.5)
Respondent residence			0.00	
Rural	128(64)	24(12)		152(76)
Urban	8(4)	40(20)		48(24)
Occupation			0.00	
Farmer	120(60)	22(11)		142(71)
Merchant	15(7.5)	39(19)		54(27)
Casual work	1(0.5)	3(1.5)		4(2)

12.3.2 *Utilization of* Awaze

Respondents (99%) had knowledge of the *Awaze* and the product is available (68%) for the respondents in each household. A high portion of respondents daily (58.5%), twice a week (8.5%), and once a month (33%) consumed *Awaze* throughout the year. Among the respondents, 97% prepare the product and about 41% of the participants can prepare it within one week. Almost two-thirds (68%) of the respondents prefer the dry season to prepare *Awaze,* and this might be because this season is preferable to get the raw material and dry the spices with sunlight in all surveyed districts. The majority (94%) of the participants responded that the use of the product is for nutritional (94%) and respondents (67%) agreed that the cost of the product is 300–350 birr per kilogram to prepare *Awaze* (Table 12.3). Spice-based food like *Awaze* was adequate for all age groups of the population because spices are natural antioxidants with their health benefits such as cardiovascular diseases, cancer, and diabetes [13]. *Awaze* is a hot condiment made with different spices and is

Table 12.3 Availability, processing method, and utilization of *Awaze* from the surveyed districts

Variable	Category	Respondent's response (percentage %)
1. Knowledge of *awaze*	Yes	198(99)
	No	2(1)
2. Availability of the product in each household	Yes	136(68)
	No	64(32)
3. Frequency of eating *awaze*	Daily	117(58.5)
	Twice a week	17(8.5)
	Once a month	66(33)
4. Preparation of the product by household	Prepare	194(97)
	Not prepare	6(3)
5. Time taken to prepare the product	One week	82(41)
	Two week	68(34)
	Three week	50(25)
6. Season to prepare the product	Dry season	136(68)
	Wet season	64(32)
7. Advantage of *awaze*	Medicinal	12(6)
	Nutritional	188(94)
8. Used as alternative source of income for farmers if it is commercialized	Yes	200(100)
	No	0
9. If the product is prepared, packed, and commercialized more	Yes	200(100)
	No	0
10. Cost of product development	300–350 birr per kg	134(67)
	350–400 birr per kg	66(33)
11. Way of utilization	With stew and *Injera*	51(25.5)
	With meat	40(20)
	With bread	10(5)
	With boiled potato	37(18.5)
	With all the above food items	62(31)

used as an Ethiopian-style ketchup served on the side to up the heat quotient of stews and soups and used to add heat to any dish.

The study respondents (58.5%) reported that the product is more prevalent in rural areas to prepare stew with meat, boiled potatoes, and Injera during coffee ceremonies in the household than in urban areas. This urban-rural difference in utilization of *Awaze* is because the use of spicy red pepper mix is prevalent in the urban area rather than *Awaze*. The respondents also replied about the consumption of *Awaze* in different ways, such as with meat (20%), for stew preparation (25.5%), with boiled potato (18.5%), and with bread (5%), utilized as part of meal for the family, and the consumption of the product in this form has also got vast recognition for its nutritional and medicinal value (Table 12.4). In this regard, 94% of

Table 12.4 Adjusted odds ratios for consumption of *Awaze* by age, respondents' residence, education, and occupation ($n = 200$)

Variables	Fully adjusted or (95% CI)	*P*- value
Age		0.056
25–40	0.426(0.159, 1.147)	
41–50	0.563(0.204, 1.553)	
51–60		
Respondents' residence		0.000
Rural	0.038(0.016, 0.090)	
Urban		
Education		0.001
None	3(0.122, 73.64)	
Primary	11.125(2.11, 58.66)	
Secondary	4.0(0.758, 21.095)	
Higher		
Occupation		0.000
Farmer	10.81(0.940, 124.51)	
Merchant	0.8(0.068, 9.45)	
Casual work		

respondents used *Awaze* for nutritional purposes and 6% for medicinal values of the product (Table 12.3).

12.3.3 Adjusted Associations

In the adjusted model, associations with *Awaze* consumption were retained with respondents' residence, education, and occupation (Table 12.4). Gender and marital status were not significantly associated with the consumption of *Awaze*, so they were not included in the final model.

Participants in the age group of 41–50 years were 43.7% less likely to have consumed the *Awaze* the day before the survey than the other age group of 51–60 years (Table 12.5). Respondents from urban were 96.2% less likely to have consumed *Awaze* the day before the survey than the respondents from rural households. Respondents at a primary education level were 11.12 times as likely to have consumed the product before the survey, compared to those at a higher level of education. Respondents from agriculturists were 10.81 times as likely to have consumed *Awaze*, compared to casual work households, while respondents from merchants were 20% less likely to have consumed the product before the survey and during the previous year, based on the adjusted model (Table 12.4).

In this study, the findings indicated that above two-thirds of the population consumed *Awaze* in the 24 h preceding the interview. The result confirmed the hypothesis that the *Awaze* consumption of the households in the rural areas and urban areas

Table 12.5 DPPH free radical scavenging activity of the collected samples of *Awaze* at different concentrations

	DPPH free radical scavenging activity (%) at different concentrations				
Sample Location	20(μl)	40(μl)	60(μl)	80(μl)	100(μl)
1. AN	11.16 ± 0.44^{b}	50.72 ± 0.00^{b}	56.18 ± 1.39^{a}	53.87 ± 0.16^{c}	59.78 ± 0.37^{a}
2. GO	39.9 ± 0.71^{a}	51.44 ± 0.053^{a}	51.1 ± 0.32^{b}	52.73 ± 0.16^{d}	61.08 ± 1.98^{a}
3. DE	42.25 ± 0.59^{a}	46.77 ± 0.00^{c}	55.35 ± 1.5^{ab}	59.22 ± 0.107^{a}	52.61 ± 1.07^{c}
4. BU	43.84 ± 0.91^{a}	46.43 ± 0.05^{d}	53.22 ± 1.82^{ab}	55.39 ± 0.05^{b}	47.72 ± 0.05^{d}
5. FO	38.76 ± 0.053^{a}	26.8 ± 0.12^{e}	53.98 ± 0.68^{ab}	55.46 ± 0.26^{b}	56.79 ± 0.10^{b}

The data were expressed as mean ± standard deviation. The different values in the same column indicate significant differences ($p < 0.05$)

studied is associated with the household and community characteristics such as age, respondents' residence, education, and occupation. The factors associated with the consumption of *Awaze* from the five districts in the Amhara region of Ethiopia were respondents' residence, education, and occupation. Respondents from urban areas were less likely to consume the *Awaze* compared to rural areas. Agricultural households were more likely to have consumed the product than merchants' households. Gender and marital status were not significantly associated with *Awaze* consumption in the adjusted model analysis.

12.3.4 Materials Used by Women for Awaze *Storage*

Regarding the storage method of *Awaze*, 24.5%, 44.5%, and 31% of the participants store the product in *Qelle (□□)*, shekella pot *(□□□□)*, and plastic pot, respectively. The product was used up to 12 months without deterioration. The respondents indicated that the product's deterioration is known by showing characteristics of growth of yeast and mold on the surface, foam on the surface, and change of taste according to the respondent's response. The formation of foam on the surface of food is an indicator of mold development and deterioration [24].

12.3.5 Antioxidant Activity, Physicochemical, and Microbiological Quality of Collected Awaze *Samples*

In the present study, the antioxidant activity was measured through the evaluation of free radical scavenging activity (DPPH) collected from different districts (Table 12.5). There was a significant difference in the percentage of inhibition at $p < 0.05$ among the collection districts. Free DPPH radical scavenging activity

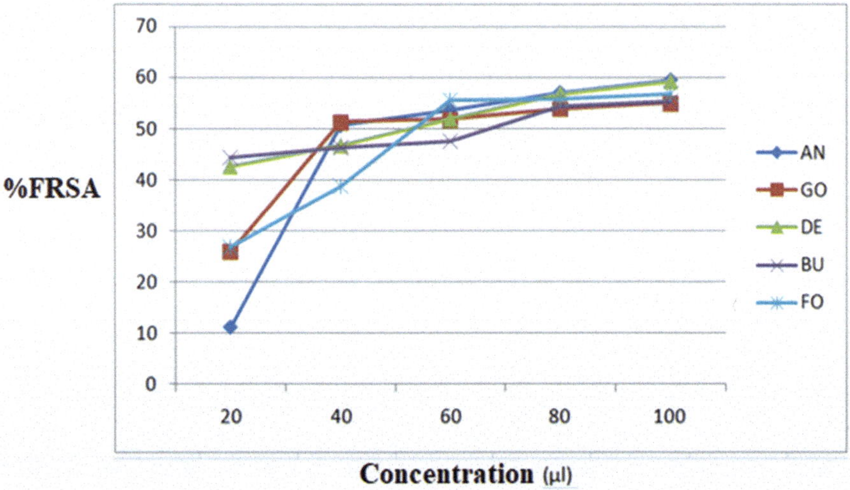

Fig. 12.3 Percentage of free radical scavenging of the methanolic extract of *Awaze* as per DPPH assay expressed as a function of concentration of the extract collected sample from Anedede, Gozamin, Dembecha, Bure, and Fogera

Table 12.6 IC_{50} value of DPPH value of collected *Awaze* sample

	IC_{50}(µg/mL)		
Sample location	60(µl)	80(µl)	100(µl)
1. AN	4.79 ± 0.24^{b}	8.97 ± 0.9^{d}	12.28 ± 0.94^{d}
2. GO	6.24 ± 1.24^{b}	9.8 ± 0.88^{d}	13.36 ± 0.52^{d}
3. DE	9.33 ± 1.53^{a}	17.53 ± 3.32^{c}	30.05 ± 2.06^{b}
4. BU	9.9 ± 0.98^{a}	22.95 ± 1.06^{b}	36 ± 2.02^{b}
5. FO	10.72 ± 0.42^{a}	28.97 ± 0.67^{a}	48.95 ± 3.18^{a}

The data were expressed as mean ± standard deviation. The different values in the same column indicate significant differences ($p < 0.05$)

ranged from 11.16% to 61.08% (% inhibition) in the *Awaze* collected samples from different districts. Therefore, we suggest that the percentage of inhibition of the paste samples is affected by samples from different districts. The various factors such as the ingredient type and storage time affect the chemical composition of the food [28].

As percentage of inhibition against concentration in Fig. 12.3 and IC_{50} values for collected samples for *Awaze* are presented in Table 12.6. According to the derived IC_{50} values (Fig. 12.3), the collected *Awaze* samples with the concentration of 60–100 µl exhibited IC_{50} values (4.79–48.98 µg/mL). Our findings agree with the previous report on IC_{50} values of 1.65–2224 µg/mL herbs and spices in Brazil [22], and also with coffee mangrove, tea mangrove, and long pepper powder IC_{50} values of 1.56–61.84 µg/mL [18].

Table 12.7 Color values of the collected samples of *Awaze* from five districts in the Amhara region, Ethiopia

Sample location	Color value				
	Lightness(L*)	Redness(a*)	Yellowness(b*)	Chroma(C*)	Hue angle(h*)
1. AN	18.25 ± 0.3^{a}	18.65 ± 0.77^{ab}	27.5 ± 1.27^{a}	33.22 ± 1.49^{a}	55.86 ± 0.12^{a}
2. GO	18.95 ± 0.49^{a}	19.35 ± 0.7^{a}	26.9 ± 1.12^{a}	33.14 ± 1.17^{a}	54.25 ± 1.10^{a}
3. DE	1.55 ± 0.35^{d}	2.5 ± 0.42^{d}	2.05 ± 0.07^{c}	3.23 ± 0.37^{c}	39.58 ± 1.8^{d}
4. BU	13.3 ± 1.13^{c}	16.75 ± 0.63^{c}	19.1 ± 1.27^{b}	25.4 ± 1.37^{b}	48.73 ± 0.70^{b}
5. FO	15.45 ± 0.63^{b}	17.2 ± 0.14^{bc}	25.1 ± 0.00^{a}	30.42 ± 0.07^{a}	55.58 ± 0.26^{a}

The data were expressed as mean ± standard deviation. The different values in the same column indicate significant differences ($p < 0.05$)

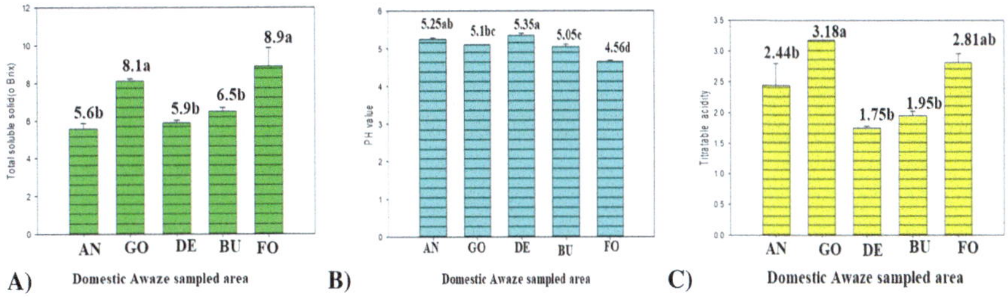

Fig. 12.4 Physicochemical property (**a**) total soluble solid (°Birix), (**b**) pH value, and (**c**) titratable acidity (%) of *Awaze* collected from five districts in Amhara region, Ethiopia

The color value of the collected *Awaze* sample was measured and presented in Table 12.7. The range of lightness (L*), redness(a*), yellowness(b*), chroma, and hue angle of *Awaze* products, were 1.55–18.95, 2.5–19.35, 2.05–27.5, 3.23–33.23 and 39.58–55.86, respectively. In the present study, the sample collected from Gozamin (GO) had the highest value of redness (19.35) while samples collected from Dembecha (DE) had the lowest value (2.5). This may be due to the limited amount of red pepper and red onion powder during preparation of *Awaze,* or the difference in storage method or processing method. In addition to the raw material, the difference in the color values for different *Awaze* products is associated with the processing practice and the microbial composition of the product [23].

The pH, titratable acidity, and total soluble solids of the collected *Awaze* sample are shown in Fig. 12.4. P^{H} is the prerequisite physicochemical parameter of fermented foods. Based on the P^{H} values, all the *Awaze* products were in the category of low acidic foods because the P^{H} content of the collected *Awaze* samples ranged from 4.65–5.35. Samples with pH value higher than 4.6 are considered as low acidic foods. The P^{H} values are dependent on the fermentation time of foods [14]. Total soluble solids for all pastes ranged from 5.6–8.9%. The mean value of total soluble solid measured in this investigation was in agreement with that of previously reported for ginger-garlic-chili paste [9]. The Titratable acidity of the collected *Awaze* sample exhibited acidity in the range of 1.75–3.18%. The difference may be due to the processing technique during household preparation and storage time of

Table 12.8 The macronutrient content in *Awaze* collected from five districts in Amhara region, Ethiopia

	Sample location				
Macronutrient	AN	GO	DE	BU	FO
Moisture content (%)	4.96 ± 0.09^{a}	4.37 ± 0.16^{b}	4.17 ± 0.16^{bc}	5.0 ± 0.24^{a}	3.69 ± 0.23^{c}
Protein content (%)	10.95 ± 1.34^{a}	9.83 ± 0.51^{ab}	6.18 ± 0.19^{b}	10.93 ± 0.98^{a}	9.27 ± 0.09^{ab}
Fat content (%)	9.15 ± 0.07^{b}	7.9 ± 0.28^{a}	5.7 ± 0.14^{c}	9.35 ± 0.21^{a}	8.35 ± 0.21^{b}
Ash content (%)	10.49 ± 0.23^{a}	7.39 ± 1.8^{b}	7.81 ± 0.31^{ab}	7.0 ± 0.46^{b}	9.05 ± 1.34^{ab}
Fiber content (%)	17.44 ± 0.10^{d}	18.03 ± 0.73^{cd}	25.85 ± 0.44^{a}	20.5 ± 0.86^{b}	19.13 ± 0.31^{bc}
Carbohydrate (%)	52.99 ± 1.03^{a}	47.53 ± 1.21^{a}	49.72 ± 0.24^{a}	52.79 ± 2.2^{a}	49.5 ± 1.37^{a}
Energy (calorie, kcal)	314.18 ± 0.60^{a}	320.31 ± 1.24^{a}	277.16 ± 3.04^{b}	316.7 ± 0.42^{a}	314.24 ± 0.9^{a}

The data were expressed as mean ± standard deviation. The different values in the same row indicate significant differences ($p < 0.05$)

the paste. The mean acidity of the collected sample was higher acidity value to than previously reported for hot red pepper paste [5].

In the collected *Awaze* products, the proximate content is presented in Table 12.8. The proximate content of *Awaze* samples was significantly different at $p < 0.05$ among the collected locations (districts). The moisture content, protein content, fat, ash, fiber, carbohydrate, and energy value for the product ranged from 3.69–5.0%, 6.18–10.95%, 5.7–9.35%, 7–10.49%, 17.44–25.85%, 47.53–52.99%, and 277.16–320.31% (kcal.), respectively. However, the current result was in agreement with the previous study of the Korean traditional red pepper paste (*gochujang)* products, indicating protein, ash, and carbohydrate content ranged from 3.4–9.3%, 5.1–21.1%, and 20.9–52.43%, respectively [20].

The Fe, Zn, K, Ca, and Mg contents of the collected samples were significantly different at $p < 0.05$ among the location (districts), as shown in Fig. 12.5, in terms of their essential mineral composition (expressed in mg/100 g). These were found within a range of 47.6 to 59.11 for Fe, 2.3 to 6.04 for Zn, 819.8 to 1468.8 for K, and 73.6 to 93.5 for Mg. Thus, it can be concluded that in addition to its nutritive profile, the presence of these elements in *Awaze* makes it useful to people prone to micronutrient deficiency. The current result for mineral contents, Fe, Zn, Ca, and K was in agreement with the sweet red pepper paste mineral content reported by [3].

The tested *Awaze* samples contained *total bacterial count* in the range of 2 to 2.7 log cfu/g (Table 12.9). In addition to *total bacterial count, yeast and mold* count in the *Awaze* sample ranged from 1.44 to 2.41 logcfu/g (Table 12.9). There was not any *coliform* development in the collected *Awaze* sample. The microbial population in the product may be dependent on the processing practice, external environment, physicochemical, and microbial profile of raw material. The *Awaze* sample from this result is safe for consumption because the *total aerobic mesophilic bacteria* equal to or more than 6 logcfu/g and *mold and yeast* equal to or more than 4 log

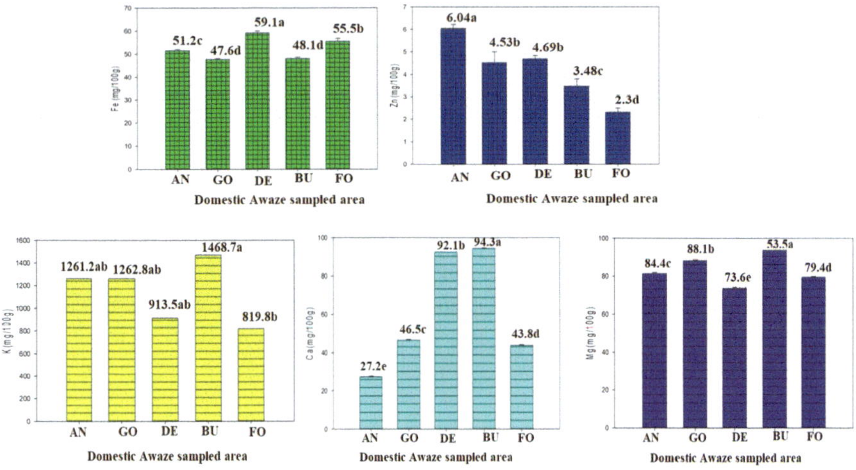

Fig. 12.5 The micronutrient content in *Awaze* collected from five districts in Amhara region, Ethiopia

Table 12.9 Microbial profile of *Awaze* collected from five districts in Amhara region, Ethiopia

	Microbial profile			
Sample location	Total bacterial plate count(logcfu/g)	Yeast and mold (logcfu/g)	*Lactic acid bacteria* (counts)	*Coliform* (logcfu/g)
1. AN	2.7 ± 0.00^{a}	2.41 ± 00^{a}	<30	ND
2. GO	2.66 ± 0.07^{b}	2.33 ± 0.05^{b}	<30	ND
3. DM	2.0 ± 0.00	1.44 ± 0.01^{e}	<30	ND
4. BU	2.69 ± 0.06^{a}	2.2 ± 0.014^{c}	<30	ND
5. FO	2.49 ± 0.01^{c}	1.9 ± 0.02^{d}	<30	ND

The data were expressed as mean ± standard deviation. The different values in the same column indicate significant differences ($p < 0.05$)

cfu/g is not acceptable based on the international commission on microbial specification for food [11]. The tasted *Awaze* samples' *total bacterial count,* and *yeast and mold* content in this investigation were less in content than that previously reported hot pepper paste [5] and *Awaze* [12].

12.4 Conclusion

Awaze is an important food product in Ethiopia. It is produced domestically at household level and not commercially. *Awaze* is a principal spice-based food with high sociocultural and economic values among the study districts at Aneded, Gozamin, Dembecha, Bure, and Fogera districts. It is consumed in different ways, as side dishes and ceremonial dishes. The study demonstrated the variation in physicochemical, nutritional, and microbial characteristics of domestic *Awaze* product, which may be due to the influence of raw materials processing methods, and fermentation conditions. The outcome of the study indicates that the microbial profile was at the safe level of *total bacterial count, yeast,* and *mold*, while in terms of nutritional content, the product had high antioxidant activity and ash content. Ethiopian *Awaze*-based food should be improved more in formulation and it has not been commercialized standardizing the ingredient use and modernizing the processing technology. Conventional processing methods for the production of traditional food types from spices need to be optimized. This study also recommends regular analysis of domestic *Awaze* products prepared at household level to ensure their safety toward consumers because it is a public health issue.

Acknowledgments The authors would like to thank the Ethiopian Institute of Agricultural Research (EIAR), Bahir Dar University- Institute of Technology for their administrative, financial, and resource support for the project.

Conflict of Interest There is no conflict of interest.

Data Availability Data will be available upon request.

References

1. AOAC, 1995. (1986). Official methods of analysis, 14th edition. *Journal of Chromatography A, 351*(3). https://doi.org/10.1016/s0021-9673(01)83549-4
2. AOAC.1990. (2003). *Association of official analytical chemists* (Vol. 1, pp. 73–80). https://doi.org/10.7312/seir17116-004
3. Atwaa, E., Ramadan, M., & Abd El-Sattar, E. (2020). Production of functional spreadable processed cheese supplemented with sweet red pepper paste. *Journal of Food and Dairy Sciences, 11*(5), 127–132. https://doi.org/10.21608/jfds.2020.102741
4. Bamdad, F., Kadivar, M., & Keramat, J. (2006). Evaluation of phenolic content and antioxidant activity of Iranian caraway in comparison with clove and BHT using model systems and vegetable oil. *International Journal of Food Science and Technology, 41*(Suppl. 1), 20–27. https://doi.org/10.1111/j.1365-2621.2006.01238.x
5. Bozkurt, H., & Erkmen, O. (2004). Effects of production techniques on the quality of hot pepper paste. *Journal of Food Engineering, 64*(2), 173–178. https://doi.org/10.1016/j.jfoodeng.2003.09.028
6. Choi, J. E., & Lee, J. H. (2017). Non-Newtonian characteristics of gochujang and chogochujang at different temperatures. *Preventive Nutrition and Food Science, 22*(1), 62–66. https://doi.org/10.3746/pnf.2017.22.1.62
7. Dessalegn, A., & Ashenafi, M. (2010). Internet journal of food safety evaluation of the probiotic properties and antibiotic resistance of lactic acid bacteria isolated from awaze, Qotchqotcha and Tef dough , traditional Ethiopian fermented foods. *Internet Journal of Food Safety, 12*(January 2010), 187–191.
8. Fagerland, M. W., & Hosmer, D. W. (2012). A generalized Hosmer-Lemeshow goodness-of-fit test for multinomial logistic regression models. *Stata Journal, 12*(3), 447–453. https://doi.org/10.1177/1536867x1201200307

9. Ghodke, S. V., More, K. D., & Chavan, D. H. (2014). Studies on standardization of ginger-garlic-chilli paste and its utilization in bread. *Food Science Research Journal, 5*(2), 100–105. https://doi.org/10.15740/has/fsrj/5.2/100-105
10. Guthausen, A., Guthausen, G., Kamlowski, A., Todt, H., Burk, W., & Schmalbein, D. (2004). Measurement of fat content of food with single-sided NMR. *JAOCS, Journal of the American Oil Chemists' Society, 81*(8), 727–731. https://doi.org/10.1007/s11746-004-0969-5
11. ICMSF, I. C. on M. S. for F. (2018). Microorganisms in foods 7: Microbiological testing in food safety management. In *Microorganisms in foods 7: Microbiological testing in food safety management* (2nd ed.). https://doi.org/10.1007/978-3-319-68460-4
12. Idris, A., Mehari, T., & Ashenafi, M. (2001). Some microbiological and biochemical studies on the fermentation of "awaze" and "datta", traditional Ethiopian condiments. *International Journal of Food Sciences and Nutrition, 52*(1), 5–14. https://doi.org/10.1080/09637480020027174
13. Jessica Elizabeth, D. L. T., Gassara, F., Kouassi, A. P., Brar, S. K., & Belkacemi, K. (2017). Spice use in food: Properties and benefits. *Critical Reviews in Food Science and Nutrition, 57*(6), 1078–1088. https://doi.org/10.1080/10408398.2013.858235
14. Kim, G.-T., Hwang, Y.-I., Lim, S.-I., & Lee, D.-S. (2000). Carbon dioxide production and quality changes in Korean fermented soybean paste and hot pepper-soybean paste. *Journal of the Korean Society of Food Science and Nutrition, 29*(5), 807–813.
15. Li, J., Zhao, F., Liu, H., Li, R., Wang, Y., & Liao, X. (2016a). Fermented minced pepper by high pressure processing, high pressure processing with mild temperature and thermal pasteurization. *Innovative Food Science and Emerging Technologies, 36*, 34–41. https://doi.org/10.1016/j.ifset.2016.05.012
16. Li, Z., Dong, L., Huang, Q., & Wang, X. (2016b). Bacterial communities and volatile compounds in Doubanjiang, a Chinese traditional red pepper paste. *Journal of Applied Microbiology, 120*(6), 1585–1594. https://doi.org/10.1111/jam.13130
17. Maharjan, D., Singh, A., Lekhak, B., Basnyat, S., & Gautam, L. S. (2012). Study on antibacterial activity of common spices. *Nepal Journal of Science and Technology, 12*, 312–317. https://doi.org/10.3126/njst.v12i0.6518
18. Miranti, D. I., Ichiura, H., & Ohtani, Y. (2018). The bioactive compounds and antioxidant activity of food products of *Rhizophora stylosa* fruit (coffee and tea mangrove). *International Journal of Forestry Research, 2018*(2), 1–6.
19. Okonkwo, C., & Ogu, A. (2014). Nutritional evaluation of some selected spices commonly used in the south-eastern part of Nigeria. *Journal of Biology, Agriculture and Healthcare, 4*(15), 97–103. http://iiste.org/Journals/index.php/JBAH/article/view/14211
20. Park, S.-Y., Kim, S., Hong, S., & Lim, S.-D. (2017). Analysis of quality characteristics of traditional and commercial red pepper pastes (gochujang). *Korean Journal of Food & Cookery Science, 33*(2), 137–147. https://doi.org/10.9724/kfcs.2017.33.2.137
21. Ramalingam, S. (2022). Physicochemical, microbial, and volatile compound characteristics of gochujang, fermented red pepper paste, produced by traditional cottage industries. *Food, 11*(3), 375.
22. Regina, L., Mariutti, B., Pizarro, G., Barreto, D. M., Bragagnolo, N., & Mercadante, A. Z. (2008). Free radical scavenging activity of Ethanolic extracts from herbs and spices commercialized in Brazil. *Brazilian Archives of Biology and Technology, 51*(December), 1225–1232.
23. Ryu, M. H., Zhang, J., Toth, T., Khokhani, D., Geddes, B. A., Mus, F., Garcia-Costas, A., Peters, J. W., Poole, P. S., Ané, J. M., & Voigt, C. A. (2020). Control of nitrogen fixation in bacteria that associate with cereals. *Nature Microbiology, 5*(2), 314–330. https://doi.org/10.1038/s41564-019-0631-2
24. Schnurer, J., Olsson, J., & Borjesson, T. (1999). Fungal volatiles as indicators. *Fungal Genetics and Biology, 217*(2–3), 209–217.

25. Shah, A. M., Tarfeen, N., Mohamed, H., & Song, Y. (2023). Fermented foods: Their health-promoting components and potential effects on gut microbiota. *Fermentation, 9*(2), 1–23. https://doi.org/10.3390/fermentation9020118
26. Sharifi, A., Niakousari, M., Maskooki, A., & Mortazavi, S. A. (2015). Effect of spray drying conditions on the physicochemical properties of barberry (Berberis vulgaris) extract powder. *International Food Research Journal, 22*(6), 2364–2370.
27. Siruguri, V., & Bhat, R. V. (2015). Assessing intake of spices by pattern of spice use, frequency of consumption and portion size of spices consumed from routinely prepared dishes in southern India. *Nutrition Journal, 14*(1), 1–9. https://doi.org/10.1186/1475-2891-14-7
28. Sora, G. T. S., Haminiuk, C. W. I., da Silva, M. V., Zielinski, A. A. F., Gonçalves, G. A., Bracht, A., & Peralta, R. M. (2015). A comparative study of the capsaicinoid and phenolic contents and in vitro antioxidant activities of the peppers of the genus capsicum: An application of chemometrics. *Journal of Food Science and Technology, 52*(12), 8086–8094. https://doi.org/10.1007/s13197-015-1935-8
29. Thangaraj, P. (2016). Proximate composition analysis. *Progress in Drug Research, 71*, 21–31. https://doi.org/10.1007/978-3-319-26811-8_5
30. Tigu, F., Assefa, F., Mehari, T., & Ashenafi, M. (2016). Probiotic property of lactic acid bacteria from traditional fermented condiments: Datta and awaze. *International Food Research Journal, 23*(2), 770–776.
31. Tsegaye, M., Ephraim, E., & Ashenafi, M. (2004). Behaviour of Escherichia coli O157:H7 during the fermentation of Datta and awaze, traditional Ethiopian fermented condiments, and during product storage at ambient and refrigeration temperatures. *Food Microbiology, 21*(6), 743–751. https://doi.org/10.1016/j.fm.2004.02.003

Chapter 13
Microwave-Assisted Hydrothermal Pretreatment of Coffee Husk: Process Parameter Optimization Using Response Surface Methodology

Gashaw Muluken Tessera, Nigus Gabbiye Habtu, Metadel Kassahun Abera, Kefyalew Dagnew Addisu, and Fantahun Woldesenbet Misganaw

13.1 Introduction

Coffee processing industries produce a lot of coffee residues. Depending on how coffee cherries are processed, the main wastes that are released into the environment are coffee pulp, coffee husk, and coffee silver skin [35]. Coffee husk is the main processing residue, it represents about 43.2% of the weight of the coffee fruit and the coffee bean represents only 38.9% of the whole fruit. Therefore, compared to processed coffee beans, more coffee husk is generated on average [13]. The coffee husk residues could be excellent, cheap substrates for the sustainable production of high-added-value biochemicals. It has a high concentration of carbohydrates and may therefore be regarded as a potential raw material for the production of fermentable sugar [31, 35]. To produce fermentable sugar from the lignocellulosic wastes, pretreatment and hydrolysis are the main duties; hence pretreatment of coffee husk biomass is a very important task. Lignocellulosic biomass is the most ubiquitous renewable source of carbon on Earth; it originates from a variety of sources, including municipal solid waste, industrial waste, crop residues, agricultural residues, and animal manures. It mostly consists of cellulose, hemicellulose, and lignin in various proportions [1]. The production of different industrially important biochemicals

G. M. Tessera (✉)
Bahr Dar University Institute of Technology, Bahr Dar, Ethiopia

Bio and Emerging Technology Institute, Addis Abeba, Ethiopia

N. G. Habtu · M. K. Abera · K. D. Addisu
Bahr Dar University Institute of Technology, Bahr Dar, Ethiopia

F. W. Misganaw
Bio and Emerging Technology Institute, Addis Abeba, Ethiopia

S. W. Fanta et al. (eds.), *Sustainable Development Research in Manufacturing, Process Engineering, Green Infrastructure, and Water Resources*, Green Energy and Technology, https://doi.org/10.1007/978-3-031-77339-6_13

from lignocellulosic material requires pretreatment, hydrolysis, and fermentation steps. To break down the recalcitrant structure of lignocellulosic biomass, the design of different pretreatment technologies is very important to release fermentable sugars in biochemical production. Microwave-based pretreatment of lignocellulosic biomass has been considered one of the most promising methods of pretreatment. The benefits of microwave-assisted pretreatment include higher reaction rates, higher yield, purity, and lower energy consumption [15, 22].

Pretreatment is to break the lignin sheath, disrupt the crystalline structure of cellulose, increase the surface accessibility for hydrolysis, promote cellulose decrystallization and partial hemicellulose depolymerization, and reduce the lignin recalcitrance in the treated biomass [29, 33]. Numerous types of pretreatment have been applied to lignocellulosic biomass to produce fermentable sugars, such as dilute acid, alkaline, organosolv, chemical oxidation, liquid hot water, microwave, steam explosion, and biological pretreatment with fungi, etc. [25].

Microwave irradiation has been used in various technological applications, including food processing, activated carbon regeneration, metals and ceramics sintering, plasma processing, polymer processing, pollution control, and pyrolysis. The use of volumetric heating, which permits quicker heating and, as a result, accelerates and intensifies chemical processes, makes microwave heating distinctive [42]. Microwave heating offers a number of advantages over conventional heating methods, such as non-contact heating, energy transfer instead of heat transfer, selective heating to target materials; lower activation temperature, faster heat transfer, volumetric heating, quick start-up and stopping, efficiently improved product yield, processing time reduction, unique material microstructure and properties, environmentally friendly process, and heating capacity increments [19, 40]. Microwaves cause a quick conversion of the electromagnetic energy into heat when they come into direct contact with microwave-sensitive materials. Microwave heat has unique properties that can be taken as advantage to improve the conversion, selectivity, and/or energy efficiency of chemical processes [20].

Microwave heating is a non-contact method that transfers heat to the product via electromagnetic waves and can transfer significant amounts of heat to the interior of the material, minimizing the effects of differential heating. This alternative method addresses the shortcomings of conventional fast heating [19]. Microwave heating could facilitate heating by allowing material to interact directly with electromagnetic energy. Several researchers now prefer microwave-assisted with other chemical methods such as acid, alkali, and ionic liquid (IL) treatments [3].

The microwave-induced reaction may occur more rapidly and efficiently at a lower bulk temperature due to the significant thermal gradient between the interior of the particle and its cold surface. This reduces processing time and energy consumption [14]. During microwave-assisted pretreatment of biomass, there is minimal generation of inhibitors, such as hydroxyl methyl furfural (HMF), furfural, weak acids (including formic acid, acetic acids, and lactic acids) and phenolic compounds [2]. For example, Ha et al. studied the microwave-assisted ionic liquid (IL) pretreatment of cotton biomass and reported dramatic reduction of degree of polymerization (DP) of cellulose [18]. Several microwave-assisted pretreatment studies were carried out with different biomasses, such as pine chips, beech chips, wheat straw, pineapple waste, and other lignocellulosic biomass [8, 27]; however, microwave-assisted hydrothermal pretreatment of coffee husk has not been sufficiently studied for cellulose recovery in the bioconversion process. The main objective of the study was to maximize the production of cellulose used as a source of fermentable sugars by optimizing the process conditions of microwave-assisted hydrothermal treatment.

13.2 Materials and Method

13.2.1 Sample Collection and Preparation

Coffee husk samples obtained from a dry de-hulling process were collected from Lake Tana in Zege (Bahr Dar, Ethiopia) local coffee market and transported to the Bahir Dar Institute of Technology research laboratory. A disk miller (FRITSC, D-55743, Idar-Oberstein, Germany) was used to reduce the particle size of the coffee husk sample. A sieve size of 2.36 mm and 1.70 mm (Retsch, AS200 Control, Germany) was used to have an average particle size of 2.03 mm. The sample was cleaned with tap water to get rid of any dirt and foreign objects, and then it was dried in an oven at 70 °C until it reached constant weight, then stored in a high-density polyethylene (HDPE) bag for the pretreatment process [23].

13.2.2 Pretreatment of Coffee Husk Biomass

Microwave-assisted hydrothermal pretreatment of coffee husk was performed using a domestic microwave oven (MW1023-02D, China) at a frequency setting of 2450 MHz. The combination of pretreatment conditions was generated from the design of experiment (DoE-V12) software (Table 13.2). Pretreatment was done at a microwave irradiation power, irradiation time, and liquid-to-solid ratio (LSR) of 400–800W, 5–10 min, and 10–15 mL/g, respectively, by modifying the procedure reported by [26]. The coffee husk sample (25 g) was submerged in distilled water based on the LSR in a 500 mL round flask with a fitted cap and positioned at the center of a rotating circular plate in the microwave oven. After pretreatment, the mixture was cooled down to room temperature. The solid fraction was separated from the liquid fraction with vacuum filtration, then the solid residue was washed using 800 ml distilled water until neutral pH. The washed solid residue was dried in an oven at 70 °C until constant weight [32] and kept in a dry HDPE plastic bag for chemical composition analysis and further study.

13.2.3 Experimental Design

Optimization of the process conditions for lignin removal and cellulose recovery was studied using a response surface methodology (RSM) of central composite design (CCD) at a rotatable α value of 1.68179 with center points of 6. The independent variables were examined at −α, 0, and +α of low, center, and high levels, respectively. The independent variables were microwave irradiation power (400 – 800 W), irradiation time (5–10 min.), and liquid-to-solid ratio (10–15 ml/g). During the employ of CCD, the second-order polynomial equation was developed to fit the experimental data and evaluate the effect of each independent variable on the responses. The optimum process condition was obtained and identified by the following general model terms:

$$y=\beta_o+\Sigma\beta_i x_i+\Sigma\beta_j x_j+\Sigma\beta_{ii} x_{ii}2+\Sigma\beta_{jj} x_{jj}2+\Sigma\beta_{ij} x_i x_j+\varepsilon \tag{13.1}$$

where, y is the response function, x_i, and x_j refer to the independent variables, β_0 is the model intercept term, β_i, and β_j are the liner effect coefficient terms, β_{ij} is the interaction coefficients between the input factors x_i and x_j, β_{ii} and β_{jj} are the coefficient of quadratic terms, x_{ii}^2 and x_{jj}^2 are the quadratic terms, and ε is the error of the model. For this study, the independent variables were coded as A, B, and C, thus the general equation could be described as follows:

$$\begin{aligned} y=&\beta o+\beta aA+\beta bB+\beta cC+\beta abAB+\beta acAC \\ &+\beta bcBC+\beta aA2+\beta bB2+\beta cC2+\varepsilon \end{aligned} \tag{13.2}$$

Each of the actual variables was investigated at five coded levels ($-\alpha$, -1, 0, +1, $+\alpha$), as listed in Table 13.1.

The coefficient of determination (R^2), analysis of variance (ANOVA), response plots, and optimization were used to examine the experimental results. The selected experimental design matrix of the average of the replicates is shown in Table 13.2.

Table 13.1 The independent variables and their coded levels used for optimization in CCD

			Coded level				
Variables	Unit	Coding	−α	−1	0	+1	+α
Microwave power	Watt	A	263.64	400	600	800	936.36
Irradiation time	min	B	3.29	5	7.5	10	11.70
Liquid-to-solid loading	ml/g	C	8.29	10	12.5	15	16.70

Table 13.2 The selected experimental design matrix of central composite design (CCD) runs with corresponding responses

	Experimental variables			Experimental responses					
Sr. No.	A: MWP	B: RT	C: LSR	Lignin content (%)		Hemicellulose content (%)		Cellulose content (%)	
	(W)	(min.)	(mL/g)	Actual	Predicted	Actual	Predicted	Actual	Predicted
1	800	10	10	26.62	27.10	48.42	47.62	12.79	12.97
2	800	10	15	29.02	29.16	46.81	46.35	17.84	15.63
3	400	5	10	28.04	28.54	50.72	49.85	10.12	10.80
4	800	5	15	28.6	28.87	48.07	47.84	10.47	9.84
5	600	3.3	12.5	28.52	28.24	48.82	49.09	10.54	11.15
6	400	5	15	26.86	27.03	48.01	47.49	13.19	11.49
7	600	7.5	12.5	29.5	29.55	46.57	46.46	14.25	14.13
8	400	10	15	29.16	29.76	48.38	46.96	14.98	14.07
9	400	10	10	28.74	29.12	48.55	47.45	11.78	10.88
10	264	7.5	12.5	29.08	28.41	45.42	47.11	10.14	11.08
11	800	5	10	28.92	28.97	50.87	50.96	10.3	9.68
12	936	7.5	12.5	28.51	28.26	47.35	47.54	10.23	11.45
13	600	11.7	12.5	29.61	28.97	44.23	45.84	14.55	16.09
14	600	7.5	16.7	29.56	29.16	46.68	47.61	10.16	12.66
15	600	7.5	8.3	29.22	28.70	49.7	50.65	10.19	9.85
16[a]	0	0	0	29.84	-	37.86	-	9.44	-

[a]control; *MWP* microwave power; *RT* Radiation Time; *LSR* liquid–to-solid ratio
NB: The microwave power of 600 and 264 watts have no adjustable scale setup, alternatively 640 and 240 watts were used, respectively, for the pretreatment of the biomass

13.2.4 Characterization of Coffee Husk Biomass

Chemical Composition Analysis of Coffee Husk The major chemical components of coffee husk biomass, such as ash, extractives, lignin (acid soluble and acid insoluble), hemicelluloses, and cellulose were determined using the National Renewable Energy Laboratory's (NREL) standard protocols [24, 36–38].

Lignin Determination . The amount of acid-insoluble and acid-soluble lignin (ASL) was determined from extractive-free coffee husks. According to NREL/TP-510-42618 procedures, total lignin was calculated by adding the acid-insoluble and acid-soluble lignin together. Using an analytical digital balance (FA2104, China), a sample of coffee husk weighing 0.300 g was put into a 125 mL conical flask along with 3 ml of H_2SO_4 (72%). Acid hydrolysis was made by placing the sample in a digital water bath (XMTD-204, China) set at 30 °C for one hour while stirring at 10-min intervals. After 1 h of acid hydrolysis, 84 ml of distilled water was added to each flask, bringing the total volume to 87 ml, and then the sample was autoclaved for 1 h at 121 °C/15 psi for the second weak acid hydrolysis step. Thereafter the hydrolysis was cooled to room temperature. By using vacuum filtration, the solid and liquid components were separated. The filtrate (liquid) part was used for acid-soluble lignin (ASL) determination using UV/VIS spectrometer (PerkinElmer Lambda 850, USA). The absorbance of the liquid fraction was measured at a recommended wavelength of 320 nm to determine ASL. The solid residue was utilized to determine acid-insoluble lignin (AIL) by drying the residue at 105 °C until it reached constant weight, calculating the acid-insoluble residue (AIR), and then ashing the residue by incinerating it at 600 °C for 4 h in a muffle furnace (Nabertherm GmbH, Bahnhafstr 20, 28865, Germany). Finally, AIL was calculated from the difference between AIR and the ash of the residue [36–38].

$$\mathrm{AIL}(\%) = \frac{W_{\mathrm{AIR}} - W_{\mathrm{ash}}}{ODW} * 100 \tag{13.3}$$

where AIL = acid-insoluble lignin, W_{AIR} = weight of acid-insoluble residue, W_{ash} = weight of ash of the acid-insoluble residue, ODW = oven dry weight

$$ASL(\%) = \frac{UV_{\mathrm{absorbance}} * V_{\mathrm{filtrate}} * D}{e * ODW_{\mathrm{sample}} * \mathrm{Pathlength}} * 100 \tag{13.4}$$

where ASL = acid-soluble lignin, *e* = absorptivity at recommended wavelength (λ = 320 nm) = 30 L/g cm. ODW = oven dry weight; *V*-filtrate = volume of filtrate = 86.73 ml; *D* = dilution.

Whereas dilution (*D*) can be determined as follows:

$$D = \frac{V_{\mathrm{sample}} + V_{\mathrm{dilluting\,solvent}}}{V_{\mathrm{sample}}}$$

Hemicellulose Determination Extractive-free coffee husk biomass (1.0 g) was transferred into a 250 mL Erlenmeyer flask with 150 mL NaOH solution (0.5 M). The mixture was heated at a reaction temperature of 80 ±3 °C for 3.5 hrs; after heating, it was cooled to room temperature and then washed with distilled water under vacuum filtration until the pH of the solution approached 7. The residue was dried at 105 °C, cooled in a desiccator, & weighed repeatedly until it reached a constant weight [7, 24]. Then the hemicellulose content was calculated by taking the weight difference before and after treatment.

$$\mathrm{Hemicellulose}(\%) = \frac{W_o - W_r}{W_o} * 100 \tag{13.5}$$

where W_o = weight before treatment, W_r = weight of the residue after treatment

Cellulose Determination The cellulose content was calculated by differences, suppose the total components of the biomass are only composed of extractives, hemicellulose, lignin, ash, and cellulose [24].

$$\text{Cellulose}(\%) = 100 - \Sigma(\text{ash}\% + \text{extractives}\% + \text{lignin}\% + \text{hemicellulose}\%) \tag{13.6}$$

FT-IR Spectroscopy Analysis Fourier-transformed infrared (FT-IR) spectroscopy was used as an analytical tool to qualitatively determine the functional group and chemical changes in the lignocellulosic material upon pretreatment. FT-IR spectra of untreated and treated coffee husk were obtained by direct transmittance using the KBr pellet technique (Shimadzu). The spectra range of 400–4000 cm^{-1} was used at a spectral resolution of 1 cm^{-1} [23]. The spectra of FT-IR were drawn by using origin software (Origin 2022).

Brunauer–Emmett–Teller (BET) Surface Area Analysis The surface area, pore volume, and pore size distribution of both untreated and pretreated samples were measured using BET analyzer (Nova 4000e, USA). The BET analyzer used nitrogen as an analysis gas. The degassing time was 8 h at a Temperature of 300 °C. The pressure tolerance for the analysis was 0.100/0.100 (ads/des) equilibrium time of 60/60 sec. and the equilibrium time was 240/240 s. (ads/des). The analysis contact time was 80.9 min [5].

13.2.5 Statistical Analysis

All the experiments were carried out in replicates and the analysis was done using Design Expert-12 and origin software (V2022) for graphical illustration.

13.3 Results and Discussion

13.3.1 Compositional Analysis of Coffee Husk

The untreated coffee husk analysis results of ash, extractives, lignin, hemicellulose, and cellulose were 4.46 ± 0.11, 18.40 ± 0.65, 29.84 ± 0.92, 37.86 ± 0.06, and 9.44 ± 0.09 w%, respectively as depicted in Table 13.3. The ash and extractive contents

Table 13.3 Chemical composition analysis of coffee husk

	Chemical composition				
Samples	Ash (w.%)	Extractives (w.%)	Lignin (w.%)	Hemicellulose (w.%)	Cellulose (w.%)
Raw coffee husk	4.46 ± 0.11	18.40 ± 0.65	29.84 ± 0.92	37.86 ± 0.06	9.44 ± 0.09
Treated coffee husk	2.23 ± 0.29	8.53 ± 0.63	28.08 ± 0.46	46.07 ± 1.02	15.09 ± 0.42

of this study are in line with the work of [16] as reported for ash (5.4w%) and extractives (18.9 w%). Another similarity report for ash (4.6w%) and extractives (17.67w%) by Baeta et al. [9]. The lignin content (29.84%) obtained in this study for raw coffee husk is in agreement with previous studies reported (29.55%) by Veiga et al. [41]. Some previous studies also reported a slight difference in the chemical composition analysis of coffee husks. For instance, de Carvalho et al. reported about cellulose (26.5 w%), hemicellulose (25.5 w%), lignin (33.5 w%), ash (0.2 w%), and extractives (6.7 w%) [12]; Gouvea et al. [17] reported cellulose (43 w%), hemicellulose (7 w%), lignin (9 w%), and minerals (3 – 7 w%); Veiga et al. [41] also showed extractives (47.38 w%), ash (7.75 w%), and lignin (29.55 w%). The study of Bekalo et al.[11] also reported cellulose (24.5 w %), hemicellulose (29.17 w %), lignin (23.70 w %), and ash (6.20 w %). In general, the variation in chemical composition reported by different scholars might be due to different factors including agronomical conditions, growth environment, crop variety, harvesting season, and coffee processing methods [28].

At the optimal process condition, the chemical composition of coffee husk was determined as shown in Table 13.3. The amount of extractives in the treated coffee husk decreased by 2.16 times that of raw coffee husk; which is 53.64 % of the extractives were removed from the treated coffee husk. The lignin percentage (28.08 wt %) of the treated coffee husk was reduced by 1.06 times that of the raw coffee husk (29.84 wt %). In hemicellulose content, a 21.68 % increment was observed in the pretreatment. The cellulose content of the treated coffee husk is 1.60 times higher than that of the raw coffee husk (9.44 wt %); this reflects that the microwave-assisted hydrothermal pretreatment improved the cellulose content by 59.64%.

13.3.2 Structural Changes in the Coffee Husk Biomass

Fourier-Transformed Infrared (FT-IR) Spectroscopy Analysis FT-IR spectroscopy is used to determine the functional groups arrangement and structural changes before and after pretreatment of the coffee husk biomass. The FT-IR is mainly used to measure the light absorption in the wavenumber range of 4,000 to 400 cm^{-1} in order to identify and quantify the presence and absence of different functional groups in the biomass (see Fig. 13.1). In the FT-IR spectra, high intensity means strong bonding and weak absorption means weaker bonding of the chemical component [34].

The strong broad spectrum represents O-H bond absorption, which is observed between 3500 and 3300 cm^{-1} for treated coffee husk and raw coffee husk (control); it is the O-H functional group in alcohol and phenol compounds. The treated coffee husk and control showed peaks at 2913 and 1319cm^{-1} which are the characteristic peaks with different intensities for C-H bond stretching and bending vibrations in cellulose and hemicellulose, respectively [4, 21] . The peak at 2851 cm^{-1} is

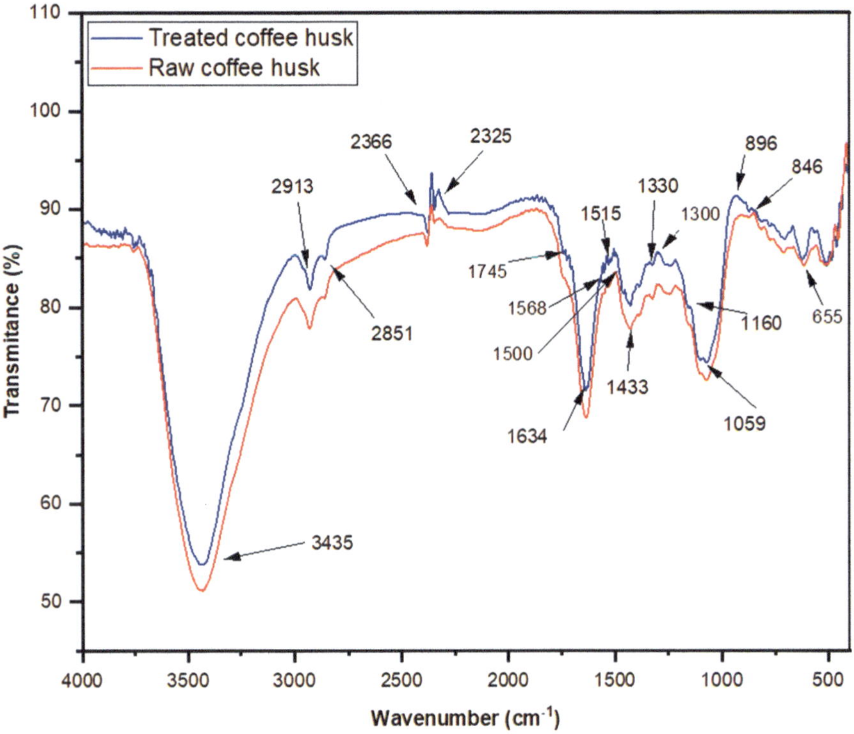

Fig. 13.1 FT-IR spectra of the treated and untreated coffee husk biomass

associated with the presence of -CH_2 groups of cellulose and hemicellulose. The peaks observed on 2350 and 2325 cm^{-1} indicated the presence of –COO group in cellulose [34].

The peaks around 1634 cm^{-1} in control and microwave-treated coffee husk were attributed to C=O stretching vibration of acetyl groups in hemicellulose compounds, as shown in Fig. 13.1 [4]. The three small peaks are observed around 1745 cm^{-1} of the C=O stretching for hemicellulose in the pretreated coffee husk but not in the raw coffee husk.

The peak observed at 1330.86cm^{-1} in the raw coffee husk disappeared after microwave treatment. The peak at 1515.6 cm^{-1} is caused by acetyl (lignin) C-O stretching and intensity reduction was observed during pretreatment. The tiny peaks between 1433 and 1568.17cm^{-1} represent the aromatic structure of lignin; the strength of the peaks dropped after pretreatment this may prove that the small amount of lignin content was removed in the course of pretreatment. The peaks visible between 1300 and 1500 cm^{-1} represent the C-H bending of hemicellulose, O-H and C-O stretching of cellulose [4]; the peaks at 1160 and 896 cm^{-1} is the C-O-C stretching at β-(1– 4) glycosidic linkages in the cellulose and hemicellulose in the pretreated coffee husk; this may state the existence of more hemicellulose and cellulose contents. The peak observed at 846 cm^{-1} in the raw coffee husk is the

Table 13.4 Surface area and porosity of the coffee husk biomass

Samples type	Surface Area (m^2/g)	Pore volume (cm^3/g)	Pore radius (nm)	Category
Untreated	567.31	0.1384	1.324	Microporous
Pretreated	529.10	0.1451	1.324	Microporous

aromatic C-H out of plane bending for lignin which is removed in pretreated coffee husk [43]; it may be an indication of the lignin removed during pretreatment of coffee husk. The other intense peaks around 655 cm^{-1} indicated the presence of C–OH Carboxyl group for cellulose [34]; the peaks revealed different intensity strengths in treated and untreated coffee husks.

Brunauer–Emmett–Teller (BET) Surface Area Analysis The surface area, pore volume, and pore size (radius) of the investigated coffee husk sample are presented in Table 13.4. The BET surface area and non-local density functional theory (NLDFT) pore volume of pretreated coffee husks are 529.095 m^2/g, 0.15 cm^3/g, and 1.3 nm, respectively. The pretreated coffee husk samples showed a change in surface area and pore volume but not in pore size compared with the untreated sample. The change in BET surface area was observed between treated and raw coffee husks. This shows the effect of microwave-assisted hydrothermal pretreatment method on the structural morphology of the coffee husks. A closely related result was reported by [30] on BET surface area (539 m^2/g) and pore volume (0.32cm^3/g) by pyrolysis pretreating the spent coffee ground for biochar. According to the investigation by Ayalew and Aragaw, they determined the pore diameter value of 1.1 nm for the activated coffee husk using chemical activation (H_3PO_4) was found to be closely similar to this work [6].

The result shows that there is a surface area reduction; this may be the result of particle agglomeration and collapse of capillary structure due to the microwave energy on the structure of cellulose. Which plays a crucial role in the cellulose-water interactions (−OH group accessibility). Such phenomena were reasonably explained in the work of [10, 39].

13.3.3 Model Fittingness Analysis

Central composite design (CCD) was applied to study the effect of microwave irradiation power, irradiation time, and liquid-to-solid ratio on lignin, hemicellulose, and cellulose content of coffee husk. The quadratic model was suggested; the linear terms, interaction terms, and quadratic terms of the models were fitted to the experimental data to obtain the regression equations of each response. Cubic model was found to be aliased. Two different tests namely sequential model sum of squares (see Annex Tables 13.9, 13.11, and 13.13) and model summary statistics (see Annex Tables 13.10, 13.12, and 13.14) were employed to decide about the adequacy of various models to represent lignin, hemicellulose and cellulose contents in the pretreated coffee husk biomass.

For the response lignin, as per the sequential model sum of squares test (see Annex Table 13.9), the p-value (< 0.05) is highly significant and the suggested model was a quadratic model, which was used for the next analysis. As per model summary statistics (see Annex Table 13.10), the quadratic model was found to have an R^2 of 0.7954, adjusted R^2 of 0.7187, and a predicted R^2 of 0.5607. The adjusted R^2 and predicted R^2 difference are less than 0.2, and are in a reasonable agreement. Therefore, the quadratic model was suggested as the best for lignin percentage prediction.

Similarly, the hemicellulose and cellulose test analyses for the fittingness of the models were performed (see Annex Tables 13.11, 13.12, 13.13, and 13.14). As per the sequential model sum of squares test (see Annex Tables 13.11 and 13.13), the models were highly significant ($p < 0.0023$ and $p < 0.0002$) and the quadratic model was suggested for both hemicellulose and cellulose content prediction, respectively.

13.3.4 Statistical Analysis of the Experimental Design

The contents of experimental responses of lignin, hemicellulose, and cellulose at different pretreatment conditions of microwave irradiation power, irradiation time, and liquid-to-solid ratio were determined, as shown in Table 13.2. Lignin content was recorded in the range of 26.62 and 29.61wt% at the pretreatment condition of (800W, 10 min, 10 mL/g) and (600W, 11.7min, 12.5ml/g), respectively. Hemicellulose content was recorded in the range of 44.23 and 50.87 wt% at the pretreatment condition of (600W, 11.7 min, 12.5ml/g) and (800W, 5 min, 10mL/g), respectively. Cellulose content was determined in the range of 10.12 and 17.84 wt% at the pretreatment conditions of (400W, 5min, 10ml/g) and (800W, 10 min., 15ml/g), respectively.

Furthermore, to analyze the effects of variables, the experimental method generated a mathematical model in the form of polynomial equation (Eqs. 13.7, 13.8, and 13.9) which represents the experimental relationship between the responses (lignin, hemicellulose, and cellulose yields) and the factors (microwave power – A, irradiation time – B, and liquid-to-solid ratio – C). The generated polynomial equation in terms of coded factors and the relationship between the responses and the factors are given in the following mathematical model equations.

$$\begin{aligned} \text{Lignin} = {} & 29.55 - 0.04\text{A} + 0.22\text{B} + 0.14\text{C} - 0.61\text{AB} + 0.36\text{AC} \\ & + 0.54\text{BC} - 0.43\text{A2} - 0.33\text{B2} - 0.22\text{C2} \end{aligned} \tag{13.7}$$

$$\begin{aligned} \text{Hemicellulose} = {} & 46.46 + 0.13\text{A} - 0.97\text{B} - 0.91\text{C} - 0.24\text{AB} - 0.19\text{AC} \\ & + 0.47\text{BC} + 0.30\text{A2} + 0.35\text{B2} + 0.94\text{C2} \end{aligned} \tag{13.8}$$

$$\begin{aligned} \text{Cellulose} = {} & 14.13 + 0.11\text{A} + 1.47\text{B} + 0.84\text{C} + 0.80\text{AB} - 0.13\text{AC} \\ & + 0.63\text{BC} - 1.01\text{A2} - 0.18\text{B2} - 1.02\text{C2} \end{aligned} \tag{13.9}$$

Analysis of Responses (Lignin, Hemicellulose, and Cellulose) Analysis of variance (ANOVA) was used to evaluate the statistical significance of the model terms for lignin, hemicellulose, and cellulose, as indicated in Tables 13.5, 13.6, and 13.7. The model is significant ($p < 0.0001$) for lignin as indicated in Table 13.5, and the main factors such as irradiation time (B), interaction terms (AB, AC, and BC), and quadratic terms (A^2, B^2 and C^2) are significant ($p < 0.05$) for lignin removal. Whereas the model terms such as microwave irradiation power (A) and liquid-to-solid ratio (C) are not significant ($p > 0.1000$).

The value of R^2 (0.7954), adjusted R^2 (0.7187), and predicted R^2 of 0.5607 clearly demonstrated that the model used to represent the relationship between the factors and the response lignin were well correlated and defined. The predicted R^2 of 0.5607 and Adjusted R^2 of 0.7187 difference less than 0.2 is in reasonable agreement. Adequate precision represents the measure of signal-to-noise ratio and a ratio greater than 4 is desirable; hence, in this case, the adequate precision was found to be 10.9468 and this is an indication of the fitness of the developed model; therefore, the model was used to navigate the points in the design spaces. Besides, the lower values of the coefficient of variance (1.60%) precisely exhibited a high degree of precision and good agreement of reliability on the experimental results [3].

The ANOVA table of hemicellulose (Table 13.6) described that the model is highly significant ($p < 0.0001$). The linear terms viz. irradiation time (B), liquid-to-solid ratio (C), and the quadratic terms (C^2) are highly significant, whereas the linear model terms such as microwave irradiation power (A), the interaction terms (AB, AC, and BC) and the quadratic terms (A^2 and B^2) are not significant ($p > 0.05$). The R^2 (0.7463), adjusted R^2 (0.6512), and predicted R^2 (0.4564) verified that the model describes the relationship between the factors and the hemicellulose. The predicted R^2 (0.4564) is in reasonable agreement with the adjusted R^2 (0.6512) due to their difference of less than 0.2. Therefore, the model was employed to navigate the points in the design spaces.

Table 13.5 ANOVA for quadratic model of the response lignin

Source	Sum of squares	df	Mean square	*F*-value	*P*-value	
Model	19.77	9	2.20	10.37	< 0.0001	Significant
A-MW power	0.0525	1	0.0525	0.2477	0.6232	
B-Time	1.28	1	1.28	6.03	0.0217	
C-L/S	0.5241	1	0.5241	2.47	0.1288	
AB	5.95	1	5.95	28.10	< 0.0001	
AC	2.02	1	2.02	9.52	0.0051	
BC	4.67	1	4.67	22.02	< 0.0001	
A^2	4.16	1	4.16	19.66	0.0002	
B^2	2.52	1	2.52	11.89	0.0021	
C^2	1.09	1	1.09	5.12	0.0329	
$R^2 = 0.7954$		Adjusted $R^2 =$ 0.7187		Predicted $R^2 = 0.5607$		
C.V (%) = 1.60						

Table 13.6 ANOVA table for quadratic model of the response hemicellulose

Source	Sum of squares	df	Mean square	*F*-value	*P*-value	
Model	73.65	9	8.18	7.85	< 0.0001	Significant
A-MW power	0.4515	1	0.4515	0.4328	0.5169	
B-Time	25.63	1	25.63	24.57	< 0.0001	
C-L/S (LSR)	22.41	1	22.41	21.48	0.0001	
AB	0.9120	1	0.9120	0.8743	0.3591	
AC	0.5852	1	0.5852	0.5610	0.4611	
BC	3.48	1	3.48	3.33	0.0803	
A^2	2.09	1	2.09	2.01	0.1694	
B^2	2.83	1	2.83	2.71	0.1126	
C^2	20.05	1	20.05	19.22	0.0002	
$R^2 = 0.7463$		Adjusted $R^2 = 0.6512$			Predicted $R^2 = 0.4564$	
Adeq Precision = 9.2587		C.V. % = 2.14				

Table 13.7 ANOVA table for quadratic model of the response cellulose

Source	Sum of squares	df	Mean square	*F*-value	*P*-value	
Model	133.10	9	14.79	7.99	< 0.0001	Significant
A-MW power	0.3214	1	0.3214	0.1735	0.6807	
B-Time	58.90	1	58.90	31.80	< 0.0001	
C-L/S (LSR)	19.16	1	19.16	10.35	0.0037	
AB	10.27	1	10.27	5.55	0.0270	
AC	0.2756	1	0.2756	0.1488	0.7031	
BC	6.28	1	6.28	3.39	0.0781	
A^2	23.10	1	23.10	12.47	0.0017	
B^2	0.7126	1	0.7126	0.3848	0.5409	
C^2	23.26	1	23.26	12.56	0.0017	
$R^2 = 0.7496$		Adjusted $R^2 = 0.6558$			Predicted $R^2 = 0.4601$	
Adeq Precision = 8.6899		C.V. % = 11.02				

The statistical significance of the model terms for cellulose was evaluated using the ANOVA table, as shown in Table 13.7. The ANOVA table for the response cellulose described that the model is significant ($p < 0.0001$). The model terms (B, C, AB, A^2, and C^2) are significantly ($p < 0.05$) influencing the cellulose yield in the pretreatment process, whereas the model terms (A, AC, BC, and B^2) have no significant ($p > 0.05$) influence on cellulose.

The R^2 (0.7496), adjusted R^2 (0.6558), and predicted R^2 (0.4601) demonstrated that the model used to represent the relationship between the factors and the response cellulose was well defined. The predicted R^2 (0.4601) is in reasonable agreement with the adjusted R^2 (0.6558) due to their difference of less than 0.2. Similarly, adequate precision analysis (8.6899) greater than 4 shows a good fitness of the developed model; therefore, the model was used to navigate the points in the design spaces [3].

13.3.5 *Effect of Process Factors*

13.3.5.1 Effect of Microwave Power, Irradiation Time, and Liquid-to-Solid Ratio on Lignin Removal

As shown in Figure 13.2a–c, the microwave power, irradiation time, and liquid-to-solid ratio have an effect on lignin removal. The increment of microwave power from 400 to 800 W and the decrement of irradiation time from 10 to 5 min. can cause an increase of lignin content on the pretreated coffee husk residue, at low microwave power versus short irradiation time and high microwave power versus long irradiation time showed decreasing of lignin content; this is the region of the highest delignification process taking place (Fig. 13.2a). The interaction at high microwave power with low liquid-to-solid ratio (Fig. 13.2b) and long irradiation time versus low liquid-to-solid ratio (Fig. 13.2b, c) are highly significant on the delignification process; the ANOVA table verified this (see Table 13.5).

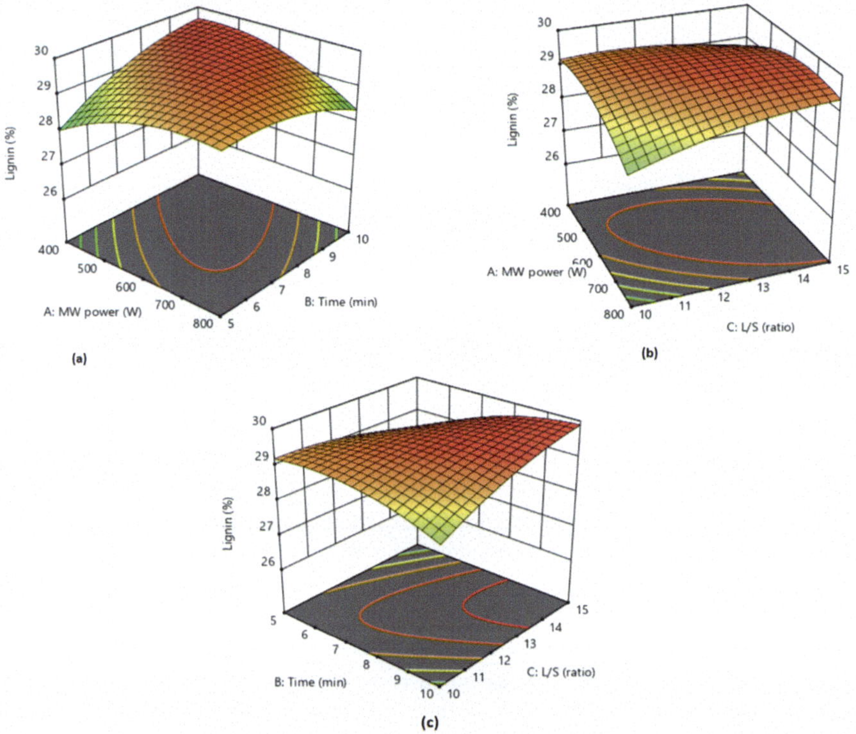

Fig. 13.2 The 3D surface plots for the effect of factors on percentage of lignin concentration (**a**) effect of microwave power (W) versus irradiation time (min.). (**b**) effect of microwave power (W) versus liquid-to-solid ratio (ml/g). (**c**) effect of irradiation time (min.) versus liquid-to-solid ratio (ml/g)

13.3.5.2 Effect of Microwave Power, Irradiation Time, and Liquid-to-Solid Ratio on Hemicellulose Content

The hemicellulose content increased gradually with a decrease in irradiation time from 10 to 5 min. and increasing the microwave power from 400 to 800W (Fig. 13.3a); on the one hand, hemicellulose content increased gradually when the microwave power increased from 400 to 800W and decreasing of the liquid-to-solid ratio from 15 to 10mL/g (Fig. 13.3b). The highest hemicellulose content was observed at the irradiation time of 5min. and liquid-to-solid ratio of 10mL/g (Fig. 13.3c). The highest percentage of hemicellulose was found at 800W, 5 min, and 10mL/g of microwave power, irradiation time, and liquid-to-solid ratio, respectively (Fig. 13.3a–c).

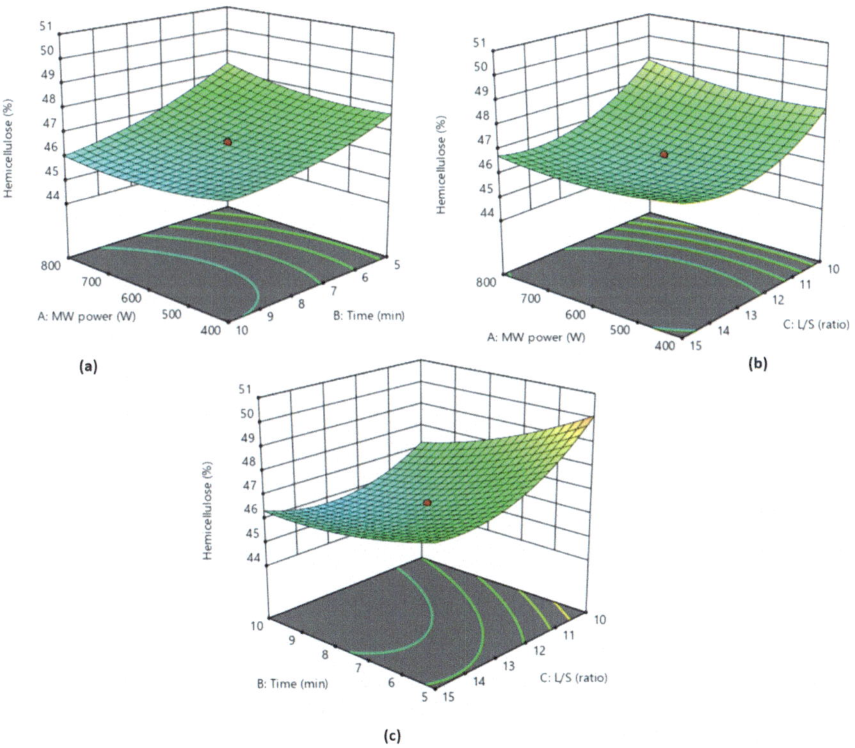

Fig. 13.3 The 3D surface plots for the effect of factors on Hemicellulose content (**a**) effect of microwave irradiation power (W) versus irradiation time (min.) (**b**) effect of microwave irradiation power versus liquid-to-solid ratio (mL/g). (**c**) effect of irradiation time (min.) versus liquid-to-solid ratio (mL/g)

13.3.5.3 Effect of Microwave Power, Irradiation Time, and Liquid-to-Solid Ratio on Cellulose Content

The yields of cellulose increased significantly as the irradiation time (B) rose from 5 to 10 min and microwave power from 400 to 800W (Fig. 13.4a). Hence, high microwave power versus long irradiation time showed a high recovery of cellulose. On the other hand, low microwave power (from 800 to 400w) versus high liquid-to-solid ratio (from 10 to 15ml/g) showed the increment of cellulose content (Fig. 13.4c). Irradiation time and liquid-to-solid ratio showed significant differences in the yield of cellulose (Fig. 13.4a–c). Whereas microwave power has a marginal effect (see Table 13.7).

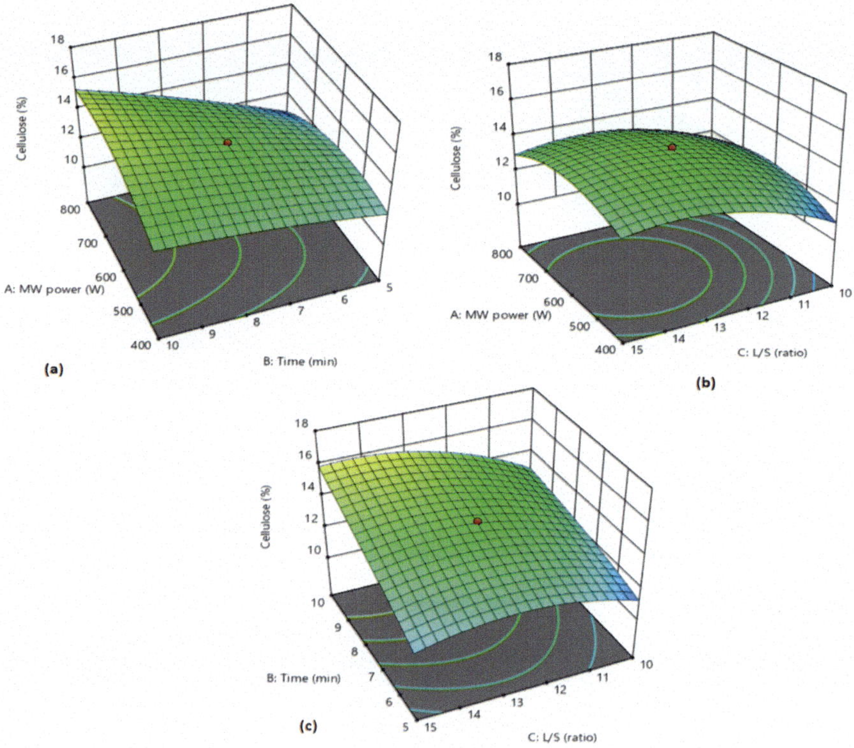

Fig. 13.4 The 3D surface plots for the effect of factors on cellulose content (**a**) effect of microwave irradiation power (W) versus irradiation time (min.), (**b**) effect of microwave irradiation power versus liquid-to-solid ratio (mL/g), (c) effect of irradiation time (min.) versus liquid-to-solid ratio (mL/g)

13.3.6 Validation of Optimal Data

Process variables optimization was performed with a target goal of minimizing lignin and maximizing cellulose yields. The optimum process condition was obtained at a microwave irradiation power of 800W, irradiation time of 10min., and liquid-to-solid ratio of 10.94 ml/g at the desirability function value of 0.585.

After experimentation, the actual percentage of lignin (28.08%), hemicellulose (46.07%), and cellulose (15.09%) were obtained (Table 13.8). The data were validated at the confirmation node and verified within the 95% prediction interval (PI) for the fittingness of the optimal results. Hence, the data mean values were accurately fitted within the prediction intervals.

13.3.7 Evaluation of Pretreatment Efficiency

The effectiveness of the microwave-assisted hydrothermal pretreatment method was evaluated by comparing the chemical composition (ash, extractives, lignin, hemicellulose, and cellulose contents) of coffee husk before and after pretreatment. The efficacy of the pretreatment method to remove lignin (the delignification process) from coffee husk was 5.9%; this result was below expectation. This may have happened due to the repolymerization of lignin fragments in the pretreatment matrix; therefore, the efficiency of the pretreatment method might have dropped.

The content of hemicellulose and cellulose in the pretreated coffee husk residues increased by 1.23 and 1.60 times, respectively. The efficiency of the pretreatment method for hemicellulose and cellulose recovery is 21.69 and 59.85%, respectively. The efficiency of the pretreatment method to remove extractive compounds and inorganic materials (ash) from coffee husk residues was 53.64 and 50%, respectively. At this optimized pretreatment condition, a large amount of organic compounds (extractives) and inorganic materials were removed; this might contribute to increasing the content of cellulose and hemicellulose. In addition, the FT-IR spectra, BET surface area, and pore volume results showed some structural modification in the pretreated coffee husk. A significant improvement in cellulose content was observed in the solid residue of coffee husk after pretreatment. Therefore, the microwave-assisted hydrothermal pretreatment method is more effective for cellulose recovery.

Table 13.8 Validation of the data obtained from optimal process conditions

Response	Experimental data	Predicted mean	Predicted median	Std Dev	*n*	SE Pred	95% PI low	Data mean	95% PI high
Lignin	28.32	27.62	27.62	0.46	2	0.39	26.80	28.08	28.43
	27.84								
Hemicellulose	45.65	46.81	46.81	1.02	2	0.88	45.00	46.07	48.62
	46.49								
Cellulose	15.51	14.08	14.08	1.36	2	1.17	11.67	15.09	16.49
	14.67								

13.4 Conclusion

The coffee husk chemical composition analysis result revealed that it contains high amounts of extractives and lignin as compared with cellulose content. Optimization of process parameter conditions was done using response surface methodology. The optimal process conditions were determined at the microwave irradiation power, irradiation time, and liquid-to-solid ratio of 800W, 10 min, and 10.94mL/g, respectively. The optimal process conditions showed an increase in pore volume of 0.145 cm^3/g and a decrease in surface area to 529.09 m^2/g. At the optimal process condition, the cellulose content was significantly improved from 9.44 to 15.07% whereas the lignin content was reduced from 29.84 to 28.08%. The pretreatment efficiency for cellulose recovery and delignification is found to be 59.85% and 5.9%, respectively. The yield of hemicellulose was improved by 21.68% during the pretreatment of microwave-assisted hydrothermal. The effectiveness of delignification process might be influenced by repolymerization of lignin fragments. Interestingly, the pretreatment method removed a large percentage of extractives (53.64%), this might be an indication of the coffee husk containing a lot of extractives. The cellulose-rich solid residue of coffee husk biomass can be used for different bioconversion processes to contribute to the circular economy concept.

Acknowledgment The authors would like to acknowledge the Bahir Dar University Institute of Technology and Bio and Emerging Technology Institute for providing the necessary laboratory facilities and resources for this study.

Annex: Statistical Tables

Table 13.9 Sequential model sum of squares for the response lignin

Source	Sum of squares	df	Mean square	*F*-value	*p*-value	
Mean vs. total	28127.28	1	28127.28			
Linear vs. mean	1.85	3	0.6179	0.8060	0.5005	
2FI vs. linear	12.64	3	4.21	10.97	< 0.0001	
Quadratic vs 2FI	**5.28**	**3**	**1.76**	**8.31**	**0.0006**	**Suggested**
Cubic vs. quadratic	0.8700	4	0.2175	1.03	0.4150	Aliased
Residual	4.21	20	0.2107			
Total	28152.13	34	828.00			

Table 13.10 Model summary statistics for the response lignin

Source	Std. dev.	R^2	Adjusted R^2	Predicted R^2	Press	
Linear	0.8756	0.0746	–0.0180	–0.2094	30.06	
2FI	0.6195	0.5830	0.4903	0.3105	17.14	
Quadratic	**0.4603**	**0.7954**	**0.7187**	**0.5607**	**10.92**	**Suggested**
Cubic	0.4590	0.8304	0.7202	0.4186	14.45	Aliased

Table 13.11 Sequential model sum of squares for the response hemicellulose

Source	Sum of squares	df	Mean square	*F*-value	*p*-value	
Mean vs. total	77520.22	1	77520.22			
Linear vs. mean	48.49	3	16.16	9.66	0.0001	
2FI vs. linear	4.98	3	1.66	0.9902	0.4122	
Quadratic vs. 2FI	**20.19**	**3**	**6.73**	**6.45**	**0.0023**	**Suggested**
Cubic vs. quadratic	7.31	4	1.83	2.06	0.1239	Aliased
Residual	17.72	20	0.8861			
Total	77618.90	34	2282.91			

Table 13.12 Model summary statistics for hemicellulose

Source	Std. dev.	R^2	Adjusted R^2	Predicted R^2	Press	
Linear	1.29	0.4913	0.4405	0.3414	64.99	
2FI	1.29	0.5417	0.4399	0.2477	74.24	
Quadratic	**1.02**	**0.7463**	**0.6512**	**0.4564**	**53.65**	**Suggested**
Cubic	0.9413	0.8204	0.7037	0.3843	60.77	Aliased

Table 13.13 Sequential model sum of squares for cellulose

Source	Sum of squares	df	Mean square	*F*-value	*p*-value	
Mean vs. total	0.0011	1	0.0011			
Linear vs. mean	0.0001	3	0.0000	6.93	0.0011	
2FI vs. linear	7.590E-06	3	2.530×10^{-6}	1.01	0.4038	
Quadratic vs. 2FI	**0.0000**	**3**	**0.0000**	**10.01**	**0.0002**	**Suggested**
Cubic vs. quadratic	0.0000	4	2.700×10^{-6}	2.80	0.0536	Aliased

Table 13.14 Model summary statistics for cellulose

Source	Std. dev.	R^2	Adjusted R^2	Predicted R^2	Press	
Linear	0.0016	0.4094	0.3503	0.2572	0.0001	
2FI	0.0016	0.4689	0.3509	0.2898	0.0001	
Quadratic	**0.0011**	**0.7641**	**0.6756**	**0.4924**	**0.0001**	**Suggested**
Cubic	0.0010	0.8488	0.7506	0.4817	0.0001	Aliased

References

1. Abo, B. O., Gao, M., Wang, Y., Wu, C., Ma, H., & Wang, Q. (2019, January). Lignocellulosic biomass for bioethanol : an overview on pretreatment, hydrolysis and fermentation processes. *Reviews on Environmental Health, 13*. https://doi.org/10.1515/reveh-2018-0054
2. Aftab, M. N., Iqbal, I., Riaz, F., Karadag, A., & Tabatabaei, M. (2019). Different Pretreatment Methods of Lignocellulosic Biomass for Use in Biofuel Production. *Biomass for Bioenergy - Recent Trends and Future Challenges*, 1–24. https://doi.org/10.5772/intechopen.84995
3. Alexander, R. A., Innasimuthu, G. M., Rajaram, S. K., Jeganathan, P. M., & Chellam Somasundarar, S. (2020). Process optimization of microwave-assisted alkali pretreatment for enhanced delignification of Prosopis juliflora biomass. *Environmental Progress and Sustainable Energy, 39*(1), 31. https://doi.org/10.1002/ep.13289

4. Amena, B. T., Altenbach, H., Tibba, G. S., & Hossain, N. (2022). Physico-Chemical Characterization of Alkali-Treated Ethiopian Arabica Coffee Husk Fiber for Composite Materials Production. *Journal of Composites Science, 6*(8), 1–10. https://doi.org/10.3390/jcs6080233
5. Awoyale, A. A., & Lokhat, D. (2021). Experimental determination of the effects of pretreatment on selected Nigerian lignocellulosic biomass in bioethanol production. *Scientific Reports, 0123456789*, 1–16. https://doi.org/10.1038/s41598-020-78105-8
6. Ayalew, A. A., & Aragaw, T. A. (2020). Utilization of treated coffee husk as low-cost biosorbent for adsorption of methylene blue. *Adsorption Science and Technology, 38*(5–6), 205–222. https://doi.org/10.1177/0263617420920516
7. Ayeni, A. O., Hymore, F. K., Mudliar, S. N., Deshmukh, S. C., Satpute, D. B., Omoleye, J. A., & Pandey, R. A. (2013). Hydrogen peroxide and lime based oxidative pretreatment of wood waste to enhance enzymatic hydrolysis for a biorefinery : Process parameters optimization using response surface methodology. *Fuel, 106*, 187–194. https://doi.org/10.1016/j.fuel.2012.12.078
8. Azelee, N. I. W., Adnan, S. A. M., Manas, N. H. A., Dailin, D. J., Ramli, A. N. M., & Illias, R. M. (2019). Assessment of microwave-assisted pretreatments for enhancing pineapple waste delignification. In *Proceedings of the 2nd International Conference on Biosciences and Medical Engineering (ICBME2019)* (p. 7).
9. Baêta, B. E. L., de Cordeiro, P. H. M., Passos, F., Gurgel, L. V. A., de Aquino, S. F., & Fdz-Polanco, F. (2017). Steam explosion pretreatment improved the biomethanization of coffee husks. *Bioresource Technology, 245*(August), 66–72. https://doi.org/10.1016/j.biortech.2017.08.110
10. Barakat, A., Mayer-Laigle, C., Solhy, A., Arancon, R. A. D., De Vries, H., & Luque, R. (2014). Mechanical pretreatments of lignocellulosic biomass: Towards facile and environmentally sound technologies for biofuels production. *RSC Advances, 4*(89), 48109–48127. https://doi.org/10.1039/c4ra07568d
11. Bekalo, S. A., & Reinhardt, H.-W. (2010). Fibers of Coffee Husk and Hulls for the Production of Particleboard. *Materials and Structures, 43*, 1049–1060.
12. de Carvalho Oliveira, F., Srinivas, K., Helms, G. L., Isern, N. G., Cort, J. R., Gonçalves, A. R., & Ahring, B. K. (2018). Characterization of coffee (Coffea arabica) husk lignin and degradation products obtained after oxygen and alkali addition. *Bioresource Technology, 257*, 172–180. https://doi.org/10.1016/j.biortech.2018.01.041
13. de Oliveira Silva e., A., Garcia, F. P., Perazzini, M. T. B., & Perazzini, H. (2023). Design and economic analysis of a pre-treatment process of coffee husks biomass for an integrated bioenergy plant. *Environmental Technology and Innovation, 30*, 103131. https://doi.org/10.1016/j.eti.2023.103131
14. Deng, H., Zhang, G., Xu, X., Tao, G., & Dai, J. (2010). Optimization of preparation of activated carbon from cotton stalk by microwave assisted phosphoric acid-chemical activation. *Journal of Hazardous Materials, 182*(1–3), 217–224. https://doi.org/10.1016/j.jhazmat.2010.06.018
15. Ethaib, S., Omar, R., Kamal, S. M., & Biak, D. R. (2015). Microwave Assisted Pretreatment.of Lignocellulosic Biomass: A Review. *Journal of Engineering Science and Technology, special Issue, 21*, 97–109.
16. Gonçalves, B. M. M., de Camillo, M. O., Oliveira, M. P., Carreira, L. G., Moulin, J. C., Neto, H. F., de Oliveira, B. F., Pereira, A. C., & Monteiro, S. N. (2021). Surface treatments of coffee husk fiber waste for effective incorporation into polymer biocomposites. *Polymers, 13*(3428), 1–22. https://doi.org/10.3390/polym13193428
17. Gouvea, B. M., Torres, C., Franca, A. S., Oliveira, L. S., & Oliveira, E. S. (2009). Feasibility of ethanol production from coffee husks. *Biotechnology Letters, 31*(9), 1315–1319. https://doi.org/10.1007/s10529-009-0023-4
18. Ha, S. H., Mai, N. L., An, G., & Koo, Y. M. (2011). Microwave-assisted pretreatment of cellulose in ionic liquid for accelerated enzymatic hydrolysis. *Bioresource Technology, 102*(2), 1214–1219. https://doi.org/10.1016/j.biortech.2010.07.108

19. Hesas, H. R., Mohd, W., Wan, A., Sahu, J. N., & Arami-niya, A. (2013). The effects of a microwave heating method on the production of activated carbon from agricultural waste : A review. *Journal of Analytical and Applied Pyrolysis, 100*, 1–11. https://doi.org/10.1016/j.jaap.2012.12.019
20. Hueso, J. L., Mallada, R., & Santamaria, J. (2022). Gas-solid contactors and catalytic reactors with direct microwave heating: Current status and perspectives. *Catalysis Today, September*, 14. https://doi.org/10.1016/j.cattod.2022.10.009
21. Ibrahim, N. A., Hadithon, K. A., & Abdan, K. (2010). Effect of fiber treatment on mechanical properties of kenaf fiber-ecoflex composites. *Journal of Reinforced Plastics and Composites, 29*(14), 2192–2198. https://doi.org/10.1177/0731684409347592
22. Inan, H., Turkay, O., & Akkiris, C. (2014). Microwave and microwave-alkali effect on barley straw for total sugar yield. *International Journal of Global Warming, 6*(2–3), 212–221. https://doi.org/10.1504/IJGW.2014.061011
23. Kandasamy, S., Muthusamy, G., Balakrishnan, S., Duraisamy, S., Thangasamy, S., Seralathan, K.-K., & Chinnappan, S. (2016). Optimization of protease production from surface-modified coffee pulp waste and corncobs using Bacillus sp . by SSF. *3 Biotech, 6*(167), 1–11. https://doi.org/10.1007/s13205-016-0481-z
24. Lin, L., Yan, R., Liu, Y., & Jiang, W. (2010). In-depth investigation of enzymatic hydrolysis of biomass wastes based on three major components : Cellulose , hemicellulose and lignin. *Bioresource Technology, 101*(21), 8217–8223. https://doi.org/10.1016/j.biortech.2010.05.084
25. López-Linares, J. C., García-Cubero, M. T., Lucas, S., González-Benito, G., & Coca, M. (2019). Microwave assisted hydrothermal as greener pretreatment of brewer's spent grains for biobutanol production. *Chemical Engineering Journal, 368*, 1045–1055. https://doi.org/10.1016/j.cej.2019.03.032
26. Lu, X., Xi, B., Zhang, Y., & Angelidaki, I. (2011). *Microwave pretreatment of rape straw for bioethanol production : Focus on energy efficiency., 102*, 7937–7940. https://doi.org/10.1016/j.biortech.2011.06.065
27. Mikulski, D., & Kłosowski, G. (2023). High pressure microwave - assisted pretreatment of softwood , hardwood and non - wood biomass using different solvents in the production of cellulosic ethanol. *Biotechnology for Biofuels and Bioproducts, 16*(19), 1–19. https://doi.org/10.1186/s13068-023-02272-9
28. Morales-Martínez, J. L., Aguilar-Uscanga, M. G., Bolaños-Reynoso, E., & López-Zamora, L. (2021). Optimization of Chemical Pretreatments Using Response Surface Methodology for Second-Generation Ethanol Production from Coffee Husk Waste. *Bioenergy Research, 14*(3), 815–827. https://doi.org/10.1007/s12155-020-10197-6
29. Mosier, N., Wyman, C., Dale, B., Elander, R., Lee, Y. Y., Holtzapple, M., & Ladisch, M. (2005). Features of promising technologies for pretreatment of lignocellulosic biomass. *Bioresource Technology, 96*(6), 673–686. https://doi.org/10.1016/j.biortech.2004.06.025
30. Mukherjee, A., Borugadda, V. B., Dynes, J. J., Niu, C., & Dalai, A. K. (2021). Carbon dioxide capture from flue gas in biochar produced from spent coffee grounds: Effect of surface chemistry and porous structure. *Journal of Environmental Chemical Engineering, 9*(5), 106049. https://doi.org/10.1016/j.jece.2021.106049
31. Murthy, P. S., & Naidu, M. M. (2012). Production and Application of Xylanase from Penicillium sp. Utilizing Coffee By-products. *Food and Bioprocess Technology, 5*(2), 657–664. https://doi.org/10.1007/s11947-010-0331-7
32. Parchami, M., Agnihotri, S., & Taherzadeh, M. J. (2022). Bioresource Technology Aqueous ethanol organosolv process for the valorization of Brewer ' s spent grain (BSG). *Bioresource Technology, 362*(August), 127764. https://doi.org/10.1016/j.biortech.2022.127764
33. Patri, A. S., Mcalister, L., Cai, C. M., Kumar, R., & Wyman, C. E. (2019). Biotechnology for Biofuels CELF significantly reduces milling requirements and improves soaking effectiveness for maximum sugar recovery of Alamo switchgrass over dilute sulfuric acid pretreatment. *Biotechnology for Biofuels, 12*(177), 1–11. https://doi.org/10.1186/s13068-019-1515-7

34. Salim, M. R., Asik, J., & Sarjadi, M. S. (2021). Chemical functional groups of extractives, cellulose and lignin extracted from native Leucaena leucocephala bark. *Wood Science and Technology, 55*(2), 295–313. https://doi.org/10.1007/s00226-020-01258-2
35. Sime, W., Kasirajan, R., Latebo, S., Mohammed, A., & Seraw, E. (2017). Coffee Husk Highly Available in Ethiopia as an Alternative Waste Source for Biofuel Production. *International Journal of Scientific & Engineering Research, 8*(7), 1874–1880. http://www.ijser.org
36. Sluiter, A., Ruiz, R., Scarlata, C., Sluiter, J., & Templeton, D. (2005). *Determination of Extractives in Biomass. Laboratory Analytical Procedure (LAP). Technical Report NREL/TP-510-42619.*
37. Sluiter, A., Hames, B., Ruiz, R., Scarlata, C., Sluiter, J., & Templeton, D. (2008). *Determination of Ash in Biomass. Laboratory Analytical Procedure (LAP). Technical Report NREL/TP-510-42622* (p. 8).
38. Sluiter, A., Hames, B., Ruiz, R., Scarlata, C., Sluiter, J., Templeton, D., & Crocker, D. (2012). *Determination of Structural Carbohydrates and Lignin in Biomass. Laboratory Analytical Procedure (LAP). Technical Report NREL/TP-510-42618* (p. 18).
39. Solhi, L., Guccini, V., Heise, K., Solala, I., Niinivaara, E., Xu, W., Mihhels, K., Kröger, M., Meng, Z., Wohlert, J., Tao, H., Cranston, E. D., & Kontturi, E. (2023). Understanding Nanocellulose-Water Interactions: Turning a Detriment into an Asset. *Chemical Reviews, 123*(5), 1925–2015. https://doi.org/10.1021/acs.chemrev.2c00611
40. Sun, J., Wang, W., Yue, Q., Ma, C., Zhang, J., Zhao, X., & Song, Z. (2016). Review on microwave – metal discharges and their applications in energy and industrial processes. *Applied Energy, 175*, 141–157. https://doi.org/10.1016/j.apenergy.2016.04.091
41. Veiga, T. R. L. A., Lima, J. T., de Dessimoni, A. L. A., Pego, M. F. F., Soares, J. R., & Trugilho, P. F. (2017). Caracterização de diferentes biomassas vegetais para produção de biocarvões. *Cerne, 23*(4), 529–536. https://doi.org/10.1590/01047760201723042373
42. Zhou, Y., Wang, W., Sun, J., Fu, L., Song, Z., Zhao, X., & Mao, Y. (2017). Microwave-induced electrical discharge of metal strips for the degradation of biomass tar. *Energy, 126*, 42–52. https://doi.org/10.1016/j.energy.2017.03.008
43. Zhuang, J., Li, M., Pu, Y., Ragauskas, A. J., & Yoo, C. G. (2020). Observation of potential contaminants in processed biomass using fourier transform infrared spectroscopy. *Applied Sciences (Switzerland), 10*(4345), 1–13. https://doi.org/10.3390/app10124345

Chapter 14
Formulation, Extraction, and Characterization of Biopesticides from Birbira (*Milletia ferruginea*) Seed and Neem (*Azadirachta indica*) Leaf

Mequanint Demeke Aynalem, Surafel Aregahegn Agdew, Ayalneh Adnew Alemayehu, Menber Berihun Nigatu, Addisu Getahun Adugna, Asmare Tezera Admassie, Kindie Bitew Worku, and Alebel Abebaw Teshager

14.1 Introduction

Pesticides are chemical substances used in agricultural practices to control pests, dis eases and weeds, and aid production and yield [1]. Synthetic pesticides are persistent and wide-spectrum compounds that bring an accumulation of pesticidal residues in the environment when they are utilized excessively in agriculture [2]. Many stable pesticide compounds can persist in the environment for a long time and they may accumulate in natural resources and enter the food chain. This can lead to increased health risks as some pesticides are toxic and can cause harm even in small amounts [3, 4]. The active ingredients present in synthetic pesticides cause severe side effects [5, 6] and lead to disturbances in the environment, causing pest-resistant organisms [7]. Contemporary.

research findings reported that synthetic insecticides, besides their high cost and environmental hazardous, penetrated into the stored seeds and reduced the germination capacity of the seed [2, 8]. Thus, there is a need to reduce the use of synthetic pesticides and promote sustainable and eco-friendly agricultural practices, botanical pesticides are environmentally friendly [8], and using plant extracts help in reducing environmental and health concerns [9]. For this reason,

M. D. Aynalem (✉) · S. A. Agdew · A. A. Alemayehu · A. G. Adugna · A. T. Admassie
A. A. Teshager
Faculty of Chemical and Food Engineering, Bahir Dar Institute of Technology, Bahir Dar University, Bahir Dar, Ethiopia

M. B. Nigatu · K. B. Worku
Faculty of Civil and Water Resources Engineering, Bahir Dar Institute of Technology, Bahir Dar University, Bahir Dar, Ethiopia

S. W. Fanta et al. (eds.), *Sustainable Development Research in Manufacturing, Process Engineering, Green Infrastructure, and Water Resources*, Green Energy and Technology, https://doi.org/10.1007/978-3-031-77339-6_14

utilizing botanical pesticides is an alternative strategy for agriculture [7] and it has lower toxicity on the environment. In addition, biopesticides affect the target pest and they do not build up resistance in target pests [10]. The application of biopesticides in Africa is still at its beginning and little information is available on adoption of biopesticides on the continent [11]. Botanical pesticides are naturally available chemicals extracted from different parts of plants such as Birbira (*Milletia ferruginea*) and neem plants which contain bioactive compounds that affect the reproductive and digestive process of important pests [8, 12, 13]. Birbira seed is rich in over 25 flavonoids, 50 isoflavonoids, 12 chalcones, and miscellaneous compounds. Rotenone is one of the dominant compounds found in the seed and stem bark of *M. ferruginea* [14]. Neem (Mahogany family Meliaceae) contains an active ingredient (azadirachtin) which has proven effectiveness as a pesticide against about 200 insect species and is non-toxic to humans [15]. This plant is rich in more than two hundred allelochemicals that provide a multiplicity of pesticidal properties [16]. Traditionally, neem leaves have been utilized for centuries to protect stored grain from pests due to its pesticidal properties [17], which can be exploited to develop an efficient pest control strategy [18]. Biopesticides are chief part of sustainable agriculture, which fulfills 2030 agenda of United Nations [19]. However, their application and availability increment and efficacy and stability improvements are made through new active substances and advanced formulations [20] which enable the pesticides to be applied in a wettable powder, soluble powder, or emulsifiable concentrate [21]. In the formulation and extraction process, the solvent type, extraction time, and method of extraction influence the toxicity effect of botanical pesticides [23, 24, 28]. The solvent with good extraction capacity, low viscosity, and higher solubility with oil is preferred, for it can penetrate plant samples inside to extract oils effectively. Such solvents are organic solvents (ethanol, hexane) which are excellent for plant oil extraction [25]. The oil extraction from plant seeds and leaves has been influenced with extraction time depending on the plant material and previous research findings reported that range from 3–8 h is a recommended extraction time [23, 28] to result in an increased yield of plant seeds and leaves oil [22]. Researches reveal that there are several extraction methods like mechanical pressing, supercritical fluid extraction, steam pressure extraction, and solvent extraction [25, 26]. Soxhlet extraction gives high yield and less turbid biopesticides, in addition to its low operating and investment costs [12, 27], which can be applied for botanical pesticide extraction [28]. Soxhlet extraction can utilize (10–30 g) of plant samples with solid-to-solvent ratios of (1:5 to 1:10), respectively [28, 29]. Proper extraction time to prevent bioactive component degradation due to temperature effect [30]. The research findings concluded that the solid-to-solvent ratio should be in the proper ranges to get the desired product in low energy and short time [31, 32]. The present research paper has investigated the formulation, extraction, and characterization of biopesticide from birbira seed and neem leaf which are available in Ethiopia throughout the year.

14.2 Materials and Methods

14.2.1 Equipment and Instrument

Sieve (SETHI) was used to determine particle size, electronic balance (FA2104) used to weigh mass, and cutter and milling machine (PULVERISETTE13) used for size reduction. Filter paper and measuring cylinder were used to separate mixtures and measure volume. Soxhlet extraction was used to extract plant oil and steam distillation was used to concentrate biopesticide. Oven (PID system M60-vf) to dry neem leaf and birbira seed and P^H meter (PHS-25) used to measure acidity or basicity of biopesticide. A distiller (DZ-5LII) was used to produce distilled water. A mixer (MS7-H550-Pro) was used to mix ingredients. Polyethylene bag was used to store the neem leaf and birbira seed powder, flask was used to store supernatant filtrate, a beaker was used to hold distilled water/extracted product, thin layer chromatography (TLC) paper used to know the purity, chromatography chamber to sample holder, water bath (HH-S) used to heat the extract for phytochemical analysis. Hydrometer is used to measure the specific gravity (SG).

14.2.2 Chemicals and Reagents

Ethanol (C_2H_5OH, 97%) used as a solvent for extraction, distilled water (*H2O*) to wash the birbira seed and fresh leaf, ethyl acetate ($C_4H_8O_2$, 92.5%), diethyl ether ($(C_2H_5)_{2^o}$, 99%) for TLC characterization and potassium bromide (KBr, 99%) for Fourier-transform infrared (FTIR) analysis of biopesticide. Birbira seed and neem leaf (air-dried) were used as raw materials.

14.2.3 Experimental Setup and Description

Raw materials were collected, prepared, and stored for further processes. Then, after drying, the birbira seed and neem leaf were ground to power with the mill. Next, a given mass of birbira seed and neem leaf powder was weighed and added to a measured volume of ethanol (solvent) and samples were soaked into the extractor, as illustrated in Fig. 14.1. Powdered samples were enclosed in a porous bag (thimble). Then, the active components from birbira seed and neem leaf powder were extracted using Soxhlet extractor with ethanol (solvent). Finally, the product was characterized with physicochemical properties, TLC, FTIR, phenolic content, and flavonoid contents.

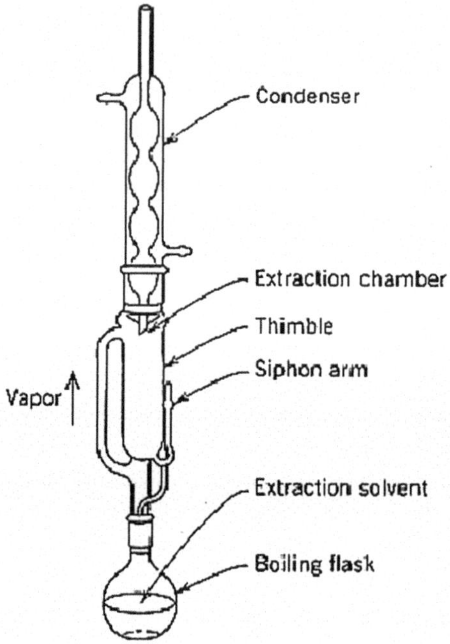

Fig. 14.1 Experimental setup of Soxhlet extraction apparatus [33]

14.2.4 Methods

14.2.4.1 Collection and Preparation of Raw Materials

A 200 g birbira seed and neem leaf were collected from the local market and Bahir Dar Institute of Technology, respectively, and then they were washed with tap water and dried to remove impurities. For the removal of moisture content birbira seed and neem leaf were dried in oven at 110 °C for 24 h so as to ensure high extraction efficiency. The samples were pulverized to fine powders with a grinding mill and the powdered samples were stored in an organic lab at Bahir Dar Institute of Technology (BiT), Bahir Dar University till further use (Fig. 14.2).

14.2.5 Formulations of Biopesticides

In an experiment, ethanol was used as a solvent for the extraction of biopesticide from birbira seed and neem leaf. In Soxhlet extraction, it is allowed to use a sample mass of (10–30 g) was extracted in the solid-to-solvent ratio of (1:5 to 1:10), respectively with the range of extraction time (3–8 h) [28, 29]. In this experiment, both extraction time (3, 4, and 5 h) [13, 23] and mixing ratios (R_1 (1:1), R_2 (2:1), R_3 (1:2),

Fig. 14.2 Raw material collection and preparation

and birbira seed to neem leaf powders) were considered as factors. After the raw materials and the solvent (ethanol) were collected and prepared for extraction, the biopesticide was formulated as birbira seed powder to neem leaf powder. Birbira seed is rich in rotenone compounds.

[14] and neem leaf contains an active ingredient (azadirachtin) [15] which provides a diversity of pesticidal properties [16].

14.2.6 Extraction and Recovery of Biopesticides

The powder samples (birbira seed and neem leaves) were proportionated as sample powder to solvent ethanol ratio (1:7.5 w/v) [28, 29] and placed into the thimble made of cotton wool and placed in the Soxhlet extractor chamber. A pure ethanol solvent was measured with a measuring cylinder and poured into the round bottom distillation flask. The apparatus was then attached to the heating mantle which was set at solvent boiling point (78 °C) and the condenser unit was connected to an overhead water tank to cool rising solvent vapor. As the solvent was being heated, then the vaporized petroleum ether passed through the condenser and condensed. The condensed solvent dropped down into the thimble containing the material to be extracted. When the liquid level filled in the thimble, automatically the liquid fell back into the round bottom distillation flask with the extracted solute into the bulk liquid. The process was sustained until the extraction times (3, 4, 5 h) were completed. After extraction, the product solution was a mixture of solvent (ethanol) and extracted solute. The supernatants were separated from solvent using simple distillation techniques and the solvent was evaporated at a temperature of 78 °C. Then, the evaporated solvent was passed through the condenser and condensed. The condensed solvent was dropped into the conical flask. The process was continued until the extracted oil was completely free from the solvent. Finally, the remaining product was biopesticide. The solvent was recovered (reused) until the operation was finished. And, the procedure was repeated till the extraction process was completed.

14.3 Raw Materials and Product Characterization

14.3.1 Determination of Moisture Content

A 5.0 gm birbira seed or neem leaves powder (W_{Before}) was taken and dried in an oven at 105 °C for 24 h. The sample was measured until the mass became constant (W_{After}). Finally, the moisture content (MC) was determined with the empirical equation.

$$\mathrm{MC}(\%) = \frac{W_{\mathrm{before}} - W_{\mathrm{after}}}{W_{\mathrm{before}}} * 100 \quad (14.1)$$

14.3.2 Determination of Percentage Yield

The yield of extraction is calculated from the relation between the extracted biopesticide weight obtained and the raw material weight used in the extraction. Sample before extraction was weighed (WBefore) and the weight of the sample after extraction and oven drying was recorded (WAfter). Finally, the percentage of biopesticide yield (BY) was calculated with the empirical Eq. (14.2) [34] as

$$\mathrm{BY}(\%) = \frac{W_{\mathrm{Before}} - W_{\mathrm{After}}}{W_{\mathrm{After}}} \quad (14.2)$$

14.3.3 Determination of Specific Gravity

After biopesticide was extracted and separated through filtration, the density of the oil was determined using specific gravity. Specific gravity is the measurement of the relative density of the sample with respect to water. The determination was carried out using AOAC 2000 standard method. The density of the extract was determined using the density bottle method. A clean and dry density bottle (measuring cylinder) of 10 ml capacity at 30 °C was weighed in grams. Then the measuring cylinder was filled with water and re-weighed at 30 °C. The extract was brought to 30 °C and the water was substituted with this extract after drying the density bottle and weighed again and the specific gravity was determined by dividing the density of the extract by the density of water. The empirical equation uses the mass of density bottle (10 ml measuring cylinder) (W_0), mass of bottle and water (W_1), and mass of bottle and sample (W_2).

$$\text{Specific gravity}(\mathrm{SG}) = \frac{W_2 - W_0}{W_1 - W_0} \quad (14.3)$$

14.3.4 Determination of P^H Values

The pH value of the extract was measured by dissolving 500 mg of extract in 10 ml at room temperature and 1 atm. The solution was placed in a beaker and the electrode of the digital PH meter treated with buffer solution was placed in the biopesticide. Finally, the acidity or basicity of the extracted biopesticide was recorded from a digital P^H meter.

14.3.5 Thin Layer Chromatography (TLC) Analysis of Biopesticide

This characterization analysis simply indicates the qualitative characteristics of the extracted compound. The sample was dropped on a TLC plate and it was placed in iodine chamber consisting of ethyl acetate to ethyl ether (5:5) ratio as mobile phase. A number of spots were detected and the retention factor (Rf) of those spots was calculated with the following empirical equation [35]:

distance traveled by component

$$R_f = \frac{\text{distance traveled by component}}{\text{distance traveled by solvent}} \tag{14.4}$$

14.3.6 FTIR Analysis of Biopesticide

For FTIR analysis, 10 mg of dried extract was mixed in 100 mg of potassium bromide and the resulting pellet was placed in FTIR spectroscope. The functional groups were analyzed for the scanning extract sample with a scan range from 400 to 4000 cm-1, and the peak values of FTIR were recorded [36].

14.3.7 Determination of Total Phenolic Content (TPC)

Folin-Ciocalteu's reagent was used for the determination of total phenolic content, with gallic acid as a standard. At 760 nm, the absorbance was measured, and results were articulated as (mg gallic acid equivalent (GAE) per g) [36, 37]. The measurements were done in triplicate. The data for total phenolic contents was expressed as mg of gallic acid equivalent weight (GAE)/g of dry mass with the empirical Eq. (14.5) [38, 39] using total phenolic content (P) calculated from the calibration curve (mg GAE/ml), sample volume (V in ml), and sample fresh weight (m in g).

$$\text{Total phenolic content}\left(\text{mg}\frac{\text{GAE}}{\text{g}}=\frac{\text{P}*\text{V}}{\text{m}}\right) \tag{14.5}$$

14.3.8 Determination of Total Flavonoids Content (TFC)

The TFC of the biopesticide was determined according to the aluminum chloride colorimetric method using quercetin as standard [40, 41]. Each sample of the extract (4 mg) was suspended in methanol (4 ml). A 0.25 ml of each suspension was transferred to a tube followed by the addition of 1.25 ml of distilled water and 0.3 ml of 5% $NaNO_3$. After 6 min, 0.150 ml of 10% aluminum chloride was added and the mixture was left for 5 min in the dark. This was followed by adding 0.5 ml of 5% sodium hydroxide and 0.275 ml water and measurement of absorbance by UV–visible spectroscopy at 500 nm wave length. The TFC in the extract was expressed as milligrams quercetin equivalent per gram of dry weight (mg QE/g) which was calculated in all extracts separately using Eq. (14.6) [39, 40], using concentration of quercetin from the calibration curve (C in mg/ml), extract sample volume (V in ml), and sample weight (m in g).

$$\text{Total Flavonoids Content}\left(\text{mg}\frac{\text{QE}}{\text{g}}=\frac{C*V}{m}\right) \tag{14.6}$$

14.3.9 Efficacy of Biopesticides on Worker Termites

The botanical biopesticides were diluted to different concentrations (25, 50, and 75% w/v) using appropriate solvents and incorporated into the termite diet to assess the mortality of worker termites with the control of commercial insecticide (roach-killer). The Petri dishes having treated filter paper with botanical biopesticides were held at temperatures of 25 ± 3 °C. All Petri dishes were kept at room temperature for 15 min, 30 min, 45 min, 60 min, and 85 min. After the treatment, each termite in the Petri dishes was carefully examined. The percent mortality rate of dead worker termites (*Macrotermes* spp.) was counted and calculated using the following empirical Eq. (14.7) [43]:

$$\text{Mortality rate}(\%)=\frac{\text{Number of Dead Termites}}{\text{Total Number of Termites}}*100\% \tag{14.7}$$

14.4 Results and Discussion

14.4.1 Raw Material Characterization

The raw materials were studied with different physicochemical properties which are important parameters to characterize the product. Moisture content of birbira seed (6.10 ± 0.125%) is low since the dry seed was collected while a fresh neem leaf has high percentage moisture content (21 ± 0.12%), which is then dried with air to assist solvent diffusion towards the leaf surface for extraction. A related research finding reported that higher moisture content blocks solvent diffusion and yields less bioactive extract as compared to those with lower moisture content [44]. Moreover, they have high organic matter (96.5 ± 0.2% of birbira seed and 96.4 ± 0.19% of neem leaf) and low ash content, which approve to extract high-yield, biodegradable, and environmentally sustainable products. In addition, these botanical plants have highly soluble properties (93.06 ± 7.08% of birbira seed and 99.71 ± 0.01% of neem leaf) which implies a lower residue left on the crop and uniformly high dispersion tendency of biopesticide on the field. Further, they have lower P^H values (4 ± 0.21 of birbira seed and 5 ± 0.25 of neem leaf) implying mildly acidic, which indicates the presence of free fatty acid. Besides, they contained low specific gravity (0.84 ± 0.12 of birbira seed and 1.34 ± 0.07 of neem leaf) which indicates less impurities are available in the biopesticide. The investigation of raw materials indicated that they contained potential physicochemical characteristics to extract bioactive organic pesticides.

14.5 Physicochemical Characteristics of Biopesticides

14.5.1 Determination of Biopesticide P^H *Values*

As illustrated in Fig. 14.3, the general trend of the biopesticide of all formulations demonstrated a decreasing tendency in P^H values from 3 h to 4 h and then a slight increase in P^H from 4 h to 5 h of extraction times. All the formulations have high P^H values (low acidity) at 3 h extraction time because the bioactive components were not completely extracted and mixed with the solvent. When the extraction time was increased to 4 h, the P^H values of all formulations were decreased (high acidity) since the bioactive components of powders were extracted and blended with the solvent. On the other hand, there was a slight increment of the P^H value in the three formulations when the extraction time was extended to 5 h because probably some bioactive phytochemicals were denatured. Generally, formulation R_2 contained the lowest P^H value (highest acidity) among all formulations, especially at 4 h extraction time. Related research findings reported similar outputs on biopesticides (up to 5.8 P^H) [19, 33]. This indicated that birbira seed and neem leaf have the potential to produce organic biopesticide.

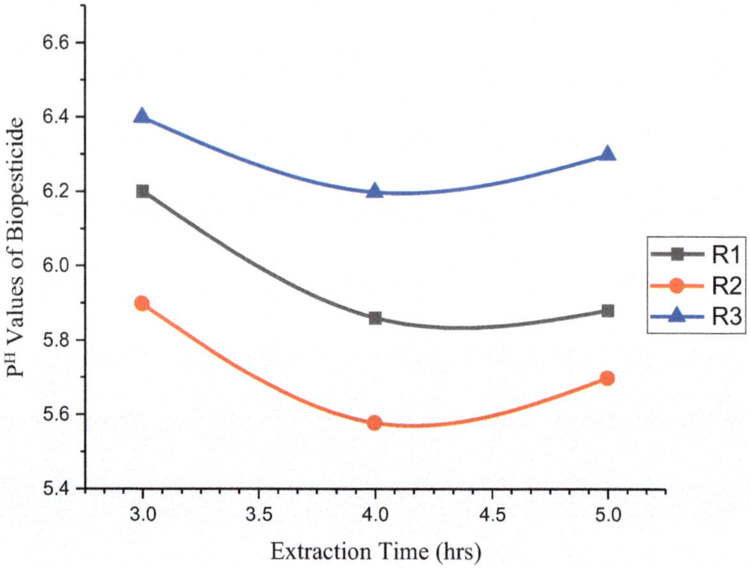

Fig. 14.3 P^H value of organic biopesticide

Table 14.1 Specific gravity (SG) of organic biopesticide

		Formulations	
Time (h)	R1	R2	R3
3	0.91 ± 0.0145	0.94 ± 0.015	0.9 ± 0.0144
4	0.96 ± 0.0154	0.97 ± 0.0155	0.95 ± 0.0152
5	0.93 ± 0.0149	0.94 ± 0.0150	0.92 ± 0.0147

14.5.2 Specific Gravity of Organic Biopesticide

As illustrated in Table 14.1, the effect of extraction time and mixing ratios on the specific gravities (SG) of all the three formulations of organic biopesticides are approximately the same (>90%) with the specific gravity of water. Among biopesticide formulations, the highest specific gravity (SG) was found at R_2 and 3 h extraction time as compared to other formulations of specific gravies. This indicates that the extracted bioactive compounds can be blended properly with water and enhance the dissociation due to diffusion and surface attraction with water during solubilization. In addition, it is important to make solution not to be easily dispersed by wind after use on the field.

14.5.3 Percentage Yield of Biopesticide

As demonstrated in Table 14.2, the effect of formulations and extraction times on the yield of biopesticide is illustrated. In all formulations of organic biopesticide (R_1, R_2, R_3), the maximum results were obtained at 4 h extraction time as compared with 3 h and 5 h of extraction time as it is observed. This result justifies that as extraction time increased from 3 h to 4 h, the amount of extract was increased. This is because the solvent has enough time to penetrate the membrane and solubilize the active component. On the contrary, when the extraction time was further increased from 4 h to 5 h of extraction time, the yield was decreased this was probably due to thermal degradation of bioactive compounds. In conclusion, according to this research the highest percent-age yield of organic biopesticide was obtained at formulation ratio R_2 and 4 h extraction time.

14.5.4 Thin Layer Chromatography (TLC) Biopesticide Analysis

The qualitative TLC analysis characterized and identified a number of useful compounds present in the biopesticide and the investigations are presented in Table 14.1. Ac cording to TLC analysis, if Rf value of pure component is matching to any Rf value of the extract, it shows the presence of that pure compound in the extract. As illustrated in Table 14.1, it shows that the Rf value is greater than 0.36 and less than 1, which indicates the product has high interaction to the mobile phase. And the main components of the product are found to be Nimbin, Azadiracthin, and rotenone, with Rf value of 0.9, 0.71, and 0.71, respectively including amino acids with Rf value of 0.64. These values are approximately the same with the standard Rf values [35]. In conclusion, TLC characterization demonstrates the potential to extract biopesticides from birbira seed and Neem leaf plants (Table 14.3).

Table 14.2 Percentage yield of biopesticide

		Formulations	
Time (h)	R1	R2	R3
3	26.9 ± 0.43	28.3 ± 0.45	25.6 ± 0.41
4	37.5 ± 0.6	40.5 ± 0.648	36 ± 0.576
5	31.6 ± 0.506	33.2 ± 0.531	30.2 ± 0.483

Table 14.3 TLC analysis of biopesticide

Number of components	Distance covered by solvent	Distance covered by component	*Rf*	Number of compounds identi fied
A	7	5	0.71	Azadirachtin and rotenone
B	7	4.5	0.64	Amino acid
C	7	6	0.9	Nimbin

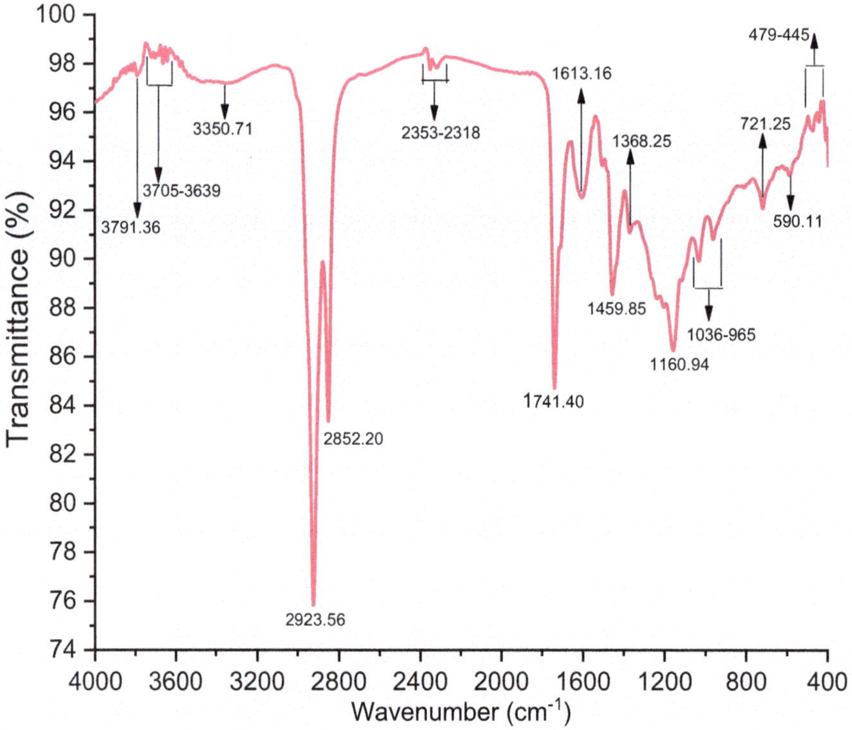

Fig. 14.4 FTIR analysis of functional groups in the biopesticide

14.5.5 Fourier-Transform Infrared Spectroscopy Analysis of Biopesticide

FTIR spectrum of ethanoic extract biopesticide of mixing ration R_2 is shown in Fig. 14.4, which aids to identify functional groups between 400 and 4000 cm^{-1} regions based on the peak values of the spectrum. The data on the peak values and the probable functional groups (obtained by FTIR analysis) are presented in Table 14.4. As illustrated in Table 14.4 which is the peak of Fig. 14.3, convincingly, FTIR detected the existence of hydroxyl groups, amine, alkane, alkynes, nitrile, carbonyl, amide, alcohol (phenol), ester, alkyl-halide, and aryl disulfide that are

Table 14.4 FTIR analysis of functional groups in the biopesticide

Absorbance (cm^{-1})	Bond/group	Functional group
3791–3639	O-H (stretch)	Hydroxyl groups
3350.71	N-H(stretch)	Amine (aliphatic)
2923–2852	C-H (stretch)	Alkane
2353–2318	C (triple bond), N stretch	Alkynes, nitrile
1741–1613	C=O(stretch)	Carbonyl and amide
1459.85–1368	C-H(bend)	Alkane
1160–1037	C-O (stretch)	Alcohol and ester
721.25	C-H (bend)	Alkane
590.11	C-I (stretch)	Alkyl-halide
479–445	S-S (stretch)	Aryl disulfide

most likely responsible for the antimicrobial properties of the biopesticide. These peak values are found in the following absorption bands of related works. The absorption regions 3700–3000 cm^{-1} is probably due to the vibrations of O-H stretching arising from hydrogen bonding [45]. The region of absorbance at 3000–2800 cm^{-1} is most likely assigned for C-H, O-H, and NH_3, which may be related to carboxylic acids, free amino acids, and phenolics [43, 44]. Carbonyl groups are located in the 1850–1540 cm^{-1} region and aromatics of strong skeletal bands in the 1600–1300 cm^{-1} region. Aliphatic structures are located in the region 909–650 cm^{-1} [48]. The vibration range is 1800–1700 cm^{-1}, which is most probably associated to C=O elongation of the ester-type carboxylic [46, 47]. Aliphatic compounds type methylene groups present stretching at 2930 cm^{-1} (asymmetric) and at 2850 cm^{-1} (symmetric). In-plane binding of C-H is located between 1275 and 1000 cm^{-1} and out-of-plane binding appears in the region of 900–690 cm^{-1} [50]. The vibration in the region of 1700–1550 cm^{-1} is typically originated from stretching band of carbonyl groups C=O and C=C [42, 49]. Additionally, a vibration approximately at 1155 cm^{-1} is a result of C–O stretching [51] while bands corresponding to organic acids (1500–1300 cm^{-1}), and phenols (1175–940 cm^{-1}) [45]. Therefore, FTIR analysis demonstrated that the extracted biopesticide is rich in bioactive compounds, which can be used as pest control for agricultural applications.

14.5.6 Determination of Phenolic Content

The total phenol composition of biopesticide extract was determined using Folin-Ciocalteu's reagent using gallic acid as the standard. The absorbance values at different concentrations of gallic acid were used to develop the calibration curve which was obtained by plotting the values of different concentrations of gallic acid versus their absorbance. The total phenolic contents were found different for the three formulations. Total phenolic content of formulation R_1 (144 ± 0.45 mg GAE/g dry extract) was found between the phenolic content of formulation R_2 (170 ± 0.36 mg

GAE/g dry extract) and formulation R_3 (36 ± 0.27 mg GAE/g dry extract). Relatively, the maximum phenol content was found at formulation R_2. Shewale & Rathod [33] found phenolic content of (6.64–13.75 mg GAE/g dry extract) from neem leaf powder using ethanol as solvent. In another study, Khamis Al-Jadidi & Hossain [42] reported that the phenol content from neem leaf was (20.8–51 mg GAE/g dry extract) with different solvents. This clearly indicated that birbira seed contained a higher amount of phenolic content than neem leaf. Thus, both plants are rich in bioactive compounds and can be utilized for botanical pesticides production for sustainable agricultural practices.

14.5.7 Determination of Flavonoid Content

In the present study, total flavonoids contents of the different formulations were determined with standard method as quercetin equivalents [41, 42]. The amount of total flavonoid contents were found different for the three formulations. The results revealed that R_2 has the highest total flavonoid contents (72.2 ± 0.65 mg QE/g dry extract) while R_3 is with the lowest total flavonoid contents (28.8 ± 0.37 mg QE/g dry extract). Additionally, R_1 (44.9 ± 0.45 mg QE/g dry extract) total flavonoid content was found between the flavonoid contents of R_2 and R_3. Several related research findings on flavonoid con- tents reported similar results with some differences [38, 39], which might possibly be due to the different methods of extraction, solvents polarities, or some volatiles active compounds denature or evaporated from the samples during processing. In general, the flavonoid content was found high when birbira seed powder increased in the formulation (R_2), which indicated that birbira seed contained high amount of flavonoid content than neem leaf has according to the findings of the present observations.

14.5.8 Efficacy of Biopesticides on Worker Termites

In the toxicity test of biopesticide formulation, different concentrations were prepared using extracted oil of mixing ratio R_2 in distilled water as (25%, 50%, and 75% w/v). The present study indicated that the percentage mortality rate of worker termites was influenced by both the concentration of biopesticide and the length of treatment time. As demonstrated in Fig. 14.5, the percentage mortality rate of worker termite increased when the treatment time was varied from 15 min to 85 min. Among various concentrations of biopesticide the lowest percentage mortality rate was observed at 25% w/v concentration while the highest percentage mortality rate was recorded at 75% w/v concentration. Moreover, the percentage mortality rate of worker termites treated with 50% w/v and 75% w/v concentrations of biopesticide from 60 min to 85 min were found to be approximately comparable to the control treatment (roachkiller) and showed excellent results in termite mortality. A related research work on neem found and reported that the neem crude extracts killed pests

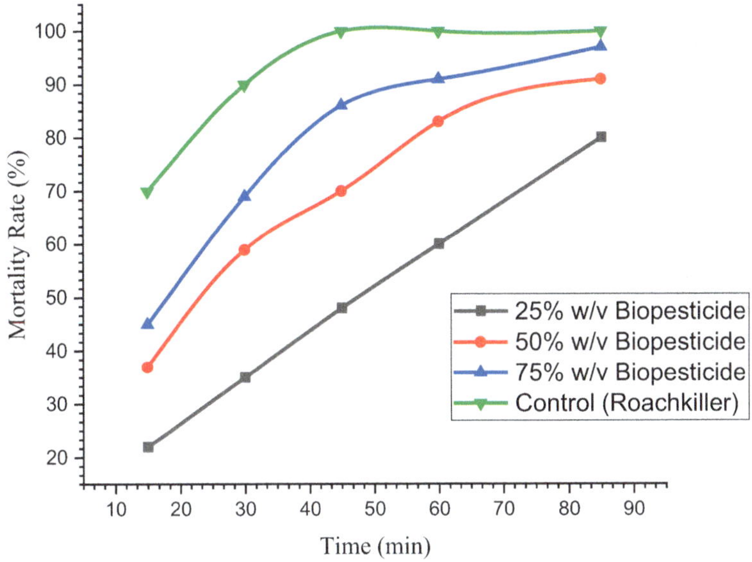

Fig. 14.5 Mortalities of worker termites treated with biopesticides

in 8 h [28] and another research finding on *M. ferruginea* reported that the birbira (*M. ferruginea*) seed crude extracts can kill insects within 24 h [14]. The combined effect of botanical compounds may increase the efficacy of the product and result in a shorter time to kill worker termites. This could be a promising solution of natural products for pest controlling in a more sustainable and environmentally friendly way. Therefore, these botanical plants can be potentially used to produce biopesticides.

14.6 Conclusions

In this chapter, raw botanic plants were investigated for their potential to extract bioactive organic pesticides. The plants were found to contain a high percentage of organic matter, highly soluble properties, lower pH value, and low specific gravity. These physicochemical characteristics showed that the plants are vital to extract bioactive organic pesticides. The formulation and characterization of biopesticides indicated they contain the lowest P^H value, highest specific gravity, and the maximum yield at R_2 formulation and 4 h extraction time. The qualitative TLC analysis identified that azadirachtin, rotenone, amino acid, and nimbin compounds are present in the biopesticide. FTIR spectrum analysis based on wavenumber demonstrated the existence of carboxylic acids, free amino acids, phenolics, organic acids, aromatics (flavonoids), and phenols that are most likely responsible for the antimicrobial properties of the biopesticide extracts. Quantitatively, the maximum phenol content (170 ± 0.36 mg GAE/g dry extract) and flavonoid contents (72.2 ± 0.65 mg

QE/g dry extract) were found at formulation R_2 with 4 h extraction time. In the efficacy of biopesticide, 97% mortality rate of worker termites was observed at 75% w/v extracted oil in 85 min while the control (roachkiller) resulted in 100% mortality rate in about 40 min. The combined bioactive compounds of these botanical plants might bring the strength of this natural biopesticide to achieve 97% mortality rate for pests. Therefore, these botanical plants can be potentially utilized to produce biopesticides.

References

1. Kumar, N., Pathera, A., et al. (2012). Harmful effects of pesticides on human health. *Annals of Agri-bio Research, 17*(2), 125–127.
2. Gilden, R. C., Huffling, K., & Sattler, B. (2010). Pesticides and health risks. *Journal of Obstetric, Gynecologic, and Neonatal Nursing, 39*(1), 103–110.
3. Lopes-Ferreira, M., et al. (2022). Impact of pesticides on human health in the last six years in Brazil. *International Journal of Environmental Research and Public Health, 19*(6), 3198.
4. Mahmood, I., Imadi, S. R., Shazadi, K., Gul, A., & Hakeem, K. R. (2016). Effects of pesticides on environment. *Plant, soil and microbe, Volume 1: Implications in Crop Science, 1*(1), 253–269.
5. Townson, H. (1992). *Public Health Impact of Pesticides Used in Agriculture, 86*(3).
6. W. O. World-Health-Organization. (1990). *Public health impact of pesticides used in agriculture* (Vol. 12, p. 5). World Health Organization.
7. Chaudhary, S., et al. (2017). Progress on Azadirachta indica based biopesticides in replacing synthetic toxic pesticides. *Frontiers in Plant Science, 8*, 1–13.
8. Getu, E. (2014). Bioefficacy of products derived from Milletia ferruginea (Hochst) baker against the bean bruchid, Zabrotes subfasciatus (bruchidae: Coleoptera) in stored beans in Ethiopia. *African Journal of Agricultural Research, 9*(37), 2819–2826.
9. Mulatu, B. (2009). Contact bioassay of an endemic plant to Ethiopia on three aphid species. *Ethiopian Journal of Biological Sciences, 6*(1), 51–62.
10. Akutse, K. S., Subramanian, S., Maniania, N. K., & Dubois, T. (2020). *Biopesticide research and product development in Africa for sustainable agriculture and food security—Experiences from the International Centre of Insect Physiology and Ecology (icipe)* (Vol. 4, pp. 1–14).
11. Olson, S. (2015). An analysis of the biopesticide market now and where it is going. *Outlooks Pest Management, 26*(5), 203–206.
12. Gebrehawaria, A. (2010). *Extraction of azadirachtin from neem seeds for biopesticide purpose.*," Doctoral Diss. ADDIS ABABA Univ (Vol. 1, pp. 82–95).
13. Vizitiu, D. E., et al. (2022). Grapevine plants management using natural extracts and Phytosynthesized silver nanoparticles. *Materials (Basel)., 15*(22), 8188.
14. Jembere, B., Getahun, D., Negash, M., & Seyoum, E. (2002). Toxicity of birbira (Milletia ferruginea) seed crude extracts to some insect pests as compared to other botanical and synthetic insecticides. In *11th NAPRECA Symposium B. Proceedings* (pp. 88–96).
15. Csurhes, S. (2016). Pest plant risk assessment Neem tree Azadirachta indica. *Department of Agriculture and Fisheries Biosecurity Queensland*, 5–6.
16. S. Koul, O., and Wahab. (2004). Neem: Today and in the new millennium. *Neem Today New Millennium, 7*(2), 125–131.
17. Koul, O., Isman, M. B., & Ketkar, C. M. (2011, Jan). Properties and uses of neem, Azadirachta indica. *Canadian Journal of Botany, 68*(1), 1–11.
18. Paul, R., Prasad, M., & Sah, N. K. (2011, Sep). Anticancer biology of Azadirachta indica L (neem): A mini review. *Cancer Biology & Therapy, 12*(6), 467–476.

19. Fenibo, E. O., Ijoma, G. N., & Matambo, T. (2021, June). Biopesticides in sustainable agriculture: A critical sustainable development driver governed by green chemistry principles. *Frontiers in Sustainable Food Systems, 5*.
20. Šunjka, D. (2022). An alternative source of biopesticides and improvement in their formulation — Recent advances. *Plants MDPI, 11*(3172), 1–13.
21. Boadu, K. O., Tulashie, S. K., Anang, M. A., & Kpan, J. D. (2011). Production of natural insecticide from neem leaves (Azadirachta indica). *Asian Journal of Plant Science and Research, 1*(4), 33–38.
22. Yadav, K. K., Garg, N., Verma, A., Kumar, S., & Trivedi, M. (2017). Optimization and extraction of oil from mango seed kernel (Mangifera indica). *Indian Journal of Agricultural Sciences, 87*(7).
23. Bekele, J. (2002). Evaluation of the toxicity potential of Milletia ferruginea (Hochest) baker againts Sitophilus zeamais (Motsch.). *International Journal of Pest Management, 48*(1), 29–32.
24. Poole, C. F., Cooke, M., & Wilson, I. D. (2000). *Encyclopedia of separation science, ten-volume set* (p. 4927). Ian Wilson.
25. Banat, F., Pal, P., Jwaied, N., & Al-Rabadi, A. (2013). Extraction of olive oil from olive cake using Soxhlet apparatus. *American Journal of Oil and Chemical Technologies, 1*(4).
26. Mahawer, S. K., et al. (2022). Extractions methods and biological applications of essential oils. *Essential Oils, 10*(112), 232–258.
27. Phong, W. N., Gibberd, M. R., Payne, A. D., Dykes, G. A., & Coorey, R. (2022). Methods used for extraction of plant volatiles have potential to preserve truffle aroma: A review. *Comprehensive Reviews in Food Science and Food Safety, 21*(2), 1677–1701.
28. Tulashie, S. K., Adjei, F., Abraham, J., & Addo, E. (2021). Potential of neem extracts as natural insecticide against fall armyworm (Spodoptera frugiperda (J. E. Smith) (Lepidoptera: Noctuidae)). Case Studies in Chemical and Environmental Engineering, 4, 100130.
29. Subramanian, S., Salleh, A. S., Bachmann, R. T., & Hossain, M. S. (2019). Simultaneous extraction and separation of oil and Azadirachtin from seeds and leaves of Azadirachta indica using binary solvent extraction. *Natural Product Sciences, 25*(2), 150–156.
30. Go, R., Kankara, S. S., Mustafa, M., Ibrahim, H. M., & Nulit, R. (2014). Effect of drying methods, solid-solvent ratio, extraction time and extraction temperature on phenolic antioxidants and antioxidant activity of Guiera senegalensis J.F. Gmel (Combretaceae) leaves water extract. *American Journal of Phytomedicine and Clinical Therapeutics, 2*(12), 1378–1392.
31. Zubairi, S., Sarmidi, M., & Malaysiana, R. A.-S. (2014). The effects of raw material particles size, types of solvents and solvent-to-solid ratio on the yield of rotenone extracted from Derris elliptica roots. *core.ac.uk, 43*(5), 707–713. Accessed: Mar. 21, 2023.
32. Nikhal, S., Dambe, P., et al. (2010). Hydroalcoholic extraction of Mangifera indica (leaves) by Soxhletion. *Citeseer, 1*(7), 30–32.
33. Shewale, S., & Rathod, V. K. (2018). Extraction of total phenolic content from Azadirachta indica or (neem) leaves: Kinetics study. *Preparative Biochemistry & Biotechnology, 48*(4), 312–320.
34. Handa, S. K., Agnihotri, N. P., & Kulshrestha, G. (1999). *Pesticide residues : Significance, management and analysis*. Research Publications & Book Pub. House. Accessed: Apr. 27, 2023.
35. Mukunda Vani, M. (2016). Identification and chemical characterization of Azadirachta Indica leaf extracts through thin layer chromatography. *International Journal of Engineering Research & Technology, 05*(02), 117–122.
36. Visveshwari, M., Subbaiyan, B., & Thangapandian, V. (2018). Phytochemical analysis, antibacterial activity, FTIR and GCMS analysis of Ceropegia juncea Roxb. *International Journal of Pharmacognosy and Pharmaceutical Research, 9*(07).
37. Hathout, A., et al. (2016). Novel Egyptian bacterial strains exhibiting antimicrobial and antiaflatoxigenic activity. *Journal of Applied Pharmaceutical Science, 6*(12), 001–010.
38. Ghareeb, M. (2009). Total phenolic contents and antioxdant activities of ficus sycomorus and azadirachta indica. *Pharmacology, 602*(3), 590–602.

39. Alam, N., & Sharma, K. R. (2020). Estimation of phenolic content, flavonoid content, antioxidant, and alpha-amylase inhibitory activity of some selected plants from Siraha District Nepal. *Asian Journal of Pharmaceutical and Clinical Research, 13*(4), 18–23.
40. Kamtekar, S., Keer, V., & Patil, V. (2014). Estimation of phenolic content, flavonoid content, antioxidant and alpha amylase inhibitory activity of marketed Polyherbal formulation. *Journal of Applied Pharmaceutical Science, 4*(9), 61–65.
41. Hossain, M. A., AL-Raqmi, K. A. S., AL-Mijizy, Z. H., Weli, A. M., & Al-Riyami, Q. (2013). Study of total phenol, flavonoids contents and phytochemical screening of various leaves crude extracts of locally grown Thymus vulgaris. *Asian Pacific Journal of Tropical Biomedicine, 3*(9), 705–710.
42. Khamis Al-Jadidi, H. S., & Hossain, M. A. (2015). Studies on total phenolics, total flavonoids and antimicrobial activity from the leaves crude extracts of neem traditionally used for the treatment of cough and nausea. *Beni-Suef University Journal of Basic and Applied Sciences, 4*(2), 93–98.
43. Abbott, W. S. (1925). A method of computing the effectiveness of an insecticide. *Journal of Economic Entomology, 18*, 265–267.
44. Orhevba, B. A., Chukwu, O., Oguagwu, V., & Osunde, Z. D. (2013). Effect of moisture content on some quality parameters of mechanically expressed neem seed kernel oil. *International Journal of Engineering Science, 2*(8), 1–7.
45. Hssaini, L., Razouk, R., & Bouslihim, Y. (2022). Rapid prediction of fig phenolic acids and flavonoids using mid-infrared spectroscopy combined with partial Least Square regression. *Frontiers in Plant Science, 13*(March), 1–16.
46. Baranowska-Wójcik, E., & Szwajgier, D. (2019). Characteristics and pro-health properties of mini kiwi (Actinidia arguta). *Horticulture, Environment and Biotechnology, 60*(2), 217–225.
47. Palmeira, L., et al. (2019). Nutritional, chemical and bioactive profiles of different parts of a Portuguese common fig (Ficus carica L.) variety Luís. *Food Research International, 126.*
48. Segneanu, A. E., Gozescu, I., Dabici, A., Sfirloaga, P., & Szabadai, Z. (2012). Organic compounds FT-IR spectroscopy, macro to Nano spectroscopy. In *National Research-Development Institute for electrochemistry and condensed matter Timisoara.*
49. Bouafif, H., Koubaa, A., Perré, P., Cloutier, A., & Riedl, B. (2008). Analysis of among- species variability in wood fiber surface using DRIFTS and XPS: Effects on esterification efficiency. *Journal of Wood Chemistry and Technology, 28*(4), 296–315.
50. B. H. Stuart, Infrared spectroscopy: Fundamentals and applications. 2005.
51. Vongsvivut, J., Heraud, P., Gupta, A., Puri, M., McNaughton, D., & Barrow, C. J. (2013). FTIR microspectroscopy for rapid screening and monitoring of polyunsaturated fatty acid production in commercially valuable marine yeasts and protists. *Analyst, 138*(20), 6016–6031.

Chapter 15
Formulation and Development of Organic Hair Shampoo from Discharged Beer in Filler Machine (Brewery Industry) and Aloe Vera Plant Gel

Mequanint Demeke Aynalem, Tegen Dagnew Tessema, Surafel Aregahegn Agdew, Tessafa Abrham Ashagrie, Belete Adane Mandie, Asmare Tezera Admassie, Smegnew Moges Mintesinot, and Desalegn Abera Mersha

15.1 Introduction

Cosmetics products are created for the purpose of cleansing, beautifying or altering appearance and enhancing attractive features when applied to the body [1]. Shampoos are widely used cosmetic products of surfactants in a suitable form to cleanse hair, to remove oils, dirt, skin particles, dandruff, and other blemish particles that gradually build up [2]. Synthetic shampoos and conditioners are cosmetics that contain harmful chemicals which cause difficulty in maintaining healthy hair. Sodium lauryl sulphate is one of the most harmful chemicals present in nearly all shampoos. When it is absorbed into the body parts or organs, it gets accumulated over long use and causes harm [3]. The researchers showed that 0.5% concentrations causes irritation; however, most products use 30% solution [4] and shampoo enters in the eyes during use, which could cause damage to the cornea and it causes hair loss as it attacks the hair follicles [5]. The related researches reported that

M. D. Aynalem (✉) · T. D. Tessema · S. A. Agdew · B. A. Mandie · A. T. Admassie · D. A. Mersha
Faculty of Chemical and Food Engineering, Bahir Dar Institute of Technology, Bahir Dar University, Bahir Dar, Ethiopia

T. A. Ashagrie
Department of Chemical Engineering, University of Gondar, Institute of Technology, Gondar, Ethiopia

S. M. Mintesinot
Faculty of Mechanical and Industrial Engineering, Bahir Dar Institute of Technology, Bahir Dar University, Bahir Dar, Ethiopia

S. W. Fanta et al. (eds.), *Sustainable Development Research in Manufacturing, Process Engineering, Green Infrastructure, and Water Resources*, Green Energy and Technology, https://doi.org/10.1007/978-3-031-77339-6_15

synthetic shampoos contained cocamidopropyl betaine which causes contact allergy in children due to augmented chemical exposures [6, 7] and diethanolamine (DEA) and N-nitrosodiethanolamine (NDELA) are potential carcinogens whose repeated use in even small doses increases the risk of cancer [6]. The formaldehyde in most shampoos as preservative (Quaternium–15) and selenium sulphide in anti-dandruff shampoos are recognized for their carcinogenic effects and allergic contact [3, 8]. Synthetic conditioners and shine-providing shampoos have silicones and silicates that form a coat around the hair strands, in which silicones accumulate in the hair and make them dull as they are non-biodegradable and difficult to wash once collected after continuous use [5]. Likewise, synthetic surfactants are added to shampoo primarily for foaming and cleansing action but their regular use leads to dryness of hair, hair loss, and irritation to scalp and eyes [9].

However, herbal formulations are considered as alternative to synthetic shampoos because they are rich in natural ingredients such as plant extracts and essential oils which have useful effects on hair [5, 10]. For this purpose, there are a number of medicinal plants that can be used in shampoo formulations [11], which are used in powdered form, crude form, purified extract form, or derivative form [12]. Medicinal plants like aloe vera are rich in bioactive compounds that dissolve dead skin cells and excess sebum which may cause clogging of hair follicles. Aloe contains salicylic acid which is an anti-inflammatory and mild anti-biotic. Aloe gel acts as an excellent moisturizer because its molecular structure is similar to keratin, a prominent protein of which skin and hair are made up, which is used to promote hair growth [13].

Brewery industries, specifically the packaging units, suffer from high mechanical down time. The key area for downtime is the filler machine, specifically the filling value, which decreases production capacity due to underfilling, overfilling problems, and uncrown ing of beer bottles. Those under-filled, overfilled, and uncrown bottles of beer are discharged immediately after the first full beer inspection [14]. However, the discharged beer contains bioactive compounds, besides ethanol, such as polyphenols (30% from hops, 70–80% from malt) which are involved in foam maintenance, physical and chemical stability, and shelf life [15, 16], and they act as antioxidants against the oxidation of ascorbic acid and unsaturated fatty acids [17]. Some related research findings reported that dried hop cones contain about 14.4% of polyphenols [18] and many extracts from barley and hops possess anti-bacterial, anti-oxidative, anti-inflammatory or anti-carcinogenic effects [19, 20]. The hop bitter acids inhibit many Gram-positive bacteria [21] and have promising molecules for cancer chemotherapy or cancer chemoprevention [22, 23]. Quercetin and ferulic acids are the two main compounds in beer that exert anti-carcinogenic effects [24]. Nowadays, the demand for herbal and organic cosmetics, like shampoos containing natural herbs and organic ingredients, is increasing and receiving more attention in public because of their high-quality properties and less side effects [25]. Shampoos containing suitable additives for hair conditioning enhancement, lubrication, and medication [13]; besides, they contain surfactants (surface active agents) plus conditioning agents, pearling agents, antimicrobial agents, colors, and fragrance [9, 13]. Evaluation of shampoos involves visual assessment and physiochemical controls such as P^H values, foaming ability, foam height, percentage of solid contents, and dirt dispersion [26]. For that reason, this study aimed to produce valuable organic hair shampoo products from discharged beer and aloe vera plant gel.

15.2 Materials and Methods

15.2.1 Equipment and Instrument

Electronic mass balance (PW 124) was used to measure the weight. P^H meter (pH 016) and Graduated cylinder (400 ml) were used to measure the mass, P^H value, and volume of the ingredients, respectively. Mixer (MS7-H550-Pro) was used to blend the ingredients; crucible, baker, pipet, and test tube were also used for holding the sample ingredient or product. Evaporating dish was used to remove the liquid part of the shampoo from the sample. Hot plate was used for heating purpose. A viscometer (Visco Star Plus) to measure viscosity and distiller (DZ-5LII) to produce distilled water.

15.2.2 Chemicals and Reagents

Citric acid (C6H8O7, 35%) was used to adjust the P^H, ascorbic acid (C6H8O6, 45%) was used to preserve the product, and bar soap was used as a source of surfactants. Discharged beer and aloe vera gel were used as raw materials. Essential oil (castor oil), deionized water, and sodium chloride (NaCl, 95%) were used as a fragrance, a medium, and viscosity adjustment, respectively.

15.2.3 Experimental Setup and Description

In the beginning, the aloe vera gel and discharged beer were collected, prepared and stored in the fridge till use. Correspondingly, surfactants, preservatives, essential oils (castor oil), and thickeners including P^H adjusting reagents were already prepared. Then and there, as illustrated in Fig. 15.1, these organized samples were blended well in the mixer with respect to their formulations and mixing speed (RPM). Finally, the product (organic shampoo) was characterized in terms of physicochemical properties.

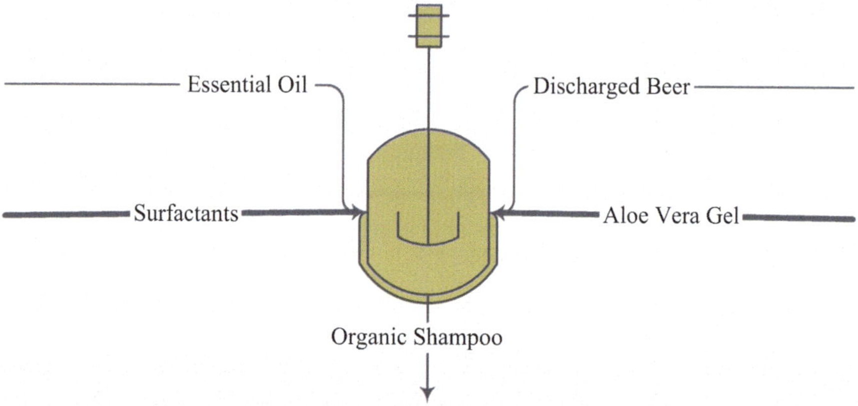

Fig. 15.1 Experimental setup of mixing process

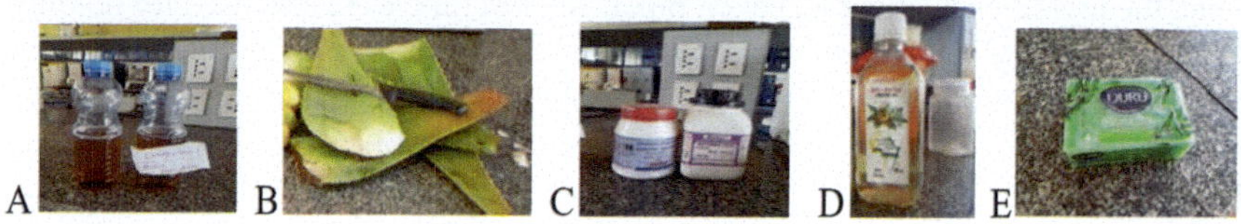

Fig. 15.2 (**a**) Discharged beer, (**b**) aloe vera plant leaf, (**c**) ascorbic acid, (**d**) essen tial oil, and (**e**) surfactant (soap bar)

15.3 Methods

15.3.1 Collection and Preparation of Raw Materials

The raw materials were collected from different areas. The discharged beer was collected directly from Dashen Brewery Share Company and it was stored in the fridge till use with plastic bottles. And aloe vera leaf was collected from the local market in Bahir Dar city and then its gel was prepared, blended with ascorbic acid, and stored in fridge for further process. Essential oil (castor oil) and surfactants (from bar soap) were bought from local markets. Correspondingly, chemicals (ascorbic acid, citric acid, sodium chloride) were collected from BiT laboratories. The soap bar was also collected locally and reduced in size to aid the solubilizing process and measured for shampoo formulation (Fig. 15.2).

15.3.2 Preparation of Aloe Vera Gel/Oil

The prepared aloe vera gel was converted into liquid form by heating at a low temperature (50 °C) in water-bath for two hours to retain thermos-sensitive bioactive ingredients in the gel (oil). Ascorbic acid was added to aloe vera concentrate to adjust the P^H within the range from 5.5 to 6.0. In organic shampoo formulation, the preserved aloe vera gel was dissolved in small amount of distilled water [27].

15.3.3 *Organic Shampoo Formulation*

All essential ingredients were measured and blended properly with different mixing speeds (5, 8, and 10 RPM) in the required quantities as shown in Table 15.1. The total amount of shampoo needed to produce for each formulation was 300 ml. To start the formulation process, the soap bar was first reduced in size and weighted (24 g, 36 g, 45 g) for each formulation (F_1, F_2, F_3), respectively. With a crucible, each measured soap bar was solubilized with distilled water (150 ml) in a hot plate at 100.5 °C for 20 min to change it to liquid phase. The hot liquid soap was put to rest to cool it down to room temperature. The measured 9.0 ml of aloe vera gel and 5.0 ml castor oil (essential oil) were poured into the liquid soap to aid the homogenization process. Then the liquid soap and oil mixture were blended well. The measured discharged beer (39 ml, 35 ml, and 32 ml) for each formulation (F_1, F_2, F_3), respectively, was heated at 78 °C for 3 min with hot plate to remove the alcohol content. The heated discharged beer was then poured into each soap and oil mixture. Following the addition of discharged beer, 1.0 g of ascorbic acid was added. Then the formulated mixtures were blended to produce shampoo. The developed shampoo was stored in a suitable container and used for further evaluations.

15.4 Product Characterization

15.4.1 *Physical Appearance (Visual Inspection)*

All developed shampoo formulations were evaluated in terms of their clarity, color, and foam-producing ability [30].

Table 15.1 Shampoo formulation

Components	F_1	F_2	F_3	Standard values (%)	
Distilled water	50	50	50	50–100	[27, 29]
Surfactants (liquid soap)	8	12	15	8–15	[27, 29]
Thickener(NaCl)	1	1	1	0–10	[27, 29]
Preservative (ascorbic acid)	0.3	0.3	0.3	0.1–3	[27, 29]
Discharged beer	39	35	32	30–50	[29]
Active ingredient (aloe vera)	3	3	3	0.1–3	[27, 29]
Fragrance (castor oil)	1.5	1.5	1.5	0.5–1.5	[27, 29]
Citric acid (P^H adjuster)	1	1	1	As needed	[27, 28]

15.4.2 Determination of P^H^

The developed formulation was diluted using distilled water to prepare a sample with 10% concentration. The prepared sample was checked for P^H using a digital P^H meter at room temperature (25 ± 2 °C) [31].

15.4.3 Determination of Percentage Solid Contents

A clean dry evaporating dish was weighed and then 4 grams of shampoo (M_1) was added to it. The dish with shampoo was weighed again to confirm the exact weight of the shampoo and then the exact weight of the shampoo was calculated. The evaporating dish with shampoo was placed on the hot plate until the liquid portion was evaporated. The weight after drying (M_2) was calculated and the percentage of solid contents was determined in the following way [32]:

$$\text{Solids Contents}(\%) = \frac{M_1 - M_2}{M_1} * 100\% \tag{15.1}$$

15.4.4 Dirt Dispersion Test

Two drops of shampoo were added in a large test tube containing 10 ml of distilled water. To this solution, one drop of India ink was added and the test tube was shaken 10 times. The amount of ink in the foam indicates the rubric such as none, light, moderate, or heavy [33] to evaluate the quality of shampoo for cleansing action.

15.4.5 Evaluation of Foaming Ability and Stability

Foaming ability is determined by using the cylinder shake method. Foam stability is evaluated by recording the foam volume after 1 min and 4 min shake tests [34].

15.4.6 Microbial Analysis of Organic Shampoo

Total Bacteria Count.

According to the American Public Health Association (APHA, 1960), this standard plate count with appropriate serial dilutions was selected given the expected total number of colonies on a plate. The standard plate count agar (PCA) was boiled, sterilized, and then cooled to 45^0c before pouring. After thoroughly mixing, the sample was serially diluted to 10^{-7} and mixed thoroughly. Then one ml of the sample from appropriate decimal dilution was placed on a Petri dish and the plate count agar medium (10 ml) was poured onto the Petri dish. The plated sample was allowed to solidify and finally incubated at 37 ^{0}c for 48 hours. The total number of colonies on plates 25 to 250 per plate was selected and the colony count [35–37].

$$\mathrm{CFU/g} = \frac{\mathrm{C}}{\mathrm{D}} \tag{15.2}$$

where C = count per plate and D = dilution

Yeast and Mold

According to American Public Health Association (APHA) (Association., 1960), yeast and mold were analyzed. Potato dextrose agar (PDA) was used for the enumeration of yeast and molds. The homogenized sample was serially diluted with 0.1% peptone water 10^{-1} to 10^{-5} enumerative (0.1 ml) from a suitable dilution and was transferred aseptically into solidified PDA plates. The sample was spread all over the surface of the plates using a sterile bent glass rod and incubated at 25^0c for five days without inverting the plate and counting colonies and calculating the result, select plates with 15 to 150 colonies with the aid of a magnifying glass fitted onto a colony counter [37]. Number of colonies per ml or g of sample (N) is expressed with the sum of all colonies on plates counted ($\sum$ C), number of plates used in the lowest dilution counted (n_1), number of plates used in the highest dilution counted (n_2), and dilution factor of lowest dilution used (d). The general formula colony count calculation is as follows:

$$\mathrm{N} = \Sigma C / v * \left[n_1 + (0.1 * n_2) \right] * d \tag{15.3}$$

15.5 Results and Discussion

15.5.1 Aloe Vera Gel Characterization

The aloe vera gel was characterized by the physicochemical properties. The refractive index of aloe vera gel (1.3378) was found within the standard range of refractive indices (1.3000–1.7000) and soluble in water [38] for shampoo formulation. The moisture content of aloe vera gel was 98.2% which is used to produce the moisturizing properties of shampoo. The P^H value of aloe vera gel was 4.86 ± 0.12 which indicates it is an acidic substance that prevents microbial attack of the product. A related research work findings on the physicochemical properties of aloe vera gel have reported similar conclusions [27]. This indicates that the gel of aloe vera plant has the potential to produce shampoo.

15.5.2 Determination of Shampoo P^H *Values*

As can be seen in Fig. 15.3, both formulations F2 and F3 showed an increasing trend from 5 RPM to 8 RPM mixing speech and then a decreasing tendency when the mixing speech was further increased from 8 RPM to 10 RPM. However, formulation F1 was observed in a slight increment of P^H values (7.0–7.05) across the graph when the mixing speed varied from 5 RPM to 10 RPM. This is due to the presence of a high amount of beer that is slightly acidic and low amount of surfactants (soap bar) also slightly basic. The pH value of F2 was found between 7.06 and 7.20 which was due to low amount of beer and high amount of soap relative to F1. The minimum pH value (7.05) was obtained at F1 and 5 RPM. The pH value of F3 also was found between 7.20 and 7.28 which is a little far from the standard pH values [39] of hair shampoo. This is due to the presence of low amount of beer and high amount of soap bar. The recommended hair shampoo has a P^H between 5 and 7 [2, 35] so that sebum is at the right balance on the hair and scalp [2]. Therefore,

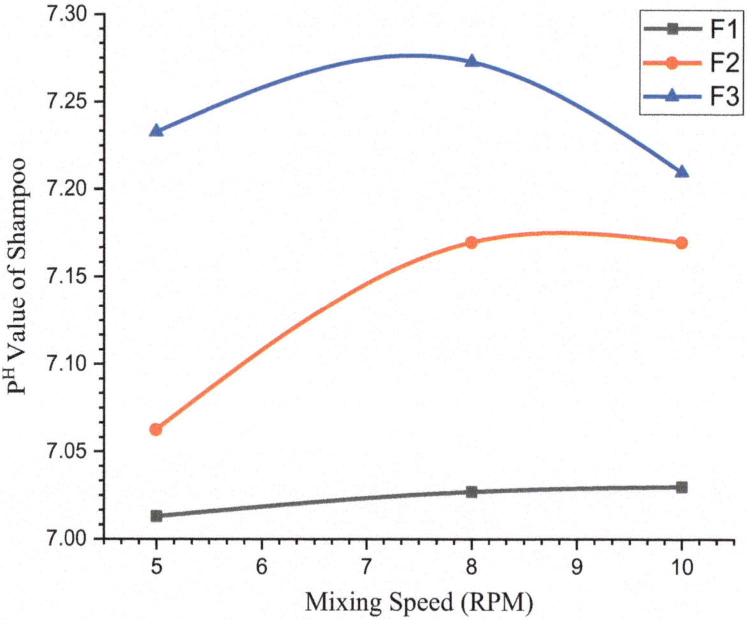

Fig. 15.3 P^H value of organic shampoo

relatively, formulation F1 is the best choice as organic hair shampoo according to the P^H values, which are important for improving and enhancing the quality of hair.

15.5.3 Foaming Ability and Foam Height

As illustrated in Fig. 15.4, the foaming ability and foam height of all formulations (F1, F2, F3) were observed highest (8.67, 4.33, and 6 mm, respectively) in the beginning at 5 RPM mixing speed and then they were detected in decreasing trend when the mixing speed is varied from 8 RPM to 10 RPM. The foam height of F1 was found to be highest across the graph when compared with the foam height of formulations F2 and F3. In general, the maximum foam height (8.67 mm) was found at formulation F1 and 5 RPM mixing speed. Foam generation is an important criterion for a good shampoo which produces adequate foam when shaken with water. Due to this, a formulation product with a higher amount of foaming height was found in formulation F1 and was chosen. The previous research findings reported that foam-generation capacity is of paramount importance to the consumer and it is an important criterion in evaluating shampoos [2, 26]. The formulated shampoo F1 can be utilized for the production and formulation of organic hair shampoo according to the foam height and foaming ability of the discharged beer from the filler machine and the prepared aloe vera gel.

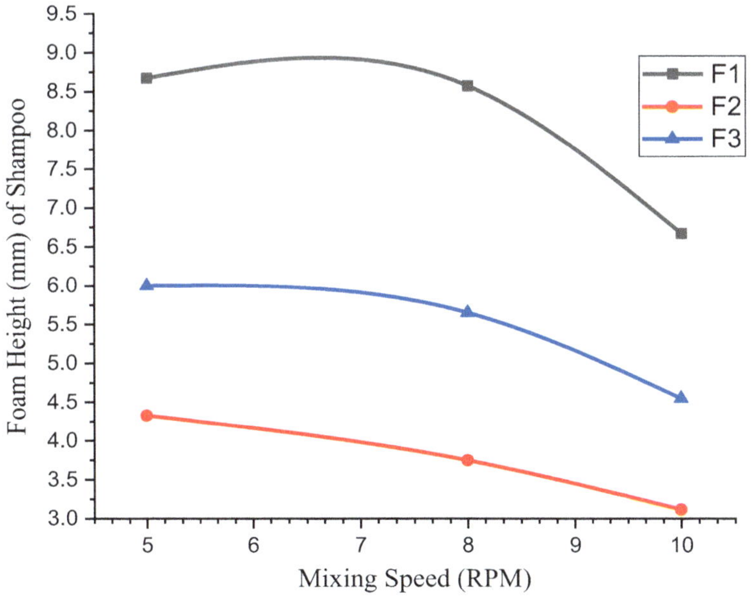

Fig. 15.4 Foam height of organic shampoo

15.5.4 Percentage of Solid Content

The percentage of solid content, as demonstrated in Fig. 15.5, generally showed an increasing trend from 5 RPM to 8 RPM and then it illustrated a decreasing tendency when the mixing speed varied from 8 RPM to 10 RPM. The highest quantity of solid content percentage was obtained in formulation F3 and the lowest percentage of solid content was observed in formulation F1 through the graph. Specifically, the solid content percentage of F1 was found between 7.853 and 8.75%, and F2 was found between 12.767 and 13.845%. Correspondingly, the solid content percentage of F3 was observed between 14.91 and 19.97%. The previous research works reported that a good shampoo should contain sufficient solid content percentages which enable be applied and rinsed out from the hair and lack of enough solids results in to quick wash away of shampoo from the hair [36]. And another research finding reported that good shampoos usually have 20%–30% solid content as it is easy to apply and rinse out from the hair [25], and in herbal shampoo formulation, the research findings obtained 3.8% solid content and concluded that it is not only safe but is also great in reducing the hair loss as well as enhancing the hair growth [33]. Formulation F3 has the highest solid content percentage because the formulation contained the highest amount of surfactant which is rich in solid content relative to the solid content of discharged beer. However, F1 has the lowest percentage

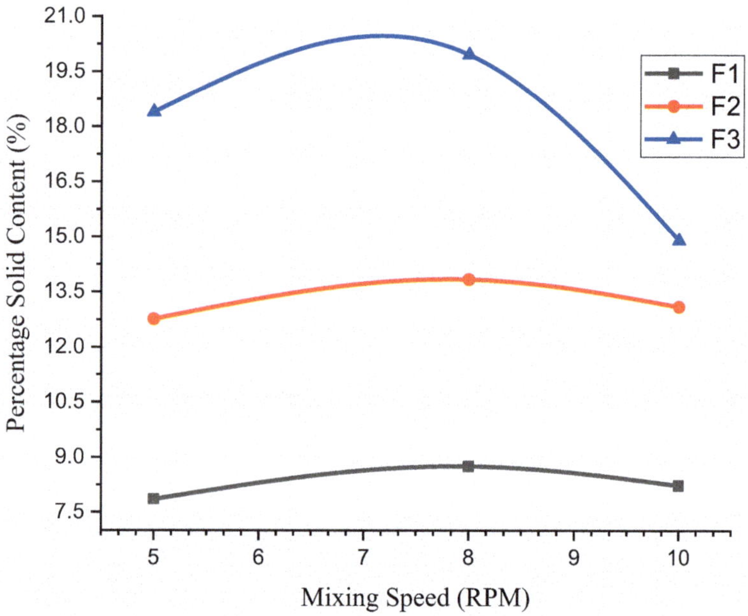

Fig. 15.5 Solid content percentage of organic shampoo

of solid content since it contained 39% discharged beer (low solid content) and 8% surfactant. But in terms of other parameters like pH value and foam height, it was found good. Thus, the organic shampoo formulation F1 can be the best choice from other formulations.

15.5.5 Dirt Dispersion

In shampoo formulation, the important criterion for evaluation of the cleansing action of shampoo is dirt dispersion test. As demonstrated in Table 15.2, the degree of dirt dispersion test for 5 minutes indicates the highest to the lowest cleansing action found in formulation F1 (moderate and none), formulation F2 (light and moderate), and formulation F3 (heavy and light). The best cleansing action was found in F1 and the poor quality of cleansing action was observed in F3. A related research finding for 3 minutes of dirt dispersion test reported that shampoos that cause Indian ink to concentrate in the foam are considered as poor quality [25]. Because ink or dirt that stays in foam is difficult to rinse away and gets re-deposited on the hair [25]. Therefore, the dirt should stay in the water portion to achieve better cleansing action [33]. According to dirt dispersion test results, F1 is the best choice because all the results of dirt dispersion at 5 minutes showed none or no ink accumulation in the solution.

15.5.6 Physical Appearance/Visual Inspection

All formulated shampoos have a good appealing physical appearance which were obviously light green in color. The clarity of the formulated shampoos was evaluated visually and observed clearly. It can be concluded that all formulations had good characteristics with respect to appearance. With respect to physical appearance, formulation F1 appeared to be the best choice from other formulations (Table 15.3).

Table 15.2 Results of dirt dispersion test for all formulations

Dirt dispersion test results			
Speed (RPM)	F1	F2	F3
5	Moderate	Light	Heavy
8	None	Light	Heavy
10	None	Moderate	Light

Table 15.3 Physical appearance of the products

		Clarity	
Speed (RPM)	F1	F2	F3
5	Clear	Clear	Slightly clear
8	Slightly clear	Clear	Clear
10	Clear	Slightly clear	Clear

15.5.7 Microbial Analysis of Organic Shampoo

As illustrated in Table 15.4, it was observed that the total bacterial count (TBC) ranged from $1.3 * 10^4$–$2.1 * 10^4$ CFU/g while the total yeast counts (TYC) ranged from $1 * 10^3$–$1.2 * 10^3$ CFU/g. A previous research finding made on microbial analysis of marketed herbal shampoo reported that 10^2–10^4 CFU/g bacterial count and 10^3–10^4 CFU /gm yeast count [35]. Besides, another research finding observed $3*10^3$–$4*10^5$ CFU/g bacterial count in shampoo products [40]. However, the microbial count of all formulations of the present study demonstrated that they were found within the reported values of microbial analysis. Therefore, the microbial analysis of the formulated organic shampoos was observed good and acceptable.

15.6 Conclusion

Nowadays, shampoo products are loaded with chemicals that have long-term effects on hair, so it is important to switch shampoo products with less chemical concentrations. This chapter used discharged beer from brewery industries and aloe vera gel, which substitute chemicals, for better use and value-added organic shampoo formulation and production. The physicochemical properties of the formulated shampoo were assessed through established procedures, demonstrating intriguing features in p^H values, foaming ability, foam height, solid content percentage, dirt dispersion, physical appearance, and microbial analysis, particularly when compared to prior observations and standard values. These parameters revealed the ideal characteristics of a shampoo devoid of any harmful chemicals and can be used as an alternative to its synthetic counterpart. Among the three formulated shampoos, F1 was found to be the most preferable formulation with 7.05 ± 0.1 P^H, 8.67 ± 0.6 *mm* foam height, $7.85 \pm 0.4\%$ solid content, and acceptable microbial count. The product analysis indicates that discharged beer and aloe vera gel are outstanding raw material sources for organic hair shampoo production.

Acknowledgments The authors would like to thank Bahir Dar University, Bahir Dar Institute of Technology for financial support. The authors honestly acknowledge laboratory technicians for their patience, friendly approach, and collaboration for the accomplishment of this work.

Table 15.4 Microbial analysis of organic shampoo

Sample	TYC (CFU/g)	TBC(CFU/g)
F1	$1.0*10^3$	$2.1*10^4$
F2	$1.1*10^3$	$1.3*10^4$
F3	$1.2*10^3$	$1.6*10^4$

References

1. Preethi, P., Padmini, K., Srikanth, J., Lohita, M., Swetha, K., & Rao, P. V. (2014). A review on herbal shampoo and its evaluation. *Asian Journal of Pharmaceutical Analysis, 3*(4), 153–156.
2. Krunali, T., Dhara, P., Meshram, D. B., & Mitesh, P. (2013). Evaluation of standards of some selected. *World Journal of Pharmaceutical Sciences, 2*(5), 3622–3630.
3. Black, G., & Howes, D. (1979). Skin penetration of chemically related detergents. *Semantic Scholar, 10*(30), 157–163.
4. Cowley, N. C., & Farr, P. M. (1992). A dose-response study of irritant reactions to sodium lauryl sulphate in patients with seborrhoeic dermatitis and atopic eczema.pdf. *Acta Dermato-Venereologica, 72*(6), 432–435.
5. Saini, R. D. (2018). Chemical shampoos vs herbal shampoos. *International Journal of Creative Research Thoughts, 6*(1), 760. [Online]. Available: www.ijcrt.org
6. Schothorst, R. C., & Stephany, R. W. (2001). Occurrence of N-nitrosodiethanolamine (NDELA) in cosmetics from the Dutch market. *International Journal of Cosmetic Science, 23*(2), 109–114. https://doi.org/10.1046/j.1467-2494.2001.00062.x
7. Belloni Fortina, A., Caroppo, F., & Tadiotto Cicogna, G. (2020). Allergic contact dermatitis in children. *Expert Review of Clinical Immunology, 16*(6), 579–589. https://doi.org/10.1080/1744666X.2020.1777858
8. Heise, D., & Bork, K. (1979). Formaldehyde in hair shampoos. *Dermatosen in Beruf und Umwelt, 27*(1), 10–12.
9. Potluri, A., Shaheda, S., Rallapally, N., Durrivel, S., & Harish, G. (2013). A review on herbs used in anti-dandruff shampoo and its evaluation parameters. *Indo American Journal of Pharmaceutical Research, 4*(1), 5–13.
10. Chavan, V. M., Kiran, K. J. T., Suryavanshi, A., & Bhor, A. S. (2019). Formulation and evaluation of herbal shampoo. *American Journal of PharmTech Research, 9*(5), 88–96. https://doi.org/10.46624/ajptr.2019.v9.i5.008
11. Mohamed Halith, S., Abirami, A., Jayaprakash, S., Karthikeyini, C., Kulathuran Pillai, K., & Mohamed Firthouse, P. U. (2009). Effect of Ocimum sanctum and Azadiracta indica on the formulation of antidandruff herbal shampoo powder. *Der Pharmacia Lettre, 1*(2), 68–76. [Online]. Available: www.scholarsresearchlibrary.com
12. Arora, P., Nanda, A., & Karan, M. (2011). Shampoos based on synthetic ingredients VIS-A-VIS shampoos based on herbal ingredients: A review. *International Journal of Pharmaceutical Sciences Review and Research, 7*(1), 41–46.
13. Mainkar, A. R., & Jolly, C. I. (2001). Formulation of natural shampoos. *International Journal of Cosmetic Science, 23*(1), 59–62. https://doi.org/10.1046/j.1467-2494.2001.00057.x
14. Simões, M., Martins, R. F., & Marmé, M. (2018). Failure analysis of a filling valve from a Brewery's beer filler. *Engineering Failure Analysis, 93*(October 2017), 87–99. [Online]. Available: https://doi.org/10.1016/j.engfailanal.2018.07.012
15. Gribkova, I. N., Eliseev, M. N., Zakharov, M. A., Kosareva, O. A., & Zakharova, V. A. (2022). Developing colloidal structure of beer by grain organic compounds. *Foods Raw Materials, 10*(2), 318–328. https://doi.org/10.21603/2308-4057-2022-2-538
16. Hough, J. S., Briggs, D. E., Stevens, R., & Young, T. W. (2018). Chemical and physical properties of beer. *Malt Brewing Science*, 776–838. https://doi.org/10.4324/9780203489239-2
17. Montanari, L., Perretti, G., Natella, F., Guidi, A., & Fantozzi, P. (1999). Organic and phenolic acids in beer. *Lwt, 32*(8), 535–539. https://doi.org/10.1006/fstl.1999.0593

18. Arranz, S., Chiva-Blanch, G., Valderas-Martínez, P., Medina-Remón, A., Lamuela-Raventós, R. M., & Estruch, R. (2012). Wine, beer, alcohol and polyphenols on cardiovascular disease and cancer. *Nutrients, 4*(7), 759–781. https://doi.org/10.3390/nu4070759
19. Chen, W., Becker, T., Qian, F., & Ring, J. (2014). Beer and beer compounds: Physiological effects on skin health. *Journal of the European Academy of Dermatology and Venereology, 28*(2), 142–150. https://doi.org/10.1111/jdv.12204
20. Jeong, H. J., Jeong, J. B., Hsieh, C. C., Hernández-Ledesma, B., & De Lumen, B. O. (2010). Lunasin is prevalent in barley and is bioavailable and bioactive in in vivo and in vitro studies. *Nutrition and Cancer, 62*(8), 1113–1119. https://doi.org/10.1080/01635581.2010.515529
21. Gerhäuser, C. (2005). Broad spectrum antiinfective potential of xanthohumol from hop (Humulus lupulus L.) in comparison with activities of other hop constituents and xanthohumol metabolites. *Molecular Nutrition & Food Research, 49*(9), 827–831. https://doi.org/10.1002/mnfr.200500091
22. Vinson, J. A., Mandarano, M., Hirst, M., Trevithick, J. R., & Bose, P. (2003). Phenol antioxidant quantity and quality in foods: Beers and the effect of two types of beer on an animal model of atherosclerosis. *Journal of Agricultural and Food Chemistry, 51*(18), 5528–5533. https://doi.org/10.1021/jf034189k
23. Van Cleemput, M., Cattoor, K., De Bosscher, K., Haegeman, G., De Keukeleire, D., & Heyerick, A. (2009). Hop (Humulus lupulus)-derived bitter acids as multipotent bioactive compounds. *Journal of Natural Products, 72*(6), 1220–1230.
24. Olson, E. R., Melton, T., Dickinson, S. E., Dong, Z., Alberts, D. S., & Bowden, G. T. (2010). Quercetin potentiates UVB-induced c-Fos expression: Implications for its use as a chemopreventive agent. *Cancer Prevention Research, 3*(7), 876–884. https://doi.org/10.1158/1940-6207.CAPR-09-0220
25. Al Badi, K., & Khan, S. A. (2014). Formulation, evaluation and comparison of the herbal shampoo with the commercial shampoos. *Beni-Suef University Journal of Basic and Applied Sciences, 3*(4), 301–305. https://doi.org/10.1016/j.bjbas.2014.11.005
26. Mainkar, A. R., & Jolly, C. I. (2000). Evaluation of commercial herbal shampoos. *International Journal of Cosmetic Science, 22*(5), 385–391. https://doi.org/10.1046/j.1467-2494.2000.00047.x
27. Pounikar, Y., Jain, P., Khurana, N., Omray, L. K., Patil, S., & Gajbhiye, A. (2012). Formulation and characterization of aloe vera cosmetic herbal hydrogel. *International Journal of Pharmacy and Pharmaceutical Sciences, 4*(Suppl. 4), 85–86.
28. Karsheva, M., & Georgieva, S. (2007). The choice of the thickener-a way to improve the cosmetics sensory properties. *Handjieva Journal University of Chemical Technology Metall, 42*(2), 187–194.
29. Yang, J. (2017). *Hair care cosmetics*. Elsevier. https://doi.org/10.1016/B978-0-12-802005-0.00036-7
30. Aghel, N., Moghimipour, E., & Raies Dana, A. (2007). Formulation of a herbal shampoo using total saponins of Acanthophyllum squarrosum. *Iranian Journal of Pharmaceutical Research, 6*(3), 167–172.
31. Tarun, J., Susan, J., Suria, J., Susan, V. J., & Criton, S. (2014). Evaluation of pH of bathing soaps and shampoos for skin and hair care. *Indian Journal of Dermatology, 59*(5), 442.
32. Salvador, A., & Chisvert, A. (2017). *Analysis of cosmetic products: Second edition* (pp. 1–606).
33. Ali, H. S., Shehab, N. A., Ahaj, E., & Mohamed, B. (2013). Formulation and evaluation of herbal cream from Ziziphus Spina leaves extract. *International Research Journal of Pharmacy, 4*(6), 44–48.
34. Klein, K. (2004). Evaluating shampoo foam. *Cosmetics & Toiletries, 119*(10), 32–35.
35. Abaas, I. S., Mohammed, H. R., & Majeed, A. H. (2013). The microbial contamination study of some herbal cosmatics products used in traditional medicine in Iraq. *Kerbala Journal of Pharmaceutical Sciences, 5*(6), 708–712.
36. Vijayalakshmi, A., Sangeetha, S., & Ranjith, N. (2018). Formulation and evaluation of herbal shampoo. *Asian Journal of Pharmaceutical and Clinical Research, 11*(Special Issue 4), 121–124.

37. Da Silva, N., Taniwaki, M. H., Junqueira, V. C. A., de Arruda Silveira, N. F., Okazaki, M. M., & Gomes, R. A. R. (2018). *Microbiological examination methods of food and water: A laboratory manual*. CRC Press.
38. Baxter, G. P., & Burgess, L. L. (1911). The refractive index of water. *Journal of the American Chemical Society, 33*(6), 893–901.
39. Kumar, A. (2014). Evaluation of prepared shampoo formulations and to compare formulated shampoo with marketed shampoos. *International Journal of Pharmaceutical Sciences Review and Research, 3*(1), 120–126.
40. Shaqra, Q. A. (2011). Microbiological quality and in use preservative capacity of shampoo preparations manufactured in Jordan. *Jordan Journal of Biological Sciences, 4*(2), 113–118.

Chapter 16
Formulation and experimental evaluation of antidandruff shampoo from leaf-extracted herbs of *Aloe vera* (Eret), *Ocimum sanctum* (Damakase), and *Withania somnifera* (Gizewa): An outlook as a herbal medicine

Tegen Dagnew Tessema, Yohannes Shetawn Wendante, Asmare Tezera Admassie, Mequanint Demeke Aynalem, Asnakew Baye Workie, Desalegn Abera Mersha, Alemu Engashu Wubie, Haile Mariam Mulusaw Minalu, Abraham Sintayehu Siyum, Bereded Gedamu Eshetie, and Behailu Bisenebt Mussie

16.1 Introduction

16.1.1 Background of the study and knowledge gaps

To date, fungal pathogens like dandruff are a big concern in the health of human skin because the chronic noninflammatory condition of the scalp is characterized by excessive scaling, itching, and redness, which is caused by *Malassezia* spp. of fungi, resulting in common dermatological conditions [1]. More specifically, dandruff is apparently

Supplementary Information The online version contains supplementary material available at https://doi.org/10.1007/978-3-031-77339-6_16.

T. D. Tessema (✉)
Faculty of Chemical and Food Engineering, Department of Chemical Engineering, Bahir Dar Institute of Technology, Bahir Dar University, Bahir Dar, Ethiopia

Department of Mechanical Engineering, Universitat Rovira i Virgili, Applied Thermal Engineering, CREVER–Research group, Tarragona, Spain
e-mail: tegen.dagnew@bdu.edu.et

Y. S. Wendante (✉) · A. T. Admassie (✉) · M. D. Aynalem · A. B. Workie · D. A. Mersha · A. E. Wubie · H. M. M. Minalu · A. S. Siyum · B. G. Eshetie · B. B. Mussie
Faculty of Chemical and Food Engineering, Department of Chemical Engineering, Bahir Dar Institute of Technology, Bahir Dar University, Bahir Dar, Ethiopia

S. W. Fanta et al. (eds.), *Sustainable Development Research in Manufacturing, Process Engineering, Green Infrastructure, and Water Resources*, Green Energy and Technology, https://doi.org/10.1007/978-3-031-77339-6_16

caused by the fungi *Malassezia furfur* and *Malassezia globosa*, which were formerly called Pittosporum, causing infections in the skin and scalp [2]. Malassezia spp. feed on oils on the scalps of most adults; dry skin, sensitivity to hair care products, contact dermatitis, and other skin conditions, such as psoriasis and eczema, are the causative factors behind dandruff and can ultimately cause embarrassment in public places and hair loss [3]. Dandruff occurs exclusively on skin in areas with high levels of sebum [4], and its symptoms mainly include itching, flaking, and scalp redness [5].

Although a variety of antidandruff shampoos are commercially available on the market for hair care treatment—with the advantage of controlling dandruff and a variety of infections, particularly those that spread through the skin—all of them include chemicals that may cause negative effects such as loss of hair, increased scaling, itching, irritation, nausea, and headache [6, 7]. In particular, diethanol amine (DEA) is a common ingredient in all chemical antidandruff shampoos, and DEA and combinations of DEA, including the widely used cocamide DEA, have been found to pose a serious health risk to customers [8]. Nitroso diethanolamine (NDELA) is a recognized carcinogen, and according to reports, when sodium lauryl sulfate (SLS) comes into contact with alkanolamines, it becomes contaminated with NDELA [9].

Over the past few decades, great emphasis has been placed on natural products, as shown by the tremendous increase in herb-based cosmetic applications. Interestingly, many herbal shampoos available nowadays contain herbal ingredients like essential oils and plant extracts [10]. Tremendous effort has been made in the field by scholars for independent applications of *Aloe vera*, *Ocimum sanctum*, and *Withania somnifera* [11–13]. However, the synergetic performance of these three herbal extracts of oils for the application of specific antimicrobial strain is limited due to the percentage of the active phytochemical components in the oil that is responsible for the antimicrobial effect. In explicit terms, these herbs have exceptional phytochemical components. *Withania somnifera* has antioxidant and anti-inflammatory properties, which can promote a healthy scalp and healthy hair. *Ocimum sanctum* has antimicrobial and anti-inflammatory properties. Finally, *Aloe vera* (olive) (eret) acts as a natural moisturizer by hydrating the scalp and preventing excessive dryness, which can contribute to the formation of dandruff flakes [11, 14, 15].

Ethiopia, in particular has a diverse range of climate and ecological conditions, as well as an enormous diversity of fauna and flora, where much of that flora makes up the backbone of traditional medicines [16]. In the Ethiopian traditional system of medicine, different plants and their components have been extensively used for treating human illnesses and persist in community experiences from time immemorial [15]. Moreover, the country has a diverse range of potentially useful medicinal plants, far more than is available in many other parts of the world, and they're used by traditional healers. In addition, local communities have been using several plants for wound treatment, specifically the oil obtained from three herbs (i.e., *Aloe vera*, *Ocimum sanctum*, and *Withania somnifera*), via strategies such as concoction, decoction, pounding, powdering, squeezing, and/or crushing these plants, as reported in [14]. However, to the best of our knowledge, the country cannot exploit its herbal resources to the level of modern drug production because of the lack of well-documented information on the performance of the phytochemical properties of plant extract, though their formulated conjugates hold promise.

Hence, this study aims to formulate and evaluate three herbal-based (i.e., *Aloe vera*, *Ocimum sanctum*, and *Withania somnifera*) antidandruff shampoos that can serve as viable substitutes for synthetic antidandruff shampoos. The herb-based shampoos have proven desirable properties such as dirt dispersion, solid content, foam stability, creating a zone of inhibition, and viscosity for promoting the growth, strengthening, and darkening of hair in comparison to the results of synthetic shampoos already on the market. Moreover, quantitative Fourier transform–infrared (FT-IR) analyses were carried out to characterize the functional properties and structure of the herbal oils before synthesizing formulated products.

16.2 Materials and Methods

16.2.1 Materials, Reagents, and Equipment Used

Sodium lauryl sulfate powder ($C_2H_{25}O_4S.Na$) 70% w/w (g) was used as detergent for cleaning action. EDTA ($C_{10}H_{16}N_2Na_2O_8$) 99.8% powders were used for chelating and as sequestering agents. Ethanol (97% w/v) was used for extracting the leaves' oils and using them as clarifying agents. Formaldehyde liquid 37–41% w/v was used as a preservative. Sodium hydroxide pellets (98% w/w) were used for neutralization; citric acid anhydrous (99.7%) for adjusting the pH value; and sodium chloride powder 98.5% w/w for the thickening agent. Distilled water was used for dilution and checking surface tension. All chemicals and reagents used in this study were analytical grades (LBS Marg, Mumbai, India).

16.2.2 Equipments and Apparatus Used

The equipment used for this experimental study were as follows: Soxhlet apparatus, a sieve, an electronic weighing balance (CAS-164,0.001 gram) for measuring the raw materials and the solid chemical used during the formulation and for checking solid content, and a pH meter for measuring the acidity and basicity of the formulated antidandruff during antifungal performance evaluation. A drying oven (M40-VF, MPM Instruments Italy: oven incubator) was used to remove the moisture from the herbal leaves and the solid content from shampoo. Tongs, an Erlenmeyer flask, a measuring cylinder, a beaker, scissors, and a rotary evaporator were used, and a viscometer (VISCO STAR + H FUNGILAB CO. Ltd., Barcelona) was used to measure the viscosity of the formulated antidandruff shampoo. A mortar and pestle (Fritsch D-55745idar, Oberstein, Germany) was used to grind the leaves into a suitable powder for the extraction process. A Jasco FT-IR 6600 spectrometer was used to record the infrared (IR) spectra. The FT-IR spectra of *Aloe vera*, *Withania somnifera*, and *Ocimum sanctum* oil were acquired by using a Jasco FT-IR 6600 spectrometer, and functional groups were determined by using IR correlation charts.

16.2.3 Collection of Plant Materials and Preparation for Extraction

Healthy, well-matured *Aloe vera* (eret), *Ocimum sanctum* (damakase), and *Withania somnifera* (gizewa) (Fig. 16.1) were collected from the northern part of Bahir Dar city, specifically Bezawit Hill, Emperor Hailsilase Palace, Ethiopia, in March 2021, and the respective herbal leaves were cleaned. Then, the herbal leaves allowed to dry exposed to the air for 1 d at 8 °C by using a dry oven. The leaves were subsequently ground into a powder by using a mortar and pestle to prepare it for the next experiment.

16.2.4 Extraction of Plant Materials

Fresh and healthy leaves of *Aloe vera* (eret), *Ocimum sanctum* (damakase), and *Withania somnifera* (gizewa) were dried in a dry oven at 80 °C for 48 h, and then 20 g of each powder was soaked in 200 mL of distilled water and ethanol (80%) for 24 h at room temperature using the aqueous alcohol extraction method employed in [17]. The crude extracts were then filtered through Whattman filter paper, and the filtrate was evaporated until dry by using a rotary evaporator under low pressure, a technique adopted from [18, 19]. The dried extract was further powdered and then dissolved in distilled water. The extracts were transferred to a glass vial and kept at 4 °C for further experiments.

16.2.5 Formulation of Herbal Shampoo

For the shampoo, 12 g (12%) of sodium lauryl sulfate was dissolved in distilled water, and 5 g of citric acid was added into the beaker containing the solution of sodium lauryl sulfate, followed by the addition of 5 g of viscosity modifiers. Sodium chloride was added to adjust the viscosity of the shampoo, and 0.5 mL of

Fig. 16.1 Herbal leaves used for the study: (**a**) *Aloe vera* (common name: eret), (**b**) *Ocimum sanctum* (common name: damakase), and (**c**) *Withania somnifera* (common name: gizewa)

Table 16.1 Composition of fillers and formulation ratios of leaf-extracted oils used for the experiment

S. No	Ingredients	F1	F2	F3	Reference/s
1	*Aloe vera*	50 mL	25 mL	25 mL	This study
2	*O. sanctum*	25 mL	25 mL	50 mL	This study
3	*W. somnifera*	25 mL	50 mL	25 mL	This study
4	SLS	12 g	12 g	12 g	[20]
5	Citric acid	0.5 g	0.5 g	0.5 g	[20]
6	NaCl	2.4 g	3 g	5 g	[20]
7	Formaldehyde	0.05 mL	0.05 mL	0.05 mL	[20]
8	EDTA	0.5 g	0.5 g	0.5 g	[20]
9	Perfume	0.5 mL	0.5 mL	0.5 mL	[20]
10	Distilled water	90 mL	90 mL	90 mL	[20]
11	NaOH	0.06 g	0.06 g	0.06 g	[20]

Note: *SLS* sodium lauryl sulfate

formaldehyde was added as a preservative. The pH of the shampoo was adjusted by using 0.6 g of sodium hydroxide, followed by the addition of 5 mL of perfumes, 0.5% dye (red), and herbal extracts (25–50%) of *Aloe vera* (local name: eret), *Ocimum sanctum* (local name: damakase), and *Withania somnifera* (local name: gizewa) in different proportions, as illustrated in Table 16.1.

16.2.6 Strains and Media Preparation

Clinical isolates of a fungal strain of *Malassezia furfur* were collected from the microbiology laboratory department of the Bahir Dar Institute of Technology, Bahir Dar University, Ethiopia. The strain was grown in Sabouraud Dextrose Broth (SDB) with olive oil to inoculate the tubes with the test ingredients and was incubated at 30 °C for 24 h.

16.2.7 Experimental Design

A statistical method for the central composite design and analysis of variance was employed to test the effect of each experimental factor on the responses. Three experimental variables, namely time, volume ratio, and the speed of mixing (rpm), were chosen for the study: volume ratio (75, 50, and 25) in mL, mixing time (2, 4, and 6) in h, and mixing speed (2.5, 5, and 10) in rpm, at three levels. The total number of experimental runs (three factors, three levels, and two replications) is illustrated in Fig. 16.2.

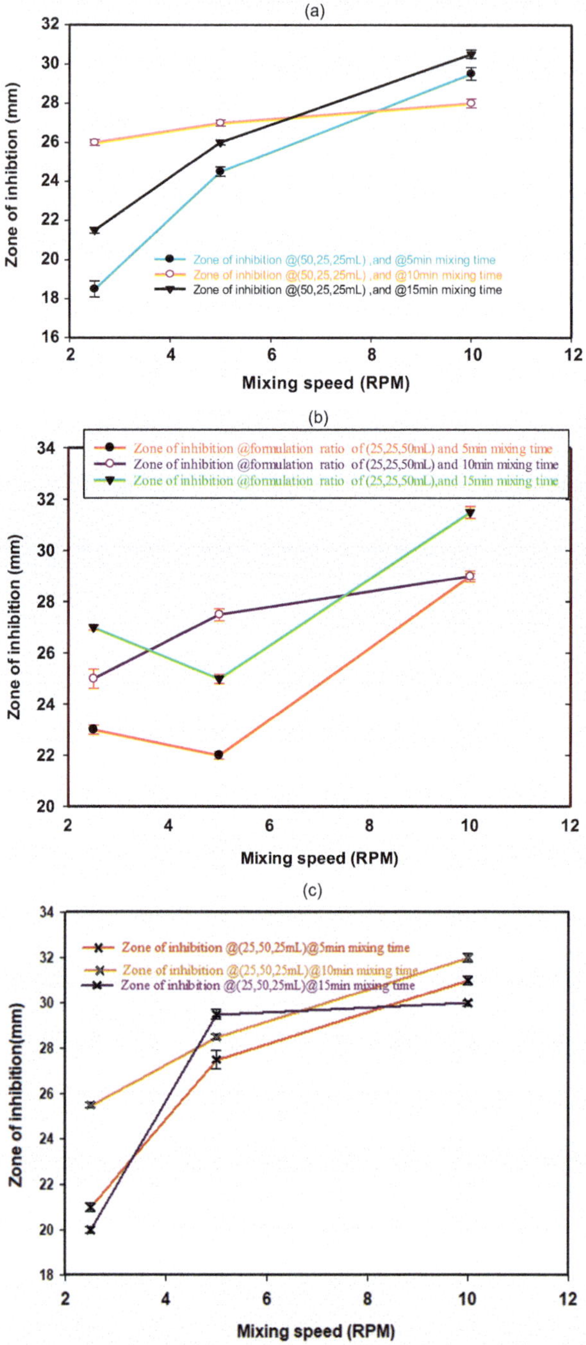

Fig. 16.2 Antifungal activity of synthesized antidandruff shampoo in terms of zone of inhibition (mm): (**a**) formulation ratios of 50, 25, and 25 mL for 5 min, 10 min, and 15 min mixing times and at 2.5, 5, and 10 rpm mixing speeds; (**b**) formulation ratios of 25, 25, and 50 mL for 5 min, 10 min, and 15 min mixing times and at 2.5, 5, and 10 rpm mixing speeds; and (**c**) formulation ratios of 25, 50, and 25 mL for 5 min, 10 min, and 15 min mixing times and at 2.5, 5, and 10 rpm mixing speeds. (Note: Data presented as mean ± standard deviation)

16.2.8 Evaluation of Antidandruff Herbal Shampoo

Organoleptic Evaluations The developed herbal antidandruff shampoos were evaluated for their organoleptic characteristics: color, odor, texture, and physical appearance. Visual inspections of physical appearance were evaluated in terms of their clarity, foamability, and fluidity [21].

16.2.8.1 Physicochemical Evaluations

pH A 10% v/v shampoo solution was constituted in distilled water, and the pH of the solution was measured by using a calibrated pH meter according to the method employed in [22, 23].

Percentage of Solid Content A clean, dry evaporating dish was weighed, and 4 g of shampoo was added to the evaporating dish. The dish and shampoo were weighed, as it was the exact weight of the shampoo, which was measured by placing the evaporating dish with shampoo onto a hot plate until the liquid portion had fully evaporated. The weight of the formulated shampoo, only (solids) after drying, was calculated according to the method reported in [24].

Dirt Dispersion Two drops of herbal shampoo were placed into a large test tube containing 10 mL of distilled water with one drop of India ink. Next, the test tube was stoppered and then shaken ten times. Afterward, the amount of ink in the foam was estimated as none, light, moderate, or heavy according to the method in [25].

Foamability and Foam Stability The cylinder shake method was employed to determine the formulated shampoo's foamability. Here, 50 mL of the 1% shampoo solution was put into a 250 mL graduated cylinder, after which the cylinder was covered with aluminum foil then shaken 10 times. The volumes of the foam contents were recorded after 1 min of shaking. The foam volume was calculated immediately after shaking the volume of foam at 1 min intervals for 4 min according to the method in [26], and their foam stability values were recorded.

Wetting Time Test The wetting ability of the surfactant was tested by using 1 inch diameter pieces of paper with a mean weight of 0.44 g. A disc was placed on the surface of 1% v/v formulated shampoo solution, and time was recorded with a stop watch. Afterward, the time required for the disc to begin to sink was noted as the wetting time of the formulated shampoo solution, a method that follows the one in [27].

Surface Tension Measurement A surface tension measurement of the 10% w/v formulated shampoo in distilled water was taken by using a stalagmometer calibrated at room temperature [28].

Rheological Property Evaluations The viscosity profiles of the produced shampoos were determined by using a VISCO STAR + viscometer set at different spindle speeds, from 0.3 to 10 rpm. The viscosity values of the shampoos were measured by using spindle R4 (5 rpm). The temperature and sample container size were kept constant during the study [29].

In Vitro Antidandruff Activity Evaluations After having used olive oil for the inoculation of the tubes containing the test ingredients, antifungal activity tests were performed using the agar diffusion method for 24 h in a broth culture medium containing the test organism and Sabouraud dextrose broth (SDB), and the tubes were incubated at 30 °C for 24 h, following the method in [30, 31].

Afterward, a loop full of broth culture was streaked across an SDB plate overlaid with olive oil to detect the presence or absence of the growth of *Malassezia* (MTCC: 1374). Here, 4 μL of antidandruff shampoo (containing active ingredients) was weighed separately in Petri plates, and molten SDB was poured in and thoroughly mixed, followed by spreading 20 μL of culture over the agar media. The plates were incubated at 30 °C for 3–5 days, following the method in [32]. All experiments were carried out in triplicate with suitable controls. Fungal cultures (e.g., *Malassezia furfur*), agar plates (e.g., Sabouraud agar), sterile pipettes, incubators, antifungal agents (test samples), sterile swabs, sterile distilled water or saline solution, sterile loops or spreaders, and Petri dishes were used for colony enumeration and zone-of-inhibition tests [33].

A plate counting of the strain of *Malassezia furfur* was carried out via the disc diffusion method by transferring 0.1 mL of the serial diluted soil sample onto SDB media and spreading it out with a sterile spreader. Next, the average value of the triplicate plates was used to determine the colony-forming units (CFU) of the *Malassezia* stains. The number of colony-forming units (CFU) was evaluated by using the formula in Eq. 16.1 (**see** Appendix 3 in the supplementary file). The result of this test was evaluated by using CFU scores, where an acceptable standard for bacterial count per plate falls within a range of 250–300 CFU, according to [34].

$$\text{Colony-forming units (CFU)} = \frac{\text{average number of colony for triplicate}}{\text{amount of sample transfered to the media}} * \text{inverse dilution factor} \tag{16.1}$$

16.2.8.2 Functional and Structural Properties of the Herbs

Leaf Extracts A Jasco FT-IR 6600 spectrometer was used to characterize the functional groups on the basis of leaf extracts' respective chemical structures. The FT-IR spectra of *Aloe vera*, *Withania somnifera*, and *Ocimum sanctum* oil were acquired by using a Jasco FT-IR 6600 spectrometer, and their functional groups were determined by using IR correlation charts.

SPSS Tools Used for Analysis The experimental data were statistically evaluated by using IBM SPSS 20 statistical software. A univariate analysis of variance (ANOVA) and a one-way ANOVA were employed at the level of significance (cut-off point: ≤ 0.05) to detect differences among the treatment means.

16.3 Results and Discussion

16.3.1 Physical Appearance and pH

An antidandruff shampoo, like any other cosmetic product, should have an appealing physical appearance. Hence, the formulated shampoos were evaluated for physical characteristics such as color and odor (Table 16.2). The pH of shampoo is an important parameter for improving the qualities of hair, minimizing irritation to the eyes, and stabilizing the ecological balance of the scalp. Hence, the pH value of antidandruff shampoo is recommended to be in the range of 5.5–7.0, which follows the current trend in promoting shampoos with lower pH values and minimizing damage to hair, according to [35, 36].

The pH values of the formulated shampoos were measured, and the values were in the range of 5.51–5.81, as depicted in Table 16.2.

16.3.2 Evaluation of Shampoo for Foamability and Foam Stability

Although foam generation has relevance to the cleansing ability of shampoos, it is important to customers and improves the efficacy of applying antidandruff agents to affected areas. Hence, the final formulation produced stable foam, and only small changes in foam volume were noted. The foam stability of a shampoo increases with increasing the percentage of plant extract, as illustrated in Table 16.3, and this due to the increased level of the foaming agent saponin, from the leaf-extract oils of *O. sanctum*, according to [10, 37].

Table 16.2 Evaluation of the physical appearance and pH of formulated shampoos

Formulation code	Physical appearance	Texture	pH
F1	Dark brown	Smooth	5.51 ± 0.02
F2	Dark brown	Smooth	5.61 ± 0.07
F3	Dark brown	Smooth	5.81 ± 0.02

Note: Data presented as mean ± standard deviation

Table 16.3 Evaluation of shampoo foamability

	Foam volume (mL)		
Time (min)	F1	F2	F3
5	149.4 ± 0.6	154.0 ± 0.64	156.4 ± 0.15
10	147 ± 0.45	149.2 ± 0.77	152.0 ± 0.65
15	144.3 ± 0.26	145.1 ± 0.7	149.1 ± 0.66

Note: Data presented as mean ± standard deviation

16.3.3 Dirt Dispersion and Percentage of Solid Content

A standard test method was employed for dirt dispersion, yielding panelist variable scores of none or slightly light, and nearly all the formulations showed that no dirt would noticed in the foam. On the other hand, the results for the percentage of solid contents (Table 16.4) showed that they were between 22.15% and 25.30%. Of the three developed formulations, F1 showed the minimum percentage of solid content, at 22.15%, indicating that F1 may be a product that is easier to wash out and preferred by customers. All the formulations in this study had panelist scores of none for dirt dispersion, which is in good agreement with the work in [21]. Hair shampoos usually have 20–30% solid contents because as such they are easy to apply and rinse out.

16.3.4 Viscosity

The viscosity of the prepared shampoo was measured by using a VISCO STAR + viscometer. A rheological evaluation, as illustrated in (Table 16.5), revealed that the viscosity of the samples gradually decreased with an increase in rpm, which affirms that the shampoo formulations were time dependent. In addition, this decrease showed that the formulated herbal antidandruff shampoo was shear thinning or pseudoplastic in nature. Moreover, different shampoos with viscosity values between 4000 and 9000 centipoise (cP) were easily manageable and acceptable. Because shampoos with viscosity levels below 1250 centipoise are not preferred by consumers and because this viscosity can easily flow out of a container, customers experience difficulty in pouring out their desired amounts and the product can come into with the eyes—as reported in [38].

Surface tension is determined by the amount of surfactant in a given antidandruff shampoo. According to the results in Table 16.6, a decrease in surface tension occurred from the mean of 45.86 for F1, that of 46.39 for F2, and that of 36.75 for F3 to the surface tension of pure water, namely 72.28 dyn/cm. This affirms that less surfactant was available for the formulated product F3, which was because *Ocimum sanctum* has less surfactant in its phytochemical components, as reported in [39].

Table 16.4 Evaluation of formulations' percentage of solid contents and their dirt dispersion

Formulation code	Percentage solid content (%)	Dirt dispersion
F1	22.15 ± 0.42	None
F2	24.43 ± 0.79	None
F3	25.30 ± 0.87	None

Note: Data presented mean ± standard deviation

Table 16.5 Viscosity of formulated antidandruff shampoos

	Viscosity (cP)		
RPM	F1	F2	F3
2.5	8682.67 ± 0.92	8287.15 ± 0.64	8867.87 ± 0.73
5	5834 ± 0.54	5827 ± 0.45	5828 ± 0.23
10	4864 ± 0.66	4640 ± 0.76	5130 ± 0.56

Note: Data presented mean ± standard deviation

Table 16.6 Wetting time and surface tension test

rpm	**Wetting time (sec)**		
	F1	**F2**	**F3**
2.5	167.5 ± 0.72	186.15 ± 0.15	192.27 ± 0.73
5	172.1 ± 0.35	189 ± 0.35	196 ± 0.33
10	173.5 ± 0.66	193.5 ± 0.6	198.2 ± 0.46
rpm	**Surface tension (dyn/cm)**		
	F1	**F2**	**F3**
2.5	54.2 ± 0.23	49.15 ± 0.25	38.5 ± 0.73
5	42.2 ± 0.44	45.52 ± 0.45	36.5 ± 0.23
10	41.5 ± 0.66	44.5 ± 0.76	35.25 ± 0.56

On the other side, the wetting ability was characterized by the amount of wetting time (sec) (Table 16.6), and mean values were determined: 171 for F1, 189.6 for F2, and 195.5 for F3. Hence, F3 had the maximum wetting time, because it contains a lower concentration of detergents. However, a low concentration of detergents in antidandruff shampoo is usually less important in the treatment microbial pathogens [40].

16.3.5 Antidandruff Activity Evaluations

The antifungal activity of the formulated shampoos was measured in terms of the zone of inhibition, which was obtained for different formulations and compared with a standard ketoconazole shampoo (2% w/v) as a positive control in the experiment. All the developed formulations exhibited antifungal activity against *Malassezia furfur*. The mean of antifungal activity was 27.398 mm (Fig. 16.2a–c),

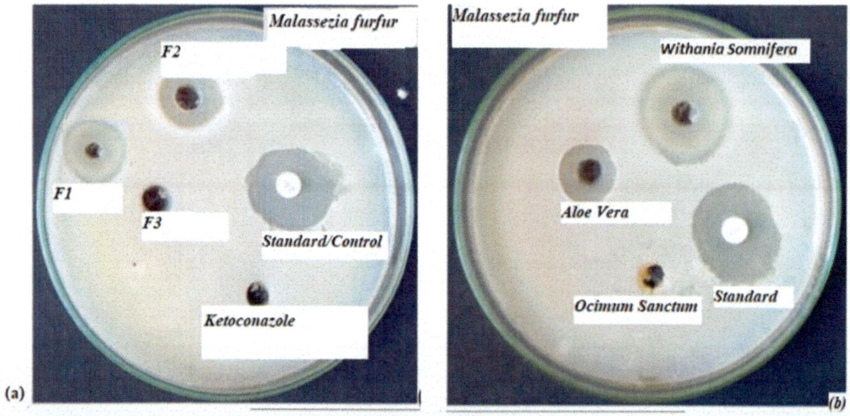

Fig. 16.3 In vitro test on the zone of inhibition for *Malassezia furfur*: (**a**) formulated antidandruff shampoos and (**b**) independent herbal antidandruff shampoos

where variation in the standard deviation ranged from 0.25 to 0.35, which indicated a small variation of antifungal performance in the group dataset. However, in comparison to the other formulations, F3 exhibited a better zone of inhibition in that its zone of inhibition (32 mm) nearly matched that of ketoconazole shampoo (36 mm, as reported in [41]) (Fig. 16.3). More importantly, for all the formulation processes, an increase in mixing speed and mixing time steadily increased their zone of inhibition. An increase in these two parameters can improve dispersion and the distribution of herbal ingredients and antimicrobial agents and can reduce the clumping of herbal ingredients, respectively [42].

The data in Fig. 16.2 were further analyzed by using statistical software (SPSS IBMS 20) and using multivariant analysis in SPSS. Table 16.5 illustrates the ANOVA test results for the dependent variable (antifungal activity). The p-value was less than 0.05 ($p < 0.05$), which affirms that the interaction with the factors had a significant effect on the dependent variable (antifungal activity) (Table 16.7).

16.3.6 Statistical analysis in One-Way ANOVA

A statistical significance difference was observed in the average antifungal activity as measured according to the formulation ratio, mixing speed, and mixing time. According to Table 16.8a,b,c, variation in the mean antifungal activity appeared between the three groups' formulation ratios (F1, F2, and F3), mixing speeds (R1, R2, and R3), and mixing times (T1, T2, and T3), and the highest level of antifungal performance was yielded by F3 (29 mm), R3 (30.33 rpm), and T3 (28.28 min) for their respective variables. More importantly, overall, differences in statistical significance appeared among all the groups of formulation ratios, mixing speeds, and mixing times in response to antifungal performance, as shown by the fact that the p-values were 0.013, 0.000, and 0.026, respectively, for the group variables.

Table 16.7 ANOVA Tests on between-subject effects, (α) = 0.05

ANOVA Tests on between-subject effects, (α) = 0.05							Remarks
Dependent variable: Antifungal activity							
Source	Type III sum of squares	Df	Mean square	F	Sig.	Partial eta squared	
Corrected model	504.565[a]	26	19.406	45.073	0.000	0.977	
Intercept	40535.560	1	40535.560	94147.108	0.000	1.000	
MS	263.787	2	131.894	306.333	0.000	0.958	√
MT	68.731	2	34.366	79.817	0.000	0.855	√
FR	80.843	2	40.421	93.882	0.000	0.874	√
MS * MT * FR	91.204	20	4.560	10.591	0.000	0.887	
Error	11.625	27	0.431				
Total	41051.750	54					
Corrected Total	516.190	53					

Note: *MS* mixing speed; *MT* mixing time; and *FR* formulation ratio
√ = significant
[a]R squared = 0.977 (adjusted R squared = 0.956)

Figure 16.3a illustrates the antifungal activity of the three formulated antidandruff shampoos in treating *Malassezia furfur*, where F3 exhibits a pronounced average antifungal performance of (33.25 mm) in comparison to the other experimental datasets and the control (ketoconazole). This could be because the formulation ratio in F3 has high *Ocimum sanctum* percentages (50%) and because *Ocimum sanctum* has high concentrations of methyl chavicol and linalool (Appendix A) among its phytochemical components, according to [11], which assessed the antifungal role of linalool and methyl chavicol by studying their respective effects on the morphological aspects of microbial *Candida* species.

For comparison, the three herbal leaf extracts of oil were evaluated for their antifungal performance values by using the same strain of *Malassezia furfur* (Fig. 16.3b), and *Ocimum sanctum* (damakase) exhibited improved performance in its zone of inhibition (34.5 mm). This might be due to the presence of methyl chavicol and linalool because these phytochemical metabolites play active roles in antimicrobial control, a finding that is inconsistent with that of previous studies [11, 43].

16.3.7 FT-IR Analysis of Oil

The functional groups in the oils extracted from *Aloe vera*, *Withania somnifera*, and *Ocimum sanctum* were determined by comparing their vibration frequencies, which fell within the spectrum range of 400–4000 cm^{-1}, and by using an FT-IR device and an IR correlation chart—as per the method employed in [44]. The peaks of the

Table 16.8 One-way ANOVA tests: (a) the zone of inhibition and formulation ratio, (b) the zone of inhibition and mixing speed, and (c) the zone of inhibition and mixing time

Descriptives								
Antifungal activity in terms of the zone of inhibition								
					95% confidence interval for mean:			
	N	Mean	Std. deviation	Std. error	Lower bound	Upper bound	Minimum	Maximum
(a)								
F1	18	26.1389	3.76831	0.88820	24.2650	28.0128	17.00	30.50
F2	18	27.0000	2.21625	0.52238	25.8979	28.1021	23.00	30.50
F3	18	29.0556	2.54887	0.60077	27.7880	30.3231	25.50	34.00
Total	54	27.3981	3.12081	0.42469	26.5463	28.2500	17.00	34.00
			Test on the homogeneity of variances					
Levene statistic		ANTI	df2	Sig.				
1.896		2	51	0.161				
One-way ANOVA								
		Sum of squares	Df	Mean square	F	Sig.		
Between groups		80.843	2	40.421	4.735	0.013		
Within groups		435.347	51	8.536				
Total		516.190	53					
Robust tests on the equality of means								
		Statistic[a]	df1	df2	Sig.			
Welch		4.849	2	32.784	0.014			
Brown-Forsythe		4.735	2	41.603	0.014			
[a]Asymptotically F distributed *F1 (50, 25, and 25), F2 (25, 25, and 50), and F3 (25, 50, and 25) for *Aloe vera*, *O. sanctum*, and *W. somnifera*, respectively								
(b)								
R1	18	25.0000	2.89523	0.68241	23.5602	26.4398	17.00	28.00
R2	18	26.8611	1.85350	0.43687	25.9394	27.7828	23.00	28.50
R3	18	30.3333	1.74052	0.41024	29.4678	31.1989	27.50	34.00
Total	54	27.3981	3.12081	0.42469	26.5463	28.2500	17.00	34.00
			Test on the homogeneity of variances					
Levene statistic		df1	df2	Sig.				
0.622		2	51	0.541				
One-way ANOVA								
		Sum of squares	Df	Mean square	F	Sig.		

(continued)

Table 16.8 (continued)

Descriptives								
Antifungal activity in terms of the zone of inhibition								
					95% confidence interval for mean:			
	N	Mean	Std. deviation	Std. error	Lower bound	Upper bound	Minimum	Maximum
Between groups		263.787	2	131.894	26.650	0.000		
Within groups		252.403	51	4.949				
Total		516.190	53					
Robust tests on the equality of means								
		Statistic[a]	df1	df2	Sig.			
Welch		28.577	2	32.869	0.000			
Brown-Forsythe		26.650	2	41.071	0.000			
[a]Asymptotically F distributed								
*R1 (2.5 rpm), R2 (5 rpm), and R3 (10 rpm)								
(c)								
T1	18	25.8056	3.96729	0.93510	23.8327	27.7784	17.00	32.50
T2	18	28.1111	2.27231	0.53559	26.9811	29.2411	25.00	34.00
T3	18	28.2778	2.32772	0.54865	27.1202	29.4353	23.50	32.50
Total	54	27.3981	3.12081	0.42469	26.5463	28.2500	17.00	34.00
			Test on the homogeneity of variances					
Levene statistic		df1	df2	Sig.				
3.107		2	51	0.053				
One-way ANOVA								
		Sum of squares	Df	Mean square	F	Sig.		
Between groups		68.731	2	34.366	3.917	0.026		
Within groups		447.458	51	8.774				
Total		516.190	53					
Robust tests on the equality of means								
		Statistic[a]	df1	df2	Sig.			
Welch		2.757	2	32.675	0.078			
Brown-Forsythe		3.917	2	38.774	0.028			

*[a]Asymptotically F distributed
*T1 (5 min), T2 (10 min), and T3 (15 min)

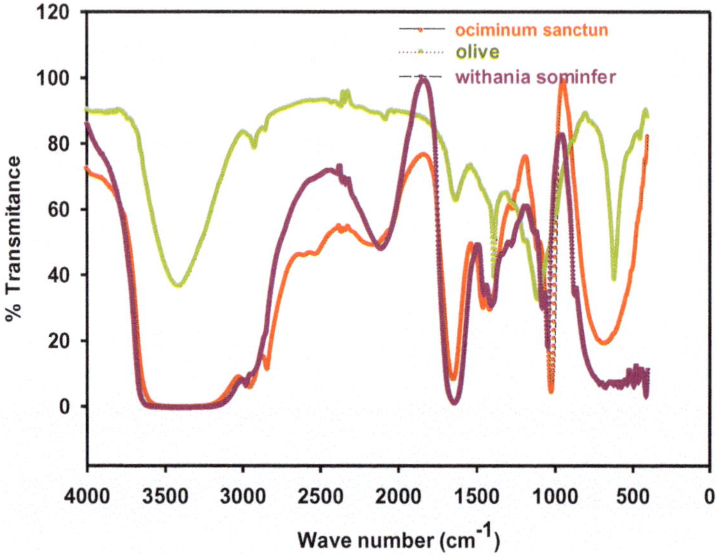

Fig. 16.4 Fourier transform–infrared spectra of *Aloe vera*, *Ocimum sanctum* (damakase), and *Withania somnifera* (gizewa)

spectrum reveal that different functional groups appeared in their phytochemical compositions, as illustrated in Fig. 16.3, where the phytochemical components are presented for each leaf-extracted oil.

The FT-IR spectra of the *Aloe era* (eret), *Ocimum sanctum* (damakase), and *Withania somnifera* (gizewa) extracts are depicted in Fig. 16.4. A notable difference between the three herbs used for this study was observed at two wave number regions: 400–2000 cm^{-1} and 2000–4000 cm^{-1}. For the single-bond area of wave number 2000–4000 cm^{-1}, five peaks were observed. The first peak, at 3562 cm^{-1}, is characterized by an internally bonded OH stretch, a tertiary alcohol OH stretch, and a phenol OH stretch, which is found in alcohol and hydroxyl compounds. The intense peak at 3081 cm^{-1} is represented as a C-H stretch and an olefinic (alkene) compound. Furthermore, the peak at 2920 cm^{-1} is characterized by the asymmetric/symmetric stretching of methylene C-H, and at 1640 cm^{-1}, olefin (alkene) appears, which is characterized by the stretching vibration of alkenyl C=C [45, 46]. The weak band at 1612 cm^{-1} and 1513 cm^{-1} can be attributed to the stretching vibration in the aromatic ring (aryl) and can also be found in C=C-C aromatic ring stretches. The band at 1442 cm^{-1} is characterized by the stretching of asymmetric/symmetric methyl C-H bonds. The peaks at 1377, 1246, and 1176 cm^{-1} are characterized by the stretching of the aromatic C-H out-of-plane bond and by C-H 1,3-disubstitution (meta). The peak at 1039 and that at 989 cm^{-1} are characterized by the stretching of the aromatic C-H in-plane bond; those at 817 and 891 cm^{-1} are characterized by the stretching of the aromatic C-H out-of-plane bond and by C-H 1,3-disubstitution (meta) [47].

16.4 Concluding Remarks

In this study, evaluations of the antifungal performance values of the formulated antidandruff shampoos made from three leaf-extracted oils were carried out. During the formulation processes, the levels of three factors, namely mixing time (min), mixing speed (rpm), and formulation ratio (mL), were varied to achieve the maximum antifungal activity in response to the zone of inhibition. According to the results of the analysis and the antifungal activity performance test using the disc diffusion method for a specific strain of *Malassezia furfur*, the formulated antidandruff shampoos were highly affected by the process parameters. Pronounced effects were yielded by varying the mixing speed, formulation ratio, and mixing time. Moreover, F3 exhibited an improved zone of inhibition because of its high concentration (50 mL) of *Ocimum sanctum* and minimal amounts of *Aloe vera* and *Withania somnifera* with the same amount of surfactant in the formulation. This suggests that a high concentration of *Ocimum sanctum* is effective as an antidandruff shampoo ingredient because it can exert a broad antifungal effect. It was also supported by the high concentration of methyl chavicol (44.63%) in its phytochemical components. Moreover, a statistical analysis of variance (univariant) and a one-way ANOVA were carried out at a 95% CI to assess the statistical differences in means among the group variables corresponding to the effects of mixing speed, mixing time, and formulation ratio on antifungal activity, and the results revealed that all factors were statistically significant. According to the results of these analyses, F3 exhibited the best formulation and can be used as an active ingredient in the development of antidandruff shampoo in modern medical systems.

Author Contributions

Tegen Dagnew Tessema conceived of the idea, designed the experiments, reviewed and edited the work, and collected and analyzed the data.

Yohannes Shitahun Wendante: designed the experiments (conceptualization), collected and analyzed the data, and wrote the first draft.

Asmare Tezera Admassie: designed the experiments (conceptualization) and collected and analyzed the data.

Mequanint Demeke Aynalem: designed the experiments (conceptualization), collected and analyzed the data.

Asnakew Baye Workie, Alemu Engashu Wubie, Desalegn Abera Mersha[1] Haile Mariam Mulusaw Minalu Abraham Sintayehu Siyum, and Behailu Bisenebt Mussie, validated the data and engaged in experimentation.

Declaration of Competing Interests The authors declare that they have no known competing financial interests or personal relationships that could have influenced the work reported in this chapter.

Acknowledgments The authors thank the Faculty of Chemical and Food Engineering, Bahir Dar Institute of Technology, Bahir Dar University, for its support in our experiments and for providing reagents. The authors apologize to all the people and organizations whose contributions to the field of herbal medicine were not adequately acknowledged in this work.

References

1. Lodha, G. (2019). Formulation and evaluation of Polyherbal shampoo to promote hair growth and provide antidandruff action. *Journal of Drug Delivery and Therapeutics, 9*(4-A), 296–300. https://doi.org/10.22270/jddt.v9i4-a.3474
2. Shukla, R., Pandey, V., Golhani, D., Shukla, A., & Rai, K. (2015). Herbal shampoo of Zizyphus oenoplia (L). *MILL from Ethanol Extract of Leaves, 2015*(July), 0–6.
3. Sawant, P. S., Sankpal, P. B., Jagtap, A. M., Gavade, A. S., & Vambhurkar, G. B. (2020). Formulation and evaluation of herbal shampoo. *Research Journal of Topical and Cosmetic Sciences, 11*(1), 01. https://doi.org/10.5958/2321-5844.2020.00001.1
4. Wijaya, W. H., Timotius, K. H., & Wijaya, J. K. (2020). Extracellular lipase of malassezia as anti dandruff drug target: A review. *Systematic Reviews in Pharmacy, 11*(8), 446–451. https://doi.org/10.31838/srp.2020.8.64
5. Pal, R. S., Saraswat, N., Wal, P., Wal, A., & Pal, Y. (2020). Preparation & Assessment of poly-herbal anti-dandruff formulation. *Open Dermatology Journal, 14*(1), 22–27. https://doi.org/10.2174/1874372202014010022
6. Ampapuram R., H. B. K, P. K. M, R. Maddileti, & A. L. G. (2019). *Powders with marketed formulation* (Vol. 9, pp. 500–503).
7. Vijayalakshmi, A., Sangeetha, S., & Ranjith, N. (2018). Formulation and evaluation of herbal shampoo. *Asian Journal of Pharmaceutical Clinical Research, 11*(Special Issue 4), 121–124. https://doi.org/10.22159/ajpcr.2018.v11s4.31713
8. Angiolella, L., Carradori, S., Maccallini, C., Giusiano, G., & Supuran, C. T. (2014). Targeting Malassezia species for novel synthetic and natural antidandruff agents. *Current Medicinal Chemistry, 24*(22), 2392–2412. https://doi.org/10.2174/0929867324666170404110631
9. Schothorst, R. C., & Stephany, R. W. (2001). Occurrence of N-nitrosodiethanolamine (NDELA) in cosmetics from the Dutch market. *International Journal of Cosmetic Science, 23*(2), 109–114. https://doi.org/10.1046/j.1467-2494.2001.00062.x
10. Rao, K. J., & Paria, S. (2016). Anti-*Malassezia furfur* activity of natural surfactant mediated in situ silver nanoparticles for a better antidandruff shampoo formulation. *RSC Advances, 6*(13), 11064–11069. https://doi.org/10.1039/c5ra23174d
11. Khan, A., et al. (2010). *Ocimum sanctum* essential oil and its active principles exert their antifungal activity by disrupting ergosterol biosynthesis and membrane integrity. *Research in Microbiology, 161*(10), 816–823. https://doi.org/10.1016/J.RESMIC.2010.09.008
12. Jasso De Rodríguez, D., Hernández-Castillo, D., Rodríguez-García, R., & Angulo-Sánchez, J. L. (2005). Antifungal activity in vitro of *Aloe vera* pulp and liquid fraction against plant pathogenic fungi. *Industrial Crops and Products, 21*(1), 81–87. https://doi.org/10.1016/J.INDCROP.2004.01.002
13. Singariya, P., Mourya, K. K., & Kumar, P. (2012, Jan). Antimicrobial activity of the crude extracts of *Withania somnifera* and Cenchrus setigerus in-vitro. *Pharmacognosy Journal, 4*(27), 60–65. https://doi.org/10.5530/PJ.2012.27.10
14. Yineger, H., & Yewhalaw, D. (2007). Traditional medicinal plant knowledge and use by local healers in Sekoru District, Jimma zone, southwestern Ethiopia. *Journal of Ethnobiology and Ethnomedicine, 3*(1), 1–7. https://doi.org/10.1186/1746-4269-3-24/FIGURES/5
15. Abbink, J. (1995). Medicinal and ritual plants of the Ethiopean southwest: An account of recent research. *Indigenous Knowledge and Development Monitor, 3*(2) Accessed: Apr. 12, 2022. [Online]. Available: https://www.ascleiden.nl/publications/medicinal-and-ritual-plants-ethiopean-southwest-account-recent-research
16. Pankhurst, R. (2001). The status and Availability of oral and written knowledge on traditional health care in Ethiopia. In *Proceedings of the National Workshop on biodiversity conservation and sustainable use of medicinal plants in Ethiopia* (pp. 92–106).

17. Handa, S. S. (2008). Extraction technologies for medicinal and aromatic plants. In *International Centre for Science and High Technology*. Accessed: Apr. 17, 2022. [Online]. Available: https://www.morningmystbotanics.com/wp-content/uploads/Extraction_technologies_for_medicinal_and_aromatic_plants.pdf#page=25
18. Shi, X. D., Yin, J. Y., Huang, X. J., Que, Z. Q., & Nie, S. P. (2018, Dec). Structural and conformational characterization of linear O-acetyl-glucomannan purified from gel of Aloe barbadensis miller. *International Journal of Biological Macromolecules, 120*, 2373–2380. https://doi.org/10.1016/J.IJBIOMAC.2018.09.005
19. Bagatharia, S. B., & Thaker, V. S. (2005) Antibacterial activity of *Aloe vera* leaf gel extracts against Staphylococcus aureus. *Indian Journal of Microbiology, 45*. Accessed: Apr. 17, 2022. [Online]. Available: https://www.researchgate.net/publication/294416621
20. Sarovar Reddy, V., & Gopinath, C. (2018, May). Formulation and evaluation of synthetic anti-dandruff shampoo. *Asian Journal of Pharmaceutical, 12*(1), S87–S91. https://doi.org/10.22377/AJP.V12I01.2045
21. Al Badi, K., & Khan, S. A. (2014). Formulation, evaluation and comparison of the herbal shampoo with the commercial shampoos. *Beni-Suef University Journal of Basic and Applied Sciences, 3*(4), 301–305. https://doi.org/10.1016/j.bjbas.2014.11.005
22. Deeksha, R. M., & Sharma, P. K. (2014). Advancement in shampoo (a dermal care product): Preparation methods, patents and commercial utility. *Recent Patents on Inflammation & Allergy Drug Discovery, 8*(1), 48–58. https://doi.org/10.2174/1872213x08666140115110727
23. Golhani, D., Pandey, V., Shukla, A., & Shukla, R. (2015). Formulation and comparative evaluation of herbal shampoo with marketed products. *Mintage Journal of Pharmaceutical and Medical Sciences, 4*(3), 3–6.
24. Kumar, A. (2014, June). *Ashok and Mali 2011.pdf*
25. Pawar, A. P., Pawar, D. N., & Dalvi, Y. V. (2019). Formulation and evaluation of Polyherbal soap. *Research Journal of Topical and Cosmetic Sciences, 10*(1), 23. https://doi.org/10.5958/2321-5844.2019.00006.2
26. Pharmaceutical, T., & Bharkatiya, M. (2019). CODEN (USA): PCJHBA Formulation and Evaluation of Herbal Shampoo. *Extract of Some Plants The Pharmaceutical and Chemical Journal, 6*(4), 74–80. Formulation and Evaluation of Herbal Shampoo by Extract of Some Plants, no. January 2019, 2021.
27. Mainkar, A. R., & Jolly, C. I. (2000). Evaluation of commercial herbal shampoos. *International Journal of Cosmetic Science, 22*(5), 385–391. https://doi.org/10.1046/J.1467-2494.2000.00047.X
28. Gaud, R. S., & Gupta, G. D. (2001). *Practical physical pharmacy* (pp. 81–105). CBS Publ. Distrib. [Online]. Available: https://scholar.google.com/scholar_lookup?title=Practicalphysicalpharmacy&publication_year=2001&author=R.S.Gaud&author=G.D. Gupta
29. Utane, R., Deo, S., & Itankar, P. (2017). Preparation of herbal shampoo (HS) by green method and their characterization. *Ijrssis, 5*(March), 254–258.
30. Vijayakumar, R., Muthukumar, C., Kumar, T., & Saravanamuthu, R. (2006). Characterization of *Malassezia furfur* and its control by using plant extracts. *Indian Journal of Dermatology, 51*(2), 145–148. https://doi.org/10.4103/0019-5154.26942
31. HiMedia. (2019). *Sabouraud Dextrose Broth (Sabouraud Liquid Medium) M033* (pp. 2–4).
32. Engel, S. M. (2011). American politicians confront the court. *American Politicians Confront the Court*, 2–3. https://doi.org/10.1017/cbo9780511994890
33. Matuschek, E., Brown, D. F. J., & Kahlmeter, G. (2014). Development of the EUCAST disk diffusion antimicrobial susceptibility testing method and its implementation in routine microbiology laboratories. *Clinical Microbiology and Infection, 20*(4), O255–O266. https://doi.org/10.1111/1469-0691.12373
34. Heidari, B. A., Badinloo, S., Ohadi, M., & Noudeh, G. D. (2016). Bioencapsulation of biosurfactant-producing bacillus subtilis (PTCC 1023) in alginate beads. *Jundishapur Journal of Natural Pharmaceutical Products, 11*(4), 33935. https://doi.org/10.17795/JJNPP-33935

35. Dias, M. F. R. G., De Almeida, A. M., Cecato, P. M. R., Adriano, A. R., & Pichler, J. (2014). The shampoo pH can affect the hair: Myth or reality? *International Journal of Trichology, 6*(3), 95–99. https://doi.org/10.4103/0974-7753.139078
36. Noudeh, G. D., Sharififar, F., Khazaeli, P., Mohajeri, E., & Jahanbakhsh, J. (2011). Formulation of herbal conditioner shampoo by using extract of fenugreek seeds and evaluation of its physicochemical parameters. *African Journal of Pharmacy and Pharmacology, 5*(22), 2420–2427. https://doi.org/10.5897/AJPP11.121
37. Jaggi, R. K., Madaan, R., & Singh, B. (2003). Anticonvulsant potential of holy basil, *Ocimum sanctum* Linn., and it s cultures. *Indian Journal of Experimental Biology*, 1329–1333.
38. Zoya, M., Bhikhu, M., & Gaurav, S. (2016). Anti-dandruff activity of synthetic and herbal shampoos on dandruff causing isolate : Malassezia. *International Journal of Applied Sciences, 2*(7), 80–85.
39. Ahmad, A., Abuzinadah, M. F., Alkreathy, H. M., Banaganapalli, B., & Mujeeb, M. (2018). Ursolic acid rich *Ocimum sanctum* L leaf extract loaded nanostructured lipid carriers ameliorate adjuvant induced arthritis in rats by inhibition of COX-1, COX-2, TNF-α and IL-1: Pharmacological and docking studies. *PLoS One, 13*(3), e0193451. https://doi.org/10.1371/JOURNAL.PONE.0193451
40. Punyoyai, C., Sirilun, S., Chantawannakul, P., & Chaiyana, W. (2018). Development of Antidandruff shampoo from the fermented product of *Ocimum sanctum* Linn. *Cosmet, 5*(3), 43. https://doi.org/10.3390/COSMETICS5030043
41. Staub, I., & Bergold, A. M. (2004). Determination of Ketoconazole in Shampoo by high performance liquid chromatography. *Acta Farmacéutica Bonaerense, 23*(3), 387–390. Accessed: Aug. 22, 2021. [Online]. Available: http://www.latamjpharm.org/trabajos/23/3/LAJOP_23_3_3_2_A6RV37R61M.pdf
42. Guo, S., Huang, W., Zhang, J., & Wang, Y. (2015). Novel inhibitor against Malassezia globosa LIP1 (SMG1), a potential anti-dandruff target. *Bioorganic & Medicinal Chemistry Letters, 25*(17), 3464–3467. https://doi.org/10.1016/J.BMCL.2015.07.005
43. Xu, J., Zhou, F., Ji, B. P., Pei, R. S., & Xu, N. (2008). The antibacterial mechanism of carvacrol and thymol against Escherichia coli. *Letters in Applied Microbiology, 47*(3), 174–179. https://doi.org/10.1111/J.1472-765X.2008.02407.X
44. Butnariu, M. V., & Giuchici, C. V. (2011). The use of some nanoemulsions based on aqueous propolis and lycopene extract in the skin's protective mechanisms against UVA radiation. *Journal of Nanobiotechnology, 9*(1), 1–9. https://doi.org/10.1186/1477-3155-9-3/TABLES/5
45. Pant, G., Nayak, N., & Gyana Prasuna, R. (2012). Enhancement of antidandruff activity of shampoo by biosynthesized silver nanoparticles from Solanum trilobatum plant leaf. *Applied Nanoscience, 3*(5), 431–439. https://doi.org/10.1007/S13204-012-0164-Y
46. Kankeaw, U., & Rawanna, R. (2015). *The study of antibacterial activity of benzimidazole derivative synthesized from citronellal*. https://doi.org/10.17706/ijbbb.2015.5.5.280-287
47. Payel Choudhury, E. (2020). Comparative antidandruff efficacy of plant extracts prepared from conventional and supercritical fluid extraction method and chemical profiling using GCMS. *The Journal of Dermatological Treatment, 33*, 989. https://doi.org/10.1080/09546634.2020.1799919

Chapter 17
Recent Advances in Microalgae Biomass Drying, Pretreatment, and Extraction of Polyunsaturated Fatty Acids (PUFAs): A Review

Tsegaye Mekuria Ayele, Metadel Kassahun Abera, Tsehayneh Geremew Yohannes, Gizachew Muluneh Amera, and Amha Belay Sahlu

17.1 Introduction

Microalgae are increasingly being viewed as a sustainable resource with enormous potential to meet the present day and future food and nutrition needs. The production of various high-value bioactive compounds, such as omega-3 and omega-6 polyunsaturated fatty acids (PUFA), has become economically and technologically feasible. The n-3 PUFA, particularly eicosapentaenoic acid (EPA) and docosahexaenoic acid (DHA), have a variety of biological functions that have been proven to have a protective effect against a variety of diseases, including cardiovascular and neurological problems [1]. As a result, over the past 20 years, the global market for

T. M. Ayele
Department of Biotechnology, College of Agriculture and Natural Resources Management Science, Debre Berhan University, Debre Birhan, Ethiopia

Faculty of Chemical and Food Engineering, Bahir Dar Institute of Technology, Bahir Dar University, Bahir Dar, Ethiopia

M. K. Abera (✉)
Faculty of Chemical and Food Engineering, Bahir Dar Institute of Technology, Bahir Dar University, Bahir Dar, Ethiopia

T. G. Yohannes
Department of Industrial Biotechnology, Institute of Biotechnology, University of Gondar, Gondar, Ethiopia

G. M. Amera
Bioinformatics and Database Management Division, Institute of Biotechnology, Bahir Dar University, Bahir Dar, Ethiopia

A. B. Sahlu
Algae4All LLC (Founder and CEO), La Quinta, CA, USA

S. W. Fanta et al. (eds.), *Sustainable Development Research in Manufacturing, Process Engineering, Green Infrastructure, and Water Resources*, Green Energy and Technology, https://doi.org/10.1007/978-3-031-77339-6_17

omega −3 PUFA has been growing quickly. However, due to a multitude of problems such as overfishing, global warming, and insufficient aquaculture productions, the supply of PUFA is currently experiencing scarcity to keep up with the rapidly rising demand. Microalgae have shown promising potential to produce PUFAs owing to their favorable omega-6/omega-3 composition [2], and short generation time of 10 days. For those reasons, the search for microalgae isolates with the prospect of high PUFAs accumulation and improvement toward their cultivation and the downstream processing technologies have become an active area of study.

The initial step in the cultivation of algae is the inoculation of the chosen algal species onto a culture medium that supplies the growth nutrients. After cultivation, algal biomass can be harvested after the culture media with the inoculum is kept at ambient conditions for a period of time [3]. Gravity sedimentation, centrifugation, filtration, flocculation, and electrolytic coagulation are the methods for harvesting microalgae biomass after cultivation [4, 5]. Since cost-effectiveness and energy efficiency during harvesting depend on the kind of strain, growth conditions, and the finished product, no single harvesting technique is appropriate for harvesting the biomass of all kinds of microalgal species [6]. Therefore, either continued improvements in the present biomass harvesting technologies or the emergence of new strain-compatible technologies are required. After harvesting microalgae biomass, it is either dried for stability and storage [7] or used as a wet slurry or cake for further processing. The significant high expenditure of microalgae PUFA production lies in the biomass drying, pretreatment, and extraction processes. Aiming to inform the technical difficulties and the cost incurring by the process to the stakeholders and to the scientific community, this review presents an overview of the methods used for drying microalgae biomass, pretreating it, and extracting PUFAS, as well as the benefits and drawbacks of the methods with overviewing the recent developments in drying, pretreatment, and extraction techniques.

17.2 Prospects of Microalgae as Alternative Source of PUFAs

The ability of microalgae to generate PUFAs is their most significant addition to the entire microalgal market [8, 9]. Long carbon chain fatty acids (> C20) are often PUFAs, whereas fatty acids with chain lengths of carbon between C14 and C20 are preferred for the production of biodiesel [10]. According to [11], the global market for ω3-PUFAs was estimated at a value of USD 2.49 billion in 2019 and is anticipated to increase by 7.7% yearly through to 2027. Martek Biosciences is the industry leader in sustainable algal omega-3 technology and the producer of goods like infant formula containing omega-3 DHA, dietary supplements, functional diets, drinks, and animal feed products.

17.2.1 Microalgae with Potential of PUFA Production

Cereal grains (rice, wheat, maize, teff), legumes (soya bean, red kidney bean), vegetable oils (sunflower, sesame, palm, olive, canola), and foods of animal origin, such as fish, beef, poultry, egg, and milk are recognized sources of PUFAs [12]. However, globally the supply of PUFAs from these sources has encountered serious scarcities due to the rising global population compounded with other global problems such as the effects of climate change, overexploitation of marine species, and low productivities of aquacultures [13]. Besides, fish is not the desirable source of omega-3 fatty acids due to the geographical and seasonal variations in quality, and as well as customer complaints about odor problems [14].

Therefore, the search for alternative sources for these essential fatty acids is becoming an immensely important issue to sustain the rising global demand for the products. Microalgae presenting as potential alternative feedstocks offer a number of advantages such as not requiring arable land for cultivation and thus not competing for land with food crops [15]; The possession of a fast growth rate and their ease to cultivate in large quantities with high photosynthesis efficiency [16]. Furthermore, several studies point out that microalgae are 20- to 40-fold more productive than oil crops, yielding 80% lipid out of the dry weight of biomass [1]. Microalgae sequester carbon dioxide from the open atmosphere and are able to be cultivated in wastewater including brackish water [17]. Therefore, microalgae present a huge and untapped reserve for PUFAs production [18, 19]. Thus, increasing PUFA production using microalgae by identifying novel species, developing and applying emergent or improved production, and processing technologies has been emerging as an active research area [20]. Microalgae species producing different PUFAs are reviewed in Table 17.1.

17.2.2 Microalgae Used at Industrial Scale Production of PUFAs

The industrial scale microalgae cultivation to produce different bio-products has expanded adequately over the last few decades [23].The development of an effective large-scale cultivation setup for the commercial production of PUFAs particularly EPA and DHA would address a significant global demand [24]. A few algal species such as Spirulina sp., *Chlorella* sp., *Dunaliella salina,* and *Haematococcus pluvialis* have been recognized as producers of nutritional supplements [25]. Production of EPA and DHA has been achieved commercially by employing the microalgae species *Phaeodactylum tricornutum*, *Nannochloropsis* sp., and *Crypthecodinium cohnii* [26]. A summary of industrial-scale microalgae PUFA products applied to food, feed, nutraceutical, and pharmaceuticals is presented in Table 17.2.

Table 17.1 Microalgae species producing different PUFAs with their content (mg/g oil) and (%W of total fatty acids)

Microalgae species	Types of PUFAs produced						References
Microalgae	ALA	SDA	EPA	DPA	DHA		Mishra [21]
Isochrysis T-iso	29 ± 4	43 ± 0	2.8 ± 0.7	–	46 ± 14		
Nannochloropsis gaditana	0.3 ± 0.03	0.3 ± 0	175 ± 12	–	–		
Nannochloropsis oculata	0.7 ± 0.1	–	193 ± 24	–	40.9 ± 0.9		
Pavlova lutheri	10.0 ± 0.3	17 ± 0.5	92 ± 2	–	40.9 ± 0.9		
Phaeodactylum tricornutum	0.8 ± 0.1	1.9 ± 0.1	111 ± 5	1.08 ± 0.01	8.3 ± 0.5		
Porphyridium cruentum	1.42 ± 0.01	–	35.6 ± 0.3	–	–		
Rhodomonas salina	92 ± 5	43 ± 2	18 ± 1	–	11.1 ± 0.8		
Tetraselmis suecica	68 ± 4	17 ± 0.5	16.3 ± 0.5	–	0.8 ± 0.1		
Thalassiosira pseudonoma	1.9 ± 0.1	20.4 ± 0.8	81 ± 2	1.82 ± 0.01	20.9 ± 0.8		
PUFA composition in some microalgae species (%W of total fatty acids)							
	ALA	GLA	ARA	EPA	DHA	Total PUFA	Karageorgou et al. [22]
Palovagyran	–	2.0	3.7	34.2	16.4	58.2	
Isocrysis galbana	9.5	6.4	3.4	10.3	33.6	64.5	
Prorocentrum minimum	1.3	0.2	13.8	14.0	28.1	62.3	
Phaeodactylum tricomutum	1.1	0.4	1.0	34.1	1.0	42.5	
Nannochloropsis oculata	Nd	Nd	9.7	21.3	Nd	36.4	
Chroomonas salina	2.6	14.2	0.9	11.9	5.2	67.9	
Schizochytrium sp. S31	0.07	0.2	0.6	1.5	29.9	50.9	
Schizochytrium limacinum OUC88	0.5	0.3	0.4	0.7	27.5	75.1	
Aurantiochytrium limacinum	Nd	Nd	0.2	21.4	0.9	66.6	
Aurantiochytrium mangrovei	Nd	Nd	0.4	1.3	35.8	49.3	
Parietichytrium sp.	Nd	Nd	5.4	8.2	22.9	54.9	
Thalassiosira pseudonana	0.2	0.1	0.3	19.3	3.9	52.6	

ALA alpha linoleic acid, *GLA* gamma linoleic acid, *SDA* Steiridonic acid, *EPA* Eicosapentanoic acid, *DPA* Docosapentanoic acid, *DHA* Docosahexanoic acid, *ARA* Arachidonic acid, dashes (–) indicate the fatty acids not detected, *nd* not determined

Table 17.2 Major industrial microalgal PUFA products that are applied to food, feed, nutraceuticals, pharmaceuticals and their producer companies [27]

Microalgae sp. of industrial PUFA producer	PUFA produced	Company	Annual Production	Applications	Function	Reference
Chlorella vulgaris, Chlorella minutissima, Virginica grossii	LA	Necton (port)	2000 tDW	Nutra	Nutritional, anticancer, anti-inflammatory, anti-aging	Kaur et al. [27]
Crypthecodinium cohnii	DHA oil	DSM (USA/NL)	240 tDHA	Pharma	Anti-inflammatory activities, improve cholesterol levels)	
Shizochytrium sp	DHA oil, ω-3, EPA, ALA	Flora health (USA)	10 tDHA	Pharma	Improved cardiovascular, brain, and eye systems	
Phaedodactylum tricornutum	EPA,	A4F (port)	–	Food	Nutrition	

A4F Algae for all, Dashes indicate data not available, *Port* Portugal, *NL* Netherlands, *tDW* tons dry weight, *tDHA* tons DHA oil, *Nutra* nutraceuticals, *Pharma* pharmaceuticals

17.3 Methods of Drying Microalgae Biomass

Harvested microalgae biomass is either stored in dried form for stability and conservation or used as a wet slurry to extract the desired cellular component. Since gathered algae slurry is perishable and can degrade quickly, drying algae before storing it is crucial [28]. Drying microalgae slurry from 95% level of moisture to a lesser moisture content of lower than 15%, creates a big financial burden which may account for up to75% of the overall processing budget [29].The varieties, advantages, and disadvantages of microalgae drying methods are summarized in Table 17.3 [30]. It has been suggested that free natural solar radiation can provide the energy needed to dry the microalgae biomass with potential of contributing significantly to the process' overall cost compensation.

17.4 Biomass Pretreatment Techniques in PUFA Extraction

In the downstream stages of microalgae cultivation, the most important and difficult phase is PUFA extraction from microalgae [10]. Prior to isolation of the desired products, the microalgae biomass is processed to break the cell wall and release the cellular contents into the cultivation medium. The unit operations undertaken to distract the cells to release their internal content are called pretreatment. A

Table 17.3 Operating principles, Pros and cons of various microalgae biomass drying methods [30]

Drying method	Operating principle	Advantages	Disadvantages
Solar drying	Solar radiation	Low energy cost	Slow operation, prone to weather conditions
Rotary drying	Rotating hot cylinder surface contacts biomass and dries	Fast operation, good sterilization	High energy cost, biomass d
Convective drying system	Heating by hot air	Effective drying	Biomass degradation
Spray drying	High-pressure water vapor contact with hot gas	Fast and efficient, used for high-value products	High energy cost, low digestibility
Cross-flow air drying	Wet biomass slurry in contact with flowing hot air	Superior cost performance, fast drying	High energy cost
Vacuum shelf drying	Heating in vacuum shelf	Fast and efficient	High cost of installment and operation
Flash drying	Wet and dried algae mixture immersed in hot gas	Quick drying	Sensitive to hot gas source and quality
Incinerator drying	Multiple incinerators	Fast, powder mass	Sophisticated process, high capital cost
Freeze drying	Warm surface heat frozen biomass	High-quality biomass, simultaneous disruption	Slow, high capital requirement, high cost for scale-up
Microwave-assisted drying	Electromagnetic radiation	Fast, controllable	High capital, safety threats

microalgal strain's unique biology and cell wall characteristics determine the best pretreatment technique to use [5]. According to [10] there are two basic groups of cell disruption techniques: mechanical and nonmechanical. Bead mill, high-pressure homogenization, lyophilization, microwave, ultrasonication, and autoclaving are a few common mechanical cell disruption techniques. Each of the techniques is briefly described here.

17.4.1 Bead Milling

In this technique, a cylindrical compartment is filled up with microalgae biomass and a large quantity of fine, high-speed glass beads. The compartment is then forced to spin at a specific speed. The suspended cells are disturbed by collision with the beads. The efficiency of the process is greatly influenced by bead size, temperature, time, cycle count, and spinning speed. The disadvantage is that it uses excessive energy and results in the release of some undesirable compounds, raising the cost of purification [10, 15].

17.4.2 A High-Speed Homogenizer (HSH)

A high-speed homogenizer (HSH) is an efficient technique that disrupts cells by hydrodynamic cavitation brought on by high-speed stirring at 10,000–20,000 rpm and shearing power at the solid–liquid interface. Although it has several limitations, such as a high-power requirement and protein disruption, this method is appropriate for large scale and only needs a brief processing period. For lipid extraction, operational periods of 30 s at 10,000 rpm have been recorded [15].

17.4.3 High-Pressure Homogenization (HPH)

High pressure (between 20 and 120 MPa) is used in the HPH method for disrupting cells in order to increase turbulence, liquid-shear stress, and friction. The efficiency of the process can be improved by optimizing the operating pressure and the quantity of homogenization passes based on the cell wall qualities. Additional factors that affect a particular energy consumption include dry cell weight concentration, microalgae species, and growth conditions [15].

17.4.4 Ultrasonication

In the ultrasonication method, intense microbubbles are created in the liquid medium. These microbubbles expand and forcefully deflate causing cavitation to occur. These bursting cavitation bubbles reach temperatures of up to 5000 K and air pressures of several orders of magnitude, pushing the release of intracellular compounds into the surrounding media and causing robust and efficient cell disruption [10, 15].

17.4.5 Microwaves

These electromagnetic waves having a frequency between 1 gigahertz and about 300 gigahertz, selectively interact with polar or dielectric molecules in suspensions of microalgae causing a rise in temperature brought on by frictional forces caused by inter- and intramolecular movement. Additionally, the availability of free water in the microalgae solution helps to improve the effectiveness of microwave-assisted extraction because when water is exposed to microwaves, it quickly approaches boiling point and swells from inside, increasing pressure. According to [10, 15], the coalition of heat and pressure created inside cells as a result of microwaves, bursts the cell wall or membrane and releases intermolecular components into the surrounding media.

17.4.6 Pulsed Electric Field (PEF)

This method induces an electric field externally across the cell wall to create the phenomenon of electroporation, which causes cell disruption. This technique is currently being used to improve the extraction of numerous important chemicals from microorganisms. This method's restriction is that the treatment solution must be ion-free, or electrically nonconductive. These microalgae would be adequately prewashed and deionized to increase the PEF cell disruption efficacy. Thus, the composition of the medium has a direct impact on the effectiveness of PEF cell disruption [10, 15]. Cell disruption via chemicals, enzymes, or osmotic shock are examples of nonmechanical techniques.

17.4.7 Enzymatic Lysis

Enzyme-assisted extraction works on the principle of using digestive enzymes to decompose the microbial cell membrane/wall and release the content into the surrounding media [10]. Enzyme-assisted extraction is a promising, eco-friendly, and nondestructive method that can protect degeneration of heat-labile biomolecules. The enzymes used vary from one microorganism to the other depending on the constituency of cell walls and membranes of the different cell types and strains [31]. This biological procedure takes place at ambient temperature. Cellulase, protease, papain, and lysozyme are typically the most researched enzymes that have been utilized to lyse the cells of many types of microorganisms. Lysozyme was used in cell disruption of *Chlorella vulgaris* to extract lipids, and it was observed that yields were 7.46-fold more compared to untreated biomass [31].

17.4.8 Osmotic Shock

Osmotic shock is a different method that causes cell disruption by causing an abrupt shift in the concentration of water across the cell membrane and causing intracellular components to leak out. To improve lipid recovery after algal cell rupture, freezing drying, and acid/base treatments are also used [10]. Fig. 17.1 gives a summary of the various biomass treatment techniques.

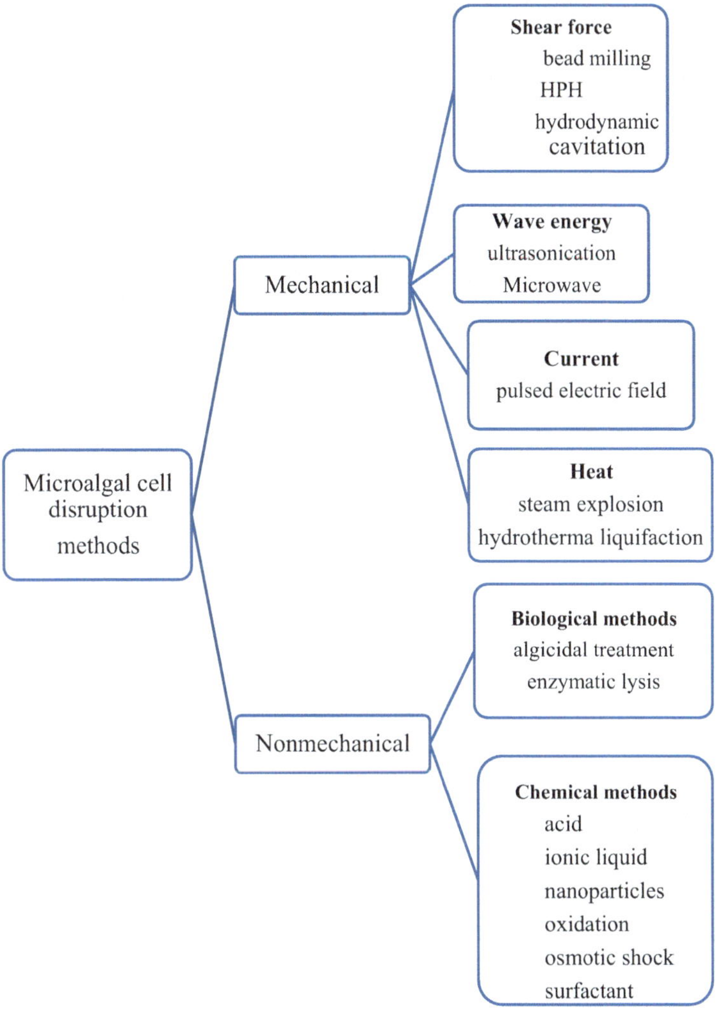

Fig. 17.1 Summary of microalgae cell disruption methods [10]

17.5 Extraction of PUFAs from Microalgae Biomass

Depending on the growing conditions, microalgae cells can gather up to 70% of their dry weight as lipids. Similar to plants, microalgae's complex cell wall structure makes it difficult to extract PUFAs from them, hence in order to obtain the greatest amount of recovery, various cell disruption techniques are used. The extracted lipid should not be damaged by the lipid extraction process, which should be quick, scalable, effective, nontoxic, and affordable [10].

17.5.1 Conventional PUFA Extraction Methods and Their Drawbacks

Because various microalgae have varying cell structures and sizes, there is no general technique for extracting PUFAs to be applied to all species [10]. The most effective conventional protocol for extracting PUFAs from microalgae is the Bligh and Dyer method that utilizes the chloroform, methanol, and water mixture. However, the employed organic solvents are poisonous and combustible, making them unsuitable for use in final products intended for food and medical applications [5]. Additionally, 89% of the total energy input and 70–75% of the overall processing cost go toward drying the biomass. As a result, it is blamed that the drying stage severely reduces the process' overall efficiency and renders it difficult for large-scale production [32].

17.5.2 Alternative Solvents

To enhance the contact of reactants in chemical processes, petroleum-derived solvents are used. However, they must either be reprocessed after each use, which often consumes a lot of energy, or their disposal into waste causes environmental pollution. Therefore, it is on the verge of stopping using these solvents or being replaced with alternatives that consume less energy and are simpler to recover. The best possibilities include solvents like water, Ionic liquids (ILs), and supercritical fluids (SCFs) [33]. Relative to their conventional ones, the typical alternative solvent should fulfill the following attributes: It must maintain the important properties of the conventional solvent that it replaces, such as polarity and reactant solubility, keep process safety, able to work at the same or moderate operating conditions, maintain or improve product quality, and be economically viable [34]. Pressurized fluid extraction (PFE), subcritical water extraction (SWE), and ionic liquid extraction are all extraction techniques that are so-called alternative solvents or green solvents [5].

17.5.2.1 Ionic Liquids for PUFA Extraction from Wet Microalgae Culture

Ionic liquids (ILs) have been described as salts with melting points between (−96 °C) and (150 °C) [35]. They typically consist of relatively bulky inorganic anions (such as chlorides, bromides, tetrafluoroborates, and hexafluorophosphates) and relatively bulky organic cations (such as pyrrolidinium, imidazolium, pyridinium, and ammonium) [36]. Special characteristics of ILs include their low vapor pressure and superior ability to dissolve both organic and inorganic substances. As a result of their low toxicity, nonflammability, and nonvolatility, ILs are frequently referred to as green solvents. They can be used either as catalyst, solvent, or support for the catalyst in the reaction medium [33]. Furthermore, by adjusting the ratio of cations to anions, ILs' physical and chemical characteristics can be tailored for different applications. These characteristics provide ILs the potential to be used in a variety of industrial operations as well as chemical reactions [4].

In order to selectively extract hydrophobic lipids from a microalgae strain *Neochloris oleoabundans* biomass intact cells, imidazolium-phosphonium ionic liquid solutions were used. Lipid recovery after cell disruption from the hydrophilic fraction containing proteins and carbohydrates showed that the ionic liquid tributyl methyl phosphonium methyl sulfate (TBP SO4) is capable of permeabilizing the fresh intact cells of *N. oleoabundans* for extracting 68% of lipids out of the cells, whereas, following cell disruption by bead milling, 80% of total proteins as well as 77% of carbohydrates could have been extracted [37] (Fig. 17.2). The effectiveness of ionic liquids to extract lipids from liquid microalgae culture was examined [38] (Fig. 17.3). The results assured that the ionic liquid (1-ethyl-3-methyl imidazolium ethylsulfate) could disrupt *C. vulgaris* microalgae cells in coalition with methanol and enabled facile recovery of lipids over a large degree of dewatered microalgae (up to 82 wt% water), within a short duration (75 min) and at room temperature, resulting in the issuing of energy saving, harmonious to aqueous environment, extraction scheme.

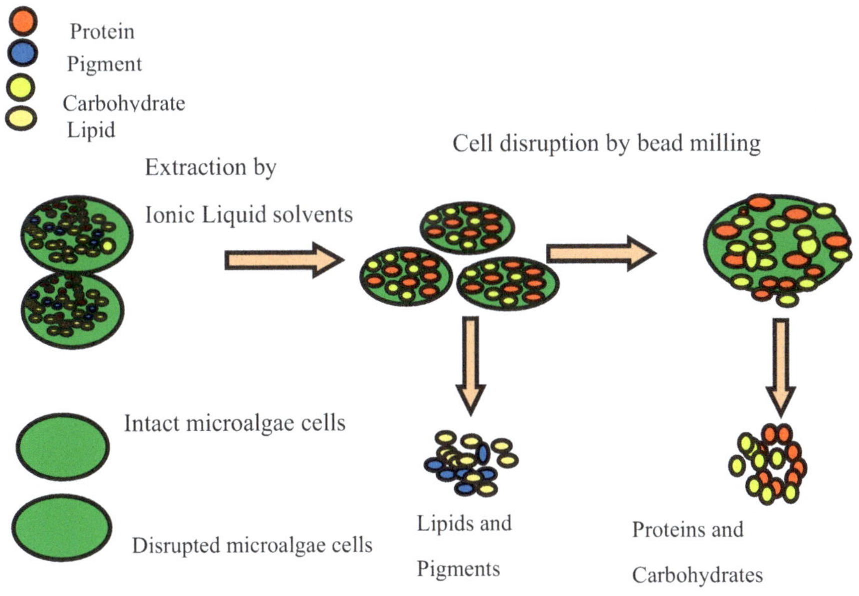

Fig. 17.2 Ionic liquid solvents enabled extraction of lipids and pigments before cell disruption, followed by recovery of proteins and carbohydrates after bead milling [37]

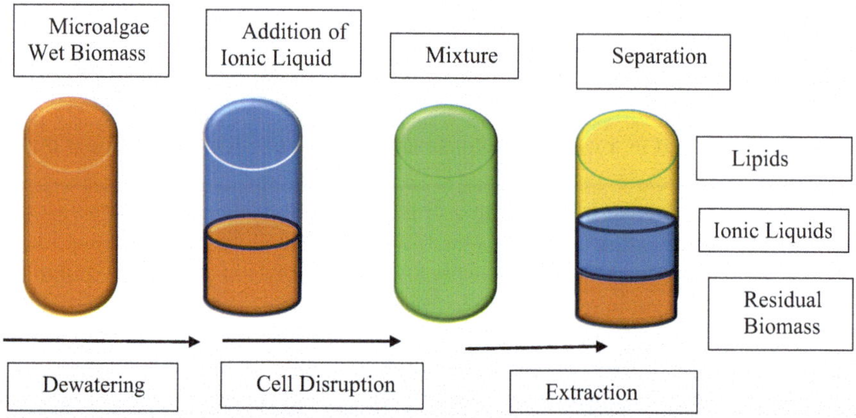

Fig. 17.3 Lipid extraction from liquid algae culture using ionic liquids [38]

17.5.2.2 Supercritical Fluid Extraction

According to [39], a mixture is characterized as being in a supercritical fluid (SCF) if it is above its critical pressure and critical temperature but below the pressure needed to cause it to solidify. Substances develop SCF condition when their critical pressure and temperature are surpassed [33]. Conducting reactions under supercritical conditions offers the chance to alter the solvent's properties by varying pressure and/or temperature, raise the solubility of the reactants and products, lower the inter-phase transport difficulties to increase reaction rates, enhance selectivity, and integrate the reaction and separation steps [40]. Several substances, such as carbon dioxide (CO_2), methanol, and ammonia, can be utilized as supercritical extractants. Because of its affordability, nontoxic nature, and lower critical temperature and pressure, which can prevent the degradation of heat-labile molecules during extraction, supercritical carbon dioxide (SC-CO_2) is the most commonly used of them. Additionally, CO_2 is gaseous at normal pressure and temperature and can be easily extracted from the extracts without the use of a solvent [41]. For supercritical CO_2 (sCO_2) extraction, dry biomass is preferred since water might generate a film barrier between the microalgae cells and the solvent CO_2 which makes lipid extraction challenging [9]. SFE extraction gives highly pure and nontoxic products (Fig. 17.4). The benefits and drawbacks of various extraction techniques are illustrated in Table 17.4. A summary of recent advancements in drying, pretreatment, and extraction techniques for microalgae biomass is provided in Table 17.5.

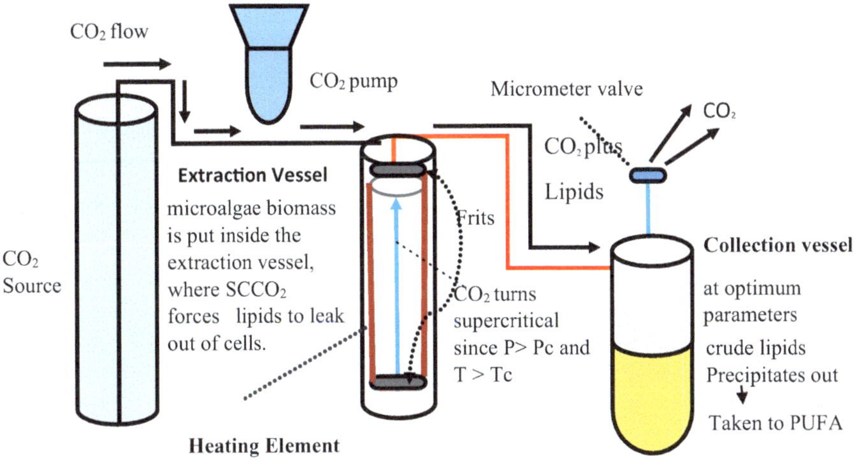

Fig. 17.4 Lipid extraction from microalgae using supercritical CO_2 [10]

Table 17.4 Advantages and disadvantages of different extraction methods [41]

Method		Advantage	Disadvantage
Traditional extraction methods	Soxhlet	Low cost, easy operation, speed of extraction	Time taking, high reagent & energy consumption
	Folch	Quick, easy to hold large number of samples	Toxic reagents are used which is harmful to human health and to the environment
	Bligh-dyer	Simultaneous lipid extraction & separation.	Extractive reagents are toxic; have fewer substitutes, and the cost is high
Emergent extraction methods	Supercritical fluids	Very good extraction efficiency, less toxic reagents	Selectivity to polarity, expensive equipment
	Pressurized lipids extraction	Separation of lipids, protect bioactive compounds, reduce energy consumption and pollution	
	Pulsed electric fields	Simple operation, no pollution, amount of sample	Adverse effect of excessive electric field
	Ultrasound-assisted extraction	Low process temperature and less energy required. High rate of extraction	The intensity and time of ultrasound need to be controlled to avoid negative effects
	Microwave-assisted extraction	Reduce extraction time and energy consumption, improve extraction efficiency	The polarity of the solvent has an impact on the extraction, unsuitable for heat-labile substances
	Ionic liquids	Low toxicity, stability, tunable physical and chemical properties	Possibility of pollution during synthesis
	Enzyme-assisted extraction	Substrate selective, pretreatment may be completed at room temperature & pressure to reduce energy use	The price of enzyme production is high. Optimizing reaction conditions is cumbersome

Table 17.5 Summary of advances in microalgae biomass drying, pretreatment, and extraction methods

Process operations	Study title and intent	Results obtained	References
Drying Microalgae biomass	Drying of algae by various drying methods: Solar conduction drying vs open sun drying	Solar conduction dryer showed better result than the open sun drying	Bheda and Shinde [28]
	Drying microalgae using an industrial solar dryer: A biomass quality assessment	Functional properties of the solar-dried biomass products proved to be suitable for food and feed products	Schmid et al. [42]
	Drying techniques of microalgal biomass: A review	Utilization of dried biomass for extracting useful metabolites from microalgae has been recommended	Kim and Kim [30]
Biomass pretreatment methods	Advanced extraction of lipids with DHA from *Isochrysisgalbana* with enzymatic pretreatment combined with pressurized liquids and ultrasound-assisted extractions	Extraction of polar lipids was performed using UAE (50 °C and 15 min) and pressurized liquid extraction (100 °C) techniques. The best omega-3 PUFA amount (33.2%), EPA (3.3%), and DHA (12.0%) were achieved with the UAE, indicating the optimized method as a practical option to produce lipids desirable for food and nutraceutical supplements	Señoráns et al. [43]
Lipid extraction methods	Supercritical carbon dioxide extraction of microalgae lipid: Process optimization and laboratory scale-up	53 °C, 500 bar, and 1.9 $gmin^{-1}$ $SCCO_2$ flow conditions resulted in 7.41% lipid extracted having negligible difference with the conventional method	Taher et al. [44]
	Extraction of microalgae-derived lipids with supercritical carbon dioxide in an industrial-relevant pilot plant	12 MPa pressure, a temperature of 20 °C, and a CO_2 to biomass ratio of 100 resulted in a high extraction efficiency of 92%	Lorenzen et al. [45]
	Production of α-linoleic acid by an oleaginous green algae *Acutodesmus obliquus* isolated from Malaysia	Highest amount of ALA could have been extracted from freshwater microalgae, *Acutodesmus obliquus* CN01 (38%) followed by *Chlorella. Vulgaris* NIES-1269 (35%), and *Chlorella* sp. *carolina*-15-2069 (17.9%)	Othman et al. [14]

UAE ultrasound-assisted extraction

17.6 Conclusions

To make microalgae-based PUFA production on an industrial scale practicable, a number of technological improvements are required, such as optimizing growing conditions to scale-up biomass and lipid productivities, Utilizing lower cost biomass drying techniques, and Facilitation of downstream procedures to reduce expenses while retaining product quality. Lipids are typically extracted using organic solvents. However, these solvents are toxic for the food and pharmaceutical applications of the products and therefore, they ought to be substituted by the so-called green solvents, ionic liquids (ILs), and supercritical fluids (SCFs). Ionic liquids are considered to be green for their reduced toxicity, nonflammability and nonvolatile properties compared to the common organic solvents. Ionic liquids could disrupt the lipid-bearing microalgae biomass and allow facilitated recovery of lipids from undried microalgae biomass, in a short time and at room temperature, bringing the development of a low-energy, water-compatible extraction scheme. Supercritical fluid extraction being suitable for dried microalgae biomass stands out as an eco-friendly extraction protocol. Because of its nontoxicity, and minimum critical temperature and pressure, which can prevent the degradation of heat-liable compounds and minimize operating expenses, SC-CO_2 is the most effective extraction technology.

References

1. Ibrahim, T., Nur, A., Suzianti, F., Noor, H., Kamaludin, W. Y., Amit, B., et al. (2023). Biological active metabolites from microalgae for healthcare and pharmaceutical industries: A comprehensive review. *Bioresource Technology, 375*, 1–14.
2. Haile, A., & Mohammed, E. (2021). Overconsumption of omega-6 polyunsaturated fatty acids (PUFAs) versus deficiency of omega-3 PUFAs in modern-day diets: The disturbing factor for their "balanced antagonistic metabolic functions" in the human body. *Journal of Lipids, 2021*(1), 8848161.
3. Jeevanandam, J., & Danquah, M. K. (2020). Dewatering and drying of algal cultures. In E. Jacob-Lopes et al. (Eds.), *Handbook of microalgae-based products and processes: Fundamentals and advances in energy, food, feed, fertilizer, and bioactive compounds* (pp. 207–224). Elsevier.
4. Liu, J., Sun, Z., & Gerken., H. (2016). *Recent advances in microalgal biotechnology*. OMICS Group eBooks.
5. Sivakumar, R., Sachin, S., Priyadarshini, R., & Ghosh, S. (2022). Sustainable production of eicosapentaenoic acid-rich oil from microalgae: Towards an algal biorefinery. *Journal of Applied Microbiology, 132*(6), 4170–4185.
6. Grima, E. M., Acién Fernández, F. G., & Robles Medina, A. (2013). Downstream processing of cell mass and products. In A. R. Hu (Ed.), *Handbook of microalgae culture: Applied phycology and biotechnology* (pp. 267–309). Wiley.
7. Chen, W., Li, T., Du, S., Chen, H., & Wang, Q. (2023). Microalgal polyunsaturated fatty acids: Hotspots and production techniques. *Frontiers in Bioengineering and Biotechnology, 11*, 1146881.
8. Terisha, N., Nodumo, Z., Dheepak, M., & Rajesh, L. (2013). Value-added products from microalgae. In F. Bux (Ed.), *Biotechnological applications of microalgae: Biodiesel and value added products* (pp. 137–159). CRC Press.
9. Van der Voort, M., Spruijt, J., Potters, J., Wolf, P., & Elissen, H. (2017). *Socio-economic assessment of algae-based PUFA production*. PUFA Chain.
10. Sharma, A. K., Chintal, V., Praveen, G., Prasher, P., & Patel, A. (2020). *Extraction and purification of PUFA from microbial biomass*. Scrivener Publishing LLC.
11. Blasio, & Balzano. (2021). Fatty acids derivatives from eukaryotic microalgae, pathways and potential applications. *Frontiers in Microbiology, Review*, 1–21.

12. Michaelsen, K. F., Dewey, K. G., Perez-Exposito, A. B., Nurhasan, M., Lauritzen, L., & Roos, N. (2011). Food sources and intake of n-6 and n-3 fatty acids in low-income countries with emphasis on infants, young children (6–24 months), and pregnant and lactating women. *Maternal & Child Nutrition, 7*, 124–140.
13. Remize, M., Brunel, Y., Silva, J. L., Berthon, J. Y., & Filaire, E. (2021). Microalgae n-3 PUFAs production and use in food and feed industries. *Marine Drugs, 19*(2), 113.
14. Othman, F. S., Jamaluddin, H., Ibrahim, Z., Hara, H., Nurul, A. Y., et al. (2019). Production of α-linolenic acid by an oleaginous green algae Acutodesmus obliquus isolated from Malaysia. *Journal of Pure and Applied Microbiology, 13*(3), 1297–1306.
15. Corrêa, W. G., Júnior, M., Martins, A., Caetano, N. S., & Mata, a. T. (2020). Microalgae biomolecules: Extraction, separation and purification methods. *PRO, 9*(10), 1–43.
16. Ajala, E. O., Ajala, M. A., Akinpelu, G. S., & Akubude, V. C. (2020). Cultivation and processing of microalgae for sustainability as a feedstock for biodiesel production. *Nigerian Journal of Technological Development, 18*(4), 322–343.
17. Udayan, A., Ashutosh, K. P., Ranjan, S., Nidhin, S., Byoung, S., Sung, J. S., et al. (2022). Production of microalgae with high lipid content and their potential as sources of nutraceuticals. *Phytochemistry Reviews, 22*(4), 833–860.
18. Bredda, E. H., Silva, M. B., Castro, H. F. D., Silva, A. P. T. D., & Rós, P. C. M. D. (2019). Microalgae as a source of functional PUFAs: A green low-cost pathway via enzymatic hydrolysis. *Journal of Advances in Biology and Biotechnology*, 1–13.
19. Khozin, G., Inna, S. L., & Boussiba, S. (2016). Microalgae as a source for VLC-PUFA production. In Y. L.-B. Y. Nakamura (Ed.), *Lipids in plant and algae development subcell biochemistry* (Vol. 86, pp. 471–510). Springer Int. Pub.
20. Ren, X., Liu, Y., Fan, C., Hong, H., Wu, W., Wang, W., & Yanwen. (2022). Production, processing, and protection of microalgal n-3 PUFA-rich oils. *Food, 11*(1215), 1–33.
21. Mishra, G. (2015). Polyunsaturated fatty acids from algae. In J. Seckbach (Ed.), *The algae world: Cellular origin, life in extreme habitats and astrobiology* (Vol. 26, pp. 467–481). The Hebrew University of Jerusalem.
22. Karageorgou, D., Bakratsas, G., & Katapodis, & Petros. (2020). Production of PUFAs as dietary and health supplements from oleaginous microalgae utilizing inexpensive renewable substrates. In A. K. Matsakas, A. K. Patel, & L. Matsakas (Eds.), *Nutraceutical fatty acids from oleaginous microalgae: A human health perspective* (pp. 77–114). Scrivener Publishing.
23. Preeti, M., Kritika, S., Ravi, P. G., & Anshu, S. M. (2020). Cultivation of microalgae for the production of bioproducts and biomolecules at an industrial level. In T. K. K. Gayen (Ed.), *Sustainable downstream processing of microalgae for industrial applications* (pp. 69–82). CRC Press.
24. Adarme-Vega, T. C., Lim, D. K. Y., Timmins, M., Vernen, F., Li, Y., & Schenk, P. M. (2012). Microalgal biofactories: a promising approach towards sustainable omega-3 fatty acid production. Review. *Microbial Cell Factories, 11*(96), 1–10..
25. Ravishankar, A., Ranga, R., & Gokare, A. (2020). Global microalgal-based products for industrial applications. In G. A. Rao (Ed.), *Handbook of algal technologies and phytochemicals; food, health and nutritional applications* (Vol. 1, pp. 267–277). CRC Press.
26. Paliwal, C., Nesamma, A. A., & Jutur, P. P. (2020). Industrial scope with high-value biomolecules from microalgae. In T. K. K. Gayen (Ed.), *Sustainable downstream processing of microalgae for industrial applications* (pp. 83–98). CRC Press.
27. Kaur, A., Gaur, S., & Adholeya, A. (2023). Functional metabolites from microalgae for multiple commercial streams. In J. Sangeetha, S. Codreanu, & D. Thangadurai (Eds.), *Microalgal biotechnology bioprospecting microalgae for functional metabolites towards commercial and sustainable applications* (pp. 149–182). CRC Press.
28. Bheda, M., & Shinde, R. (2018, September 11–14). Drying of algae by various drying methods. In *21st international drying symposium* (pp. 1791–1798).
29. Bi, & He. (2020). Biodiesel from microalgae. In E. Jacob-Lopes, M. M. Maroneze, M. I. Queiroz, & L. Q. Zepka (Eds.), *Handbook of microalgae-based processes and products:*

Fundamentals and advances in energy, food, feed, fertilizer, and bioactive compounds (pp. 329–371). Academic Press.
30. Kim, M., & Kim, K. (2022). Drying techniques of microalgal biomass: A review. *Journal of Chemistry and Applied Chemical Engineering, 33*(2), 145–150.
31. Hrishikesh, A., Tavanandi, A., Chandralekha, D., & Raghavarao, K. S. (2020). Sustainable pre-treatment methods for downstream processing of harvested microalgae. In T. K. K. Gayen (Ed.), *Sustainable downstream processing of microalgae for industrial applications* (pp. 100–135).
32. Pôjo, V., Tavares, T., Xavier, F., & Malcata. (2021). Processing methodologies of wet microalga biomass toward oil separation: An overview: Review. *Molecules, 26*, 1–26.
33. Adam, H. (2013). Green chemistry in oscillatory baffled reactors. In K. B. Harvey (Ed.), *Process intensification of green chemistry* (pp. 157–175). Wiley.
34. Beh, C. C., & Mammucari, N. R. (2017). Neoteric media as tools for process intensification. *IOP Conf Ser Mater Sci Eng, 2016*, 1–12.
35. Davis, J. H. (2003). Ionic liquids derived from natural products and other novel chemistries. In R. Rogers et al. (Eds.), *Green industrial applications of ionic liquids* (pp. 385–402). Unversity of South Alabama: Kluwer Academic Publishers.
36. Cañadas, R., González, M., González, E. J., Díaz, I., & Rodríguez, M. (2020). Overview of neoteric solvents as extractants in food industry: A focus on phenolic compounds separation from liquid streams. *Food Research International, 136*, 109558.
37. Desai, R. K., Fernandez, M. S., Wijffels, R. H., & Eppink, M. H. (2019). Mild fractionation of hydrophilic and hydrophobic components from Neochloris oleoabidans using ionic liquids. *Frontiers in Bioengineering and Biotechnology, 7*, 284.
38. Orr, V. C., Plechkova, N. V., Seddon, K. R., & Rehmann, L. (2016). Disruption and wet extraction of the microalgae Chlorella vulgaris using room-temperature ionic liquids. *ACS Sustainable Chemistry & Engineering, 4*(2), 591–600.
39. Jessop, P. G., & Leitner, W. (1999). Supercritical fluids as media for chemical reactions. In *Chemical synthesis using supercritical fluids* (pp. 1–36). https://doi.org/10.1002/9783527613687.ch1
40. Adams, D. J., & Tavener, S. J. (2005). *Chemistry in alternative reaction media* (Vol. 3). Wiley. https://doi.org/10.1002/0470869666
41. Zhou, J., Min, W., Jorge, A., Saraiva, A. P., Martins, C. A., Pinto, M. A., et al. (2022). Extraction of lipids from microalgae using classical and innovative approaches. *Food Chemistry, 384*, 1–15.
42. Schmid, B., Navalho, S., Schulze, P. S., Walle, S. V.-D., Royen, G. V., Schüler, L. M., et al. (2022). Drying microalgae using an industrial solar dryer: A biomass quality assessment. *Food*, 1–20.
43. Señoráns, M., Castejón, N., Javier, F., & Señoráns. (2020). Advanced extraction of lipids with DHA from Isochrysis galbana with enzymatic pre-treatment combined with pressurized liquids and ultrasound assisted extractions. *Molecules, 25*, 1–14.
44. Taher, H., Al-Zuhair, S., Al-Marzouqi, A. H., & Haik, Y. (2014). Supercritical carbon dioxide extraction of microalgae lipid: Process optimization and laboratory scale-up. *The Journal of Supercritical Fluids, 86*, 57–66.
45. Lorenzen, J., Igl, L., Nadine, T., Marlene, S. A., Igl, F. Q., Andrea, S., et al. (2017). Extraction of microalgae derived lipids with supercritical carbon dioxide in an industrial relevant pilot plant. *Bioprocess and Biosystems Engineering, 40*, 911–918.

Chapter 18
Effect of Spontaneous Sedimentation on the Chemical Composition of Compressed Water Hyacinth Juice as a Pretreatment for Anaerobic Digestion: Comparison of Different Growing Areas

Pranshu Bhatia, Anas Hijazi, Shinichi Akizuki, Nigus Gabbiye Habtu, Shinjiro Sato, and Tatsuki Toda

18.1 Introduction

Water hyacinth (WH), *Eichhornia crassipes*, is an invasive aquatic weed native to the Amazon in South America and is now proliferating all over the world [1]. WH is the worst aquatic plant on the earth with fast productivity capacity, doubling its population within 2 weeks [2, 3]. WH, due to its inherentt nature, can grow on both soil/land and water which makes it a major threat to the environment [4]. This nature of WH interferes with nature habitats causing numerous issues including clogging slow-flowing streams, polluting waterbodies, and land degradation [5, 6]. The weed's appearance on Lake Tana, the largest Lake in Ethiopia, was declared in 2011 [3]. Over the last few decades, several programs of WH management (removal and/or spread control) implemented across the world required high financial and logistical mobilization with no satisfactory results [7–9].

Creating valuable products from WHs is desirable to compensate for its expensive harvesting costs. To achieve this, our research group developed a novel

P. Bhatia (✉)
Faculty of Science and Engineering, Soka University, Tokyo, Japan
e-mail: bhatiapranshu@soka.gr.jp

A. Hijazi · S. Sato · T. Toda
Graduate School of Science and Engineering, Soka University, Tokyo, Japan

S. Akizuki
Institute of Plankton Eco-engineering, Soka University, Tokyo, Japan

N. G. Habtu
Faculty of Chemical and Food Engineering, Bahir Dar Institute of Technology, Bahir Dar, Ethiopia

S. W. Fanta et al. (eds.), *Sustainable Development Research in Manufacturing, Process Engineering, Green Infrastructure, and Water Resources*, Green Energy and Technology, https://doi.org/10.1007/978-3-031-77339-6_18

treatment process (Japanese Patent No. 7204263). This method involves crushing and compressing the WH, recovering biogas from compressed water hyacinth juice (WHJ) through up-flow anaerobic sludge blanket (UASB) treatment by anaerobic digestion (AD), and producing biochar from compressed solid residue for fuel and soil conditioning. The AD effluent can be used for purposes such as growing nutritional microalgae and vegetables using hydroponics. This process may be useful, especially in developing countries that face difficulties in financing the annual harvest.

Using WHJ for the AD process can be a great potential for the creation of fast and efficient gas recovery. Past researchers used WHJ and have proven its potential to achieve fast and stable operation within the AD reactor. Different studies on batch experiments [10, 11] or continuous reactors using UASB were conducted [12–14]. WHJ is relatively a new substrate and not many studies are performed till now. The compression process leads to a high suspended solids (SS) concentration remaining in the liquid fraction. These solids are not quickly degradable and may pose multiple operational issues such as the clogging of the feed pipe and the accumulation of solids in the reactor which leads to a slow degradation process, unstable operation and the need for cleaning and restarting the AD operation [5]. For instance, using compressed WHJ had its rate limitations when digested in the UASB reactor, research showed [12] a severe performance decrease when digesting WHJ at a rate faster than 3.2 days hydraulic retention time (HRT). Pushing the treatment rate past this point required higher system complexity or sacrificing the degradation efficiency. Also, our previous experiment confirmed that using WHJ with high SS (around 9000 mg L^{-1}) clogged the UASB and caused a washout of the microorganisms due to floating, negatively affecting the reactor performance [15]. In addition, further solubilization of WHJ has the potential to produce even faster pretreatments with higher biomass utilization and improved biogas recovery potential. Hence, the use of pretreatments is essential in achieving these goals.

Spontaneous sedimentation is an economic pretreatment method that easily separates the WHJ into two obvious phases: the supernatant, which is the equivalent of filtered WHJ utilized here as the experimental substrate, and the sediments collected at the bottom of the container. Sedimentation pretreatment is also known by different names like batch settling time, gravitational sedimentation and liquid-solid separation [16, 17]. Some studies use different types of equipment such as tilted plane screens, 2-stage separators and belt press separators which can increase the cost related to the pretreatment [18]. In this study, sedimentation pretreatment performed does not require any equipment, manual effort, chemical additions etc. and it can work effectively to reduce the SS concentration in the WHJ as planned making it a low-cost pretreatment. Furthermore, it can make the recovery of the sediments, the removed parts, much easier in comparison with mesh filtration techniques. However, this method reduces the amount of accessible nutrients, particularly carbon, by excluding more sediments as time passes [19]. Therefore, sedimentation time needs to be evaluated and optimized to ensure efficient SS reduction for processes like AD while maintaining sufficient nutrients for competitive biogas potential. The main purpose of sedimentation pretreatment is to remove SS, inhibitors by retaining the organic fraction and other nutrients.

Therefore, the objective of this study was to evaluate the effect of sedimentation pretreatment on the chemical composition of three different types of compressed water hyacinth juice. The purpose of this study was to remove the suspended solids and retain the organic fraction and nutrients throughout the sedimentation pretreatment. The effect of sedimentation pretreatment on the carbon mass balance was conducted to reach a deeper understanding. Finally, a comparison of different studies using WHJ as a substrate was also performed. This study can further help to optimize the WHJ as a substrate for a subsequent biological process. To our best knowledge, this is the first study using WHJ from three different sources as a substrate for sedimentation pretreatment.

18.2 Material and Methods

18.2.1 Substrate Collection

WH samples were harvested from three different sampling sites in Japan. WH was categorized as floating-, rotten-, and ground-type WH. The "floating-type" WH was harvested from a lake in Gifu, Japan. The "rotten-type" WH was collected from Suijo Park, Gyoda, Saitama, Japan. The rotten-type WH was also floating but the WH plant was already rotten. The "ground-type" WH was collected from a paddy field near the Doyo-no-mori Hurusato Otone road station in Kazo City, Saitama, Japan. For ground-type WH, the roots were cut to prevent soil contamination, and the part above the roots was collected. All three types of WH were crushed (2–5 cm) and compressed using a milling machine (RSC-3500, R-ing Co. Ltd., Japan) and a screw dewatering machine (Dash-1, R-ing Co. Ltd., Japan). Consequently, the solid and liquid fractions were separated, and the liquid fraction (WHJ) was collected and stored at +4 °C for experiment use.

18.2.2 Sedimentation Pretreatment

Sedimentation pretreatment was performed by leaving the WHJ to stand at room temperature (25 ± 1 °C) for 38 h (Fig. 18.1). The purpose for choosing 38 h was to have maximum removal of SS at the bottom of the cylinder for all three types of WHJ and to have the sedimentation time as short as possible due to the self-degradation of the WHJ (flocs formation). The sedimentation treatment was performed by using a 500 mL volumetric cylinder in duplicate to imitate the actual scalable condition. After 38 h, the supernatant fraction and sedimented fraction

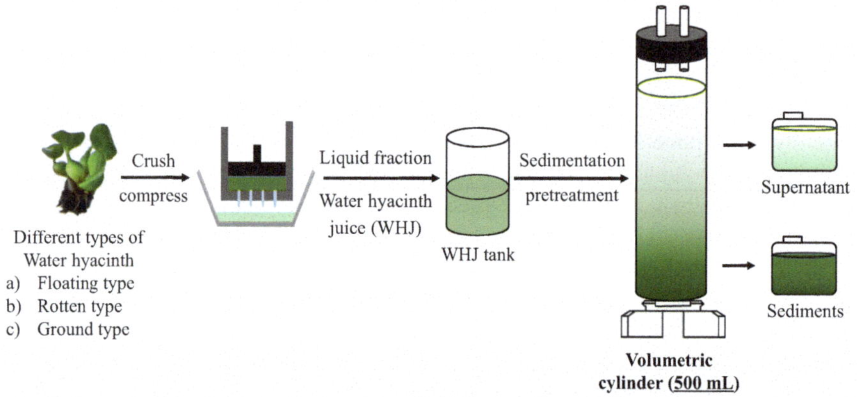

Fig. 18.1 Experimental setup of the sedimentation pretreatment of water hyacinth juice (WHJ)

were obtained and used for further analysis. The mouth of the flasks was sealed with a parafilm to prevent water loss due to evaporation.

18.2.3 Analytical Methods

The pH, salinity, SS, VSS, DOC, carbon content and phenolic compounds were measured for whole WHJ, supernatant and sediments fraction. For the nutrient cation analysis, Na^+, NH_4^+, K^+ and Mg^{2+} were measured to evaluate the effect of sedimentation pretreatment on the whole WHJ and the supernatant fraction.

The pH was measured using a pH meter (Seven Compact™ pH/Ion meter, S220, Mettler Toledo, Japan) and salinity was measured using a digital salt meter (ES-421, Atago U.S.A). SS and VSS were determined following the standard methods of the American Public Health Association (APHA, 1998) and measured along different settling times of 0.5, 3, 9 and 38 h.

DOC was measured using a high-temperature combustion method (TOC-L and TNM-L; Shimadzu, Japan). The samples were filtered using a 0.45 μm glass fiber filter (Advantec, GC-50, 47 mm, Japan) before performing the analysis. The machine itself measures the triplicates of the samples to account for errors.

Carbon in the WHJ and its components was measured using (2400 CHNS/O Series II System (110 V), Perkin Elmer, USA). The WHJ, supernatant and sedimented liquid fraction were dried at 105 °C for 24 h [20]. After that, the respective fraction was converted to a fine powder using a mortar and stored in an air-tight container before the analysis.

Phenolic compounds in the WHJ, supernatant, and sedimented liquid fraction were measured according to the Folin-Ciocalteau method [21] by conducting the tests in triplicates to account for errors.

Cations (Na^+, NH_4^+, K^+, Mg^{2+}) were measured using high-performance liquid chromatography system (HPLC, Shimadzu, Japan) with a conductometric detector (CDD-10AVP; Shimadzu, Japan) at 40 °C, equipped with HPLC columns (an IC YS-50 column for cation analysis; Showa Denko, Shodex Japan) [22, 23]. The samples were filtered using a 0.45 μm pore size glass fiber filter before the analysis. The calibration of the standard is shown in the supplementary figure (Fig. 18.2).

18.2.4 Calculations

SS removal efficiency and sedimentation efficiency were calculated according to the following equations (Eqs. 18.1 and 18.2). The removal efficiency was calculated at different settling times of 0.5, 3, 9 and 38 h, respectively.

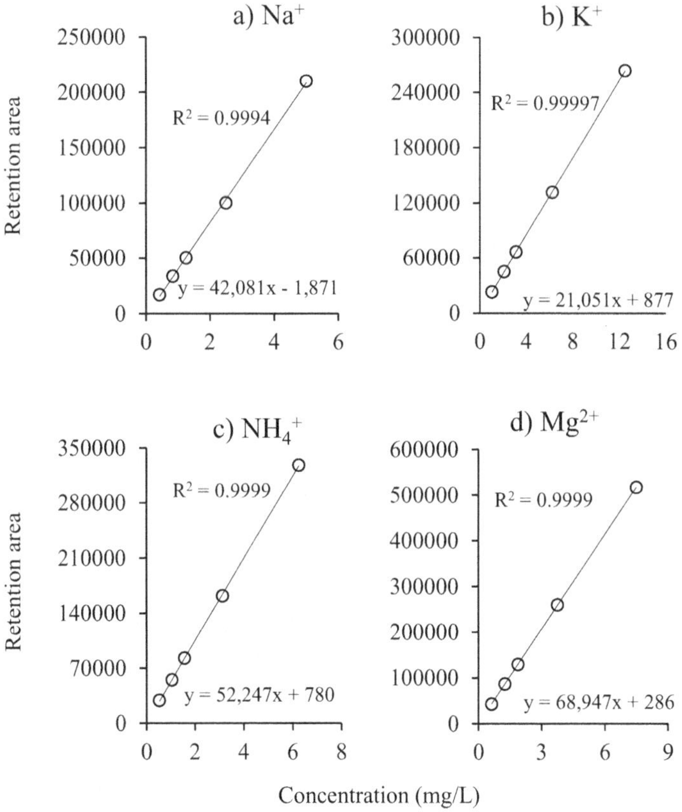

Fig. 18.2 Calibration curves for different cations used in this study

$$\text{Removal efficiency}\,(\%) = \frac{\text{SS}(\text{sediments})(\text{g}\,\text{L}-1)\times\text{Volume}(\text{sediments})(\text{L})}{\text{SS}(\text{whole WHJ})(\text{g}\,\text{L}-1)\times\text{Volume}(\text{whole WHJ})(\text{L})} \tag{18.1}$$

$$\text{Sedimentation efficiency}\,(\%) = \frac{\text{Sedimentation volume}(\text{mL})\,\text{after 38h}}{\text{Total volume}(\text{mL})} \tag{18.2}$$

18.3 Results and Discussions

18.3.1 Effect of Sedimentation Pretreatment on the Suspended Solids

The SS and its removal efficiency were calculated after the sedimentation pretreatment for different settling times of 0.5, 3, 9 and 38 h, respectively (Fig. 18.3). Firstly, in Fig. 18.3a, the SS and the removal efficiency of the floating-type WHJ are shown. The SS of the untreated WHJ was 5.0 ± 0.1 g L^{-1}, which kept on decreasing in the supernatant with the increase in the settling time of the sedimentation experiment. The value of the untreated WHJ was quite similar to the WHJ used by Liu et al. (4363–4785 mg L^{-1}) [12]. The study used filter bag squeeze filtration which led to the SS decrease which increased the cost of the process. However, in this study, the SS concentration in the supernatant after 0.5 h. 3 h, 9 h and 38 h was 4.6 ± 0.1, 4.2 ± 0.2, 4.1 ± 0.1 and 3.0 ± 0.1 g L^{-1}, respectively. The SS concentration of the supernatant had a negative relationship with the SS removal efficiency. The removal efficiency increased with the increase in the settling time ($y = 6.57 \ln(x) + 28.22$, $n = 4$, $R^2 = 0.92$, $p < 0.05$) as the SS concentration was decreased. The SS removal efficiency ranged from 26.7 ± 0.2% at 0.5 h to 54.4 ± 0.4% at 38-h settling time.

For the rotten-type WHJ, the SS concentration in the untreated WHJ was 6.1 ± 0.2 g L^{-1} which was drastically decreased after the sedimentation pretreatment. The SS concentration in the supernatant ranged from 3.2 ± 0.0 to 1.9 ± 0.0 g L^{-1} at different settling times. The rotten-type WHJ had readily available degraded compounds which led to a drastic decrease in the SS concentration. The SS removal efficiency was quite higher compared to the floating-type WHJ. The removal efficiency after 0.5, 3, 9 and 38 h were 76.3 ± 1.3, 79.4 ± 2.2, 79.7 ± 2.0 and 76.8 ± 0.8%, respectively. Even though the SS removal efficiency was quite high, the effect of settling time was not significant after 3 h. A previous study mentioned that due to a combination of hindered settling and increased density and viscosity of the juice, a large content of small particles causes an increased viscosity of the liquid and thereby leading to no change in the removal rate [24, 25]. The settling time of 0.5 and 3 h removed a significant amount of SS from the WHJ followed by a stable phase.

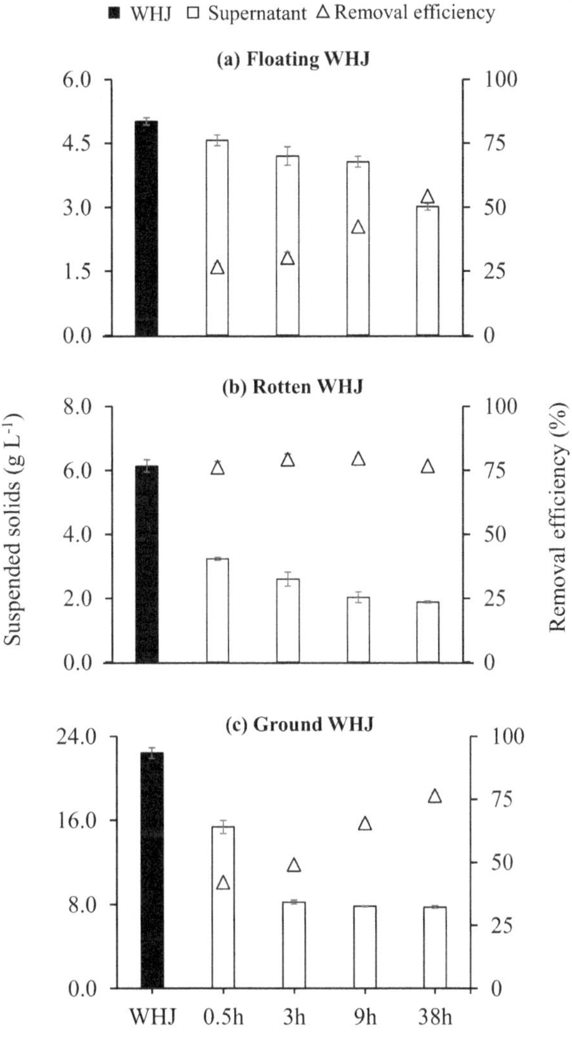

Fig. 18.3 The suspended solids (SS) concentration at different sampling times (0.5, 3, 9, 38 h) during sedimentation pretreatment relationship with the SS removal efficiency (%) on the different types of water hyacinth juice. Removal efficiency was calculated by using Eq. 18.1. (**a**) Floating WHJ. (**b**) Rotten WHJ. (**c**) Ground WHJ

Finally, for the ground-type WHJ, the SS of the untreated WHJ was the maximum having a value of 22.4 g L^{-1}. The high value of this WHJ is due to the habitat of the ground-type WHJ. As WH was growing on the soil, the nature of WHJ was different from the above-mentioned floating and rotten type. WH growing in the freshwater/pond is known for its high moisture content and low total solids (TS) [26] whereas when growing on terrestrial the SS was drastically increased. The sedimentation pretreatment helped to remove the SS from this WHJ giving values ranging from 15.4 ± 0.6, 8.2 ± 0.2, 7.8 ± 0.0 and 7.8 ± 0.1 g L^{-1} at the respective settling time. The SS removal rate ranged from 42.13 ± 0.9 at 0.5 h to 76.41 ± 1.1 at 38 h after the sedimentation pretreatment. The SS removal rate kept on increasing with the increase in the settling time ($y = 8.28 \ln(x) + 45.37$, $n = 4$, $R^2 = 0.95$, $p < 0.05$).

From the results above for all three types of WHJ, the sedimentation pretreatment seems to be quite effective making all three substrates quite attractive for further usage in different treatment processes requiring lower SS substrates. The settling time of 38 h seems quite effective in removing the SS concentration from the WHJ giving a clear supernatant fraction. Around 38 h it was seen that the layer of supernatant had some flocs formation proving the self-degradation of the substrate. Therefore, increasing the time by more than 38 h could affect the supernatant chemical properties. In previous studies, it was shown that settling time increases the separation efficiency [27, 28]. The floating-type WHJ was collected from a freshwater lake which led to lower SS in the untreated WHJ compared to the rotten and the ground type giving a sedimentation efficiency of 9.6 ± 0.2% (Fig. 18.4). The rotten-type WHJ was also growing in a pond but the WH plant was already rotten which means that some degraded matter was released into the sediments during the sedimentation pretreatment leading to the maximum SS removal efficiency (Sedimentation efficiency = 19.9 ± 0.4%). In a report, it was mentioned that rotting of WH in moist air in loosely sealed containers seems to promote the most uniform and rapid decomposition, presumably controlled by the growth of aerobic bacteria. Reduction in mass up to 40% occurred during the rotting process as is proven in the experiments at Biberton HMP. This is a great advantage as in turn it reduces the amount of digestion needed, either digestion time or digestion chemical [20]. Finally, for the ground-type WHJ, due to the growing conditions of the WH plant, the sedimentation efficiency was 14.45 ± 0.4%. The WH was growing on soil and was in its adult phase as the flowers were already shed. Even though the roots were removed before crushing and compressing the WH weed, some soil must have been transferred to the juice. These results can be compared with the SS removal rate as discussed before. The sedimentation efficiency results were in accordance with the

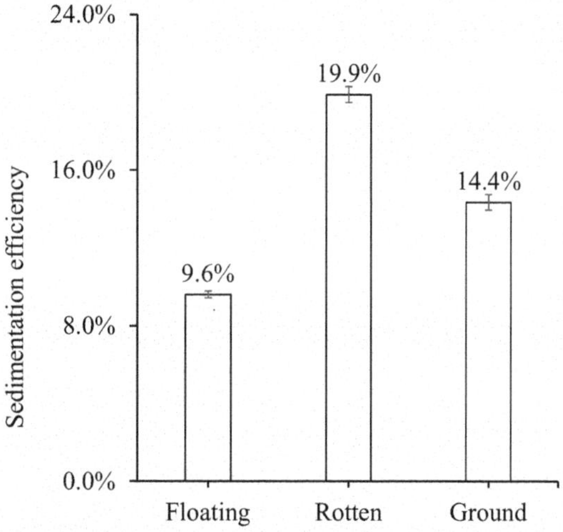

Fig. 18.4 The sedimentation efficiency by sedimentation pretreatment on the different types of water hyacinth juice

SS removal rate. The sedimentation efficiency was maximum for the rotten type followed by the ground type and finally the floating type (Fig. 18.4).

18.3.2 *Effect of Sedimentation Pretreatment on the Chemical Composition of Different Types of WHJ*

Table 18.1 shows the different chemical parameters of the untreated WHJ and supernatant after 38 h for each type of WHJ. Table 18.1a shows the chemical parameters of the floating-type WHJ. The pH of the untreated WHJ was 5.16 ± 0.0 which decreased slightly after the sedimentation pretreatment to 4.88 ± 0.0. Similarly, the salinity also had a slight change after the sedimentation pretreatment changing from

Table 18.1 Chemical composition of the untreated water hyacinth juice and the supernatant after the sedimentation pretreatment

A. Floating-type WHJ		
Parameter	**Whole WHJ**	**Supernatant**
pH	5.16 ± 0.0	4.88 ± 0.0*
Salinity	0.52 ± 0.0	0.56 ± 0.0*
DOC	1385 ± 8.7	1259 ± 9.6
Phenolic compounds	206.6 ± 1.6	208.7 ± 0.9
Na^+	116.7 ± 5.5	133.3 ± 3.8*
NH_{4+}	106.8 ± 5.0	122.6 ± 1.4*
Mg^{2+}	180.2 ± 12.0	150.3 ± 7.7
K^+	2440.5 ± 98.4	2266.3 ± 22.4
B. Rotten-type WHJ		
pH	7.31 ± 0.0	6.65 ± 0.0*
Salinity	0.24 ± 0.0	0.25 ± 0.0
DOC	1052 ± 16.8	1197 ± 16.0
Phenolic compounds	141.6 ± 0.6	147.6 ± 0.7*
Na^+	54.8 ± 1.1	31.7 ± 1.3*
NH_4^+	113 ± 0.5	65 ± 10.9*
Mg^{2+}	61 ± 6.5	28 ± 1.8*
K^+	310.8 ± 6.6	141.1 ± 14.8*
C. Ground-type WHJ		
pH	4.32 ± 0.0	4.16 ± 0.0*
Salinity	0.68 ± 0.0	0.71 ± 0.0
DOC	6897 ± 48.5	7788 ± 18.4
Phenolic compounds	638.8 ± 23.3	301.2 ± 2.7*
Na^+	420.5 ± 21.5	435.9 ± 14.8
NH_4^+	66.9 ± 2.5	107.3 ± 7.6*
Mg^{2+}	325.5 ± 6.6	346.8 ± 5.6*
K^+	2973.5 ± 132	3057 ± 54.1

* p values ≤0.05, all values are in mg L^{-1} except pH and salinity

0.52 to 0.56. The decrease in the pH was influenced by the increase in the salinity. The pH and salinity were not affected by the sedimentation pretreatment. For the DOC concentration, the value in the untreated WHJ and the supernatant was 1385 ± 8.7 and 1259 ± 9.6 mg L^{-1}, respectively. The decrease in DOC concentration over a period of 38 h was not significant. But some dissolved carbon might have been transferred to the sediments during the pretreatment as mentioned by previous studies [29, 30]. In a study by Li et al. [30] using rice straw hydrolysate the COD removal efficiency was 5.22% after 12 h of sedimentation. The study mentioned that the low removal efficiency was due to the organics dissolved in the hydrolysate being mainly composed of low-molecular-weight.

For the rotten-type WHJ, the pH was relatively higher when compared with other types of WHJ. The pH was 7.31 ± 0.0 for the untreated WHJ and decreased to 6.65 ± 0.0 after the sedimentation pretreatment. The salinity was not changed a lot; however, quite lower as shown in Table 18.2b. During the sedimentation pretreatment of the rotten-type WHJ, flocs/scum formation was observed on the layer on the WHJ which could be due to its rotten nature. In a study by Imchen et al. [31] it was observed that when the salinity of the water was low, the DO concentration increased which led to the rotting of plants. Previous studies using sugarcane juice also reported the formation of flocs which could lead to high turbidity [32]. Finally, for the DOC concentration, the value of DOC increased from WHJ to the pretreated supernatant (1052 ± 16.8 to 1197 ± 16.0 mg L^{-1}). This increase can be owed to solubilization at room temperature of the substrate with time passing [33].

The pH and salinity of the untreated ground-type WHJ were 4.32 ± 0.0 and 0.68 ± 0.0. The values changed after the pretreatment to 4.16 ± 0.0 and 0.71 ± 0.0. Similar to the floating- and rotten-type WHJ, the salinity and the pH values did not change a lot after the sedimentation pretreatment. The DOC value was quite high when compared to the floating- and rotten-type WHJ. The value of the untreated WHJ was 6897 ± 48.5 mg L^{-1} and the value increased to 7788 ± 18.4 mg L^{-1} after the sedimentation pretreatment. The DOC for the supernatant was around six times more than the other types of pretreated supernatant. This can be owed to the nurturing environment for "ground" WHJ which receives more nutrients than the floating types due to growing in soil. The effect of the same type of aquatic weed growing in water and land has not been studied till now. During sedimentation, a large amount of soluble substances in the plant residues dissolved quickly, which led to an increase in organic matter.

The purpose of sedimentation pretreatment was not only to maximize the removal of SS from the WHJ but also to make the proportion of ions and organic matter appropriate for subsequent treatment, like AD or enzymatic hydrolysis. In Fig. 18.5, the percentage changes in the supernatant after the sedimentation treatment were observed on cations and phenolic compounds. It can be seen that for the floating-type WHJ, the values did not change a lot after the sedimentation pretreatment. The Na^+ and NH_4^+ increased significantly after the sedimentation pretreatment. The Na^+ was increased from 116.7 ± 5.5 to 133.3 ± 3.8 mg L^{-1} giving a 14.2% increase whereas NH_4^+ increased from 106.8 ± 5.0 to 122.6. ± 1.4 mg L^{-1} giving a 14.8% increase. These results indicate that for the floating type, WHJ sedimentation

Table 18.2 Comparison of different parameters using water hyacinth juice as substrate (WHJ)

WHJ type	Pretreatment type	pH[a]	SS	VSS	COD	References
Floating	Crushing + Filter bag squeeze filtration	7.0–7.5	4363–4785	2475–2647	15,630–23,500	[12, 14]
Ground	Crushing + compressing + Mesh filtration	4.02 ± 0.1	12,878 ± 3139	–	29,617 ± 369	[11]
Floating	Chop + grind + squeezed + Microorganisms' mixture	6.8	10,760 (TS)	3455 (VS)	5960	[42]
Floating	Crushing + compressing + sedimentation	4.88 ± 0.0	3012.0 ± 82.9	1439.5 ± 31.3	1259 ± 9.6 (DOC)	This study
Floating (rotten)	Crushing + compressing + sedimentation	6.65 ± 0.0	1882.5 ± 31.3	1306.7 ± 31.3	1197 ± 16.0 (DOC)	This study
Ground	Crushing + compressing + sedimentation	4.16 ± 0.0	7751.4 ± 112.9	5426.0 ± 38.4	7788 ± 18.4 (DOC)	This study

[a]All pretreatments values are in mg L^{-1} except pH

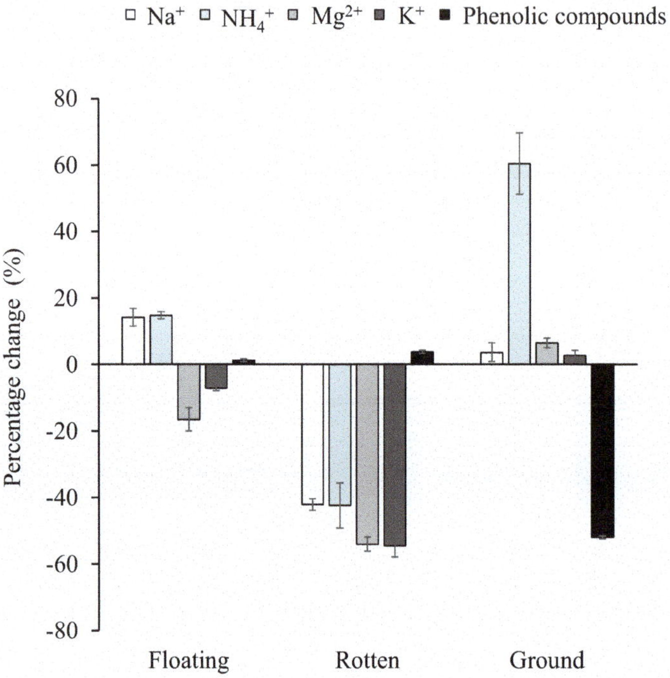

Fig. 18.5 The effect of sedimentation pretreatment on the chemical parameters in the supernatant fraction of different types of water hyacinth juice

pretreatment seems effective as no loss or addition of nutrients and phenolic compounds was seen.

For the rotten-type WHJ, it can be seen that except for the phenolic compounds, all the cations were significantly decreased after the sedimentation pretreatment. All the ions were reduced by almost 50% after the sedimentation pretreatment. The Na^+ was reduced from 54.8 ± 1.1 to 31.7 ± 1.3 mg L^{-1}, NH_4^+ from 113 ± 0.5 to 65 ± 10.9 mg L^{-1}, Mg^{2+} from 61 ± 6.5 to 28 ± 1.8 mg L^{-1} and K^+ from 310.8 to 141 mg L^{-1}. Also, the value of all the ions was quite low when compared to floating- and ground-type WHJ. In a study when the effect of rotting was evaluated on six different aquatic weeds, it was seen that after the decaying process was finished the NH_3-N and TP were transferred to the water [34]. Similarly in this study during the sedimentation of the untreated WHJ, the juice kept on decaying at a slower rate which led to the release of nutrients to the sediments. Also, after the sedimentation pretreatment, there is excessive loss of ions which are required for subsequent processes like AD. This higher loss of nutrients can negatively affect the further biological processes [35–37].

Finally, for the ground-type WHJ except NH_4^+, Mg^{2+} and phenolic compounds, no other ions changed significantly after the sedimentation pretreatment. It can be seen that NH_4^+ increased by 60.4% (66.9 ± 2.5 to 107.3 ± 7.6 mg L^{-1}), probably because some organic nitrogen had been hydrolyzed. Whereas the phenolic compounds were reduced to almost half after the pretreatment (638.8 ± 23.3 to 301.2 ± 2.7 mg L^{-1}).

The main reason for the change in the percentage of the ions can be the pH of the supernatant, and the nature of the growing area of the substrate. For example, previous studies mention that the pH of the growing area affects the availability of the nutrients of the plants. For the floating-type WHJ, the pH was around 4.88 ± 0.0 meaning an acidic environment, which means most of the nutrients were available for the WH growth. But once the pH turns alkaline most of the elements can become inactivated and will be no longer available to the plants [38]. This is the reason for quite low values of the ions in the untreated WHJ for the rotten type. Thereafter, the rotting of the plant led to an even higher reduction of these ions after the sedimentation pretreatment which can negatively affect the further process [31, 34]. For the ground type, the NH_4^+ was increased in the supernatant by 60.4% after the sedimentation pretreatment. Most K and NH_4^+ are dissolved and recovered in the supernatant [39, 40]. Even for the floating-type WHJ, the NH_4^+ was increased by 14.8% after the sedimentation pretreatment. Also, another reason for the increase of some compounds can be the separation of the solids. The sedimented WHJ may contain lower concentrations of these compounds, which were removed through sedimentation, leaving behind a higher concentration of these compounds in the supernatant.

To know how the carbon was recovered in the WHJ components after the sedimentation pretreatment, the carbon mass balance was performed. Figure 18.6 shows the carbon mass balance in the untreated WHJ and the components after the sedimentation pretreatment. Figure 18.6a shows that for the floating-type WHJ, 52.61% of carbon was recovered in the supernatant while 46.44% percent was transferred to the sediments. The floating-type WHJ had the minimum loss of carbon (0.95%) making the sedimentation pretreatment effective for the floating-type WHJ.

For the rotten-type WHJ, the carbon content recovered in the supernatant was 31.84% and 55.77% was transferred to the sediments. The rotten-type WHJ had the minimum amount of carbon recovered and the maximum amount of loss (12.39%) after the sedimentation pretreatment. As mentioned earlier due to the degrading ability, the organic fraction can be easily transferred to the settle down and get mixed with the sediments. Zhang et al. [34] mentioned that after the rot of different aquatic weeds, the COD increased in the water body. After decomposition, the COD concentration of water bodies containing submerged plants increased at 20 and 30 °C. Initially, a substantial proportion of soluble compounds in the plant residues dissolved fast, increasing the amount of organic matter and other reducing substances in the water body. Li et al. [30] mentioned that the organics that dissolved in the hydrolysate quickly were mainly composed of low-molecular-weight. The carbon mass balance for the ground-type WHJ revealed that 40.50% was recovered in the pretreated supernatant while 50.94% was transferred to the sediments with an overall loss of 8.56%. The higher loss of carbon after the sedimentation pretreatment can negatively affect processes like AD as it can reduce biogas production since carbon is the main element required for biogas production [41]. Even though

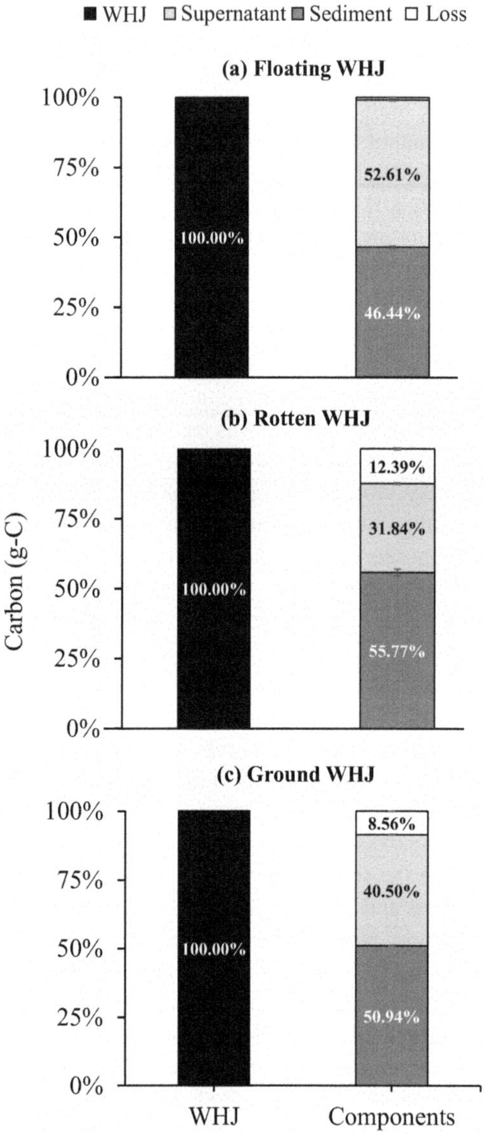

Fig. 18.6 The distribution of carbon content of untreated water hyacinth juice and water hyacinth juice components (supernatant, sediment and loss) after 38 h during the sedimentation pretreatment. (**a**) Floating WHJ. (**b**) Rotten WHJ. (**c**) Ground WHJ

the carbon concentration was high in the ground-type WHJ, the mass balance revealed that the best carbon recovery was seen in the floating-type WHJ.

18.3.3 Comparison of Different Types of Water Hyacinth Juice

As mentioned WHJ is relatively a new substrate which is been used mainly for AD [11–14] and the solid extracted is being used for biochar or pellet production [42, 43]. We tried to compare all the studies that used WHJ and presented them in the form of a table. Table summarizes the various physio-chemical parameters of the type of WHJ used. From Table , it can be seen that the WHJ juice used in this study had the minimum SS concentration depending upon its growing type. Sedimentation pretreatment helped to remove the SS from all three types of WHJ. Normally the WHJ juice studies used in Table were all from the floating sources with only Maruyama et al. [11] and this study using the ground type. The pH of both the ground types WHJ were quite similar (4.16 ± 0.0 and 4.02 ± 0.1). Maruyama et al. [11] used WHJ on the sequential batch AD reactor (SBR) and found that initially the control reactor had a stable biogas production but after day 33, the methane yield started to decrease. One reason can be the undegraded fraction that accumulated in the reactor as the SS of the WHJ was quite high (12.8 g L^{-1}). Therefore, by using the sedimented ground-type WHJ this problem can be probably solved and achieve a stable AD. The pH value for the floating type in this study was 4.16 ± 0.0 which is lower than often reported but can be due to the habitat difference when compared to free-floating water hyacinth weeds [13]. In a study by Liu et al. [13], when the HRT was lower than 3.2 days, the methane yield was drastically reduced, pushing the treatment rate past this point required higher system complexity or sacrificing the degradation efficiency. Even though the initial substrate pH was in the range of 7.0–7.5, the AD was affected. Therefore, the sedimentation pretreatment of the different types of WHJ can be a potential solution to this problem. The AD process can have a low lag phase and high methane yields due to the substrate's easily biodegradable properties in comparison with the WHJ that has the sediments, which, in turn, have more solids that contain more-difficult-to-digest biomass. This is especially important for continuous AD operations or enzymatic hydrolysis that require low SS concentration for uninterrupted stable operation [5]. In addition, this method can be a great alternative to the affordable pretreatment method since it doesn't require any external energy when compared to heat-focused options.

18.4 Conclusion

The study focused on evaluating the effectiveness of a sedimentation pretreatment method on different types of WHJ. While no specific economic analysis was conducted, it is worth noting that the sedimentation pretreatment method employed in this research is inherently low cost. The method's simplicity and independence from electric and specialized equipment make it a cost-effective approach for the preliminary treatment of WHJ. The conclusion drawn is as follows:

SS Removal: The floating-, rotten- and ground-type WHJ reduced the SS concentration in the pretreated supernatant by 54.3, 76.8 and 76.5%, respectively, after 38 h of sedimentation pretreatment.

Organic Fraction: The DOC fraction after the sedimentation pretreatment was evaluated in all three types of WHJ. The carbon mass balance revealed the maximum recovery in the order of floating-, ground- and rotten-type WHJ due to their habitat conditions.

Nutrient Retention: The floating and ground types had the maximum values of nutrient retention, making them easily applicable to the subsequent processes. However, for the rotten type, nutrient retention was negatively related to the sedimentation pretreatment.

Future Work

Our results demonstrate that these pretreated WHJ could be used in a process that requires low SS in the substrate. The effect of sedimentation pretreatment on the physio-chemical composition of different types of WHJ revealed that sedimentation pretreatment is well-suited for removing SS from the WHJ. However, the effect of these parameters on subsequent processes like AD or enzymatic hydrolysis needs to be evaluated to suggest which type of WHJ can be used.

Acknowledgments We are deeply grateful to all three sampling sites for the donation of the water hyacinths. This research was funded by the Japan Science and Technology Agency (JST)/Japan International Cooperation Agency (JICA), Science and Technology Research Partnership for Sustainable Development (SATREPS) through the project for Eco-Engineering for Agricultural Revitalization toward Improvement of Human Nutrition (EARTH): Water Hyacinth to Energy and Agricultural Crops (grant number: JPMJSA2005).

References

1. Villamagna, A. M., & Murphy, B. R. (2010). Ecological and socio-economic impacts of invasive water hyacinth (*Eichhornia crassipes*): A review. *Freshwater Biology, 55*, 282–298. https://doi.org/10.1111/J.1365-2427.2009.02294.X
2. Saleh, H. M. (2016). Biological remediation of hazardous pollutants using water hyacinth- a review. *Journal of Biotechnology, 2*, 80–91.
3. Damtie, Y. A., Berlie, A. B., & Gessese, G. M. (2022). Impact of water hyacinth on rural livelihoods: The case of Lake Tana, Amhara region, Ethiopia. *Heliyon, 8*, e09132. https://doi.org/10.1016/j.heliyon.2022.e09132
4. Harun, I., Pushiri, H., Amirul-Aiman, A. J., & Zulkeflee, Z. (2021). Invasive water hyacinth: Ecology, impacts and prospects for the rural economy. *Plants, 10*. https://doi.org/10.3390/PLANTS10081613
5. Malik, A. (2007). Environmental challenge vis a vis opportunity: The case of water hyacinth. *Environment International, 33*, 122–138. https://doi.org/10.1016/J.ENVINT.2006.08.004

6. Ferrer, I., Palatsi, J., Campos, E., & Flotats, X. (2010). Mesophilic and thermophilic anaerobic biodegradability of water hyacinth pre-treated at 80 °C. *Waste Management, 30*, 1763–1767. https://doi.org/10.1016/J.WASMAN.2009.09.020
7. Crossetti, L. O., Bicudo, D. D. C., Bini, L. M., Dala-Corte, R. B., Ferragut, C., & de Mattos Bicudo, C. E. (2019). Phytoplankton species interactions and invasion by Ceratium furcoides are influenced by extreme drought and water-hyacinth removal in a shallow tropical reservoir. *Hydrobiologia, 831*, 71–85. https://doi.org/10.1007/S10750-018-3607-Y/FIGURES/7
8. Wanda, F. M., Namukose, M., & Matuha, M. (2015). Water hyacinth hotspots in the Ugandan waters of Lake Victoria in 1994–2012: Implications for management. *African Journal of Aquatic Science, 40*, 101–106. https://doi.org/10.2989/16085914.2014.997181
9. Djihouessi, M. B., Olokotum, M., Chabi, L. C., Mouftaou, F., & Aina, M. P. (2023). Paradigm shifts for sustainable management of water hyacinth in tropical ecosystems: A review and overview of current challenges. *Environmental Challenges, 11*, 100705. https://doi.org/10.1016/J.ENVC.2023.100705
10. Kist, D., Pérez-Elvira, S. I., & Monteggia, L. (2018). Evaluation of the steam explosion pretreatment upon the anaerobic digestion of water hyacinth biomass: Influence on liquid and solid fractions. *American Scientific Research Journal for Engineering, Technology, and Sciences*, 298–317.
11. Maruyama, D. O., Akizuki, S. I., Sekine, M., Fujita, A., Habtu, N. G., Sato, S., & Toda, T. (2023). Enhancement of water hyacinth juice treatment in an anaerobic sequential batch reactor with coffee husk–derived biochar. *Bioenergy Research, 1*, 1–10. https://doi.org/10.1007/S12155-023-10593-8/FIGURES/6
12. Liu, J., Wang, C., Wu, K., Tang, Z., Peng, S., Huang, J., Li, F., Zhao, X., Yin, F., Yang, B., Liu, J., Yang, H., & Zhang, W. (2020). Comparison of long-term energy efficiency and microbial community dynamics of different reactors in response to increased loadings of water hyacinth juice. *Science of the Total Environment, 744*, 140812. https://doi.org/10.1016/j.scitotenv.2020.140812
13. Liu, J., Tang, Z., Wang, C., Wu, K., Song, Y., Wang, X., Zhang, Z., Zhao, X., Yang, B., Piao, M., Yin, F., & Zhang, W. (2021). Novel technique for sustainable utilisation of water hyacinth using EGSB and MCSTR: Control overgrowth, energy recovery, and microbial metabolic mechanism. *Renewable Energy, 163*, 1701–1710. https://doi.org/10.1016/j.renene.2020.10.093
14. Liu, J., Wang, C., Wu, K., Huang, L., Tang, Z., Zhang, C., Wang, C., Zhao, X., Yin, F., Yang, B., Liu, J., Yang, H., & Zhang, W. (2020). Novel start-up process for the efficient degradation of high COD wastewater with up-flow anaerobic sludge blanket technology and a modified internal circulation reactor. *Bioresource Technology, 308*, 123300. https://doi.org/10.1016/j.biortech.2020.123300
15. Bhatia, P., Fujiwara, M., Ban, S., & Toda, T. (2020). Effect of steam explosion pre-treatment on methane generation from *Ludwigia grandiflora*. *Biomass and Bioenergy., 142*, 105771. https://doi.org/10.1016/J.BIOMBIOE.2020.105771
16. Aguirre-Villegas, H. A., Larson, R. A., & Sharara, M. A. (2019). Anaerobic digestion, solid-liquid separation, and drying of dairy manure: Measuring constituents and modeling emission. *Science of the Total Environment, 696*, 134059. https://doi.org/10.1016/J.SCITOTENV.2019.134059
17. Deng, L., Chen, Z., Yang, H., Zhu, J., Liu, Y., Long, Y., & Zheng, D. (2012). Biogas fermentation of swine slurry based on the separation of concentrated liquid and low content liquid. *Biomass and Bioenergy., 45*, 187–194. https://doi.org/10.1016/J.BIOMBIOE.2012.06.004
18. Møller, H. B., Lund, I., & Sommer, S. G. (2000). Solid-liquid separation of livestock slurry: Efficiency and cost. *Bioresource Technology, 74*, 223–229. https://doi.org/10.1016/S0960-8524(00)00016-X
19. Beccari, M., Majone, M., Riccardi, C., Savarese, F., & Torrisi, L. (1999). Integrated treatment of olive oil mill effluents: Effect of chemical and physical pretreatment on anaerobic treatability. *Water Science and Technology, 40*, 347–355. https://doi.org/10.1016/S0273-1223(99)00383-2

20. Bhatia, P., Fujiwara, M., Salangsang, M. C. D., Qian, J., Liu, X., Ban, S., Myojin, M., & Toda, T. (2021). Effect of semi-continuous anaerobic digestion on the substrate solubilisation of lignin-rich steam-exploded *Ludwigia grandiflora*. *Applied Sciences, 11*, 4452. https://doi.org/10.3390/app11104452
21. Singleton, V. L., Orthofer, R., & Lamuela-Raventós, R. M. (1999). Analysis of total phenols and other oxidation substrates and antioxidants by means of folin-ciocalteu reagent. *Methods in Enzymology, 299*, 152–178. https://doi.org/10.1016/S0076-6879(99)99017-1
22. Sekine, M., Yoshida, A., Akizuki, S., Kishi, M., & Toda, T. (2020). Microalgae cultivation using undiluted anaerobic digestate by introducing aerobic nitrification-desulfurization treatment. *Water Science and Technology, 82*, 1070–1080. https://doi.org/10.2166/WST.2020.153
23. Sekine, M., Akizuki, S., Kishi, M., Kurosawa, N., & Toda, T. (2020). Simultaneous biological nitrification and desulfurization treatment of ammonium and sulfide-rich wastewater: Effectiveness of a sequential batch operation. *Chemosphere, 244*, 125381. https://doi.org/10.1016/J.CHEMOSPHERE.2019.125381
24. Henriksen, K., Berthelsen, L., & Matzen, R. (1998). Separation of liquid pig manure by flocculation and ion exchange part 2: Pilot-scale system. *Journal of Agricultural Engineering Research, 69*, 127–131. https://doi.org/10.1006/JAER.1997.0230
25. Henriksen, K., Berthelsen, L., & Matzen, R. (1998). Separation of liquid pig manure by flocculation and ion exchange part 1: Laboratory experiments. *Journal of Agricultural Engineering Research, 69*, 115–125. https://doi.org/10.1006/JAER.1997.0229
26. Sivasankari, B., & Ravindran, D. (2016). A study on chemical analysis of water hyacinth (Eichornia crassipes), water lettuce (Pistia stratiotes). *International Journal of Innovative Research in Science, Engineering and Technology, 5*(10), 17566–17570.
27. Ndegwa, P. M., Zhu, J., & Luo, A. (2001). Effects of solid levels and chemical additives on removal of solids and phosphorus in swine manure. *Journal of Environmental Engineering, 127*, 1111–1115. https://doi.org/10.1061/(ASCE)0733-9372(2001)127:12(1111)
28. Converse, J. C., & Karthikeyan, K. G. (2002). Nutrient and solids separation of flushed dairy manure by settling. In *2002 ASAE annual meeting* (p. 1). American Society of Agricultural and Biological Engineers.
29. Song, Z., Williams, C. J., & Edyvean, R. G. J. (2000). Sedimentation of tannery wastewater. *Water Research, 34*, 2171–2176. https://doi.org/10.1016/S0043-1354(99)00358-9
30. Li, Y., Qi, B., & Wan, Y. (2020). Separation of monosaccharides from pretreatment inhibitors by nanofiltration in lignocellulosic hydrolysate: Fouling mitigation by activated carbon adsorption. *Biomass and Bioenergy, 136*, 105527. https://doi.org/10.1016/J.BIOMBIOE.2020.105527
31. Imchen, T., Ezaz, W., & Sawant, S. S. (2020). Potential contribution to the nutrient pool due to the decomposition of *Eichhornia crassipes* under different salinity in an aquatic ecosystem – A laboratory study. *Regional Studies in Marine Science, 35*, 101222. https://doi.org/10.1016/J.RSMA.2020.101222
32. Rein, P. (2016). *Cane sugar engineering* (2nd ed., pp. 91–113).
33. Patel, V., Desai, M., & Madamwar, D. (1993). Thermochemical pretreatment of water hyacinth for improved biomethanation. *Applied Biochemistry and Biotechnology, 42*, 67–74. https://doi.org/10.1007/BF02788902
34. Zhang, F., Li, Z., Tang, L., Wang, J., Shen, C., He, X., Li, R., Feng, J., & Li, N. (2022). Effects of submerged macrophyte decomposition on water quality. *Nature, Environment and Pollution Technology, 21*, 1049–1058. https://doi.org/10.46488/NEPT.2022.V21I03.011
35. Mir, M. A., Hussain, A., & Verma, C. (2016). Design considerations and operational performance of anaerobic digester: A review. *Cogent Engineering, 3*. https://doi.org/10.1080/23311916.2016.1181696
36. Fujiwara, M., Koyama, M., Akizuki, S., Ban, S., & Toda, T. (2022). Influence of lignocellulosic components on the anaerobic digestibility of aquatic weeds: Comparison with terrestrial crops. *Industrial Crops and Products, 178*, 114576. https://doi.org/10.1016/J.INDCROP.2022.114576

37. Sun, S., Sun, S., Cao, X., & Sun, R. (2016). The role of pretreatment in improving the enzymatic hydrolysis of lignocellulosic materials. *Bioresource Technology, 199*, 49–58. https://doi.org/10.1016/j.biortech.2015.08.061
38. Canna. (n.d.). An interactive relationship: Plant and substrate influence on pH and vice versa | Canna Gardening USA . https://www.cannagardening.com/interactive_relationship_plant_and_substrate_influence_ph_and_vice_versa. Accessed 24 July 2023.
39. Masse, L., Massé, D. I., Beaudette, V., & Muir, M. (2005). Size distribution and composition of particles in raw and anaerobically digested swine manure. *Transactions of ASAE, 48*, 1943–1949. https://doi.org/10.13031/2013.20003
40. Massé, D. I., Croteau, F., & Masse, L. (2007). The fate of crop nutrients during digestion of swine manure in psychrophilic anaerobic sequencing batch reactors. *Bioresource Technology, 98*, 2819–2823. https://doi.org/10.1016/J.BIORTECH.2006.07.040
41. O'Sullivan, C., Rounsefell, B., Grinham, A., Clarke, W., & Udy, J. (2010). Anaerobic digestion of harvested aquatic weeds: Water hyacinth (*Eichhornia crassipes*), cabomba (*Cabomba Caroliniana*) and salvinia (*Salvinia molesta*). *Ecological Engineering, 36*, 1459–1468. https://doi.org/10.1016/j.ecoleng.2010.06.027
42. Hudakorn, T., & Sritrakul, N. (2020). Biogas and biomass pellet production from water hyacinth. *Energy Reports, 6*, 532–538. https://doi.org/10.1016/J.EGYR.2019.11.115
43. Tassapak, W. S. A. S. S. (2022). Hydrothermal carbonization of compressed water hyacinth: Effects of operation parameters on energy conversion and characterization of products. *プランクトン工学研究, 2*, 44–55.

Chapter 19
Treatment of Water Hyacinth Juice in an Upflow Anaerobic Digester: Process Stability and Quality of Digestate's Macronutrients

Shinichi Akizuki, Pranshu Bhatia, Nigus Gabbiye Habtu, Tatsuki Toda, and Shinjiro Sato

19.1 Introduction

Water hyacinth (*Pontederia crassipes*, also referred to as *Eichhornia crassipes*) is a well-known, globally persistent, and invasive aquatic macrophyte. In particular, water hyacinth has been naturalized in more than 60 countries, including countries in Africa, Australia, Central America, Latin America, India, and New Zealand, owing to its rapid proliferation and growth [1, 2]. The invasion of water hyacinth evidently causes severe environmental and economic complications, including biodiversity loss, micro-habitat formation for disease vectors, hindrance to water transportation, increased evapotranspiration, and fishing-related problems [3]. Although several countries are continuously trying to control overgrowth using various approaches (e.g., mechanical harvesters, chemical herbicides, and biological agents) [4], the growth rate of water hyacinth still exceeds their removal/management rate [5].

Previous studies have focused on controlling the expansion of the areas invaded with water hyacinths. Recently, the focus has shifted to the valorization methods for water hyacinth [6] with several techniques, including its use for bioenergy

S. Akizuki (✉)
Institute of Plankton Eco-engineering, Soka University, Tokyo, Japan
e-mail: akizuki@soka.ac.jp

P. Bhatia
Faculty Science and Engineering, Soka University, Tokyo, Japan

N. G. Habtu
Bahir Dar Institute of Technology, Bahir Dar University, Bahir Dar, Ethiopia

T. Toda · S. Sato
Graduate School of Science and Engineering, Soka University, Tokyo, Japan

S. W. Fanta et al. (eds.), *Sustainable Development Research in Manufacturing, Process Engineering, Green Infrastructure, and Water Resources*, Green Energy and Technology, https://doi.org/10.1007/978-3-031-77339-6_19

production (e.g., biogas, bioethanol, and charcoal briquettes) [7], raw materials (e.g., furniture and handicrafts) [8], organic fertilizers [9], and animal feed [10]. Biogas recovery via wet anaerobic digestion is considered one of the most efficient methods as it is a simple energy-saving process that can be used to treat biomasses with high water content, such as water hyacinths. In addition, the nutrient-rich liquid fraction of the digestate and the microalgae biomass can be useful nutrient sources for valuable products, which can be further used in cosmetics, chemical products, animal feed, and nutritional health supplements [11]. Although the anaerobic digestion of water hyacinth alone has a small economic advantage, its combination with high-value microalgae production as the driving force could potentially be an economically feasible and attractive valorization alternative.

The direct use of water hyacinth as the feedstock for traditional digesters presents several challenges, as follows: (1) the necessity of physical pretreatment (chopping or beating to break the lignin component) of the biomass to enhance digestion and remove the air entrapped in the tissue and prevent it from floating in the digester and (2) the requirement of a large digester to conform to the long hydraulic retention time (HRT) [3, 9]. To overcome these hurdles, several attempts have been reported to increase the anaerobic degradability and shorten the HRT required for biogas production by chemical pretreatment, such as using sulfuric acid [12] or a two-stage digester [13]. However, compared to a conventional single digester, the additional chemical costs and complexity of the two-stage digester operation can be bottlenecks for the realization of the technology.

As a promising alternative approach, a technique involving high-rate anaerobic digestion, including the crushing and compression of water hyacinth into a juice and solid residue, has been proposed [14]. Water hyacinth juice (WHJ can be degraded more easily than whole water hyacinth plant, resulting in the absence of water hyacinth floating and a shorter HRT. Furthermore, a smaller digester is used than when using the plant itself. The solid residue can be valorized as silage or organic fertilizers owing to its ideal water content (<70%) [15, 16]. Reports on anaerobic digestion for WHJ treatment, including batch [17, 18] and continuous [14, 15, 19] experiments, have been gradually increasing in recent years. In particular, there is a strong interest in the effective utilization of the anaerobic digestate. Several studies have evaluated the composition of the digestate derived from different substrates, such as food waste [20], sewage sludge [21], and animal manure [22]. However, to the best of our knowledge, WHJ is yet to be investigated. When lignocellulosic biomass is used as a substrate for anaerobic digestion, the digestate has a large amount of magnesium because of its core molecule of chlorophyll. Ionic magnesium (Mg^{2+}) is known to coprecipitate with various ions, including ammonium (NH_4^+), potassium (K^+), and phosphate (PO_4^{3-}) [23], which are essential for microalgae growth.

This study aims to evaluate the process stability of anaerobic digestion in terms of the organic removal efficiency, biogas, and methane (CH_4) production rates in an upflow anaerobic digester continuously fed with WHJ. In addition, the quality of the digestate's macronutrients, including Mg^{2+}, NH_4^+, K^+, sodium (Na^+), calcium (Ca^{2+}), and PO_4^{3-}, was analyzed from the perspective of its use for microalgae cultivation.

19.2 Materials and Methods

19.2.1 Preparation of Substrate and Seed Sludge

Water hyacinth was obtained from a quagmire near the Doyo-no-mori Hurusato Otone road station in Kazo City, Saitama, Japan. The roots with attached soil were cut and only the part above the roots was collected to prevent soil contamination. The water hyacinth was immediately transported to Soka University after the collection. Subsequently, it was crushed and squeezed on the same day using a milling machine (RSC-3500, R-ing, Japan) and a screw dewatering machine (Dash-1, R-ing Co. Ltd., Japan), respectively. The obtained WHJ was filtered through a 106 μm mesh; thereafter, it was used as a substrate for the digester. The characteristics of the substrate were as follows: pH = 4.2 ± 0.1; total solids (TS) = 35.1 ± 0.4 g L^{-1}; volatile solids (VS) = 26.0 ± 0.4 g L^{-1}; total suspended solids (TSS) = 14.1 ± 3.5 g L^{-1}; total volatile suspended solids (VSS) = 11.7 ± 2.9 g L^{-1}; chemical oxygen demand (COD) = 29.6 ± 0.4 g-COD L^{-1}; total organic carbon (TOC) = 9.60 ± 0.57 g-TOC L^{-1}; total nitrogen (TN) = 0.24 ± 0.04 mg-N L^{-1}. Mesophilic anaerobic sludge was collected from a full-scale anaerobic digester that treats dewatered sewage sludge at the Hokubu Sludge Treatment Center, Yokohama City, Kanagawa, Japan. The collected sludge was stored for a few days at 37 ± 1 °C to remove residual organic compounds from the sludge; thereafter, it was used as the seed sludge.

19.2.2 Reactor Operation and Experimental Description

An upflow anaerobic digester with an effective volume of 7.8 L (internal diameter of 100 mm) was used as the reactor (Fig. 19.1). First, 7.8 L of the seed sludge was added to the reactor. Thereafter, the substrate was continuously fed into the bottom inlet of the reactor, with a constant HRT of 8 days, thereby leading to an organic loading rate (OLR) of 3.70 ± 0.05 g-COD L^{-1} day^{-1} (= 1.20 ± 0.07 g-TOC L^{-1} day^{-1}). The substrate was then diffused into the reactor using inert alumina beads (diameter = 10 mm). The resulting effluent was passed through a U-shaped pipe connected to the upper portion of the reactor and was stored in an effluent (or digestate) storage tank. The amount of biogas produced through anaerobic digestion was measured using a wet gas meter (W-NK-0.5B, Shinagawa, Japan) and stored in an aluminum gas bag (GL Science, Japan). The reactor temperature was maintained at 37 ± 1 °C in a walk-in, temperature-controlled laboratory and was operational for 100 days.

On day 57 of the stable reactor operation period, samples were taken from three different sampling points (i.e., substrate tank, middle sampling port of the reactor, and U-shape pipe, referred to as "inlet," "middle," and "outlet," respectively) to evaluate the variations in nutrient concentration while the substrate passed through the reactor. Dissolved total nitrogen (DTN), Na^+, K^+, Mg^{2+}, Ca^{2+}, and PO_4^{3-} were selected for analysis as they contribute to microalgal growth as macronutrients. The

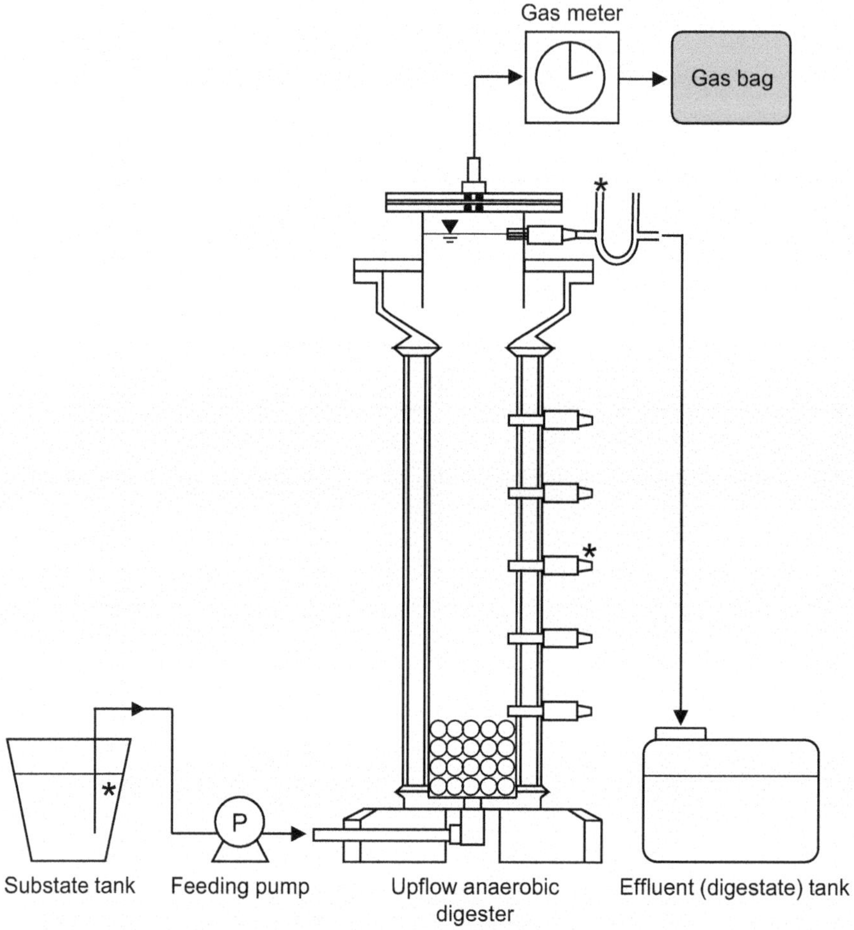

Fig. 19.1 Configuration of an upflow anaerobic digester. The asterisks (*) indicate the points at which the samples were taken on day 57

"outlet" values were considered as digestate component values; furthermore, the values were used for comparison with those obtained from the microalgae culture medium in Sect. 19.3.3.

19.2.3 Analytical Parameters

The pH, TSS, TOC, and dissolved organic carbon (DOC) of the influent and effluent were measured during the experimental period. The pH was measured using a pH meter (SevenCompact™ pH/Ion meter, S220, Mettler Toledo, Japan), and TSS was determined in accordance with the standard methods of the American Public Health

Association [24]. TOC and DOC were measured using a total organic carbon analyzer (TOC-L CPH/CPN, Shimadzu, Japan). The COD of the substrate was determined colorimetrically (DR-3900, Hach, USA) following dichromate digestion. The biogas composition in the aluminum gas bag was determined using a gas chromatograph (GC-14B, Shimadzu, Japan) equipped with a packed column (Micropacked-ST, Shimadzu, Japan) and a thermal conductivity detector. The injector and detector temperatures were maintained at 100 and 200 °C, respectively. Argon gas was used as the carrier gas at a flow rate of 10 mL min^{-1}.

The DTN was measured on day 57 using a total organic carbon analyzer, as mentioned previously. Cations (NH_4^+, Na^+, K^+, Mg^{2+}, and Ca^{2+}) were measured using high-performance liquid chromatography at 40 °C, equipped with high-performance liquid chromatography columns. PO_4^{3-} was measured in accordance with the Hach method 8114, in which orthophosphate reacts with molybdate in an acidic medium to produce a mixed phosphate/molybdate complex. Multication Standard Solution III (Fujifilm, Japan) was used as the standard solution for the cation measurement, and Anion Mixture Standard Solution I (Fujifilm, Japan) was used as the standard solution for the PO_4^{3-} measurement.

19.2.4 Calculations

The TOC removal efficiency was calculated using Eq. (19.1) as follows:

$$\text{TOC removal}(\%) = \left(1 - \frac{\text{TOC}_{\text{ef}}}{\text{TOC}_{\text{in}}}\right) \times 100 \tag{19.1}$$

where TOCin and TOCef refer to the influent and effluent TOC concentrations, respectively.

The carbon mass balance during the stable operation period was calculated using Eq. (19.2):

$$\text{Input} - \text{C} = \text{CH4} - \text{C} + \text{CO2} - \text{C} + \text{Effluent} - \text{C} + \text{Biomass} - \text{C} + \text{Others} - \text{C} \tag{19.2}$$

where Input-C, CH_4-C, CO_2-C, Effluent-C, and Biomass-C refer to the amounts of carbon derived from the WHJ-fed, CH_4, and CO_2 produced, effluent, and converted biomass, respectively. The biomass yield was assumed to be 1.4% of the carbon produced during substrate gasification [25]. Others-C was calculated by subtracting CH_4-C, CO_2-C, Effluent-C, and Biomass-C from Input-C, which were mainly undegraded WHJ residues.

19.3 Results and Discussion

19.3.1 Process Stability and Reactor Performance

The overall performance of the reactor in terms of the pH, total suspended solid (TSS) and volatile suspended solid (VSS) removal, CH_4 production rate, and total organic carbon (TOC) removal is summarized in Fig. 19.2. Throughout the experimental period, the mildly acidic pH (4.13–4.44) of the influent consistently increased to a neutral or mildly alkaline range (7.07–9.19) in the effluent (Fig. 19.2a). The increase in the pH to a near-neutral value indicates the stable and optimum conditions for anaerobic microbial reactions [25]. For the TSS and VSS removal, relatively low removal efficiency was observed during the first 20 days (Fig. 19.2b), which is likely ascribed to the partial washout of the dispersed inoculum. After 20 days, the TSS and VSS removal sharply increased and stabilized at 66.3% ± 13.7% and 76.8% ± 10.6%, respectively. It is worth noting that the influent and effluent VSS/TSS ratios are different; the influent value (0.84 ± 0.04) was higher than the effluent value (0.59 ± 0.13) (Fig. 19.3). These results indicate that the suspended materials in the influent were mainly organic, whereas the effluent potentially contained inert suspended solids [26] and/or inorganic precipitates [27]. The CH_4 content was found to be 70.4% ± 2.9%, which was maintained throughout the experimental period (Fig. 19.2c). The CH_4 production rate increased during the first

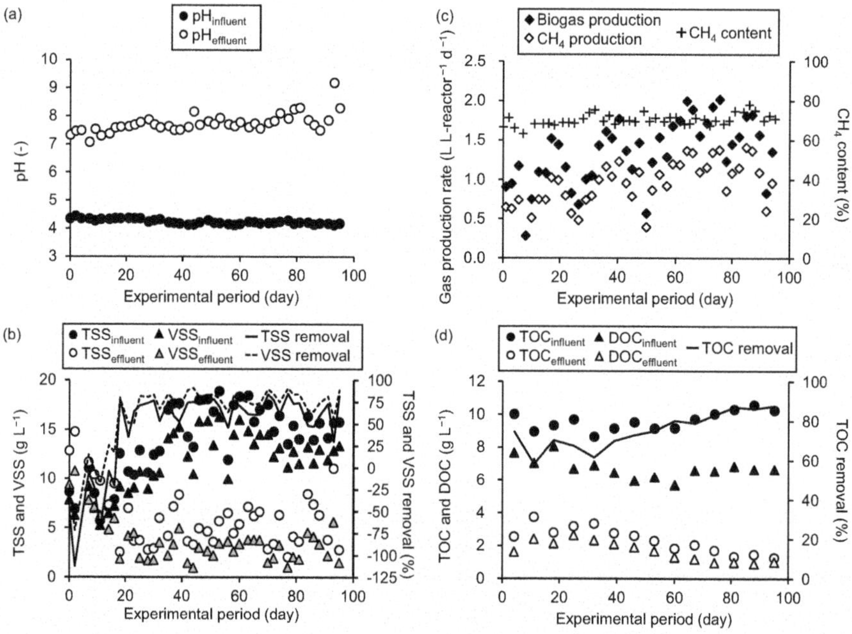

Fig. 19.2 Overall anaerobic digestion process performance. (**a**) pH variation; (**b**) TSS and VSS variations; (**c**) Biogas and CH4 production rates; (**d**) TOC and DOC variations

Fig. 19.3 Variation of VSS/TSS ratio

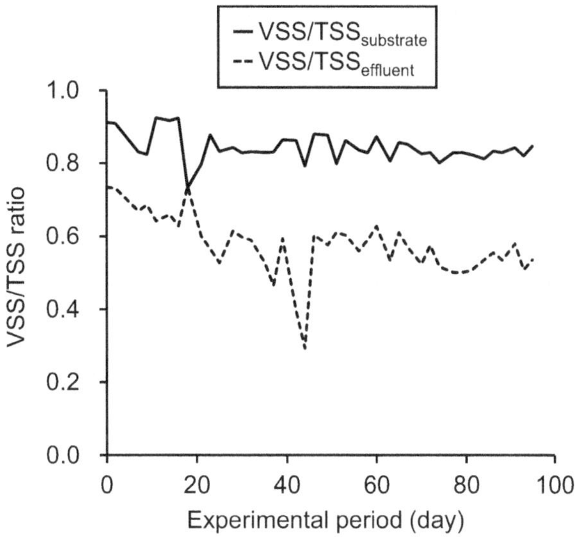

Table 19.1 Fed-carbon mass balance during stable operation period (unit: %)

CH_4	CO_2	Effluent TOC	Biomass	Others
42.3 ± 9.8	17.4 ± 4.5	19.9 ± 5.0	0.6 ± 0.1	19.8 ± 14.0

month and stabilized at 1.08 ± 0.25 L L-reactor^{-1} day^{-1} thereafter, resulting in the CH_4 yield of 332 ± 77 mL g-VS^{-1}. Although the TOC removal was relatively unstable with a fluctuating trend during the first 30–40 days (Fig. 19.2d), it gradually stabilized with the operation time to an average value of 80.0% ± 6.5% and remained constant thereafter. The results shown in Figs. 19.2a–d indicate the stability of the reactor and its good performance after approximately 40 days.

The carbon mass balance results showed that 59.7% of the input carbon was converted to biogas (CH_4 = 42.3% ± 9.8%; CO_2 = 17.4% ± 4.5%) with the balance allocated to the effluent (19.9%) and biomass growth (0.6%) and the remaining carbon was regarded as mainly undegraded WHJ residue (19.8%) (Table 19.1). The obtained biogas conversion efficiency was higher than that reported in a previous study (<40%, based on the chemical oxygen demand) from the anaerobic digestion of untreated water hyacinth [28]. Furthermore, the percentage of the "Others" (unallocated carbon) was lower in this study than that in a previous study (about 25%). Although it is difficult to simply compare the mass balance results because they depend on the type of water hyacinth used (i.e., whether it is floating or rooted in the soil), the results of this study suggest that WHJ obtained from the crushing and squeezing pretreatment enhanced the anaerobic biodegradability.

The poor mixing abilities of the upflow reactor used in this study resulted in the accumulation of a certain amount of residue at the bottom of the reactor. Although the process instability was not observed for approximately 100 days, the sludge (i.e., a mixture of microorganisms and residues) may need to be periodically withdrawn from the bottom, especially during long-term operation exceeding 1 year. Moreover, the residue accumulation can have adverse effects, including lower performance efficiency because of the decrease in microbial concentration. In contrast, the residues accumulated in the reactor may decompose over time, as suggested by the gradually increasing trend of the TOC removal rate after approximately 40 days (Fig. 19.2d), indicating that the microorganisms in the reactor may have acclimatized to the WHJ.

Our previous study treated the squeezed WHJ collected from the same location in an anaerobic sequential batch reactor (SBR) with an HRT of 10 days [29]. The results of the previous study demonstrated that the process was interrupted by 40 days due to the accumulation of volatile fatty acids. In SBRs, the process proceeds by agitation; however, the acid production associated with the WHJ degradation through hydrolysis and acetogenesis, and biogas conversion through methanogenesis are not balanced, making it prone to break down. Therefore, upflow anaerobic digesters are more suitable than SBRs and continuously stirred tank reactors for the WHJ used in this study.

Table 19.2 summarizes the comparison of the anaerobic digestion treatment performance of the water hyacinth biomass and WHJ reported in previous studies with the results of this study. Compared to reported untreated water hyacinths [17, 18, 28], our findings exhibited a high CH_4 yield and carbon removal efficiency despite the relatively short HRT, confirming the effectiveness of the employed pretreatment. Moreover, a higher CH_4 yield was achieved in this study compared with the other WHJ studies [15]. This difference could be attributed to the differences in the reactor type, operating conditions, such as HRT, organic loading rate, and temperature, as well as the characteristics of the water hyacinth. For instance, the lower metanization efficiency in the study by Liu et al. [15] than that in this study can be attributed to their relatively short HRT with high organic loading conditions. In addition, the water hyacinths used by Liu et al. [15] were collected from a lake, whereas those used in this study with rigid roots and stems were collected from a swamp. Such differences in the habitat may also have affected the CH_4 yield of the WHJ.

19.3.2 Changes in the Macronutrient Concentration After Passing Through the Reactor

Figure 19.4 shows the changes in the macronutrient concentration in the upflow reactor on day 57 of the experiment. Different trends were noted in the concentration changes based on the nutrient type. For instance, K^+ and Na^+ did not change considerably before and after treatment (Fig. 19.4c, f), whereas the concentration of the other components (DTN, NH_4^+, PO_4^{3-}, Ca^{2+}, and Mg^{2+}) decreased considerably. In particular, DTN, PO_4^{3-}, and Mg^{2+} decreased sharply from the "inlet" to the "middle." In anaerobic digesters, much of the organic nitrogen in the substrate is mineralized to NH_4^+ owing to hydrolysis. As the organic nitrogen in the WHJ was quickly

Table 19.2 Comparison of different studies on anaerobic digestion of water hyacinth or WHJ

Substrate type	Reactor type	Substrate carbon content (g-COD or g-TOC L^{-1})	HRT[a] (d)	OLR (g-COD or g-TOC L^{-1} d^{-1})	CH_4 yield (mL g-VS^{-1} or g-COD^{-1})	Carbon removal (%COD or TOC)	References
WH	Batch	–	35.0	–	ca. 175 (VS)	–	[17]
WH	Batch	5.96 (COD)	60.0	–	237 (vs)	–	[18]
WH	Batch	–	60.0	–	230–410 (VS)	–	[19]
WH	Single tank two-stage digester	–	20.0	3.75 (COD)	146.4 (COD)	72.5 (COD)	[28]
Pretreated WH[b]	Single tank two-stage digester	–	20.0	3.75 (COD)	340.8 (COD)	82 (COD)	[28]
WHJ	UASB	15.6–23.5 (COD)	2.0	8.85 (COD)	150 (COD)	86 (COD)	[15]
WHJ	MIC	15.6–23.5 (COD)	0.9	17.93 (COD)	210 (COD)	82 (COD)	[15]
WHJ	UAD	29.6 (COD) 9.6 (TOC)	8.0	3.7 (COD) 1.20 (TOC)	332 (VS) 291 (COD)	89 (COD) 80 (TOC)	This study

Single tank two-stage digester: structure with a partition in a single tank and an agitator on one side only; *WH* water hyacinth, *WHJ* water hyacinth juice, *UASB* upflow anaerobic sludge blanket, *MIC* modified international circulation, *UAD* upflow anaerobic digester

– not described

[a] For batch tests, the value indicates experimental period

[b] WH was pretreated at 90 °C for 1 h inside hot air oven

mineralized to NH_4^+, the decrease in DTN from the "inlet" to the "middle" can be almost substitutable for the decrease in NH_4^+. The reason for the decrease in NH_4^+ throughout the anaerobic digester can be ascribed to the gasification by volatilization. However, if the pH is near neutral (7–8) and the temperature is 37 °C, the percentage of volatilizable free ammonia (NH_3) in the total ammonia nitrogen is less than 10% [30], inhibiting the nitrogen removal by volatilization. The observed simultaneous decreases in the concentrations of the DTN (assumed to be NH_4^+), PO_4^{3-}, and Mg^{2+} can be ascribed to the coprecipitation via crystal formation. For instance, when Mg^{2+}, NH_4^+, and PO_4^{3-} coexist in a weakly acidic to mildly alkaline pH range (i.e., 4–9), the formation of magnesium ammonium phosphate hexahydrates (MAP, $MgNH_4PO_4{\cdot}6H_2O$, also known as struvite), magnesium phosphate ($Mg_3(PO_4)_2$), and magnesium hydrogen phosphate ($MgHPO_4$) proceeds [31]. In this study, the average influent and effluent pH values are 4.24 ± 0.08 and 7.86 ± 0.36, respectively (Fig. 19.2a). The pH in the "middle" is within this range, supporting the

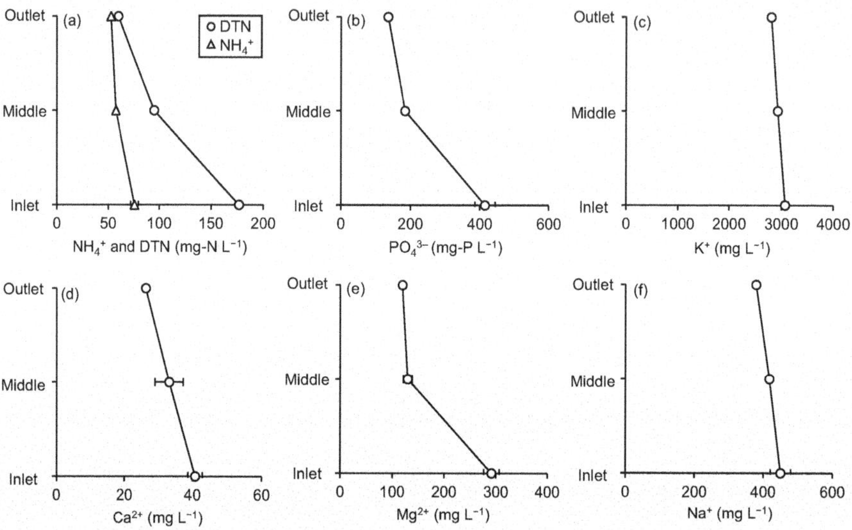

Fig. 19.4 Macronutrient concentration changes by passing through the reactor. (**a**) DTN and NH_4^+; (**b**) PO_4^{3-}; (**c**) K^+; (d) Ca^{2+}; (**e**) Mg^{2+}; (**f**) Na^+

possibility of crystallization in the reactor. Trimagnesium phosphate in two states of hydration ($Mg_3(PO_4)_2{\cdot}22H_2O$ and $Mg_3(PO_4)_2{\cdot}8H_2O$) can also be formed as crystals; however, they have a low precipitation rate and have never been identified in the pH range of 6–9 [32].

The decrease in the Mg^{2+}, DTN, and PO_4^{3-} concentration from the "inlet" to "middle" were 161 mg L^{-1}, 82 mg-N L^{-1}, and 230 mg-P L^{-1}, respectively, corresponding to 6.64, 4.53, and 7.43 mM in molar terms, respectively. Mg, N, and P react at a molar ratio of 1:1:1 in the MAP formation; however, the observed decreases did not match this ratio, indicating the likelihood of the formation of non-MAP crystals, such as $Mg_3(PO_4)_2$ and $MgHPO_4$. Alternatively, potassium struvite ($KMgPO_4{\cdot}6H_2O$), instead of MAP, has been reported to precipitate for the low NH_4^+ concentration [33]. In addition, calcium hydrogen phosphate ($CaHPO_4$) is formed when Ca^{2+} and PO_4^{3-} coexist in the pH range of 6–10 [34]. As Ca^{2+} and PO_4^{3-} gradually decreased from the "inlet" to "outlet" in this study, it is possible that other crystals, such as $CaHPO_4$, were formed. These results suggest that WHJ contains high concentrations of macronutrients that tend to coprecipitate during the anaerobic digestion process, resulting in crystal formation and accumulation in the digester. If the large number of crystals continues to accumulate in the digester, the consequent formation of a large mass may result in pipe/valve blockage and heat transfer reduction in the reactor because of scale formation [35]. In addition, considering the microalgae cultivation using the digestate, crystallization may reduce the amount of nutrients essential for cultivation, such as N and P. Nevertheless, the concentration of each nutrient and its balance are important for microalgal cultivation.

19.3.3 Evaluation of the Digestate Quality as an Alternative to the Microalgae Cultivation Medium

Figure 19.5 shows a radar chart comparing the nutrient concentrations of the digestate (i.e., effluent) on day 57 with those of the traditional artificial culture media for microalgae cultivation. Bold basal medium (BBM); C medium, which is suitable for green microalgae, such as *Chlorella*, *Scenedesmus*, and *Haematococcus*; and Spirulina–Ogawa–Terui (SOT) medium, which is suitable for blue-green microalgae *Arthrospira platensis* (Spirulina) were used as the artificial media. The media compositions were taken from the "Media list" of the National Institute for Environmental Studies, Japan [36]. The DTN, PO_4^{3-}, K^+, Ca^{2+}, Mg^{2+}, and Na^+ concentrations in the digestate were 60 mg-N L^{-1}, 136 mg-P L^{-1}, 2807 mg L^{-1}, 26 mg L^{-1}, 120 mg L^{-1}, and 379 mg L^{-1}, respectively, which exceeded those in the BBM and C medium (Fig. 19.5a, b). Meanwhile, the N and Na concentrations in the digestate were considerably lower than those in the SOT medium (Fig. 19.5c). The high Na concentration in the SOT medium was derived from the $NaHCO_3$ added to adjust the pH to 9–10.5 [37], which provides a carbonate source that can be converted to CO_2, playing a role in the photosynthetic process; however, a high Na concentration is not essential as a nutrient source. Consequently, the digestate has the potential to be used as an alternative to BBM and C medium. However, this study evaluated only the macronutrients and not the essential micronutrients, such as Fe, Mn, and Zn. In addition, undiluted digestate is often black in color, which reduces light permeability and may inhibit microalgae growth when combined with a high NH_4^+ concentration [38, 39], especially in comparison with artificial media that commonly use nitrate as an N source. Further experiments are needed to evaluate the availability of the WHJ-derived digestate in actual microalgae cultivation experiments.

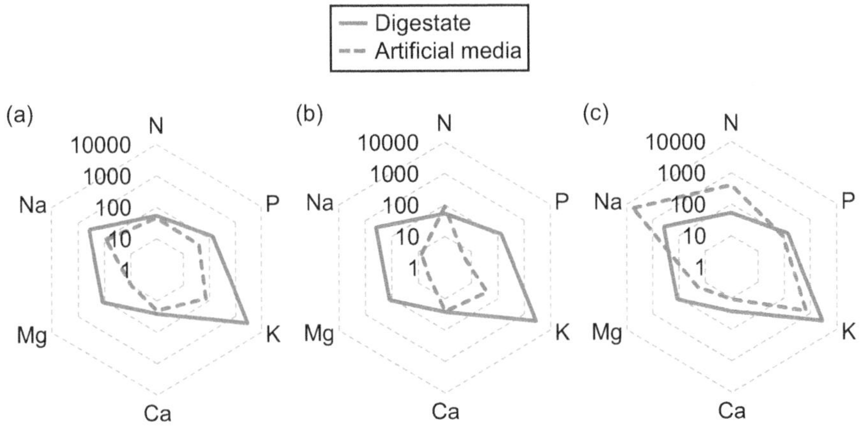

Fig. 19.5 Comparison of macronutrient concentrations in the digestate and artificial microalgae cultivation media. (**a**) BBM; (**b**) C; (**c**) SOT (Unit: mg/L)

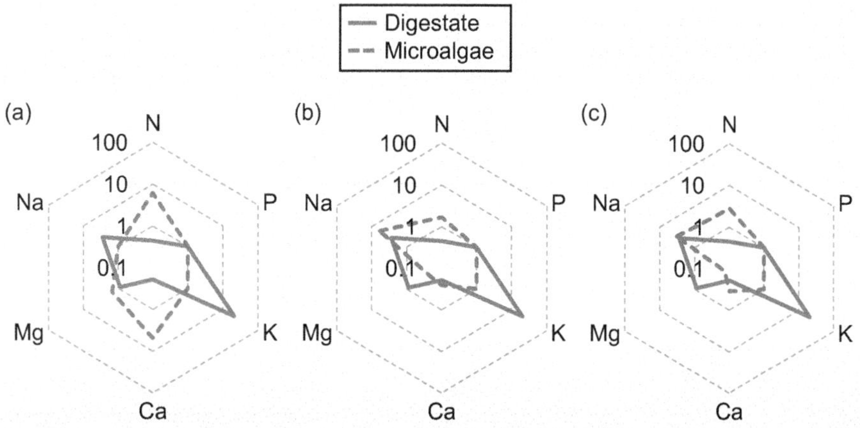

Fig. 19.6 Comparison of relative macronutrient ratio in the digestate and microalgae biomass (with $P = 1$). (**a**) *Chlorella*; (**b**) *Haematococcus*; (**c**) *Arthrospira platensis*

Another important aspect is the comparison of the nutrient balance (i.e., the ratio of each nutrient) in the digestate and microalgae biomass. For instance, if the nutrient balance in the digestate differs from that in the biomass, a certain nutrient may remain in the effluent after cultivation, often requiring further posttreatment to meet the discharge standards. The nutrient balance in the digestate was compared to that in the different microalgae biomasses: *Chlorella*, *Haematococcus*, and *A. platensis* [40] (Fig. 19.6). Compared with the balance in microalgae biomass, the N (0.44, when $P = 1$) and Ca (0.19) ratios in the digestate were lower than that of the other nutrients regardless of microalgae type. Therefore, we believe that if the microalgae are grown using this digestate, they will eventually become N- or Ca-limited materials and stop growing, leaving other nutrients in the effluent. In such cases, the nutrient balance should be adjusted by adding N and Ca as the reagents. Although the nutrient balance is not optimal, the WHJ-derived digestate has the potential to be used as a microalgae culture medium because it is rich in macronutrients necessary for microalgal growth. Further, the nutrient composition of the digestate varies based on the environmental conditions in which the water hyacinths are harvested. Therefore, further research is needed regarding this in the future.

19.4 Conclusions

This study evaluated the performance of WHJ treatment in a continuous upflow anaerobic digester and the digestate quality from the perspective of its use as a microalgae cultivation medium. The following conclusions were drawn:

(i) The anaerobic digestion process was stabilized after 40 days, achieving an average TOC removal efficiency of 80.0% ± 6.5% and CH_4 yield of 332 ± 77 mL g-VS^{-1} during the stable operation period.
(ii) A sharp decrease in the WHJ-derived DTN, PO_4^{3-}, and Mg^{2+} was observed after passing through the digester, indicating the formation of crystals, such as MAP and $Mg_3(PO_4)_2$.
(iii) The nutrient concentrations in the digestate were comparable to or higher than those in the BBM and C media used for *Chlorella*, *Scenedesmus*, and *Haematococcus*.
(iv) Compared to the nutrient balance in the microalgae biomass, the relative ratios of N and Ca in the digestate were lower than the other nutrients. Therefore, the adjustment of the nutrient balance by adding N and Ca sources as the reagents would be effective.

With the overgrowth of water hyacinths becoming a social and environmental global problem, research on the proper treatment and effective utilization of WHJ is becoming increasingly important. In particular, further investigation of microalgae culture studies employing digestate obtained from WHJ should be explored.

Acknowledgments We are deeply grateful to the Hokubu Sludge Treatment Center in Yokohama, Japan, for donating the anaerobic digestion sludge. This research was funded by the Japan Science and Technology Agency (JST)/Japan International Cooperation Agency (JICA), Science and Technology Research Partnership for Sustainable Development (SATREPS) through the project for Eco-Engineering for Agricultural Revitalization toward Improvement of Human Nutrition (EARTH): Water Hyacinth to Energy and Agricultural Crops (grant number: JPMJSA2005).

References

1. Yan, S. H., & Guo, J. Y. (2017). *Water hyacinth: Environmental challenges, management and utilization*. CRC Press.
2. Jafari, N. (2010). Ecology and socio-economic utilization of water hyacinth (*Eichhornia crassipes* Mart Solms). *Journal of Applied Sciences and Environmental Management, 14*(2), 43–49.
3. Nimule, P. E., Kumolu-Joh, C. A., & Anetekhai, M. A. (2011). The invasive aquatic macrophyte, water hyacinth {*Eichhornia crassipes* (Mart.) Solm-Laubach: Pontedericeae}: Problems and prospects. Research journal of environmental. *Science, 5*(6), 509–520.
4. Hill, M. P., & Coetzee, J. A. (2008). Integrated control of water hyacinth in Africa. *EPPO Bulletin, 38*, 452–457.
5. Julien, M. H., Griffiths, M. W., & Stanley, J. N. (2001). *Biological control of water hyacinth 2: The moths Niphograpta albiguttalis and Xubida infusellus: Biologies, host ranges, and rearing, releasing and monitoring techniques for biological control of Eichhornia Crassipes*. Australian Centre for International Agricultural Research.
6. Karouach, F., Bakrim, W. B., Ezzariai, A., Sobeh, M., Kibret, M., Yasri, A., Hafidi, M., & Kouisni, L. (2022). A comprehensive evaluation of the existing approaches for controlling and managing the proliferation of water hyacinth (*Eichhornia crassipes*): Review. *Frontiers in Environmental Science, 9*, 767871.

7. Rezania, S., Ponraj, M., Din, M. F. M., Songip, A. R., Sairan, F. M., & Chelliapan, S. (2015). The diverse applications of water hyacinth with main focus on sustainable energy and production for new era: An overview. *Renewable and Sustainable Energy Reviews, 41*, 943–954.
8. Patel, S. (2012). Threats, management and envisaged utilizations of aquatic weed *Eichhornia crassipes*: An overview. *Reviews in Environmental Science and Bio/Technology, 11*, 249–259.
9. Malik, A. (2007). Environmental challenge vis a vis opportunity: The case of water hyacinth. *Environment International, 33*(1), 122–138.
10. Lu, J., Fu, Z., & Yin, Y. (2008). Performance of a water hyacinth (*Eichhornia crassipes*) system in the treatment of wastewater from a duck farm and the effects of using water hyacinth as duck feed. *Journal of Environmental Sciences, 20*(5), 513–519.
11. Uggetti, E., Sialve, B., Latrille, E., & Steyer, J. P. (2014). Anaerobic digestate as substrate for microalgae culture: The role of ammonium concentration on the microalgae productivity. *Bioresource Technology, 152*, 437–443.
12. Manigandan, S., Preveenkumar, T. R., Anderson, A., Maryam, A., & Mahmoud, E. (2023). Benefits of pretreated water hyacinth for enhanced anaerobic digestion and biogas production. *International Journal of Thermofluids, 19*, 100369.
13. Kivaisi, A. K., & Mtila, M. (1998). Production of biogas from water hyacinth (*Eichhornia crassipes*) (Mart) (Solms) in a two-stage bioreactor. *World Journal of Microbiology and Biotechnology, 14*, 125–131.
14. Liu, J. F., Yan, L. X., Wang, Q., Wang, C., Zhao, X. L., Wang, C. M., Liu, J., Yang, H., Yin, F., & Zhang, W. D. (2016). Experimental study on dry fermentation of water hyacinth residue by anaerobic digestion. In *Proceedings of the 2016 international conference on energy development and environmental protection.*
15. Liu, J., Wang, C., Wu, K., Tang, Z., Peng, S., Huang, J., Li, F., Zhao, X., Yin, F., Yang, B., Liu, J., Yang, H., & Zhang, W. (2020). Comparison of long-term energy efficiency and microbial community dynamics of different reactors in response to increased loadings of water hyacinth juice. *Science of the Total Environment, 744*, 140812.
16. Gunnarsson, C. C., & Petersen, C. M. (2007). Water hyacinths as a resource in agriculture and energy production: A literature review. *Waste Management, 27*, 117–129.
17. Ngan, N. V. C., Viet, K. H., Cu, N. D., & Phong, N. H. (2011). Biogas production of pig manure with water hyacinth juice from batch anaerobic digestion. In M. A. Stewart & P. A. Conclanis (Eds.), *Environmental change and agricultural sustainability in the Mekong Delta* (pp. 355–369). Springer.
18. Hudakorn, T., & Sritrakul, N. (2020). Biogas and biomass pellet production from water hyacinth. *Energy Reports, 6*, 532–538.
19. Liu, J., Tang, Z., Wang, C., Wu, K., Song, Y., Wang, X., Zhang, Z., Zhao, X., Yang, B., Piao, M., Yin, F., & Zhang, W. (2021). Novel technique for sustainable utilisation of water hyacinth using EGSB and MCSTR: Control overgrowth, energy recovery, and microbial metabolic mechanism. *Renewable Energy, 163*, 1701–1710.
20. Opatokun, S. A., Yousef, L. F., & Strezov, V. (2017). Agronomic assessment of pyrolysed food waste digestate for sandy soil management. *Journal of Environmental Management, 187*(1), 24–30.
21. Parmer, K. R., & Ross, A. B. (2019). Integration of hydrothermal carbonization with anaerobic digestion: Opportunities for valorisation of digestate. *Energies, 12*(9), 1586.
22. Hung, C. Y., Tsai, W. T., Chen, J. W., Lin, Y. Q., & Chang, Y. M. (2017). Characterization of biochar prepared from biogas digestate. *Waste Management, 66*, 53–60.
23. Marti, N., Bouzas, A., Seco, A., & Ferrer, J. (2008). Struvite precipitation assessment in anaerobic digestion process. *Chemical Engineering Journal, 141*, 67–74.
24. American Public Health Association (APHA). (2005). *Standard methods for the examination of water and wastewater* (21st ed.). AHPA.
25. Speece, R. (1996). *Anaerobic biotechnology for industrial wastewaters*. Archae Press.
26. Metcalf, E., & Eddy, M. (2014). *Wastewater engineering: Treatment and resource recovery*. McGraw-Hill Education.

27. Syutsubo, K., Onodera, T., Choeisai, P., Khodphuvieng, J., Prammanee, P., Yoochatchaval, W., Kaewpradit, W., & Kubota, K. (2013). Development of appropriate technology for treatment of molasses-based wastewater. *Journal of Environmental Science and Health, Part A: Toxic/Hazardous Substances and Environmental Engineering, 48*(9), 1114–1121.
28. Barua, V. B., & Kalamdhad, A. S. (2019). Biogas production from water hyacinth in a novel anaerobic digester: A continuous study. *Process Safety and Environmental Protection, 127*, 82–89.
29. Maruyama, D. O., Akizuki, S., Sekine, M., Fujita, A., Habtu, N. G., Sato, S., & Toda, T. (2023). Enhancement of water hyacinth juice treatment in an anaerobic sequential batch reactor with coffee husk-derived biochar. *Bioenergy Research.*
30. Anthonisen, A. C., Loehr, R. C., Prakasam, T. B. S., & Srinath, E. G. (1976). Inhibition of nitrification by ammonia and nitrous acid. *Journal of the Water Pollution Control Federation, 48*, 835–852.
31. Darwish, M., Aris, A., Puteh, M. H., Abideen, M. Z., & Othman, M. N. (2016). Ammonium-nitrogen recovery from wastewater by struvite crystallization technology. *Separation & Purification Reviews, 45*, 261–274.
32. Musvoto, E. V., Wentzel, M. C., & Ekama, G. A. (2000). Integrated chemical-physical processes modelling. II. Simulating aeration treatment of anaerobic digester supernatants. *Water Research, 34*, 1868–1880.
33. Schuiling, R. D., & Andrade, A. (1999). Recovery of struvite from calf manure. *Environmental Technology, 20*, 765–768.
34. Recillas, S., Rodríguez-Lugo, V., Montero, M. L., Viquez-Cano, S., Hernandez, L., & Castaño, V. M. (2012). Studies on the precipitation behavior of calcium phosphate solutions. *Journal of Ceramic Processing Research, 13*(1), 5–10.
35. Schnaars, K. (2023). Operator essentials: What every operator should know about anaerobic digestion. *Water Environment and Technology Magazine*, https://www.wef.org/globalassets/assets-wef/direct-download-library/public/operator-essentials/wet-operator-essentials-anaerobic-digestion-dec12.pdf. Last accessed 16 May 2023.
36. National Institute for Environmental Studies (NIES). (2023). BBM, C, SOT in "Media list." NIES collection Microbial Culture Collection, https://mcc.nies.go.jp/02medium-e.html. Last accessed 16 May 2023.
37. Arvanitis, A., Sotiroudis, T., Nerantzis, E., Fournadzhieva, S., & Koultsiakis, E. (2004). Mass culture of the microalga spirulina using geothermal fluids in Greece – Antioxidant activities of spirulina powder extracts. In *Proceedings of the international conference on geothermal energy*.
38. Marcilhac, C., Sialve, B., Pourcher, A. M., Ziebal, C., Bernet, N., & Béline, F. (2014). Digestate color and light intensity affect nutrient removal and competition phenomena in a microalgal-bacterial ecosystem. *Water Research, 64*, 278–287.
39. Collos, Y., & Harrison, P. J. (2014). Acclimation and toxicity of high ammonium concentrations to unicellular algae. *Marine Pollution Bulletin, 80*(1–2), 18–23.
40. Batista, A. P., Gouveia, L., Bandarra, N. M., Franco, J. M., & Raymundo, A. (2013). Comparison of microalgal biomass profiles as novel functional ingredient for food products. *Algal Research, 2*(2), 164–173.

Chapter 20
Sustainable Energy Catalyzed Rural Entrepreneurship for Smallholder Farmers in Ethiopia Through a Living Lab Approach

Tena Tegbar

20.1 Introduction

Despite the fact that living laboratories (LL) are popular, the concept is little understood [1]. A number of researchers define LL as real or virtual environments, or interaction spaces, in which stakeholders collaborate to create new solutions to complex societal problems [2, 3]. Similar to this, LL are also defined as networks that integrate both user-centered research and open innovation [4–7]. Other scholars emphasize LL as a methodology for user engagement [8, 9]. The European Network of Living Labs (ENoLL) defines LL as "user-centered, open innovation ecosystems based on systematic user co-creation approach, integrating research and innovation processes in real life communities and settings" [1]. In general, the definitions of living laboratories are broader and somewhat diffuse. Nevertheless, the following elements tend to be core features of a living laboratory: (i) experimental approaches in a real-life context, (ii) participation and user involvement, and collaboration and co-production of knowledge. From the conceptual perspective, there are four types of LL. For instance, Leminen [10] classifies LL as: utilizer-driven, enabler-driven, provider-driven, and user-driven. Whereas, the ENoLL classifies LL as urban and rural living labs, research-driven living labs (research-focused), living test-bed (provider-focused), and living labs as a service (for SME's and startups) [1]. However, in practice, most living labs combine several types, although their focus is on one of the above [1].

Living labs differ from other labs in many ways [11]. The existing differences are summarized in Table 20.1.

Living labs operate as intermediaries among citizens, research organizations, companies, cities and regions for joint value co-creation, rapid prototyping, or

T. Tegbar (✉)
Bahir Dar Institute of Technology, Bahir Dar University, Bahir Dar, Ethiopia

S. W. Fanta et al. (eds.), *Sustainable Development Research in Manufacturing, Process Engineering, Green Infrastructure, and Water Resources*, Green Energy and Technology, https://doi.org/10.1007/978-3-031-77339-6_20

Table 20.1 Comparison of living lab against other labs

Lab research (user labs)	Action research	Living lab
Controlled environment	Real-world setting, yet typically confined to an organization or department	Controlled environment real-world setting, yet real-world setting involving multiple stakeholders from multiple organizations and their interaction
Limited, clearly assigned role of users	Not specific about user role	Active role of users as co-innovators; exposing technology to the creative and destructive energies of the users; facilitating dynamics of collective action
Designed for replicability	Active (social and political) role of researcher in the research setting	Multi-disciplinary research teams actively involved in the research settings, confronted with the technical, social and political dynamics of innovation, at times even driving the agenda
Designed for observation of outcome	The researchers observe and take part in the creation of an outcome	Joint collaboration to create a desired outcome

Source: Klein et al. [11]

validation to scale up innovation and businesses [12–14]. In other words, they are real-world, naturalistic settings used to test or evaluate concepts and/or technologies [12–14]. They facilitate learning by bridging the laboratory and the real world through open, uncontrolled, yet focused public experimentation [11]. These kinds of labs encourage multiple stakeholders with diverse backgrounds and interests to collaborate on a common goal [15]. They are proposed as a solution [16] to the imperfect connections between academia, policy, industry, and societal sectors, which are the root causes of innovation blocking [17]. In other words, living labs unblock the mutual flow of knowledge between university and business [18, 19]. Overall, living labs, in contrast to a traditional laboratory, operate in a real-life context with a user-centric approach. The physical and/or organizational boundaries of a living lab are defined by purpose, scope, and context. The scope, aims, objectives, duration, actor involvement, degree of participation, and boundaries of a living laboratory are open for definition by its participants. A living laboratory could thus be established on a street, in a house, within an organization, or include a whole city or industry, depending on the project.

Understanding the significance of living labs, the Bahir Dar Institute of Technology has endeavored to use the concept for the development of rural entrepreneurship in order to bring about significant change in Ethiopia's rural areas where there are untapped opportunities and resources. It is known that Ethiopia is one of the countries with a large renewable energy potential, particularly solar energy. East Africa is the blessed part of Africa in terms of solar energy potential, which is projected to be more than 200,000 TWh/year, accounting for one-third of Africa's potential [20]. Due to Ethiopia's proximity to the equator, its annual average irradiance is projected to be around 5.2 kWh/m2/day [21, 22]. Water is yet another abundant resource. Annual renewable groundwater resources are estimated to be 30 billion cubic meters [23]. Around 48% of the country's total 11 million hectares of land eligible for irrigation could be irrigated utilizing groundwater resources [24]. It also has a climate suitable for farm production with long sunshine hours. This demonstrates the existence of opportunities and resources. However, the rural community, which accounts for 80% of the overall population according to the 2020 World Bank report [25], is primarily reliant on rain-fed agricultural operations. Moreover, unlike other countries, there is a lack of strategy to promote entrepreneurship in rural areas, despite the fact that there is opportunity. The country is too late in providing a policy framework for entrepreneurship. Its national entrepreneurship strategy was formulated in 2019 to encourage economic growth and increase young employability across the country [26]. However, there is a lack of evidence on how it will be implemented in the context of rural Ethiopia. The Bahir Dar Institute of Technology has devised a novel approach, known as the living lab, to transform the status quo of the rural community by utilizing the available opportunities and resources on the ground. The institute has built a "Sustainable Energy and Entrepreneurship Development (SEED)" living lab in practice. Therefore, the objectives of this study are to: (i) comprehend the characteristics of BiT's living lab using the literature concepts highlighted above and (ii) evaluate the effect of the living lab on smaller holders who were encouraged to engage in entrepreneurial activity during the dry season, which lasted nearly from November to June in the majority of Ethiopia.

20.2 Research Methodology

This study is based on the empirical evidence of living lab established by the Bahir Dar University, Bahir Dar Institute of Technology (BiT) and its immediate beneficiaries. The living lab consists of an integrated system of solar photovoltaics (PV) equipped with storage batteries and charge controller that has a design capacity of 25kw, a bore hole of 160 m equipped with a submarine pump with a design flow rate of 10 L/sec and placed at a depth of 120 m to lift the water; 300 m^3 capacity water reservoir and its distribution channels to reach the water to the plots of smallholders. The living lab has a drinking water supply system for the nearby rural community. But, this is not considered in this study. Hence, immediate beneficiaries of the living lab would mean the smaller holders that participated in solar irrigation practice and generated income from crops and vegetables after receiving tailor-made entrepreneurship training. Geographically, the living lab is located at Dangishita rural village, which is accessible at 10.5 km departing from Dangila to Northwards, a town located 78 km on the main highway that runs from Bahir Dar to Addis Ababa. To achieve the research objective, primary data is collected with different tools and methods including, field visits, observation, and semi-structured questionnaires described in the subsequent sections. Firstly, interviews with key project participants are conducted using a semi-structured questionnaire and by observing the living lab at the site. Secondly, the expenditure, total revenue, and net revenue of the

Table 20.2 Template prepared to collect data of the smaller holders involved in the BiT living lab in the year 2014 E.C /2022(GC)

S. no.	Name of smallholder	Gender	Crop type	Total expense for seed and fertilizer (ETB)	Total Revenue (ETB)	Net revenue (ETB)
1						
2						
3						
.						
.						
.						
n						
Total						

smaller holders engaged in the production of income-generating crops and vegetables are collected by interviewing the smaller holders (Table 20.2).

To achieve the first objective of the research a qualitative analysis with comparison technique is employed. To achieve the second objective, a descriptive analysis method with an Excel spread sheet is used. Descriptive method is used to analyze the data. Such a method is used to convert raw data into a form that will make it easy to understand and interpret, i.e., rearranging, ordering, and manipulating data to provide insightful information about the provided data. For this particular case, bar chart is used since it is the best for presenting a single data series.

20.3 Results and Discussion

Based on the results of key project participants' interviews, the BiT SEED living lab fits the integrated model, which is characterized by: (i) a complete process from energy development to entrepreneur creation, (ii) the creation of a network around a project, (iii) the collection and use of information and co-creation of knowledge within the network, and (iv) piloting a project for regional and national implementation. This living lab differs from other living labs categorized by Leminen [10] and the ENoLL [1] in several important dimensions. Above all, the Bit's living lab is notable for being constructed in the countryside. Second, it is intended to foster entrepreneurship in rural areas. Despite its uniqueness, it possesses the essential features of a living lab. Its primary goal is to solve a deep-rooted problem that is widespread in Ethiopia, namely, a lack of capacity to utilize available potential and resources to improve the living standards of Ethiopia's rural areas, which is similar to the goal of other living labs established to solve grand societal problems. It is also established in collaboration with major players, such as the Technical University of Munich and other local stakeholders, as are other living labs mentioned in the literature. Furthermore, the living lab was developed and is run by applying suitable techniques and methods. First and foremost, a site was chosen in accordance with the project documentation developed by the University of Munich and its partners, including the BiT, in which the living labs are implemented. Following site selection, a competent institute undertook a hydro-geological study by outsourcing the task. The installation and commissioning of the solar station were then given to a qualified company through a bid process. Finally, it has delegated greater responsibility for the living lab's safety to the community after providing tailored training in entrepreneurship, irrigation management, and conflict resolution. This aids in the comprehension of the complicated processes of prototyping, validating, and refining solutions. The results of the quantitative data collected to assess how the living lab affects the project beneficiaries are shown later.

During the first phase of the living lab, approximately 28 households (21 male; 5 female) were encouraged to grow crops and vegetables on 4 hectares of land. The results of the interviews, which were mainly done to figure out the amount of expenditure, gross income, and net income of these smallholders in Ethiopian currency (Birr), are provided in Table 20.3.

The smallholders primarily grew potatoes, as shown in the table above. Based on information collected from the living lab project team members, the land area is estimated to be four hectares. The smallholders generated an economic activity worth Birr102, 820.00 while spending Birr 28,688.00 on this small plot of land. Moreover, all of the smallholders achieved a net profit from their efforts. In other words, regardless of the amount, they were all profitable. The average net profit was around Birr 2612. Lalem Mebrat, the best smallholder, was made Birr 15,250, over six times the average profit. Wudu Asemie and Ayine Amare achieved the lowest profit, Birr 600, which is approximately 4.5 times less than the average income. Figure 20.1 depicts the individual performance of the small holdings using a bar graph.

Finally, the quantitative study is based on data gathered solely through interviews with smallholders. Each smallholder's plot of land was not measured. Furthermore, the cultivation method, irrigation method, and fertilizer usage were not measured. As a result, such crucial elements will be taken into account in future investigations.

20.4 Conclusions and Recommendations

The living laboratory approach of BiT has demonstrated how untapped sustainable energy and resources—groundwater, land, and workforce—are combined to undertake economic activities that generate money. In the stated approach, all of the smallholder farmers were able to earn a net profit from their farming efforts during the dry season, when they were inactive previous to the establishment of the living lab. In general, the revenue generated by the smallholders in the living lab's initial trial is highly encouraging. This research thus leads us to the conclusion that if sustainable energy, specifically solar energy, is utilized, groundwater can be used to irrigate farmer-owned plots of land in the entire country. The use of the concept in

Table 20.3 Interview results of the beneficiaries of the BiT's living lab

S. no.	Name of smallholder	Gender	Crop type	Total expenditure for seed and fertilizer (ETB)	Total revenue (ETB)	Net income (Birr)
1	Wubet Zewdie	M	Potato	800	1600	800
2	Wudu Asemie	M	Potato	1000	1600	600
3	Ayine Amare	M	Potato	900	1500	600
4	Aderaw Amare	M	Potato	1000	1900	900
5	Wudu Azanaw	M	Potato	500	1200	700
6	Enyew Azanaw	M	Potato	600	2300	700
7	Siraw Ayalkibet	M	Potato and wheat	3700	16,000	12,300
8	Begizew Enyew	M	Potato	1070	2400	1330
9	Lalem Mebrat	M	Potato	2750	18,000	15,250
10	Antehunegn Mebrat	M	Potato	500	1600	1100
11	Chale Desalegn	M	Potato	500	1650	1150
12	Mengistu Zelalem	M	Potato	1000	2400	1400
13	Manaylush Dagnaw	F	Potato	1200	2450	1250
14	Wubante Eyasu	M	Potato	792	4800	4008
15	Enkuaye Mersha	F	Potato	530	2000	1470
16	Tagele Mebrat	M	Potato	430	1950	1520
17	Asrat Zewude	M	Potato	2600	8000	5400
18	Wubet Adugna	M	Potato	520	1670	1150
19	Melak Ejigu	M	Potato	440	2000	1560
20	Ayanaw Enyew	M	Potato	1000	4800	3800
21	Asaye Fenta	M	Potato	600	1700	1100
22	Fikr Takele	F	Potato	1100	2300	1200
23	Fentahun Teshager	M	Potato	1120	3200	2080
24	Menber Siraw	M	Potato	1000	4400	3400
25	Wudie Zewude	F	Potato	440	2000	1560
26	Mulu Defersha	F	Potato	396	2400	2004
27	Ayalkbet Shiferaw	M	Potato	1500	4500	3000
28	Zemenu Siraw	M	Potato	700	2500	1800
Total				28,688.00	102,820.00	73,132.00

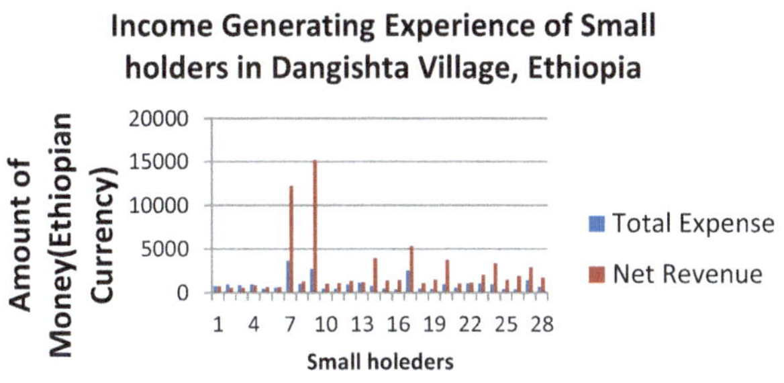

Fig. 20.1 The economic performance of smallholders at individual level

a broader context could alter production practices in Ethiopia's rural areas. They can cultivate in a different way. Finally, the approach will help to foster rural entrepreneurship in the country's rural areas. Hence, further research is recommended to improve understanding of the processes of prototyping, validating, and refining solutions while taking into account the next phases of the living lab: activities carried out to create small-scale enterprises in manufacturing, ago-processing, and services in Ethiopia's rural areas. The results of the recommended future study could aid in the formulation of lessons learned from the full implementation of the living lab of BiT, making it easier to replicate the living lab approach to entrepreneurship development at the regional and national levels.

Acknowledgments The authors would like to express their gratitude to the DAAD and Bahir Dar University for their financial and material support of the Sustainable Energy, Entrepreneurship, and Development project, which was made possible through collaboration between the Technical University of Munich and Bahir Dar University.

References

1. Purola, A., Santonen, T., Haapaniemi, H., Hirvikoski, T., Nevmerzhitskaya, J. (2020). *Living labs concepts and implementation plan for CIRC4Life-project*. European Commission Horizon. https://ec.europa.eu/research/participants/documentIds=080166e5c1e. Last accessed 23 Apr 2023.
2. Karaseva, V., Seffah, A., & Porras, J. (2015). A social-media-based living lab: An incubator for human-centric software engineering and innovation. In *International conference on software and system process 2015, ICSSP* (pp. 194–198).
3. Higgins, A., & Klein, S. (2011). I*ntroduction to the living lab approach*. Accelerating Global Supply Chains with IT-Innovations. https://doi.org/10.1007/978-3-642-15669-4_2.
4. Frößler, F., Rukanova, B., Klein, S., Higgins, A., Tan, Y., & Kelly, S. (2018). The first (beer) living lab: Learning to sustain network collaboration for digital innovation. In K. Riemer, S. Schellhammer, & M. Meinert (Eds.), *Proceedings on collaboration in the digital age: How technology enables individuals, teams and businesses* (pp. 227–247). Springer.
5. Tang, T., Wu, Z., Hämäläinen, M., & Ji, Y. (2012). From web 2.0 to living lab: An exploration of the evolved innovation principles. *Journal of Emerging Technologies in Web Intelligence, 4*(4), 379–385.

6. Schuurman, D., Lievens, B., de Marez, L., & Ballon, P. (2012). Towards optimal user involvement in innovation processes: A panel-centered living lab-approach. In: *12th proceedings of PICMET: Technology management for emerging technologies*. Canada.
7. Schumacher, J., & Niitamo, V. (2008). *European living labs: A new approach for human centric regional innovation*. Wissenschaftlicher Verlag.
8. Fulgencio, H., Lefever, H., Katzy B.: Living lab: Innovation through pastiche (a research linking disparate and discorded ontology). In: Editor Paul, C., Miriam, C. (eds.) Conference 2012, IIMC(2012).
9. de Leon, M. P., Eriksson, M., Balasubramaniam, S., & Donnelly, W. (2006, March). Creating a distributed mobile networking testbed environment-through the living labs approach. In *2nd international conference on testbeds and research infrastructures for the development of networks and communities*. TRIDENTCOM 2006 (pp. 5). IEEE.
10. Leminen, S. (2015). Q&A what are living labs. *Technology Innovation Management Review, 5*(9), 29–35.
11. Klein, S., Higgins, A., & Rukanova, B. (2011). *Network collaboration models* (pp. 255–270). Springer.
12. Kusiak, A. (2007). The University of Iowa Innovation: The living laboratory perspective. *Computer-Aided Design and Applications, 4*(6), 863–876.
13. Schumacher, J., & Feurstein, K. (2007). *Living labs – A new multi-stakeholder approach to user integration*. In Presented at the 3rd international conference on interoperability of enterprise systems and applications (I-ESA'07), Funchal, Portugal.
14. Niitamo, V., Kulkki, S., Eriksson, M., & Hribernik, K. (2006). State-of-the-art and good practice in the field of living labs. In: *Proceedings of the 12th international conference on concurrent enterprising: Innovative products and services through collaborative networks* (pp. 349–357). Milan, Italy.
15. Lucassen, I., Klievink, B., & Tavasszy, L. (2014). A living Lab framework: Facilitating the adoption of innovations in international information infrastructures. In: *International conference on Transport Research Arena* (TRA). Paris.
16. Canzler, W., Engels, F., Rogge, J. C., Simon, D., & Wentland, A. (2017). From "living lab" to strategic action field: Bringing together energy, mobility, and information technology in Germany. *Energy Research and Social Science, 27*, 25–35.
17. Burbridge, M. (2017). If living labs are the answer–what's the question? A review of the literature. *Procedia Engineering, 180*, 1725–1732.
18. Grotenhuis, F. D. (2017). Living labs as service providers: From proliferation to coordination. *Global Business and Organizational Excellence, 36*(4), 52–57.
19. Voytenko, Y., McCormick, K., Evans, J., & Schliwa, G. (2016). Urban living labs for sustainability and low carbon cities in Europe: Towards a research agenda. *Journal of Cleaner Production, 123*, 45–54.
20. Mondal, M. H. A., Gebremeskel, A. G., Gebrehiwot, K., & Ringler, C. (2018). *Ethiopian universal electrification development strategies*. International Food Policy Research Institute.
21. IEA Homepages. (2023). https://www.icafrica.org/fileadmin/documents/Knowledge/Energy/AfricaEnergyOutlook-IEA.pdf. Last accessed 21 Mar 2023.
22. Hailu, A., & Kumsa, D. (2021). Ethiopia renewable energy potentials and current state. *Journal of AIMS Energy, 9*(1), 1–14.
23. Berhanu, B., Seleshi, Y., & Melesse, A. (2014). Surface water and groundwater resources of Ethiopia: Potentials and challenges of water resources development. In *Nile River basin: Ecohydrological challenges, climate change and hydro politics*.
24. Schmitter, P., Kibret, K. S., Lefore, N., & Barron, J. (2018). Suitability mapping framework for solar photovoltaic pumps for smallholder farmers in sub-Saharan Africa. *Applied Geography, 94*, 41–57.
25. World Bank. (2020). *Ethiopia regional poverty report: Promoting equitable growth for all regions*. https://openknowledge.worldbank.org/handle/10986/34805
26. ENA Homepage. (2023). https://www.ena.et/web/eng/w/en_8792. Last accessed 20 Mar 2023.

Chapter 21
Contract Time Determination and Work Item Scheduling in Public Building Projects: Case Study in Bahir Dar

Befekadu C. Gashe, Asregidew K. Woldesenbet, Solomon M. Belay, and Habtamu E. Endalew

21.1 Introduction

The construction industry is one of the major industry sectors that play vital roles in a nation's socioeconomic development through the construction of residential housing, commercial and retail buildings, and industrial plants and through the replacement, maintenance, and restoration of infrastructure and public facilities [1]. The industry generates employment opportunities and wealth and creates a built environment for the well-being of residents through forward and backward links with other industry sectors. Typically, the industry designs, builds, operates, and maintains projects of the required quality and within the allocated budget (cost) and schedule [2].

To meet these criteria, every project is part of a contract, as a binding agreement between parties involved in the project. This contract includes specific and general

B. C. Gashe (✉)
Faculty of Civil and Water Resource Engineering, Bahir Dar Institute of Technology, Bahir Dar University, Bahir Dar, Ethiopia

Department of Structural for Engineering and Architecture, University of Naples Federico II, Naples, Italy

A. K. Woldesenbet

Ethiopian Institute of Architecture, Building Construction, and City Development, Addis Ababa University, Addis Ababa, Ethiopia
S. M. Belay
Faculty of Civil and Water Resource Engineering, Bahir Dar Institute of Technology, Bahir Dar University, Bahir Dar, Ethiopia

H. E. Endalew
School of Architecture and Construction Technology and Management, University of Gonder, Institute of Technology, Gonder, Ethiopia

S. W. Fanta et al. (eds.), *Sustainable Development Research in Manufacturing, Process Engineering, Green Infrastructure, and Water Resources*, Green Energy and Technology, https://doi.org/10.1007/978-3-031-77339-6_21

contractual agreements, which contain details on start and completion dates, schedules, responsibilities, and expectations [3]. Through practice, however, contracts often go beyond their allocated contract time for a number of reasons, including a lack of coordination, the unavailability of materials, project uncertainties, and project uniqueness [4]. According to [5], contract time is an essential and critical element of project management. Contract time includes incentive milestone dates that may warrant liquidation for damage, termination for default, and other punitive measures associated with unsatisfactory contract performance. A set contract time is nothing but a schedule showing the milestones and early project completion dates. Although such a schedule seems straightforward, executing, administering, and closing out a project is not that simple, because these require various interrelated activities. Typically, a delay in one activity can affect the other like a domino effect unless the activities are outside of a project's crucial direction. Given the aforementioned factors, contractors know how difficult completing a project on time is [6].

The construction industry in Ethiopia has indicated that building projects suffer from delays during the completion period, leading to cost overruns and inconvenience to public users, clients, and various other stakeholders [7]. For these reasons, this book makes some assertions about the shortcomings related to a lack of proper methods and procedures in determining project contract time and the scheduling of work items before construction commences. In addition, problems such as poor data-collection and management procedures for records by agencies emerge, where these agencies face difficulties in calibrating up-to-date productivity rates to help them set a more realistic contract time [8]. Today, many public building projects in Bahir Dar are characterized by delays to completion dates, leading to cost overruns and inconvenience to public users, clients, and various other stakeholders. These claims, however, are not supported by sufficient empirical evidence. Hence, the research in this book investigates procedures to determine contract time and to clearly map out the variables affecting the efficiency of construction projects in Bahir Dar and its surroundings.

This study aims to investigate the procedures used to determine a contract time procedure for public building projects by different institutions. It aims to ultimately indicate what major variables lead to delays in project completion. Also, it aims to assess the accuracy of project contract time estimates provided to contractors by clients (or owners) of public building projects by evaluating whether they are properly calculated or scheduled according to work items on the bill of quantities.

21.2 Methodology

21.2.1 Study Design

This research adopted an exploratory type (content analysis) of strategy for qualitative research and quantitative analysis on various scenarios to assess contract time determinations and the scheduling of work items in building construction.

The study utilized a three-phase approach:

1. Phase I—It reviews the various literature on the subject area of scheduling practices, current practices, and methods to determine contract time that are used by the construction industry in both developed countries and developing countries.
2. Phase II—Based on the information acquired from literature review, a survey questionnaire was designed.
   (a) A questionnaire was developed for consultant and contractor organizations. From the questionnaires given to consultants, the current practice or standard procedure used to determine contract time in construction projects was analyzed. The factors or circumstances that lead to the appropriate allocation of contract time, which itself leads to delays in the project, and the factors that critically affect contract time in the construction phase were assessed. The questionnaires distributed to contractor organization aimed to determine the productivity rates of controlling or critical activities for public building construction. The questionnaires' survey data were collected from resident engineers, contract administration engineers, and office engineers.
   (b) Projects were classified on the basis of the project type, and controlling or critical activities were identified for different types of projects.
   (c) According to the data that were collected from the questionnaire, an analysis was carried out. In the analysis, a regression model was developed to estimate the new productivity rate for controlling and critical activities at different contractor organizations.
   (d) A case study was conducted to estimate the duration (working days) of selected projects on the basis of estimated quantities and predicted production rates; next, the duration was entered into Microsoft Project, which scheduled the project to determine the contract time of the projects as a number of working days.
3. Phase III—A recommendation was made for a contract time determination procedure on the basis of the outcomes of Phases I and II.

21.2.2 Study Area

This research focused on the determination of contract time procedures for public building construction and factors that critically affect contract time during the construction phase. The projects that were considered in this research are physical public building projects at Bahir Dar University.

21.2.3 Sample Size/Research Population

The total population of contractors involved in physical projects at Bahir Dar University is 19, and the total population of consultant organizations is 40.

Sampling is the process of selecting a number of study units from a defined study population. Often, research focuses on a large population for practical reasons, but only some of its members can be included in the investigation. Hence, a sample was drawn from the total population by using the sample size formula in [9], which has a 94% confidence level (Eq. 21.1):

$$n = n' / \left(1 + \left(n' / N\right)\right) \tag{21.1}$$

where

- N = the total number of populations.
- n = the sample size from a finite population.
- n' = the sample size from an infinite population = $\frac{S2}{V2}$, where S is the variance of the population elements and V is a standard error of sampling a population (usually, $S = 0.5$ and $V = 0.06$).

For the 19 contractors,

$n' = \frac{S2}{V2} = \frac{0.52}{0.06} = 69.44.$

$N = 15$, the total number of contractors.

$n = 69.44 / (1 + (69.44 / 19)) = 15.$

For the 40 consultants,

$n' = \frac{S2}{V2} = \frac{0.52}{0.06} = 69.44.$

$N = 40$, the total number of consultants.

$n = 69.44 / (1 + (69.44 / 40)) = 25.$

As much as possible, attempts were made to draw representative samples from the study populations. Table 21.1 presents the number of sample contractors and consultants to whom the questionnaire was given.

The questionnaires were directed to those consultant professionals who have experience with contract administration and contract time determination systems, and the contractors were mainly office engineers who have experience with

Table 21.1 Respondent distribution rate

Name of party	Questionnaires distributed	Number of respondents	Response rate (%)
Consultants	25	17	68
Contractors	15	8	53.3
Total	40	25	62.5

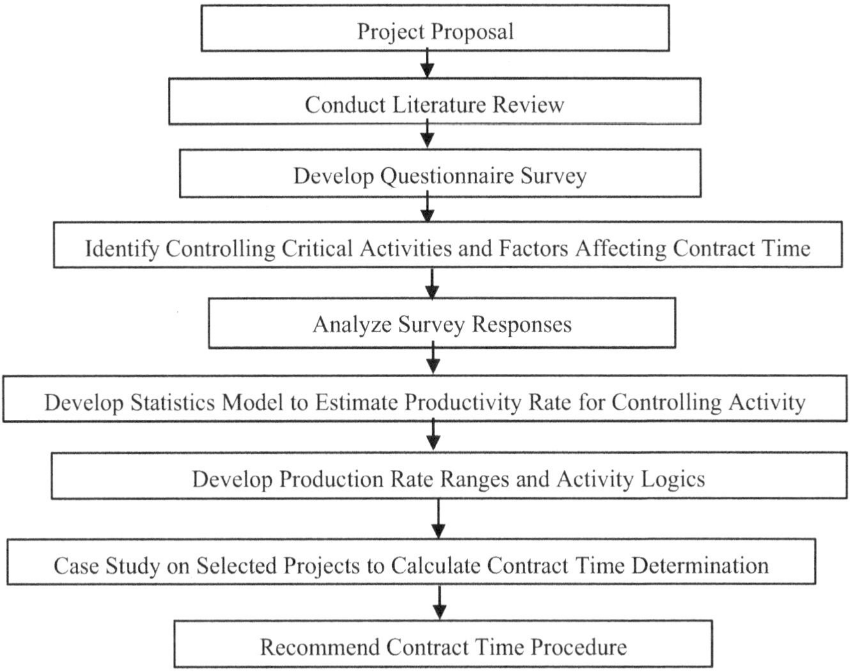

Fig. 21.1 Flowchart of study design

productivity rates for different controlling activities. The contractors included were all category 1 and general contractors. The respondents included in the survey comprised a total of 25, of which eight were contractors and 17 consultants.

The study design of this research is illustrated via a flowchart, as shown in Fig. 21.1.

21.3 Results and Discussion

21.3.1 Current Practice of Contract Time Determination Procedures

According to the respondents, eight consultants—or 47.05% of the consultants—have no standard procedure to determine contract time, whereas the rest (nine)—or 52.95% of consultants—do have a standard procedure to determine contract time.

Of the 52.95% of respondents, six consultants use a standard empirical formula to determine contract time. This empirical formula depends on project costs and different factors that affect construction, such as accessibility, material availability, labor and equipment availability, and weather conditions, as shown in Eq. 21.2:

$$\text{Contract time} = \left[\sqrt[3]{\text{project cost}/10\wedge 6}\right](\text{factors}) * 30 \tag{21.2}$$

where

Project cost is the total cost of the project.
Factors are within a range of 1 to 5, a range that is based on the hardship of the site, the accessibility of material delivery, and labor and equipment availability.

Three consultants use in-house templates in Microsoft Excel to estimate production rates and determine contract time. Although the empirical formula is a better system than no system, determining how the equation was derived (no one can justify the accuracy) and how the factor ratings were evaluated was difficult.

The overall current practice in determining contract time is based more on project costs and the productivity rate. More than half of consultant organizations use an empirical formula, which is based on project costs.

This empirical formula is limited by its being based on preliminary project costs and by its not considering cost inflation or other factors, which hinders estimations of a realistic contract time. On the contrary, some consultant organizations use productivity rates to estimate contract time.

21.3.2 Accuracy of Contract Time

According to the survey, 23.5% of consultant and contractor respondents believe that the contract time given by clients/owners or consultants for contractors are reasonable and allows them to execute projects without delay. However, the majority of consultant and contractor respondents (76.5%) believe that the contract time awarded by the client or consultant is not realistic or reasonable, leading to project delays.

Experts have identified different circumstances or factors to explain the inappropriate allocation of contract time. These typically include the following: Clients are too ambitious to quickly complete projects to provide public or business services; contractors make false promises to their clients that they will not exceed the contract agreement; and some consultants use an empirical formula that is based solely on project costs, without giving careful consideration to all activities, when they determine contract time.

Table 21.2 Top-five factors affecting contract time

No.	Factors affecting contract time	RII	Rank
1.	Design change by owners or variation order	0.760	1st
2.	Material delivery time	0.688	2nd
3.	High inflation or increase in price	0.664	3rd
4.	Weather conditions for substructure and superstructure	0.648	4th
5.	Defective design by designers during the construction phase	0.616	5th

21.3.3 Factors Affecting Contract Time

According to [10], contract time can be estimated by using different methods or techniques. However, contract time remains inaccurate unless it factors in project-specific factors and adjustments due to the unique nature of construction projects [11]. By using the data from the survey responses, the factors affecting the contract times given by consultants and contractors were identified and ranked [12]. Respondents were requested to rank the factors by using a five-point Likert scale. The five-point Likert scale was then converted to define a value on the Relative Importance Index (RII) for each individual factor by using Eq. 21.3:

$$RII = \sum \mathrm{W} / \left(\mathrm{H}^{*}\, \mathrm{N} \right) \tag{21.3}$$

Table 21.2 shows the top-five factors that affect contract time according to a consultant and contractor survey.

21.3.4 Linear Regression Models for Productivity Rate

Eight public building contractors participated in this study. Importantly, the study focuses on public projects, in particular physical projects at Bahir Dar University. In the questionnaire, 22 civil works that fall along the critical path, which is based on selected projects, were selected, including earthwork, concrete work, formwork, masonry works, hollow concrete block works, plastering works, and painting works. Contractors were expected to fill the productivity rates of these selected activities and their respective workforce assignments. In this assignment, contractors were given three scenarios—featuring low, medium, and maximum productivity rates and workforce assignments, respectively—for their project sites. These three scenarios are based on several assumptions and interviews with experts.

1. Scenario 1: Low productivity rates on construction sites are typically experienced during religious holidays (not calendar holidays), where workers are not present on-site. These include days such as the 12th, 19th, and 21st of every month, which are Saint Michael, Saint Gabriel, and Saint Mariam holidays, respectively. In addition, low workforce values are recorded during poor weather conditions (e.g., raining for half a day to a full day).

2. Scenario 2: Medium productivity values are assumed according to the regular or average productivity rate data used for construction projects, assuming only minor obstacles.
3. Scenario 3: The maximum workforce is assumed to be present on time-constrained construction projects where activities must be finished within a short period of time. In this scenario, contractors utilize as many daily laborers and other skilled personnel, such as masons, carpenters, and plasterers, to expedite the schedule, assuming that the site conditions are favorable.

This study considers labor or workforce as the only factor in estimating productivity rates, excluding equipment such as mixers, vibrators, and other machinery. The 22 activities fall along the critical path needed to finish buildings according to the selected projects. Thus, as part of developing linear regression models, in this data analysis, the productivity rate is considered as the dependent variable, whereas the workforce is taken as the independent variable.

The survey yielded 24 observations for the dependent and independent variables, for each of the 22 activities. Therefore, the study yielded a total of 528 observations.

The linear relationship for developing a regression model can be calculated by using Eq. 21.4:

$$Y = \beta o + \beta_1 X_1 + \beta_2 X_2 \tag{21.4}$$

where

Y is the dependent variable (the productivity rate for the selected activities).
X_1 and X_2 are the known independent variables (including the workforce, such as daily laborers, masons, carpenters, plasterers, etc.).
$\beta_{0,}$ β_1, and β_2 are partial regression coefficients.

In this study, a stepwise regression analysis was carried out to investigate the relationship between the independent variables and the dependent variable. By using Microsoft Excel, the line was regressed to fit the sample data and run computations, where the values of the coefficient of the regression are in Eq. 21.4. Next, the coefficient of determination (R^2) was used for each activity to measure the goodness of fit of the models derived from the empirical data. The coefficient of determination (R^2) can be calculated by using Eq. 21.5:

$$R^2 = 1 - \frac{\text{Sum of Squared Errors}}{\text{Total Sum of Squares}} \tag{21.5}$$

The predicted productivity rate models for the selected critical civil work activities, along with their coefficient of determination, are summarized in Table 21.3.

Table 21.3 Summary of predicted productivity rate models for selected critical civil work activities

No.	Selected civil work activities	Predicted productivity rate according to linear regression	R^2
1.	Site clearance	Productivity rate = 1.39 + 6.68 (daily laborer)	0.89
2.	Bulk excavation	Productivity rate = 2.32 + 1.212 (daily laborer)	0.93
3.	Backfill work	Productivity rate = 1.92 + 1.24 (daily laborer)	0.81
4.	Hard coring 25 cm thick	Productivity rate = 5.48 + 1.45 (daily laborer) + 7.04 (mason)	0.93
5.	Lean concrete	Productivity rate = 1.75 + 3.12 (daily laborer) – 1.81 (mason)	0.98
6	Footings	Productivity rate = 0.68–0.15 (daily labor) + 6.68 (mason)	0.75
7.	Foundation column	Productivity rate = 0.0735 + 0.562 (daily laborer) – 0.086 (mason)	0.66
8.	Grade beam	Productivity rate = −0.812 + 0.82 (daily laborer) – 0.18 (mason)	0.79
9.	Elevation column	Productivity rate = −0.52 + 0.24 (daily laborer) + 1.18 (mason)	0.70
10.	Floor slab formwork	Productivity rate = 0.183 + 0.834 (daily laborer) – 0.69 (ass. carpenter) – 0.23 (carpenter)	0.93
11	Floor slab	Productivity rate = – 2.19 + 2.28 (daily laborer) + 0.25 (mason)	0.92
12.	Top tie beams	Productivity rate = −0.34 + 0.39 (daily laborer) + 1.88 (mason)	0.90
13.	Stone masonry substructure	Productivity rate = 2.31–2.41 (daily laborer) + 10.95 (mason)	0.78
14.	Stone masonry superstructure	Productivity rate = – 0.47 + 1.52 (daily laborer) – 1.89 (mason)	0.88
15	20 cm thick HCB	Productivity rate = −2.46 – 3.42 (daily laborer) + 15.79 (mason)	0.76
16.	15 cm thick HCB	Productivity rate = −0.43 + 6.24 (daily laborer) – 4.77 (mason)	0.86
17	10 cm thick HCB	Productivity rate = −2.58 + 0.25 (daily laborer) + 8.23 (mason)	0.81
18	Fixing EGA sheet joining	Productivity rate = 2.90 + 8.90 (daily laborer) + 11.46 (carpenter)	0.78
19	2nd coat plastering	Productivity rate = −2.61 + 7.68 (daily laborer) – 3.13 (plasterer)	0.66
20	Fine-coat plastering	Productivity rate = −3.34 – 5.72 (daily laborer) + 29.74 (plasterer)	0.73
21	Cement screed	Productivity rate = 5.27 + 0.185 (daily laborer) + 12.00 (mason)	0.93
22	Painting	Productivity rate = 9.41 + 30.43 (daily laborer) – 35.82 (painter)	0.85

HCB Hollow Concrete Block; *EGA* Elegant Galvanized Aluminium

21.3.5 Model Mean Validation

Tables 21.3, 21.4, and Fig. 21.1 compare the developed models' productivity on the basis of its mean value with the Ministry of Urban Housing Development's productivity rates, which are used along with the average workforce assignment as benchmarks or for validation. The figure and tables test the validity of the modeled productivity rates with those of the Ministry of Urban and Housing Development. According to a *t*-test, the difference between the two productivity rate means was examined, and the results yielded a value of 0.141. This result is close to zero, and the difference between theirs means is insignificant. Thus, the predicted productivity rate value of the regression model does not diverge significantly from the observed productivity rate given by the Ministry of Urban and Housing Development, as shown on Table 21.4.

Table 21.4 Comparison of modeled predicted productivity rate with productivity rates from the Ministry of Urban and Housing Development

No.	Activity	Workforce assignment	Productivity rate from MoUHD	Modeled predicted productivity rate
1	Site clearance	1 daily laborer	10 m^2/day	8.07 m^2/day
2	Bulk excavation	1 daily laborer	1.5 m^2/day	3.53 m^2/day
3	Backfill work	1 daily laborer	2 m^2/day	3.16 m^2/day
4	Hard coring 25 cm thick	1 mason + 4 daily laborers	18 m^2/day	18.32 m^2/day
5	Lean concrete	1 mason + 6 daily laborers	20 m^2/day	18.66 m^2/day
6	Footings	1 mason + 6 daily laborers	5.5 m^3/day	6.46 m^3/day
7	Foundation column	1 mason + 6 daily laborers	2.75 m^3/day	3.36 m^3/day
8	Grade beam	1 mason + 6 daily laborers	5.5 m^3/day	3.93 m^3/day
9	Elevation column	1 mason + 6 daily laborers	2.75 m^3/day	2.1 m^3/day
10	Floor slab formwork	1 carpenter + 1 assistant carpenter + 2 daily laborers	0.5 m^2/day	0.9 m^2/day
11	Floor slab	1 mason + 6 daily laborers	10 m^2/day	11.74 m^2/day
12	Top tie beams	1 mason + 6 daily laborers	5.5 m^3/day	3.88 m^3/day
13	Stone masonry substructure	1 mason + 3 daily laborers	5 m^3/day	6.03 m^3/day
14	Stone masonry superstructure	1 mason + 3 daily laborers	3 m^3/day	2.2 m^3/day
15	20 cm thick HCB	1 mason + 2 daily laborers	10 m^2/day	6.49 m^2/day

Table 21.4 (continued)

No.	Activity	Workforce assignment	Productivity rate from MoUHD	Modeled predicted productivity rate
16	15 cm thick HCB	1 mason + 2 daily laborers	10 m^2/day	7.28 m^2/day
17	10 cm thick HCB	1 mason + 2 daily laborers	8 m^2/day	6.15 m^2/day
18	Fixing EGA sheet joining	1 carpenter + 2 daily laborers	30 m^2/day	32.16 m^2/day
19	2nd coat plastering	1 plasterer + 2 daily laborers	10 m^2/day	9.62 m^2/day
20	Fine-coat plastering	1 plasterer + 2 daily laborers	14 m^2/day	14.97 m^2/day
21	Cement screed	1 mason + 2 daily laborers	14 m^2/day	17.64 m^2/day
22	Painting	1 painter + 2 daily laborers	30 m^2/day	34.45 m^2/day

HCB Hollow Concrete Block; *EGA* Elegant Galvanized Aluminium

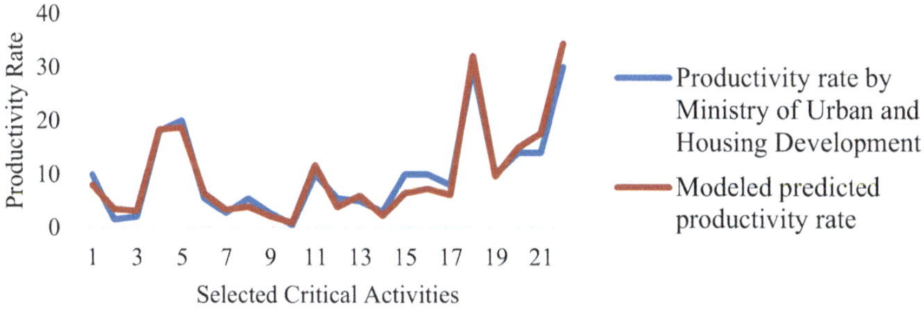

Fig. 21.2 Modeled predicted productivity rates and productivity rates given by the Ministry of Urban and Housing Development for the selected critical activities

Table 21.5 Mean difference between modeled predicted productivity rate and productivity rate given by the Ministry of Urban and Housing Development

	Productivity rate from Ministry of Urban and Housing Development	Modeled predicted productivity rate	Mean difference
Mean	9.90	10.05	0.140

For the selected critical activities, the modeled predicted productivity rates and the productivity rates given by Ministry of Urban and Housing Development are shown in Fig. 21.2.

The mean difference between the modeled predicted productivity rate and the productivity rate given by the Ministry of Urban and Housing Development is shown in Table 21.5.

21.4 Case Study

A case study was conducted on two building projects to estimate the duration of activities by using estimated quantities, lists of quantities, and the production rates yielded by the prediction models. On the basis of the productivity rates, the duration is calculated and entered into Microsoft Project to schedule the project and thus determine the contract time of projects as a number of working days [13].

21.4.1 General Overview of Selected Projects

The building projects selected for this case study are two physical projects located in Bahir Dar University. Project 1 is EITEX LOT III, a student cafeteria, found on the premises of the Ethiopian Institute of Textile and Fashion Design, and project 2 is Poly Library LOT 1, found on the campus of the Institute of Technology, which are shown in Tables 21.6 and 21.7, respectively.

Table 21.6 Duration of activities determined by using average workforce assignment data and the modeled productivity rate for project 1

No.	Activity	Average workforce assignment	Modeled productivity rate	Production rate	Total quantities	Total duration
1	Site clearance	30 DL	PR = 1.39 + 6.68 (DL)	202 m^2/day	5400 m^2	27 days
2	Bulk excavation	69 DL	PR = 2.32 + 1.212 (DL)	86 m^3/day	5500 m^3	64 days
3	Lean concrete	5 M + 31 DL	PR = 1.75 + 3.12 (DL) – 1.81 (M)	89 m^2/day	1280 m^2	15 days
4	Footing concrete	4 M + 28 DL	PR = 0.68–0.15 (DL) + 6.68 (M)	23 m^3/day	841 m^3	37 days
5	Foundation column	5 M + 32 DL	PR = 0.0735 + 0.562 (DL) – 0.086 (M)	18 m^3/day	120 m^3	7 days
6	Backfill work	61 DL	PR = 1.92 + 1.24 (DL)	78 m^3/day	5759 m^3	80 days
7	Stone masonry substructure	5 M + 15 DL	PR = 2.31–2.41 (DL) + 10.95 (M)	21 m^3/day	305 m^3	15 days
8	Stone masonry superstructure	5 M + 16 DL	PR = – 0.47 + 1.52 (DL) – 1.89 (M)	14 m^3/day	160 m^3	12 days
9	Grade beams	6 M + 30 DL	PR = −0.812 + 0.82 (DL) – 0.18(M)	23 m^3/day	231 m^3	10 days
10	Ground elevation column	6 M + 33 DL	PR = −0.52 + 0.24 (DL) + 1.18 (M)	14 m^3/day	172 m^3	13 days

(continued)

Table 21.6 (continued)

No.	Activity	Average workforce assignment	Modeled productivity rate	Production rate	Total quantities	Total duration
11	Hard coring 25 cm thick	5 M + 20 DL	PR = 5.48 + 1.45 (DL) + 7.04 (M)	69 m^2/day	5141 m^2	87 days
12	Floor slab formwork	26 C + 25 Ass. C + 53 DL	PR = 0.183 + 0.834 (DL) – 0.69 (ass. C) – 0.23(C)	22 m^2/day	1880 m^2	86 days
13	Floor slab	7 M + 41 DL	PR = – 2.19 + 2.28 (DL) + 0.25 (M)	93 m^2/day	474 m^3	17 days
14	1st floor elevation columns	6 M + 33 DL	PR = −0.52 + 0.24 (DL) + 1.18 (M)	14 m^3/day	86 m^3	6 days
15	Top tie beams	6 M + 34 DL	PR = −0.34 + 0.39 (DL) + 1.88 (M)	24 m^3/day	196 m^3	8 days
16	Fixing EGA sheet joining	9 C + 19 DL	PR = 2.90 + 8.90 (DL) + 11.46 (C)	275 m^2/day	4565 m^2	17 days
17	20 cm thick HCB	8 M + 17 DL	PR = −2.46 – 3.42 (DL) + 15.79 (M)	66 m^2/day	2108 m^2	32 days
18	15 cm thick HCB	9 M + 18 DL	PR = −0.43 + 6.24 (DL) – 4.77 (M)	69 m^2/day	152 m^2	3 days
19	10 cm thick HCB	8 M + 16 DL	PR = −2.58 + 0.25 (DL) + 8.23 (M)	67 m^2/day	130 m^2	2 days
20	2nd coat plastering	10 Pl + 20 DL	PR = −2.61 + 7.68 (DL) – 3.13 (Pl)	120 m^2/day	10,734 m^2	90 days
21	Fine-coat plastering	9 Pl + 19 DL	PR = −3.34 – 5.72 (DL) + 29.746 (Pl)	156 m^2/day	8878 m^2	57 days
22	Painting	7 P + 14 DL	PR = 9.41 + 30.43 (DL) – 35.82 (P)	185 m^2/day	11,038 m^2	60 days
23	Cement screed	12 M + 25 DL	PR = 5.27 + 0.185 (DL) + 12.00 (M)	154 m^2/day	3215 m^2	21 days

PR Productivity Rate; *DL* Daily Laborer; *M* Masson; *C* Carpenter; *Pl* Plasterer; *P* Painter; *Ass.C* Assistance Carpenter

Table 21.7 Duration of activities determined by using average workforce assignment data and the modeled productivity rate for project 2

No.	Activity	Average workforce assignment	Modeled productivity rate	Production rate	Total quantities	Total duration
1	Site clearance	30 DL	PR = 1.39 + 6.68 (DL)	202 m^2/day	3200 m^2	16 days
2	Bulk excavation	69 DL	PR = 2.32 + 1.212 (DL)	86 m^3/day	3836 m^3	45 days
3	Lean concrete	5 M + 31 DL	PR = 1.75 + 3.12 (DL) – 1.81 (M)	89 m^2/day	1020 m^2	12 days
4	Footing concrete	4 M + 28 DL	PR = 0.68–0.15 (DL) + 6.68 (M)	23 m^3/day	838 m^3	37 days

(continued)

Table 21.7 (continued)

No.	Activity	Average workforce assignment	Modeled productivity rate	Production rate	Total quantities	Total duration
5	Foundation column	5 M + 32 DL	PR = 0.0735 + 0.562 (DL) – 0.086 (M)	18 m^3/day	92 m^3	6 days
6	Backfill work	61 DL	PR = 1.92 + 1.24 (DL)	78 m^3/day	4183 m^3	54 days
7	Stone masonry substructure	5 M + 15 DL	PR = 2.31–2.41 (DL) + 10.95 (M)	21 m^3/day	240 m^3	12 days
8	Stone masonry superstructure	5 M + 16 DL	PR = – 0.47 + 1.52 (DL) – 1.89 (M)	14 m^3/day	189 m^3	14 days
9	Grade beams	6 M + 30 DL	PR = –0.812 + 0.82 (DL) – 0.18 (M)	23 m^3/day	82 m^3	4 days
10	Ground elevation column	6 M + 33 DL	PR = –0.52 + 0.24 (DL) + 1.18 (M)	14 m^3/day	38 m^3	14 days
11	1st floor formwork	26 C + 25 Ass. C + 53 DL	PR = 0.183 + 0.834 (DL) – 0.69 (Ass. C) – 0.23(C)	22 m^2/day	1702.5 m^2	74 days
12	1st floor slab and beams	7 M + 41 DL	PR = – 2.19 + 2.28 (DL) + 0.25 (M)	93 m^2/day	315 m^3	23 days
13	1st floor elevation column	6 M + 33 DL	PR = –0.52 + 0.24 (DL) + 1.18 (M)	14 m^3/day	45 m^3	4 days
14	2nd floor formwork	26 C + 25 Ass. C + 53 DL	PR = 0.183 + 0.834 (DL) – 0.69 (Ass. C) – 0.23(C)	22 m^2/day	1725.54 m^2	79 days
15	2nd floor slab and beams	7 M + 41 DL	PR = – 2.19 + 2.28 (DL) + 0.25 (M)	93 m^2/day	330.76 m^3	24 days
16	2nd floor elevation column	6 M + 33 DL	PR = –0.52 + 0.24 (DL) + 1.18 (M)	14 m^3/day	35 m	3 days
17	3rd floor formwork	26 C + 25 Ass. C + 53 DL	PR = 0.183 + 0.834 (DL) – 0.69 (Ass. C) – 0.23(C)	22 m^2/day	1806.66 m^2	82 days
18	3rd floor slab and beams	7 M + 41 DL	PR = – 2.19 + 2.28 (DL) + 0.25 (M)	93 m^2/day	339 m^3	25 days
19	3rd floor elevation column	6 M + 33 DL	PR = –0.52 + 0.24 (DL) + 1.18 (M)	14 m^3/day	40 m^3	3 days

(continued)

Table 21.7 (continued)

No.	Activity	Average workforce assignment	Modeled productivity rate	Production rate	Total quantities	Total duration
20	4th floor formwork	26 C + 25 Ass. C + 53 DL	PR = 0.183 + 0.834 (DL) – 0.69 (Ass. C) – 0.23(C)	22 m^2/day	1828.54 m^2	83 days
21	4th floor slab and beams	7 M + 41 DL	PR = – 2.19 + 2.28 (DL) + 0.25 (M)	93 m^2/day	222.7 m^3	16 days
22	4th floor elevation column	6 M + 33 DL	PR = −0.52 + 0.24 (DL) + 1.18 (M)	14 m^3/day	34 m^3	3 days
23	5th floor formwork	26 C + 25 Ass. C + 53 DL	PR = 0.183 + 0.834 (DL) – 0.69 (Ass. C) – 0.23(C)	22 m^2/day	1897 m^2	87 days
24	5th floor slab and beams	7 M + 41 DL	PR = – 2.19 + 2.28 (DL) + 0.25 (M)	93 m^2/day	337.3 m^3	25 days
25	Hard coring 25 cm thick	5 M + 20 DL	PR = 5.48 + 1.45 (DL) + 7.04 (M)	69 m^2/day	1537 m^2	27 days
26	20 cm thick HCB	8 M + 17 DL	PR = −2.46 – 3.42 (DL) + 15.79 (M)	66 m^2/day	5234 m^2	79 days
27	15 cm thick HCB	9 M + 18 DL	PR = −0.43 + 6.24 (DL) – 4.77 (M)	69 m^2/day	121 m^2	3 days
28	10 cm thick HCB	8 M + 16 DL	PR = −2.58 + 0.25 (DL) + 8.23 (M)	67 m^2/day	603 m^2	9 days
29	2nd coat plastering	10 Pl + 20 DL	PR = −2.61 + 7.68 (DL) – 3.13 (Pl)	120 m^2/day	25,173 m^2	210 days
30	Fine-coat plastering	9 Pl + 19 DL	PR = −3.34 – 5.72 (DL) + 29.746 (Pl)	156 m^2/day	21,784 m^2	139 days
31	Painting	7 P + 14 DL	PR = 9.41 + 30.43 (DL) – 35.82 (P)	185 m^2/day	25,173 m^2	136 days
32	Cement screed	12 M + 25 DL	PR = 5.27 + 0.185 (DL) + 12.00 (M)	154 m^2/day	6047 m^2	40 days

PR Productivity Rate; *DL* Daily Laborer; *M* Masson; *C* Carpenter; *Pl* Plasterer; *P* Painter; *Ass.C* - Assistance Carpenter; *HCB* Hollow Concrete Block

Workforce assignment key:

DL—daily laborer	C—carpenter	Pl—plasterer
M—mason	Ass. C—assistant carpenter	P—painter

Tables 21.6 and 21.7 show the total duration of each activity on the basis of the average workforce assignment, the modeled productivity rates, and the total quantities of work items.

The total duration for project 1 and that for project 2 are estimated from Tables 21.5 and 21.6, respectively, which are entered into Microsoft Project to schedule the project and determine the contract time of the project as a number of working days. This case study was carried out by using Microsoft Project 2007. In this study, the number of working days each week is 6, and the number of working hours each day is 8. This study considers calendar days off, which are Christian and Muslim calendar holidays and other governmental holidays.

According to the Microsoft Project output for project 1, the time elapsed is 231 working days for substructure work and 304 for superstructure work, for a total of 535 working days. When we compare the original contract time, which was estimated via an empirical formula as 420 working days, to the new contract time, which in this case study was 535 working days, we find a 115 working day difference.

According to the Microsoft Project output for project 2, the time elapsed is 166 working days for substructure work and 438 for superstructure, for a total of 604 working days. When we compare the original contract time, estimated via an empirical formula as 475 working days, to the new contract time, which in this case study was 604 working days, we find a 129 working day difference.

Through survey responses from and interviews with engineers responsible for determining contract time using an empirical formula, which was described in Eq. 21.1, we learned that they had no clue about the derivation or updating of this empirical formula. They also mentioned that this empirical formula is not updated in response to currency inflation. Accordingly, we can conclude that the original contract time estimation using this empirical formula for this project is not sufficient to finish construction, so this project will be delayed due to insufficient contract time and other factors that affect contract time during the construction phase.

21.5 Conclusion

According to our findings, some consultant organizations have a standard procedure to determine contract time, using an empirical formula. This empirical formula depends on project costs and factors affecting contract time. During our interview analysis, we found that consultant engineers had no clue about the derivation of this empirical formula, and they also said that no adjustment to or updates on the project cost are made to account for currency inflation. For these reasons, realistic contract times are not being calculated by this empirical formula.

On the accuracy of the contract times that are given by clients and consultant organizations, the majority of consultants and contractors believe that these contract times are not reasonable and instead lead to construction delays. Different circumstances affect the determination of nonaccurate contract times: Typically, clients are too ambitious to quickly complete projects to provide public or business services; contractors make false promises to their clients that they will not exceed the contract agreement; and some consultants use an empirical formula that is based solely on project costs, without giving careful consideration to all activities, when they determine contract time.

Factors affecting contract time during the construction phase were identified, assessed, and ranked on the basis of the survey responses. The factors affecting contract time are ranked as follows in descending order: design changes made by owners or variation orders, material delivery times, high inflation or increased prices, weather conditions for the substructure and superstructure, and defective designs made by designers during the design phase.

A linear regression model was developed to predict the new productivity rate for the selected critical civil work activities by using the data obtained from contractor organizations. In this analysis, the productivity rate was considered as the dependent variable, whereas workforce was taken as the independent variable. Afterwards, a case study was carried out by using modeled productivity rates, the average workforce (determined from contractor data), and the total quantities of materials to estimate the duration of each critical activity. Next, the duration was entered into Microsoft Project, which scheduled the project to determine the contract time of the project as a number of working days, and through this analysis and case studies, a new procedure for determining contract times for building projects was introduced.

According to the two case studies, the original contract times given by clients to contractors are shorter than the new contract times, which are drawn from Microsoft Project. Thus, we can conclude that the original contract times, which were estimated via an empirical formula, were not sufficient to finish construction.

We conclude that realistic contract times are not drawn from the practices and methods that are used to determine contract times. We therefore recommend that the clients and consultants of building projects give realistic and reasonable contract times to contractors and that consultant organizations use updated production rates for contract time determination systems. Changes in orders or works are the major factors that affect contract time, so clients or owners should emphasize that contractors not change orders, and they should allot additional time to contractors if the contractors change an order.

This research should be supplemented by conducting studies on the following:

- The method and procedure for time extension in the event of delays to the project
- The method and procedure for contract time determination for highway works and other horizontal projects
- Modeling production rates by using other machinery and equipment, such as earthwork machinery and concrete work machines as the independent variable, and by considering the factors affecting contract time to better predict productivity rates.

References

1. Mechegiaw, L. (2012). *Performance study of lowest bidder bid awarding system in public construction projects*. PhD diss., MS thesis, Addis Ababa Univ, Ethiopia.
2. Akadiri, P. O., Chinyio, E. A., & Olomolaiye, P. O. (2012). Design of a sustainable building: A conceptual framework for implementing sustainability in the building sector. *Buildings, 2*(2), 126–152.
3. Lahdenperä, P. (2012). Making sense of the multi-party contractual arrangements of project partnering, project alliancing and integrated project delivery. *Construction Management and Economics, 30*(1), 57–79.
4. Saeed, S. A. A. (2009). *Delay to projects-cause, effect and measures to reduce/eliminate delay by mitigation/acceleration*. PhD diss., British University in Dubai.
5. Scott, S., & Harris, R. A. (2004). United Kingdom construction claims: Views of professionals. *Journal of Construction Engineering and Management, 130*(5), 734–741.
6. De Reyck, B., & Leus, R. (2008). R&D project scheduling when activities may fail. *IIE Transactions, 40*(4), 367–384.
7. Zewdu, Z. T., & Aregaw, G. T. (2015). Causes of contractor cost overrun in construction projects: The case of Ethiopian construction sector. *International Journal of Business and Economics Research, 4*(4), 180–191.
8. Taylor, T. R. B., Goodrum, P. M., Brockman, M., Bishop, B., Shan, Y., Sturgill, R. E., & Hout, K. (2013) *Updating the Kentucky contract time determination system.*
9. Kish, L., & Frankel, M. R. (1974). Inference from complex samples. *Journal of the Royal Statistical Society Series B: Statistical Methodology, 36*(1), 1–22.
10. Atreya, S. K. (2007). *Development of an improved system for contract time determination.* Oklahoma State University.
11. Gondy, C., & Hildreth, J. (2007). *Contract time determination guidelines.*" .
12. Yadeta, A. E. (2014). Assessing the impact of variation orders on public building projects in Addis Ababa. Unpublished Masters Project, Addis Ababa: Addis Ababa University.
13. Guntuk, C. R., & Koehn, E. (2009). Construction productivity and production rates: Developing countries. In *Challenges, opportunities and solutions in structural engineering and construction* (pp. 709–714). CRC Press.

Chapter 22
Assessment of Selected Physicochemical Parameters of Selamco Dam Water, South Gondar, Ethiopia

Menber Berihun Nigatu, Mengistu Berihun Nigatu, Minaleshewa Atlabachew, Mequanint Demeke Aynalem, and Kindie Bitew Worku

22.1 Introduction

Water quality is vital for human, animal, industrial, and agricultural survival. Agriculture success is greatly dependent on the quality of water used in a farming area. The use of hazardous quality water affects agriculture land/soil and damages crop output in a variety of ways [1]. Good quality crops can be grown when irrigation water is of good quality [2]. The characteristics of irrigation water that characterize its quality vary depending on the source of the water. Water from various sources contains a substantial proportion of unwanted or undesirable substance dissolved in water as a result of natural or man-made activities [3]. It is not an easy task to say water is good or not, scientific measurements are necessary. Depending on their requirements, water may be tested for a few features or for a variety of natural substances and pollutants. Several physical, chemical, and biological characteristics define the nature and degree of water contamination [4]. Irrigation water's chemical components may have a direct influence on plant growth through toxicity or nutrient deficiency, or they may have an indirect effect by modifying the plants' availability to nutrients [5]. Heavy metals are elements with a density larger than 5 g/cm^3 [6]

M. B. Nigatu (✉) · M. B. Nigatu · K. B. Worku
Faculty of Civil and Water Resources Engineering, Bahir Dar Institute of Technology, Bahir Dar University, Bahir Dar, Ethiopia

M. Atlabachew
Faculty of Civil and Water Resources Engineering, Bahir Dar Institute of Technology, Bahir Dar University, Bahir Dar, Ethiopia

Department of Chemistry, College of Science, Bahir Dar University, Bahir Dar, Ethiopia

M. D. Aynalem
Faculty of Chemical and Food Engineering, Bahir Dar Institute of Technology, Bahir Dar University, Bahir Dar, Ethiopia

S. W. Fanta et al. (eds.), *Sustainable Development Research in Manufacturing, Process Engineering, Green Infrastructure, and Water Resources*, Green Energy and Technology, https://doi.org/10.1007/978-3-031-77339-6_22

that cause a variety of diseases and, with the exception of low tolerated dosages [7]. Because trace metals are harmful, determining the levels of metallic trace elements in water and soil is important [8]. The physical, biological, and chemical features of soil are directly and indirectly impacted by irrigation water quality and quantity, particularly in arid and semi-arid locations where irrigation is regrettably necessary [9]. There are seasonal and yearly variations in water quality and quantity because of different factors [10]. Water quantity change has a direct impact on irrigation scheme production in the short term, and water quality has a longer-term impact on productivity [11]. It is necessary to test the irrigation water prior to selecting the place and crops to be produced [12]. The determination of physical and chemical characteristics of the irrigation water have particular importance. The types of ions present in the irrigation water can affect the soil's physical and mechanical qualities, including the degree of soil particle dispersion, aggregate stability, soil structure, and permeability [13]. In Ethiopia, it is common to utilize water for irrigation purpose without checking the quality of water and proper treatment. Although there is a large gap in a systematic water quality characterization of water bodies in Lake Tana basin [14], some previous studies show that there is a daily decrease in water quality of Lake Tana [15] which is caused by different natural and anthropogenic activities in its sub basins. Ribb reservoir which is found in the sub basin of Lake Tana was characterized by low water quality status [16]. However, this research [16] did not assess the heavy metal concentration in the reservoir. Selamco River is one of the Tributaries of Ribb River. It originates from Debre Tabour Town where there is an accumulation of solid and liquid wastes dumping. Those dumps may lead to the contamination of the river in its downstream. Downstream of the town, in Selamco Kebele, there is an artificial reservoir used for irrigation purpose. Therefore, this study is aimed to determine the concentration of some selected physicochemical parameters in the reservoir due to budget restrictions. These are pH, temperature, electrical conductivity (EC), turbidity and total dissolved solids (TDS) from the physical parameters and heavy metals like chromium, manganese, nickel, zinc, cadmium, and lead from chemical parameters.

22.2 Methodology

22.2.1 Study Area Description

The study area is found in Amhara National Regional State, South Gondar Zone at the command area of Selamco Kebele in Farta district as shown in Fig. 22.1. Selamco micro earth dam was constructed by Co-SERAR in 2005. The dam is 21 m high and 216 m long with a top/crest width of about 4.5 m. Its range is categorized under medium size reservoirs based on height.

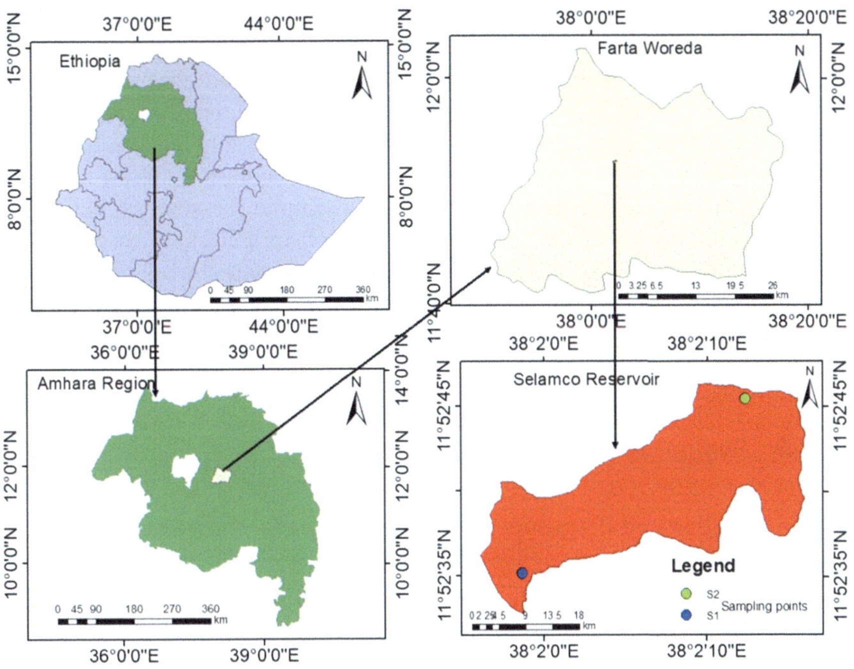

Fig. 22.1 Study area location

22.2.2 *Apparatus and Chemicals*

Apparatuses like plastic bottle (polyethylene), micropipettes, measuring cylinder, conical flask, beaker, volumetric flask, funnel, pH meter (model tochpro II), electrical conductivity/TDS meter (model CE 470cond meter 01189), thermometer, Whitman no.40 filter paper, hotplate hood, inductively coupled plasma optical emission spectroscopy (ICP-OES) (8000, perkin Elmer, Germany) were used to conduct the experiment. And chemicals used are Nitric acid, HNO3 (69%) (Spectrosol, BDH, England), and standard solutions of heavy metals Mn, Ni, Zn, Cd, Cr (VI), and Pb.

22.2.3 *Sample Collection*

For chemical analysis, polyethylene bottles were used to hold the water samples. The prewashed bottles were initially rinsed three times with water sample. Two points were carefully chosen to get representative water for analyzing selected physicochemical parameters as well as some trace metal concentrations of Selamco irrigation water at places where water enters the dam and is reserved (inlet) and water leaves out from the dam (outlet). At each sampling location, a composite water sample was obtained by mixing the grab samples taken from the reservoir's top, middle, and bottom at a depth of 1 meter by dipping bottles directly into the dam and handling them by rope. Hence, water samples from the two sampling sites were collected separately. About 1 L of composite water samples were collected twice, that is both in the dry and wet seasons at each sampling site. So the results in this study could be regarded as seasonal. The first sampling session was before the rainy season during the dry season in the last week of April 2021. The second sampling session was after rainfall during the wet season on the first week of July 2021. The samples were given a letter and number designation, such as A1 for the inlet of the dry season, B1 for the outlet, A2 for the inlet, and B2 for the exit of the wet season.

22.2.4 *Sample Preparation*

Sample Digestion and Procedure of Analysis During sample collection, physical parameters like temperature, pH, conductivity, TDS, and turbidity were measured using portable instruments. Using a thermometer, the water temperature was measured in situ and recorded in degrees Celsius. Using a field conductivity meter, electrical conductivity and TDS were assessed. Using a portable digital pH meter, pH was determined. A portable turbidity meter was used to measure the turbidity. In the digestion of heavy metals, the water sample was prepared by taking 50 mL of each water sample pipette out to a 100 mL conical flask then 5 mL of conc. HNO3 (69%) was added to each sample and watch glass was used to cover the flask and kept in a hood to avoid contamination. Then the solution was boiled at 80–95 °C for 2 h and evaporated on a hotplate to about 20 mL until it became light-colored, and a clear solution was obtained [17]. Then all samples were taken away and cooled at room temperature. The solution was then filtered using filter paper, and the filtrate was then transferred to a 50 mL volumetric flask and diluted to the proper strength using distilled water. Parallel digestions of a reagent blank were also conducted using the same digestion settings as the samples. The digestion procedure of spiked and unspiked water samples was similar. ICP-OES was then used to analyze the digested and filtered samples (both spiked and unspiked), blank samples, and standard solutions to determine the concentration of metals in the sample.

Standard Preparation and Calibration of the Instrument For the creation of working standards and intermediate standards, standard stock solutions of 1000 mg/L were employed. The dilution procedure was used to prepare the intermediate standards. In order to quantify the concentration of metals in a water sample, the working standard solutions were additionally created by properly diluting the intermediate standards with distilled water [18]. To create the working solutions for this investigation, intermediate standard solution (10 mg/L) was first made from stock standard solution (1000 mg/L). Using a micropipette, 1 mL of the stock solution was transferred into a 100 mL standard volumetric flask to create the 10 mg/L intermediate standard solution. The solution was then thoroughly mixed and diluted with deionized water to the required level before being used to prepare the working standards through serial dilution. For each metal, working standards of 0.05, 1.05, 2.05, 3.05, 4.05, and 5.05 mg/L were prepared.

22.2.5 Method Validation

Recovery Test In order to assess the validity of the digesting process and analysis procedure, two experiments, one spiked and one none spiked were conducted. In the water samples of spiked samples known concentrations of metals were added. The standard solutions of 1.62 mg/L from 1000 mg/L stock solution of each metal were added to 50 mL of water samples. Both the spiked and non-spiked samples underwent meticulous digestion under the same circumstances. Then, each digested sample was put into a 50 mL volumetric flask and diluted with distilled water to the appropriate mark. ICP-OES was then used to determine the solutions' metal concentration. The percent recovery was calculated by the standard formula.

$$\%\text{R} = \frac{\text{CSP} - \text{Cunsp}}{\text{sp}} * 100 \tag{22.1}$$

where

% R = recovery
Csp = concentration recorded in spiked sample.
Cunsp = concentration recorded in unspiked sample.
Sp = spike value (known concentration added to the sample).

Method Detection Limit (MDL) From a blank sample, the method detection limit (MDL) for each metal analysis using ICP-OES was determined. 5 mL of HNO3 (69%) were digested without the sample for the control samples. Deionized water was used to dilute the digested blank samples into a 50 mL volumetric flask for metal analysis. Then the MDL of each element is computed using three times the standard deviation of the blank sample concentration determined [19].

22.2.6 Instrumentation

An inductively coupled plasma optical emission spectrometer (ICP-OES) (Perkin Elmer Optima 8000) (Table 22.1), which is used in contemporary laboratories for metal analysis in a range of applications, is the instrument used for the analysis. To create aerosols from an aqueous sample, a nebulizer was used. The aerosols are transported to the inductively coupled plasma which is a high temperature zone

Table 22.1 Fast optima 8000 DV ICP-OES instrumental conditions

ICP_OES parameters (Optima 8000)	
Power	1500 watt
Plasma gas flow	8.0 L/min
Auxiliary gas flow	0.2 L/min
Nebulizer gas flow	0.7 L/min
Pump speed	1.0 mL/min

(8000–10,000 °C). The analytes are heated to various atomic and/or ionic states, which results in distinctive optimum emissions (light). Based on their various wavelengths, these emissions are divided, and their intensities are measured. The intensities are proportional to the analyte concentrations in the aqueous sample. By contrasting the emission intensity of an unknown sample with that of a standard sample, an external multi-point linear standardization is used for the quantification. From single- and multiple-element primary standard solutions, multi-element calibration standard solutions are created [20].

22.2.7 Data Analysis

Assessment of water quality parameters is done in two seasons which is in dry and wet seasons (winter and summer). To see whether there is a significant difference in values of the water quality parameters or not in these seasons, the commonly used statistical analysis method called t-test is used. It is useful when comparison is done between two groups.

22.3 Analysis of Metals

22.3.1 Validation of Experimental Procedures

The validation of experimental processes aids in determining the efficiency of the experimental study, which is demonstrated by some parameters [21]. Some experimental approaches were seen to be successful in determining the level of a few selected trace metals in water samples from Selamco irrigation water. MDL, recovery test (% recovery), coefficient of variation using standard deviation of triplicate sample analysis (CV), linear range, and plotting calibration curve with their coefficient of determination (R^2) were seen for the analysis of six selected trace metals in samples of Selamco irrigation water.

Coefficient of Determination (R2) It is used to infer the linearity of an analytical procedure [22]. Standard solutions of 0.05, 1.05, 2.05, 3.05, 4.05, and 5.05 in mg/L are utilized to determine the calibration curve for each metal. As shown in Table 22.2,

Table 22.2 Correlation coefficients for calibration curves

Metal	Regression equation ($y = mx + b$)	Coefficient of determination (R^2)
Manganese	$y = 10{,}430{,}000 \times -305{,}069$	0.999
Chromium	$y = 1{,}436{,}000 \times -8167$	0.999
Nickel	$y = 485{,}000 \times -1262$	0.999
Zinc	$y = 36{,}203 \times -1474$	0.998
Lead	$y = 76{,}742 \times -730$	0.999
Cadmium	$y = 76{,}328 \times -13{,}970$	0.999

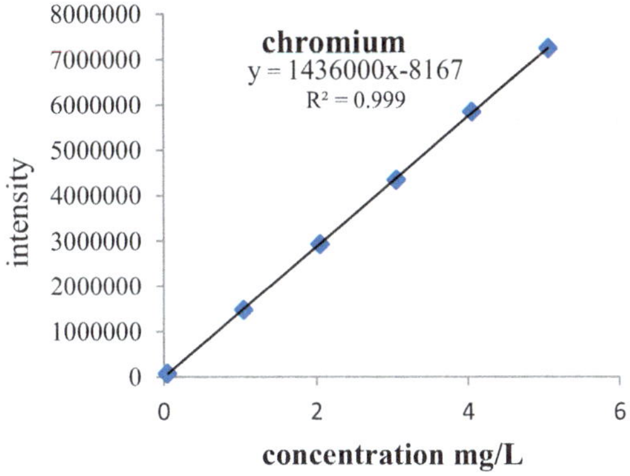

Fig. 22.2 Calibration curve for chromium

all the coefficients of determinations of the six trace metals are above 0.99 which is deemed a good fit to the linearity of calibration curve plotted analytical methods. To estimate the element's unknown concentration, a calibration curve is used. Individual calibration curve of each metal was obtained from the instrument by using a series of solutions of known concentrations.

The following figures (Figs. 22.2, 22.3, 22.4, 22.5, 22.6, and 22.7) show Calibration curves of working standards using ICP-OES for different metals.

Method Detection Limit (MDL) It is the lowest concentration of the analyte that can be detected in a measurement procedure. This value was obtained by multiplying the standard deviation of the blank by three [19].

$$\text{MDL} = 3 \times \text{Sd} \tag{22.2}$$

where

MDL = method detection limit,
Sd = Standard deviation of blank sample.

Limit of Quantification (LOQ) It is the minimum concentration level at which a measurement is meaningful quantitatively. The typical definition of the LOQ is 10 times the determined concentration of the blank sample's standard deviation [19]. In this study, the limit of quantification (LOQ) was determined from a blank sample

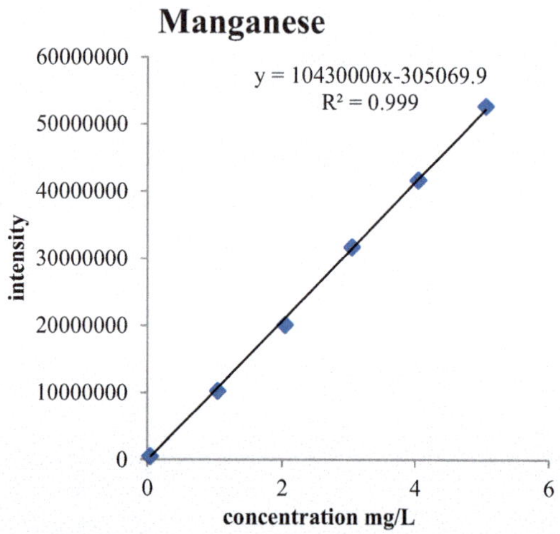

Fig. 22.3 Calibration curve for manganese

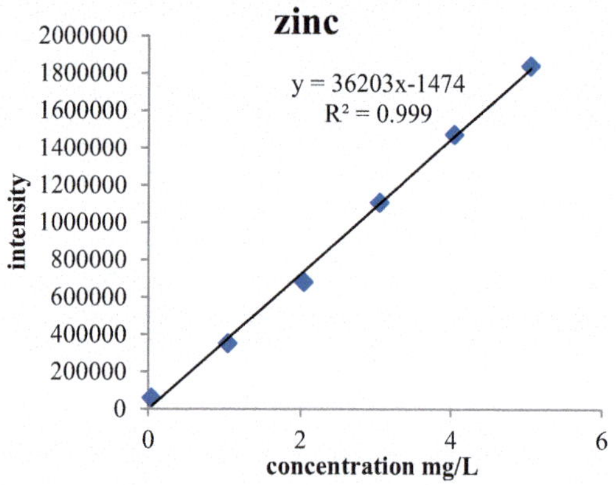

Fig. 22.4 Calibration curve for zinc

that underwent the identical digestion process. The results are displayed in Table 22.3.

$$\mathrm{LOQ} = 10 \times \mathrm{Sd} \tag{22.3}$$

where

LOQ = limit of quantification,
Sd = Standard deviation of blank sample.

From this study, MDL lies in between 0.000 and 0.0036 mg/L and the limit of quantification lies between 0.000 and 0.012 mg/L. Both the MDL and LOQ are lower enough to detect and quantify trace metals in the real sample [19].

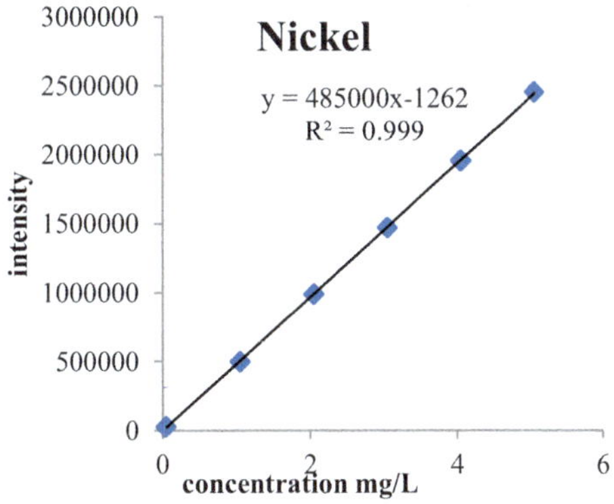

Fig. 22.5 Calibration curve for nickel

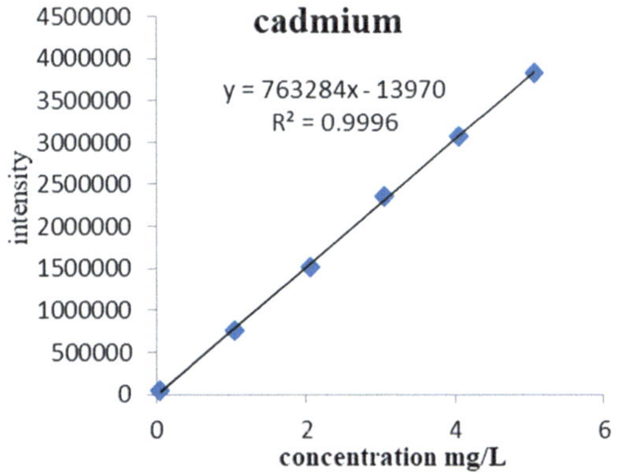

Fig. 22.6 Calibration curve for cadmium

Recovery (%) By examining the digests of spiked samples, the optimized method's efficiency and accuracy were assessed. The percent recoveries obtained for the analyzed metals are shown in Table 22.4.

As a result, the range of metal recoveries in the spiked samples was found to be between 79.19% and 98.3%. Except for zinc, all of the recovery values fell within the permissible 80–120% range for metal analysis [23]. These findings show that the suggested approach was precise and confirms that the method of analysis for each metal was acceptable.

Coefficient of Variation (CV) It is one of the measures of precision measurement of an analysis, which is determined by calculating the percent relative standard deviation or coefficient of variation [22]. The majority of the CV values are found to be less than 5%, confirming the accuracy of the measurement.

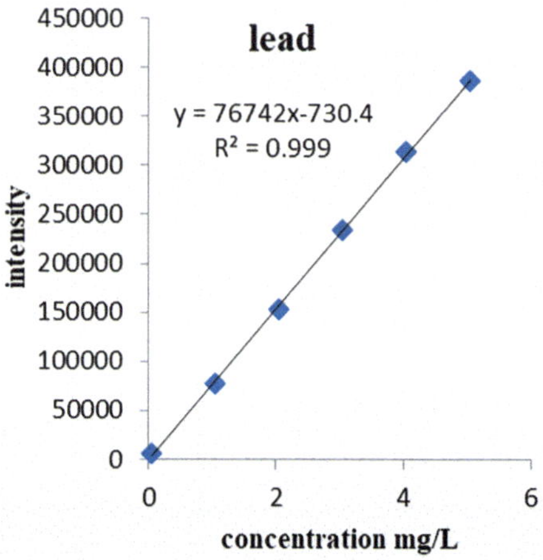

Fig. 22.7 Calibration curve for lead

Table 22.3 Method detection limit and quantification limit for determination of metals in water samples

Metal	Cr	Mn	Pb	Cd	Zn	Ni
MDL (mg/L)	0.0021	0.0006	0.0036	0.000	0.0012	0.0012
LOQ (mg/L)	0.007	0.002	0.012	0.000	0.004	0.004
IDL (mg/L)	0.0061	0.0014	0.042	0.0027	0.0059	0.015

Hint: *MDL* method detection limit, *LOQ* limit of quantification, *IDL* instrumental detection limit

22.4 Result and Discussion

22.4.1 Physical Parameters

For the water in the Selamco dam, the physical parameters including temperature, pH, EC, TDS, and turbidity were determined (see Fig. 22.8). The mean values and standard deviation of the data were obtained and are given in Table 22.5 after analysis using a field measuring instrument in both the dry and wet seasons.

Due to various activities that can affect changes in surface water temperature, aquatic bodies naturally exhibit daily and seasonal temperature shifts. Temperature has a significant impact on both the physical and chemical properties of water, the creatures that live there, and the aquatic environment as a whole. In this study, the temperature values at Selamco Reservoir A1, B1, A2, and B2 sampling locations were obtained as 23.260, 22.460, 20.60, and 21.60 °C, respectively. In comparison, the dry season had greater temperatures. Seasonal variation caused this alteration, which may have been brought on by the earth's surface warming [24]. These results were found to be within the allowable limit [25]. As a result, the water in the dam is generally at a temperature that is good for aquatic ecosystems.

pH The dissolved chemical components and biological activities in a solution control pH, which expresses how strongly an acidic or basic character of a solution is

Table 22.4 Recovery results after spiking with standard solutions

Metal	Spike value(mg/L)	Concentration recorded in unspiked sample (mg/L)	Concentration recorded in spiked sample (mg/L)	Recovery (%)
Cr	1.62	0.331 ± 0.0017	1.725 ± 0.0126	86.04
Mn	1.62	0.143 ± 0.0002	1.58 ± 0.0147	88.7
Ni	1.62	0.229 ± 0.0018	1.687 ± 0.0109	90
Zn	1.62	0.772 ± 0.0071	2.055 ± 0.0356	79.19
Cd	1.62	0.012 ± 0.000	1.605 ± 0.0269	98.3
Pb	1.62	0.023 ± 0.0002	1.498 ± 0.0046	91.04

Hint: Concentration of spiked samples is average of the analyzed sample ± standard deviation

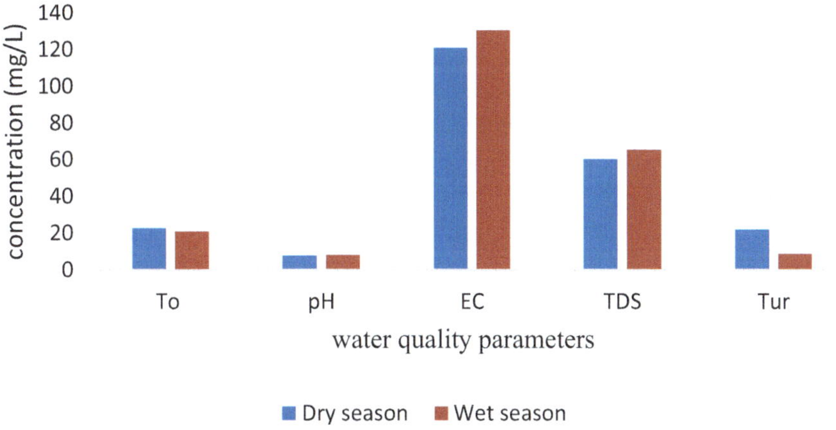

Fig. 22.8 Physical parameters at dry and wet season in Selamco irrigation water dam

present. It is typically observed for evaluations of the health of the aquatic ecosystem, drinking water sources, industrial discharges, and surface water runoff [26]. The research area's pH value indicates that it is mildly alkaline, that is A1, B1 A2, and B2 were found to be 7.93, 7.89, 8.34, and 8.21, respectively. These values fall within the permitted range as set forth in the Food and Agriculture Organization of the United Nations (FAO) irrigation guidelines. Due to floods and runoff which contains alkaline salts, the pH value of the water in the wet season differed slightly from that of the dry season.

Electrical Conductivity (EC) It is a measurement of the ability of water to carry electrical current. It represents the entire amount of dissolved salts [27]. One of the most crucial factors affecting crop growth is conductivity. The results reported at each sampling site fell within the FAO-established parameters for permissible/allowable irrigation water use. The fact that the EC of the wet season is relatively higher than that of the dry season may be because of more electrolyte compounds enter the dam through runoff during the rainy season.

Table 22.5 Results of physicochemical parameters in Selamco reservoir at dry and wet season (mean ± sd)

Parameter	Dry season		Wet season		FAO guideline (1994)
	A1	B1	A2	B2	
Temperature (°C)	23.26 ± 0.40	22.46 ± 0.41	20.6 ± 0.15	21.6 ± 0.30	–
pH	7.93 ± 0.04	7.89 ± 0.01	8.34 ± 0.08	8.21 ± 0.02	6.0–8.5
EC(μs/cm)	118.9 ± 9.00	122.9 ± 3.68	130.86 ± 2.39	129.9 ± 3.64	3000
TDS(ppm)	58.8 ± 5.68	61.6 ± 1.76	66.1 ± 1.32	65.4 ± 1.87	2000
Turbidity(NTU)	22.93 ± 0.56	20.93 ± 0.35	11.13 ± 0.50	6.57 ± 0.34	–

Hint; A1 = inlet sample, B1 = outlet sample, A2 = inlet sample and B2 = outlet sample

Total Dissolved Solid (TDS) TDS increases turbidity, which has an impact on the water's aesthetic value. High levels of TDS in water samples increase the density of the water, impair its appropriateness as a source of irrigation and drinking water, affect the osmoregulation of freshwater species, and decrease the solubility of gases. The TDS laboratory results for this specific study region were below the FAO-established maximum limits.

Turbidity It is an indicator of how cloudy the water is; high turbidity could mean that bacteria, viruses, and parasites are present, which could cause symptoms like nausea, cramps, diarrhea, and headaches [28]. Turbidity values of sampling sites A1, B1, A2 and B2 were obtained to be 22.93NTU, 20.93 NTU, 11.3 NTU, and 6.57 NTU, respectively as shown in Table 22.2 of the same document, and can be concluded that it is quite good for irrigation purpose.

Spatiotemporal Variations in the Physical Water Quality Parameters The spatiotemporal fluctuations between water samples of mean value in the dry and rainy seasons were investigated using statistical methods. For the comparison of two groups, the p-value of the t-test or statistical significance is usually used [29]. To compare the mean values of the various irrigation water quality indicators during the wet and dry seasons, a paired t-test was used in this study.

pH, EC, and turbidity showed a significant difference between the mean value of the dry and rainy seasons ($p < 0.05$) according to paired t-test results. However, temperature and TDS did not differ significantly at a 95% confidence level ($p > 0.05$), as shown in Table 22.6.

Table 22.6 Comparisons of physical parameters of the Selamco irrigation water between dry and wet seasons with its p-value

Parameter	Dry season mean ± sd	Wet season mean ± sd	P-value	Remark
Temperature (°C)	22.86 ± 0.56	21.1 ± 0.707	0.11	Ns
pH	7.91 ± 0.028	8.27 ± 0.092	0.033	S
EC(μs/cm)	120.9 ± 2.83	130.38 ± 0.67	0.043	S
TDS(ppm)	60.2 ± 1.97	65.7 ± 0.5	0.06	Ns
Turbidity(NTU)	21.93 ± 1.41	8.85 ± 3.22	0.0343	S

Hint: *S* significant, *Ns* not significant according to paired t-test value

22.4.2 Analyzed Concentration of Metals

ICP-OES was used to calculate the amount of heavy metals present in the Selamco irrigation water during the dry and rainy seasons, and the results are shown in Table 22.7 in mg/L. All heavy metal concentrations were discovered to be lower during the wet season than in the dry season, with the exception of the concentration of lead. This could be a result of the diluting impact from the reservoir's outfall runoff and heavy rain during the rainy season.

22.4.3 Discussions for the Heavy Metal Results

Chromium (VI) The levels were over the FAO's acceptable limit for irrigation purposes in both sites (inlet and outlet) during the dry season but below the permissible limit in both sites during the wet season (Table 22.8). Chromium can be harmful in large doses [30]. Large doses of chromium inhibits the uptake of water and mineral to plants and also inhibits the respiration and photosynthetic processes [31]. Uptake of chromium containing crops with unacceptable limit can led to gastric cancer risks [32].

Manganese It is found naturally in numerous sources of surface and ground water as well as in soils that could erode into these waters. However, a lot of manganese poisoning in some places' water is also a result of human activity. Except for the values observed at the dam's outlet during the dry season, all readings in the current investigation of the Selamco dam at sites A1, B1, A2, and B2 were below the FAO-permitted level. Overexposure to manganese can lead to neurotoxic consequences [33].

Nickel Concentration of 0.214, 0.212, 0.022, and 0.014 mg/L were obtained in sampling sites of A1, B1, A2, and B2, respectively. Therefore, the results in this study were comparable and greater in some extent with the permissible limit set by FAO but higher than United States Environmental Protection Agency (USEPA)

Table 22.7 Trace metals concentrations in the water sample from Selamco reservoir (Mean ± Sd) in mg/L

	Dry season		Rainy season		Average value of dry and rainy season	
Metals	A1	B1	A2	B2	Dry season	Wet season
Cr	0.323 ± 0.009	0.317 ± 0.018	0.014 ± 0.0007	0.008 ± 0.006	0.32 ± 0.004[a]	0.011 ± 0.0008[b]
Mn	0.135 ± 0.002	0.285 ± 0.052	0.133 ± 0.018	0.032 ± 0.029	0.21 ± 0.027[a]	0.082 ± 0.009[b]
Ni	0.214 ± 0.012	0.212 ± 0.017	0.022 ± 0.002	0.014 ± 0.007	0.213 ± 0.001[a]	0.018 ± 0.004[b]
Zn	0.369 ± 0.036	0.531 ± 0.233	0.184 ± 0.048	0.091 ± 0.118	0.45 ± 0.114[a]	0.137 ± 0.065[a]
Cd	0.010 ± 0.000	0.010 ± 0.000	0.005 ± 0.000	0.005 ± 0.000	0.01 ± 000 [a]	0.005 ± 000 [b]
Pb	0.017 ± 0.003	0.014 ± 0.004	0.249 ± 0.0007	0.207 ± 0.048	0.015 ± 0.002[a]	0.228 ± 0.029[b]

Hint: Values in the same row of the last two column that are followed by different letters (a and b) are significantly different ($p < 0.05$) by paired t-test. A1 = inlet sample in dry season, B1 = outlet sample in dry season, A2 = inlet sample in rainy season, and B2 = outlet sample in rainy season

Table 22.8 International guidelines of heavy metals water quality (mg/L)

Metal	FAO (1985)	USEPA (1990)
Cr	0.1	0.1
Mn	0.2	0.3
Ni	0.2	0.1
Zn	2	5
Cd	0.01	0.005
Pb	5	0.003

limit Table 22.8. The higher presence of nickel in diet can lead to nickel-induced carcinogenesis and allergy in humans [34].

Zinc It is a crucial component of the human diet. Low and high concentrations of zinc can have negative health effects [35]. The recorded value of zinc in all sampling sites of A1, B1, A2, and B2 were below the permissible limit of the guideline set by FAO as shown in Table 22.7.

Cadmium Human actions, such as agricultural practices, can contribute to higher Cadmium levels [36]. Cadmium metal of this work in all sampling sites of Selamco reservoir results at A1, B1, A2, and B2 were 0.01, 0.01, 0.005, and 0.005 mg/L, respectively. Based on the finding, the value in dry season requires attention because it is almost in the maximum limit which can toxify kidney and cause bone demineralization with long-term exposure [37].

Lead It is a substance that is present in trace amounts in a number of foods, most notably fish, which is subjected to industrial pollution [30]. In this study, the concentration of lead in all sites of A1, B1, A2, and B2 of Selamco reservoir were under the acceptable limit of FAO for irrigation purposes as shown in Tables 22.7 and 22.8 of the same document.

Spatiotemporal Variations of Heavy Metal Concentrations in Selamco Irrigation Water During Dry and Wet Seasons Some water sources' quality may change considerably over time or during specific times, such as dry or wet seasons [38]. Therefore, it is advised to take more than one sample at various times. The amount of heavy metals contained in this sample varied significantly between the dry and wet seasons. With the exception of lead, the concentration of heavy metals was higher during the dry season than during the wet season. This may be due to high evaporation rate during the dry season which resulted in higher concentrations of heavy metals in the reservoir. Whereas, there is dilution in wet season by rain water and subsequent outflow of reservoir water [39]. But for the case of lead the researchers expected that there is an unforeseen inflow of lead-containing influents to the reservoir in the rainy season, because of high runoff.

The results showed that (Table 22.7) the mean concentrations of metals ranked (low to high) Cd < Pb < Mn < Ni < Cr < Zn during dry season, whereas the concentration of heavy metals during wet season was in the following order of increasing value Cd < Cr < Ni < Mn < Zn < Pb.

22.5 Conclusion

This study was generally focused on assessing some physicochemical properties of Selamco artificial reservoir. Physical parameters were measured in situ and chemical parameters (heavy metals) were measured in laboratory using ICP_OES. Based on the finding, physical parameters were found to be within the permissible limit both in dry season and wet season based on the FAO guideline. From the heavy metals measured, three metals which are Cr, Mn, and Ni which are toxic to humans and have long-term cumulative effects on human health were found to be above the permissible limit in dry season. Actually, the water is utilized during dry season for irrigation purpose, and therefore special attention should be given to the upstream of the reservoir to prevent the entry of toxic metal-containing materials.

Acknowledgments The authors of this paper are grateful to the South Gondar Zone Agricultural office for their help in providing the data we want. We also acknowledge Bahir Dar University for financial support to conduct this research.

References

1. Pimentel, D., et al. (1987). World agriculture and soil erosion. *Bioscience, 37*(4), 277–283. https://doi.org/10.2307/1310591
2. Waterer, D. (2003). Yields and economics of high tunnels for production of warm-season vegetable crops. *HortTechnology, 13*(2), 339–343. https://doi.org/10.21273/horttech.13.2.0339
3. Yates, D. N., Sieber, J., Purkey, D. R., & Huber-Lee, A. (2005). WEAP21 – A demand- , priority- , and preference-driven water planning model part 1: Model characteristics. *Water International, 30*(4), 487–500.
4. Jonnalagadda, S. B., & Mhere, G. (2001). Water quality of the odzi river in the eastern highlands of Zimbabwe. *Water Research, 35*(10), 2371–2376. https://doi.org/10.1016/S0043-1354(00)00533-9
5. Ayers, R. S., & Westcot, D. W. (1985). *Water quality for agriculture* (Vol. 29, p. 174). Food and agriculture organization of the United Nations.
6. Mkadmi, Y., et al. (2018). Study of the impact of heavy metals and physico-chemical parameters on the quality of the wells and waters of the Holcim area (oriental region of Morocco). *Journal of Materials and Environmental Science, 9*(2), 672–679. https://doi.org/10.26872/jmes.2018.9.2.74
7. Ali, H., Khan, E., & Ilahi, I. (2019). Environmental chemistry and ecotoxicology of hazardous heavy metals: Environmental persistence, toxicity, and bioaccumulation. *Journal of Chemistry, 2019*, 1. https://doi.org/10.1155/2019/6730305
8. Lal, R., & Stewart, B. A. (2020). *Soil processes and water quality*. CRC Press. https://doi.org/10.1201/9781003070184-1
9. Chemura, A., Kutywayo, D., Chagwesha, T. M., & Chidoko, P. (2014). An assessment of irrigation water quality and selected soil parameters at Mutema irrigation scheme, Zimbabwe. *Journal of Water Resource and Protection, 06*(02), 132–140. https://doi.org/10.4236/jwarp.2014.62018
10. Yidana, S. M., Sakyi, P. A., & Stamp, G. (2011). Analysis of the suitability of surface water for irrigation purposes: The southwestern and Coastal River Systems in Ghana. *Journal of Water Resource and Protection, 03*(10), 695–710. https://doi.org/10.4236/jwarp.2011.310080

11. Almas, A., & Scholz, M. (2015). Agriculture and water resources crisis in Yemen: Need for sustainable agriculture. *Journal of Sustainable Agriculture, 53*(4), 1060–1067. https://doi.org/10.1080/19443994.2014.880156
12. Ragab, R. (2002). A holistic generic integrated approach for irrigation, crop and field management: The SALTMED model. *Environmental Modelling & Software, 17*(4), 345–361. https://doi.org/10.1016/S1364-8152(01)00079-2
13. Mujere, A. M. N., Chazovachii, B., & Chifodya, G. (2011). Evaluating factors influencing the variation of irrigated wheat yields. *Journal of Sustainable Development in Africa, 13*(4), 177–188.
14. Goshu, G., Koelmans, A. A., & de Klein, J. J. M. (2017). Water quality of Lake Tana basin, Upper Blue Nile, Ethiopia. A review of available data. In *Social and ecological system dynamics: Characteristics, trends, and integration in the Lake Tana Basin*, Ethiopia, 127–141.
15. Mulatu, C. A., Crosato, A., Langendoen, E. J., Michael, M., & Mcclain, M. E. (2020). Long-term effects of dam operations for water supply to irrigation on downstream river reaches. The case of the Ribb River, Ethiopia. *International Journal of River Basin Management, 19*, 1–51. https://doi.org/10.1080/15715124.2020.1750421
16. Andualem, Y., Diress, M., Tenagashawu, Y., & Molla, H. (2023). Evaluation of the physicochemical and microbiological current water quality status of Ribb reservoir, South Gondar, Ethiopia. *Sustainable Water Resources Management, 9*(1), 1–18. https://doi.org/10.1007/s40899-022-00801-2
17. Zhang, C. (2007). *Fundamentals of environmental sampling and analysis*. Wiley.
18. Smith, J. C., Butrimovitz, G. P., & Purdy, W. C. (1979). Direct measurement of zinc in plasma by atomic absorption spectroscopy. *Clinical Chemistry, 25*(8), 1487–1491. https://doi.org/10.1093/clinchem/25.8.1487
19. Shrivastava, A., & Gupta, V. (2011). Methods for the determination of limit of detection and limit of quantitation of the analytical methods. *Chronicles Young Sci., 2*(1), 21. https://doi.org/10.4103/2229-5186.79345
20. Boss, C. B., & Fredeen, K. J. (2004). *Concepts, instrumentation and techniques in inductively coupled plasma optical emission spectroscopy* (pp. 1–22). Perkin Elmer.
21. Bruce, G. R., & Gill, P. S. (1999). Estimates of precision in a standard additions analysis. *Journal of Chemical Education, 76*(6), 805–807. https://doi.org/10.1021/ed076p805
22. Guideline, ICH Harmonised Tripartite. (2005). Validation of analytical procedures: Text and methodology. *Q2 (R1), 1*(20), 05.
23. Alaska Department of Environmental Conservation Division of Water. (2011). *Generic Tier 2 quality assurance project plan for water quality monitoring sampling and analysis Activities*.
24. Bayingana, G., Mupenzi, C., & Maniragaba, A. (2021). Potential effect assessment of agricultural activities on water quality of rivers in Rwanda: 'Case of Muvu mba river in Nyagatare District,'. *Environmental Agricultural Research, 1*, 42–53.
25. World Health Organization. (1989). Guidelines for drinking-water quality, *35*(5), 307–312: https://doi.org/10.1248/jhs1956.35.307.
26. Kumar, M., Kumari, K., Ramanathan, A., & Saxena, R. (2007). A comparative evaluation of groundwater suitability for irrigation and drinking purposes in two intensively cultivated districts of Punjab, India. *Environmental Geology, 53*(3), 553–574. https://doi.org/10.1007/s00254-007-0672-3
27. Devi, S., & Premkumar, R. (2012). Physicochemical analysis of groundwater samples near industrial area, Cuddalore District, Tamilnadu, India. *International Journal of ChemTech Research, 4*(1), 29–34.
28. Teklay, A., & Amare, M. (2015). Water quality characteristics and pollution levels of heavy metals in Lake Haiq, Ethiopia. *Ethiopian Journal of Science and Technology, 8*(1), 15. https://doi.org/10.4314/ejst.v8i1.2
29. Miller, J. N., & Miller, J. C. (2010). *The quality of analytical measurements*. Available. http://www.tandfonline.com/doi/abs/10.1198/tech.2004.s248

30. Al-Imarah, F. J. M., Al-Nagar, G. A., & Abdulhadi Al-Hatem, Z. (2019). Heavy metals in well's waters within southern Basrah province/Iraq. *Journal of Physics Conference Series, 1294*(7), 072018. https://doi.org/10.1088/1742-6596/1294/7/072018
31. Mathur, S., Kalaji, H. M., & Jajoo, A. (2016). Investigation of deleterious effects of chromium phytotoxicity and photosynthesis in wheat plant. *Photosynthetica, 54*(2), 185–192. https://doi.org/10.1007/s11099-016-0198-6
32. Tseng, C., Lee, I., & Chen, Y. (2019). Science of the total environment evaluation of hexavalent chromium concentration in water and its health risk with a system dynamics model. *Science of the Total Environment, 669*, 103–111. https://doi.org/10.1016/j.scitotenv.2019.03.103
33. Miah, M. R. et al. (2021). HHS Public Access. https://doi.org/10.1016/j.neuint.2020.104688.The.
34. Zambelli, B., Uversky, V., & Ciurli, S. (2016). PT SC. *BBA: Proteins Proteomics, 1864*, 1714. https://doi.org/10.1016/j.bbapap.2016.09.008
35. Nazir, R., et al. (2015). Accumulation of heavy metals (Ni, cu, cd, Cr, Pb, Zn, Fe) in the soil, water and plants and analysis of physico-chemical parameters of soil and water collected from Tanda dam Kohat. *Journal of Pharmaceutical Sciences and Research, 7*(3), 89–97.
36. Cattani, I., Romani, M., & Boccelli, R. (2008). Effect of cultivation practices on cadmium concentration in rice grain. *Agronomy for Sustainable Development, 28*(2), 265–271. https://doi.org/10.1051/agro:2007033
37. Bernard, A. (2008). Cadmium & its adverse effects on human health. *Indian Journal of Medical Research, 128*(4), 557–564.
38. Delpla, I., Jung, A. V., Baures, E., Clement, M., & Thomas, O. (2009). Impacts of climate change on surface water quality in relation to drinking water production. *Environment International, 35*(8), 1225–1233. https://doi.org/10.1016/j.envint.2009.07.001
39. Rehman, H. U., Ismail, M., Khan, R. U., & Qureshi, Z. (2015). Heavy metals analysis of Bannu dam ' s and Damai stream during breeding season of fishes heavy metals analysis of Bannu dam ' s and Damai stream during breeding season of fishes. *World Applied Sciences Journal*. https://doi.org/10.5829/idosi.wasj.2015.33.12.101170

Chapter 23
Evaluating Flood Vulnerability in Bahir Dar City Through an Indicator-Based Method

Elias S. Leggesse, Walelign A. Derseh, Afework A. Simegn, and Mengiste A. Meshesha

23.1 Introduction

Floods have become more frequent worldwide, and associated risks are projected to increase as the global climate and land use change, the urban population grows, and existing infrastructure ages [1–4]. Even though there have been significant efforts to prevent floods, the resulting damage to people and property remains high, with floods making up 34% of all natural disasters worldwide in quantity wise and 40% in terms of losses [5, 6]. It can be assumed that the statistics for developing countries are even worse [7], as flood hazard assessment "research in developing countries lags significantly behind although it is developing rapidly" [2].

A study assessing the history and spatial extents of flood events in Ethiopia from the 1960s to the present [8], has shown a general, decade by decade, increase in flood frequency, characterized by spatial and temporal variation. With expected increased runoff and extreme flows as a result of climate change and urbanization, the drainage infrastructure developed for the current demand is likely to fall short of the expected service level in the future [9–12]. In some regions of Ethiopia, including Bahir Dar and Dire Dawa, it is deemed inevitable to experience flooding during the main rainy season, which lasts from June to September [7, 13, 14]. The International Federation of Red Cross and Red Crescent Societies (IFRC) reported that within a month (August to September) 2020 alone, heavy rains triggered floods, affecting more than 200,000 people and claiming eight lives [15].

Bahir Dar is facing frequent pluvial flooding, which is expected to worsen due to climate change, urban expansion, poor infrastructure, and land use changes. The flooding, even if it is shallow in depth, is deteriorating the city's infrastructure,

E. S. Leggesse · W. A. Derseh (✉) · A. A. Simegn · M. A. Meshesha
Faculty of Civil and Water Resources Engineering, Bahir Dar Institute of Technology, Bahir Dar University, Bahir Dar, Ethiopia

S. W. Fanta et al. (eds.), *Sustainable Development Research in Manufacturing, Process Engineering, Green Infrastructure, and Water Resources*, Green Energy and Technology, https://doi.org/10.1007/978-3-031-77339-6_23

particularly the transportation network, and disrupting social and economic activities. The low-lying downtown areas, where the majority of the city's population and main commercial activities are situated, are typically affected. The flooding imposes exorbitant costs on the city's fragile economy and exacerbates poverty levels. The most likely causes of flooding in Bahir Dar include lake level rise, poor drainage infrastructure, groundwater level rise, an already saturated soil receiving high rainfall in the late rainy season, or a combination of all. Monitoring data from the Ministry of Water and Energy indicates the water level of Lake Tana frequently exceeds its maximum operational level, signifying that the lake is currently experiencing higher than normal water levels [16]. The construction of a weir at Lake Tana's outlet intended to prevent the maximum level of the lake from exceeding its historical high and protect surrounding areas from flooding has also inadvertently contributed to the rise of the lake level, causing severe flooding, along its shores, including in the city [17].

Flood vulnerability assessment which is the main part of flood risk assessment aims at identifying, selecting, and localizing receptors (i.e., elements potentially at risk) that could be potentially subjected to floods and therefore exposed to adverse consequences in a flood zone. By understanding the vulnerability of an area or population to flooding, risk management plans can be developed to reduce the potential impact of flood events [18, 19]. There are several methods for assessing vulnerability, each with its own unique description, theoretical framework, variables, and methodology [20]. The most widely accepted methods can be grouped into four categories: curve method, disaster loss data method, computer modeling methods, and indicator-based methods [19, 20]. The indicator-based method is particularly popular in flood vulnerability studies [19, 21–23] and is preferred by policymakers due to its ability to provide a clear image of vulnerability across space [19, 24]. It allows to prioritize the measures and plans for risk response in specific regions, and it is adaptable to use available data to provide a logical image of a place's vulnerability.

Hence, in this study, a flood vulnerability assessment of the City of Bahir Dar was conducted by employing the indicators-based method. The authors identified the indicators responsible for the spatial variation of flood vulnerability, examined the spatial distribution of physical, social, and economic vulnerability to flooding, and clustered the vulnerability indicators to reveal the overall flood vulnerability across the city. The finding of the study provides guidance for policymakers and land managers to reduce flood risk in the most exposed areas.

23.2 Description of The Project Area

The City of Bahir Dar, located in the Amhara National Regional State (ANRS) of Ethiopia, is a regional capital that comprises six urban sub-cities, four satellite towns, and several rural administrative areas. It lies on the southern shore of Lake Tana and is bisected by the Abay River (also known as the Blue Nile River), at approximately 11°37′00″ N and 37°25′00″ E. It has an altitudinal range of 1786–1886 meters above sea level with higher elevation terrain surrounding the city that forms a drainage pattern of natural water courses that flow toward the city. Bahir Dar is one of the fastest-growing cities in Ethiopia [25–28], with an estimated population of more than 582,034 in 2018, which under a high variant scenario is projected to reach a population of 1.35 and 4.0 million by the years 2035 and 2063 respectively [29]. Bahir Dar is also among the cities with the highest proportions of economically active population in the urban Ethiopia (72%) in urban Ethiopia, most of it (82%) employed in the formal economy and the remaining (18%) employed in the growing informal sector [30]. The city's economy is heavily dependent on services and retail/wholesale trade sectors creating 60% of the jobs followed distantly by the manufacturing sector only at 15% [29].

This study concentrates on the six downtown sub-cities that constitute the Bahir Dar Proper (Fig. 23.1) as they are the most vulnerable in terms of physiographic factors such as insufficient surface slope, high groundwater table, inward drainage pattern, exposure due to the concentration of economic activities in these areas, and susceptibilities of infrastructure and people due to the existence of slums and poorly planned stormwater management infrastructure. Furthermore, more than 56% of the

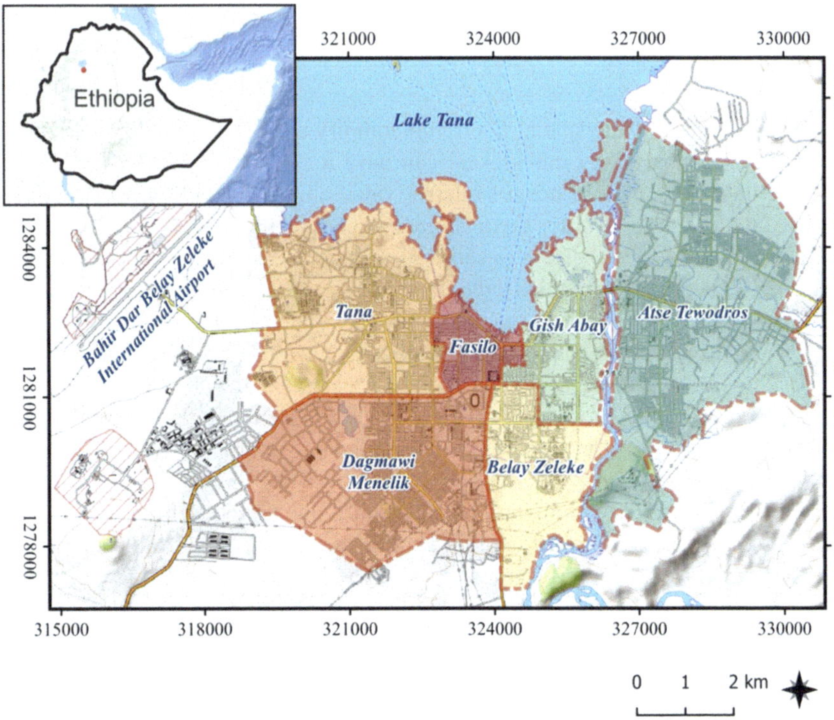

Fig. 23.1 The study area—Bahir Dar's six urban sub-cities (OpenStreetMap [31] is used as a base map)

Table 23.1 Population data and size of sub-cities

Spatial extent	Area (Km^2)	Population	Density (people/km^2)
Downtown (study area)	64.91	325,781	5019.01
Dagmawi Menelik sub-city	14.04	89,538	6378.30
Tana sub-city	13.59	78,880	5803.93
Atse Theodros sub-city	22.28	66,288	2974.57
Fasilo sub-city	2.27	32,569	14,356.43
Belay Zeleke sub-city	6.57	32,092	4880.99
Gish Abay sub-city	6.15	26,414	4293.35
Satellite towns (4 in number)	314.16	33,409	106.34
Rural administrative units	2448.36	222,844	91.02
Total municipality	2827.43	582,034	205.85

municipality's total population lives in these six sub-cities with Fasilo, Dagmawi-Minilik, and Tana sub-cities having the highest concentration, with an approximate population density of 14,356, 6,378, and 5,804 inhabitants per square kilometer, respectively (see Table 23.1) followed by Belay Zeleke, Gish Abay, and Atse Theodros sub-cities (Fig. 23.1).

23.3 Methodology

The bounding box of the spatial flood vulnerability analysis, or the spatial extent over which the analysis is undertaken, can have a considerable influence on the resulting indices [32]. Measures of relative vulnerability depend on the bounding box used to delimit the study. Often, choices of geographic bounds are determined by data availability or stakeholder needs rather than by the dynamics of the system under investigation [19]. In this study, the building blocks (property parcels delineated by local streets) were used as the smallest spatial units to compute the physical, economic, and social vulnerability indices, and the total study area was bounded by the downtown urban territory of the City of Bahir Dar. This approach was arranged to make efficient use of the extensive data collected at building block and parcel/household levels for the city's structure plan preparation as input data for the selected indicators.

Data for this study was obtained from a socioeconomic household survey conducted by the Bahir Dar City Structure Plan Preparation Project Office, as part of a Structure Plan Preparation Project for the city. The survey, which provided the data needed to analyze the indicators listed in Table 23.3, conducted as a structured interview, using a standardized questionnaire. This method of collecting indicator data was recommended by previous researchers [33, 34] as the most suitable to obtain the required data. The survey involved a larger sample size than typically used in similar studies, with a total of 1,003 households surveyed (see Table 23.2) compared to the sample size employed in previous similar studies [33]. Additionally, the

Table 23.2 Household survey attributes

Attribute	Value	Remark
Total number of Households surveyed	1003	The coordinates of households were recorded to consider and understand the spatial variation of responses
Total number of questions	88	The questions were categorized into eight sections based on thematic areas
Questions used as input data for the vulnerability indicators in this study	21	This study used a total of 24 indicators, with three additional indicators derived from GIS data

Table 23.3 Selected indicators of physical, social, and economic vulnerability; indicators marked with an asterisk (*) are derived from sources other than direct household survey responses

Dimension	Weight	Indicators	Score	Code
Physical	37%	1. Green space per building area*	0.22	PhyGR
		2. Building wall material	0.12	PhyMat
		3.Number of stories	0.13	PhyStor
		4. Age of the building	0.13	PhyAge
		5. Building roof material	0.12	PhyRoof
		6. Building footprint*	0.08	PhyFoot
		7. Building floor material	0.05	PhyFloor
		8. Land use	0.15	Phyuse
Social	28%	9. Population density*	0.13	SoHHs
		10. Female male ratio	0.12	SoFemR
		11. Female-headed household	0.12	SoFemH
		12. Percentage of age group >60/65	0.14	SoAg60
		13. Percentage of age group <18 years old	0.1	SoAg18
		14. Household head's education level	0.11	SoEduH
		15. Percentage of educated people	0.11	SoEdP
		16. Percentage of people with employment	0.07	SoEmpP
		17. Flooding experienced in the past 10 years	0.11	SoFP
Economic	35%	18. Dependency ratio (to total household size)	0.13	EcoDep
		19. Monthly household income in ETB	0.11	EcoInc
		20. Occupation of household head	0.17	EcoOcc
		21. A household that has taken out a loan in the last 10 years	0.16	EcoLoan
		22. Household-holding basis	0.10	EcoBasis
		23. Livelihood options of the household head	0.12	Ecolcopt
		24. Household with means of transportation	0.21	EcoTrans

survey recorded the geographic locations of the households, enabling the inclusion of the entire downtown area within the study's spatial extent and providing insights into spatial nuances.

The city's susceptibility to flooding was analyzed from physical, social, and economic perspectives, utilizing a combination of indicators, whose indices were sourced from the household survey data and additional geospatial data. To establish a standardized metric for measuring vulnerability, we defined vulnerability dimensions and indicators as follows:

1. Vulnerability dimensions: These are the three aspects of vulnerability – Physical, Economic and Social. The vulnerability indicators are categorized into either of these three dimensions. Each dimension was assigned a normalized weight based on its perceived significance, ensuring that the overall combined vulnerability index would appropriately reflect the relative importance of each dimension.
2. Vulnerability indicators: These are specific factors within each vulnerability dimension, identified as indicators of vulnerability. Each indicator was assigned a score representing its relative impact on vulnerability. These scores were used to weigh the numerical values obtained from the household survey and geospatial data.

The weight of the dimensions and the scores of the indicators within the dimensions were assigned based on collected expert opinions. The experts were requested to assign weights to the physical, social, and economic dimensions of vulnerability based on their perceived importance, as well as to score the relevance of indicators within each dimension. By calculating the average and normalized weights of dimensions, expressed as a percentage, and scores of indicators expressed as a rating, we ensured a balanced and informed assessment.

23.3.1 Vulnerability Dimensions and Indicators

The multifactorial nature of flooding in an urban context needs a comprehensive vulnerability assessment, which should incorporate the diverse factors. For this study, only three dimensions of vulnerability were considered: physical vulnerability, social vulnerability, and economic vulnerability. Within the three dimensions, several indicators were selected through a literature review and based on the availability of baseline socioeconomic data. Indicators such as age thresholds for vulnerable age groups and household income ranges for low-, middle- and high-income categories were further examined with respect to local context and adjusted accordingly.

To calculate Physical Vulnerability Index (PhyVI), we used eight indicators including the proportion of buildings with poor construction material per building block, the proportion of buildings with long years since construction, the proportion of green space per building block, the number of stories of the building and the general land use classification (GLUC) (see Table 23.3). We selected these indicators because they are perceived as key criteria to determine the susceptibility or safety of an area. For example, buildings with better materials, newer construction, lower height, and more green space would have lower vulnerability, as they have more ability to withstand the effects of floods and heavy rains. We also assumed that areas with less population density, such as institutional, industrial, agricultural, and open public spaces, would have lower priority for flood risk reduction.

To calculate Socio-Economic Vulnerability Index (SoVI), we used nine indicators involving population density, age distribution, percentage of females per household, percentage of female-headed households, percentage of elders, household size, experience of flooding, unemployment rate, and literacy rate. We selected these indicators because we assumed that groups with less physical and social resources, such as children, women, elders, large families, unemployed people, and uneducated people, would have higher vulnerability. We also assumed that people with no previous flood experience would have limited awareness and preparedness for flood risk.

For the Economic Vulnerability Index (EcoVI), indicators such as households with a large percentage of unemployed people are used. That is because, people who depend on others for financial aid will have to bear more economic burdens during floods and are considered more vulnerable under the economic perspective. Similarly, households that live in rooms on a rental basis will also face high vulnerability. People with fewer livelihood options also suffer from flood events. And finally, people without private car transportation systems will be more vulnerable as they will require more time to evacuate and will be exposed to flood events for a longer time.

23.3.2 *Quantification of Vulnerability Dimensions and Indicators*

The significance of vulnerability dimensions and indicators depends on local contexts and personal judgment. To obtain a bias-free assessment of the dimensions and indicator's significance, we sought the opinions of a panel comprised of 20 water professionals. These professionals were chosen for their knowledge over a randomly selected sample of residents who may not have comprehended the concept of susceptibility to disaster. To ensure a balanced weight assignment, participants were instructed to distribute the weights across the three dimensions in a manner that the sum would be 100%. The final weights of the dimensions were calculated by taking the average of the weights the participants assigned to each dimension.

Similarly, for the indicators, the 20 participants were asked to rate them on a scale of 1–10, with higher scores assigned to more affecting factors. For instance, people living on the ground floor of multistory buildings are more prone to flooding than those on higher floors. Similarly, those in poorly constructed buildings are more likely to be impacted by floods than those living in well-constructed buildings. Survey participants were instructed to assign higher scores to the "number of floors" indicator than to the "building condition" indicator, if they believed living in a poorly constructed building is less concerning compared to living on the ground floors. Subsequently, we calculated the averages of the ratings provided by the participants for each indicator.

To ensure comparability across the physical, social, and economic dimensions, we normalized the average scores of each indicator using the Min-Max technique (Eq. 23.1). This normalization process transformed the ratings by scaling the average values of the indicators between 0 and 1. Once normalized, feature scaling (Eq. 23.2) was used to ensure that the sum of all indicator scores in each dimension would be a unity. By employing this normalization step, we could fairly compare the different dimensions and their respective contributions to the combined vulnerability.

$$X_{\text{norm}} = \frac{X - X_{\min}}{X_{\max} - X_{\min}} \tag{23.1}$$

$$X_{\text{adjusted}} = \frac{X_{\text{norm}}}{\sum_{i=1}^{n} X_{norm,\,i}} \tag{23.2}$$

Where X is the average score given to a single indicator, $X_{\min}$ is minimum average indicator score, $X_{\max}$ is the maximum average indicator score, X_{norm} is the normalized indicator score, and X_{adjusted} is the scaled indicator value after feature scaling. To ensure consistency with the indicator scoring, the original values of the indicators obtained from household surveys and geospatial data were also transformed onto a 1–10 scale based on vulnerability levels, which allowed us to compute the indices. These indicator indices, measured at the building block spatial resolution, were then multiplied by the corresponding indicator scores and dimension weights to generate physical-, social-, economic-, and combined-dimension vulnerability maps (Fig. 23.2).

Out of the 24 indicators used, the data of 21 was adapted from responses obtained from the spatially attributed household survey (Table 23.2) conducted by Bahir Dar City Structural Plan Preparation Project Office [30]. A comprehensive discussion on the standardization of criteria used to transform the original responses of the household survey into vulnerability indices of the selected indicators is further elaborated in the assessment framework section (Sect. 23.3.3). The indices of the remaining three indicators, namely the proportion of green space per building, building footprint, and population density, were derived from demographic and geospatial data collected during the preparation of the city's structure plan. Population density was calculated at the sub-city level (see Table 23.1). These geospatial data, including Easting (X), Northing (Y), and Altitude (Z) coordinates of property boundaries, introduced additional variability to the vulnerability map by incorporating a building block-based spatial nuance. Subsequently, the indices of the corresponding indicators were aggregated at the building block resolution by employing zonal statistics.

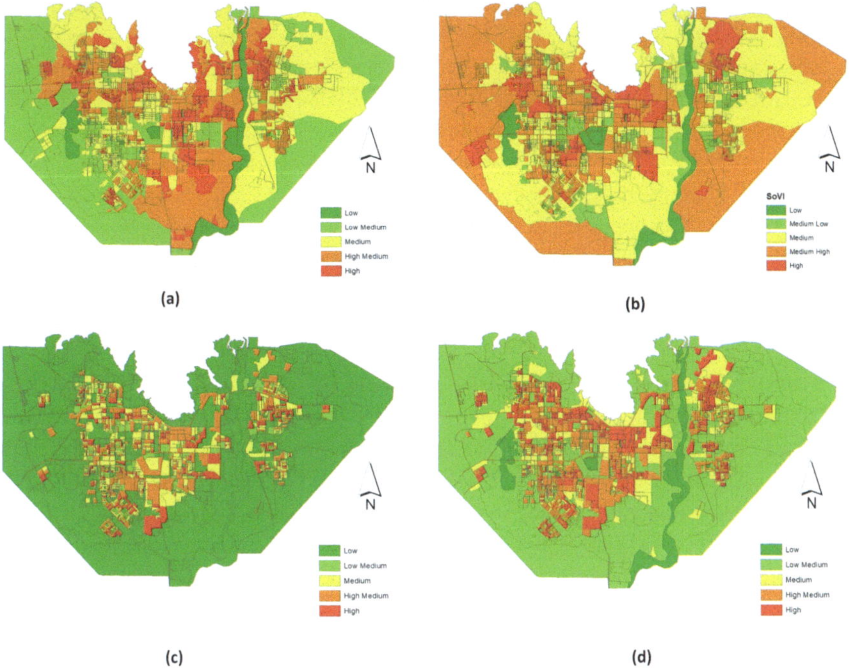

Fig. 23.2 Vulnerability maps of Bahir Dar depicting different levels of vulnerability—scaled as low, low-medium, medium, high-medium, and high. The maps represent (**a**) physical vulnerability, (**b**) social vulnerability, (**c**) economic vulnerability, and (**d**) combined vulnerability encompassing all factors

23.3.3 Assessment Framework

The main purpose of indicator-based vulnerability assessments (IBVAs) is to assess the aspects of the system that define vulnerability within certain domains of the case study area. Most of the time, these aspects are first assessed separately and then aggregated to compute combined vulnerability "indices" or "scores" [35]. The combined vulnerability or the overall vulnerability of the city is the weighted sum of the three dimensions of the vulnerability.

The following methodological processes were followed to assess the three vulnerability dimensions and then the overall/combined vulnerability of the city.

1. The hierarchical structure of the vulnerability assessment model was established: this step aimed to develop the hierarchical structure of the decision-making criteria as combined, dimension-based, and indicator-based.
2. Standardization of the criteria: at this step, the numerical scales of physical, social, and economic vulnerability indicators are standardized/transformed into

a common scale. The original values of the indicators have been transformed to 1–10 based on the vulnerability level, to compute the indices. The values closer to 1 signify low vulnerability, whereas values closer to 10 denote high vulnerability. Each variable was further divided into classes depending on its characteristics: for example, the nature of the response was divided into two classes (yes or no response), three classes, four classes, and five classes, as required. With literature support, these classes were framed to demonstrate the degree of variation, as much as possible, in that particular variable. In dual classes, the values were 1 and 10. The indicators with three classes were assigned the values 3, 7, and 10; for four classes, the values were 2, 5, 7, and 10; and for five classes, the values were 2, 4, 6, 8, and 10. Thus, the composite index for each component fell between 1 and 10.

3. Criteria weighting: in this step, the relative scores of the physical, social, and economic vulnerability indicators were set. The weights of the vulnerability dimensions and the score of each indicator were identified from the expert survey. In ArcGIS, a standardization method of equal interval was used to standardize the numerical value of each vulnerability indicator on a scale from 1 to 10.
4. Decision rules and the mapping of combined vulnerability scenarios: the final overall flood vulnerability of the city was mapped by weighted linear combination (WLC) aggregation methods. This study used an expert survey scoring technique to allocate values to classes of phenomena for each indicator and formulate indices based on the equation.

$$\text{Physical vulnerability : PhyVI} = \sum_{i}^{8} \text{PhVI}_i \text{S}_i \tag{23.3}$$

$$\text{Similarly, Social vulnerability : SoVI} = \sum_{i}^{9} \text{SoVI}_i \text{S}_i \tag{23.4}$$

$$\text{Economic vulnerability : EcoVI} = \sum_{i}^{7} \text{EcoVI}_i \text{S}_i \tag{23.5}$$

Then, the combined/overall flood vulnerability (FVI) of the city was mapped using the following equation.

$$FVI = \sum_{i}^{3} \text{W}_i VI_i \tag{23.6}$$

where PhyVIi, SoVIi, and EcoVIi are the numerical values of the individual physical, social, and economic indicators at a building block level, respectively. PhyVI, SoVI, and EcoVI are the physical, social, and economic vulnerability dimensions. Si's are the scores for each indicator in each dimension. Wi is the weight of each vulnerability dimension, VI_i is the three vulnerability dimensions and FVI is the Flood Vulnerability Index.

Table 23.4 Flood vulnerability ranking [36]

Index range	Descriptions
1–2	Low vulnerability to floods
2–4	Medium-low vulnerability to floods
4–6	Medium vulnerability to floods
6–8	Medium-high vulnerability to floods
8–10	High vulnerability to floods

The functional relationship of indicators and overall/combined vulnerability determines the level of vulnerability at each land use (unit of analysis). Table 23.4 shows how the values of the indicators were further transformed into five-level functional relationships.

23.3.4 Spatial Pattern of Vulnerability Indices

In geospatial analysis, hot spot analysis is often employed to map regions of statistically significant clusters of high and low occurrences of events. Hot spot analysis usually delineates three zones within the sphere of significance of an event: the hot spot, cold spot, and not significant. In this study, GIS analysis of the Getis-Ord Gi* statistic index was employed for the hot spot analysis to categorize the combined vulnerability of physical, economic, and social dimensions in the City of Bahir Dar into hot spots, sub-hot spots, sub-cold spots, and cold spots. The tool in GIS works by looking at the neighborhood features of a selected feature and proportionally comparing the local sum with the sum of values in the whole feature [37]. With that, a *hot spot* indicates an intense clustering of features with high indices; a *cold spot* denotes an intense clustering of features with smaller indices; and the *not significant* denotes areas that cannot be either a clustering of features with high or low values as compared to what would be the result of random chance of occurrences. Each analyzed unit of building blocks obtained a G_i* score by comparing within the context of neighboring land use features (Eqs. 23.7 and 23.8).

$$G_i^* = \frac{\sum_{j=1}^{n} w_{i,j} x_j - \bar{X} \sum_{j=1}^{n} w_{i,j}}{S \sqrt{\frac{\left[n \sum_{j=1}^{n} w^2_{i,j} - \left(\sum_{j=1}^{n} w_{i,j} \right)^2 \right]}{n-1}}} \tag{23.7}$$

$$S = \sqrt{\frac{\sum_{j=1}^{n} x_j^2}{n} - \left(\bar{X}\right)^2} \tag{23.8}$$

where x_i and x_j are the values of neighborhoods i and j, W_{ij} is the spatial weight matrix, and n is the total number of features (building blocks). G_i^* can map hot and cold spots.

23.4 Result and Discussion

Indices for each dimension were calculated using the methodology described in the previous section. The results for each dimension are separately described, followed by an overall explanation of combined vulnerability.

The analysis and assessment of the Flood Vulnerability Index (FVI) of the City of Bahir Dar identified parts of the city with comparatively high, moderate, and less vulnerable places across the city based on three vulnerability components physical, social, and economic perspectives (Figs. 23.2a and 23.3b, c) and the overall FVI of the city (Fig. 23.2d).

23.4.1 Physical Vulnerability (PhyVI)

In the physical vulnerability assessment, eight indicators were used to show the city's susceptibility to flood events, and the results of this analysis are presented (Fig. 23.2a). Blocks colored red were identified as the most food-vulnerable regions based on the assessment. While the physical vulnerability was low in suburban areas, it was high in urban centers. This was due to several factors, including a low proportion of green space, a high proportion of aged houses and poor building material condition, and a high proportion of ground floor or villa housing type and connected houses. In contrast, Kebele 14 (western part) had a lower physical vulnerability due to its percentage of multistory housing, good building material, and relatively higher green space area in the southern part.

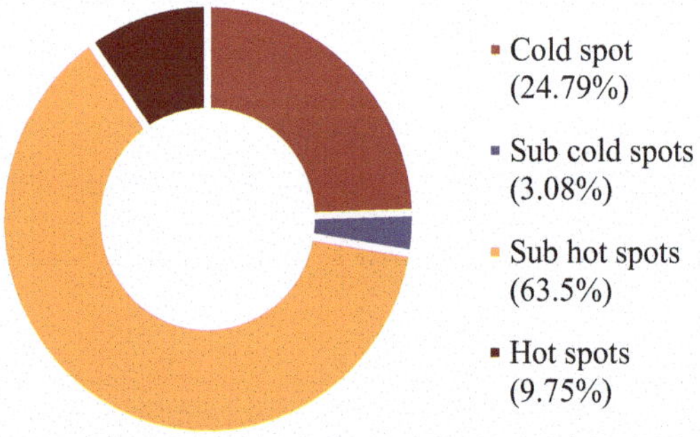

Fig. 23.3 The proportion of combined VI scores in the City of Bahir Dar by % of the area

23.4.2 Social Vulnerability (SoVI)

For the social vulnerability analysis of the city, nine indicators were identified and analyzed. The results of this analysis are presented (Fig. 23.2b), with red indicating high social vulnerability and green indicating low vulnerability to flood events. Areas with a high percentage of unemployment, a high percentage of female-headed households, a high percentage of vulnerable age groups (greater than 60 and less than 18), a low percentage of education level, and an absence of past flood experience were found to have higher to medium-high social vulnerability. These areas were located around the suburban areas of the city and in the inner part of the city and Diaspora Sefere. This suggests that these populations may be more at risk during flood events due to their socioeconomic status and lack of prior experience with floods.

23.4.3 Economical Vulnerability (EcoVI)

In the analysis of economic vulnerability, seven indicators were examined, and the results are shown (Fig. 23.2c). Factors such as high monthly income, household occupation, house-holding basis, and the presence of means of transportation were found to contribute to low economic vulnerability. High economic vulnerability, depicted by the color red, was associated with households that had low income, rental house-holding basis, limited livelihood options, and a lack of readily available transportation. The results showed that economic vulnerability was low in the outskirts of the city and relatively higher in the inner city.

23.4.4 Combined Vulnerability (ComVI)

The combined vulnerability index (Fig. 23.2d) incorporates physical, social, and economic vulnerability dimensions. The results showed that the inner center of the city, where there is high population density, high unemployment, and poor building condition, had a higher combined vulnerability index. In contrast, less developed areas in the outskirts of the city had a lower combined vulnerability index. This suggests that more developed areas in the inner city may be more vulnerable to flood events due to their socioeconomic and physical characteristics.

In deriving the vulnerability indices, a conservative approach was followed, as demonstrated by the high vulnerability observed in most parts of the city in the maps presented earlier. In particular, the social vulnerability map (Fig. 23.2b) and economic vulnerability map (Fig. 23.2c) showed that most parts of the city were highly vulnerable. This could be partly attributed to a shortage of data. For example, vulnerable age groups over 60 and below 18 were used, despite most flood

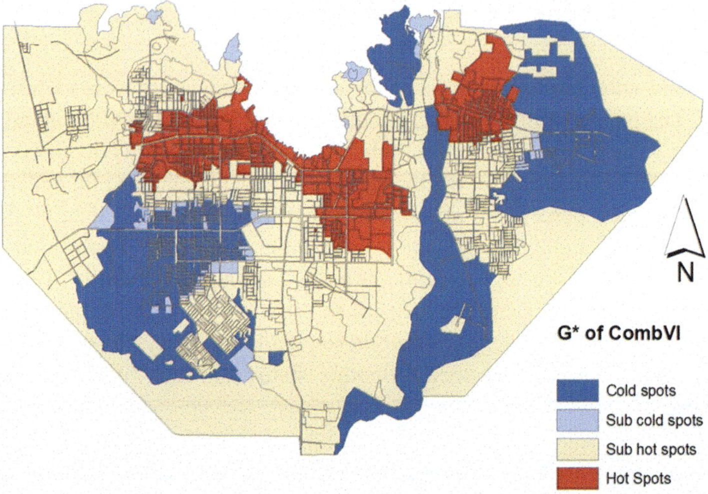

Fig. 23.4 G* of combined vulnerability

vulnerability studies using percentages of ages over 65 and under 14. For female-headed households, the highest vulnerability score was assigned. Additionally, since there were only binary data (yes/no) for indicators such as the percentage of people with experience of flooding, the percentage of unemployed households, and the occupation of households, these indicators resulted in higher social and economic vulnerability scores for most areas. This lack of variation in these indicators resulted in higher combined vulnerability scores (Fig. 23.2d) for most parts of the inner city.

The results of the hot spot analysis (Fig. 23.4) show that most parts of central Bahir Dar, in the Fasilo Subcity, Belay Zeleke Subcity, Tana Subcity, and Atse Tewodros Subcity (see Fig. 23.1) were characterized by significantly high spots. In contrast, the southern part of Kebele 14, Dona Kebele, and the eastern part of Kebele 11 were characterized by cold spots, accounting for 24% of the city. The remaining 63% of the area was characterized by sub-hot spots, while only 3% was characterized by sub-cold spots.

23.5 Conclusions and Recommendations

Planning for flood risk mitigation should take into account multiple flooding sources and scenarios, including future climates and urbanization rates. It should also consider a range of alternative solutions, both traditional and novel, to deal with uncertain future conditions and their possible consequences and damages.

Effective adaptation measures can only be implemented when the nature of the impact is well understood. For urban planning and disaster preparedness, it is important to conduct a quantitative assessment of the increase in flood hazards. However, the results of such studies should be treated only as indicators of a range of possibilities. Nevertheless, such attempts are crucial for realistic adaptation planning.

The results suggest that to decrease the flood risk of the city, a wide variety of measures should be taken. These include developing a policy framework for low-impact development (LID), creating job opportunities and alternative livelihood options, formulating structural measures, increasing drainage capacity, and adapting other flood mitigation measures such as flood zoning, green infrastructure, and large-scale implementation of rainwater harvesting facilities. The development of integrated flood management plans and strategies to minimize flood hazards and build flood resilience is paramount. This involves evaluating both traditional and novel flood protection solutions in terms of their effectiveness in reducing flood extent, depth, duration, flow velocity, and associated damages.

References

1. Hossain, M. N. (2015). Analysis of human vulnerability to cyclones and storm surges based on influencing physical and socioeconomic factors: evidences from coastal Bangladesh. *International Journal of Disaster Risk Reduction, 13*, 66–75.
2. Zhu, W., et al. (2023). A quantitative analysis of research trends in flood hazard assessment. *Stochastic Environmental Research and Risk Assessment, 37*(1), 413–428.
3. Stott, P. (2016). How climate change affects extreme weather events. *Science, 352*(6293), 1517–1518.
4. Lehner, B., et al. (2006). Estimating the impact of global change on flood and drought risks in Europe: A continental, integrated analysis. *Climatic Change, 75*, 273–299.
5. Lyu, H.-M., et al. (2019). Perspectives for flood risk assessment and management for mega-city metro system. *Tunnelling and Underground Space Technology, 84*, 31–44.
6. Petit-Boix, A., et al. (2017). Floods and consequential life cycle assessment: Integrating flood damage into the environmental assessment of stormwater best management practices. *Journal of Cleaner Production, 162*, 601–608.
7. Ketema, A., & Dwarakish, G. (2021). Climate change impacts on water resources in Ethiopia. climate change impacts on water resources: hydraulics, water resources and coastal. *Engineering*, 47–58.
8. Mamo, S., Berhanu, B., & Melesse, A. M. (2019). Chapter 29 – Historical flood events and hydrological extremes in Ethiopia. In A. M. Melesse, W. Abtew, & G. Senay (Eds.), *Extreme hydrology and climate variability* (pp. 379–384). Elsevier.
9. Mailhot, A., & Duchesne, S. (2010). Design criteria of urban drainage infrastructures under climate change. *Journal of Water Resources Planning and Management, 136*(2), 201–208.
10. van Duin, B., et al. (2021). Toward more resilient urban stormwater management systems – Bridging the gap from theory to implementation. *Frontiers in Water*, ***3***, 3.
11. Bibi, T. S., & Kara, K. G. (2023). Evaluation of climate change, urbanization, and low-impact development practices on urban flooding. *Heliyon, 9*(1), e12955.
12. Zhou, Q. (2014). A review of sustainable urban drainage systems considering the climate change and urbanization impacts. *Water, 6*(4), 976–992.
13. DREF, Plan 2010-2011. (2010). International Federation of Red Cross and Red Crescent Societies: Geneva.

14. Akola, J., Binala, J., & Ochwo, J. (2019). Guiding developments in flood-prone areas: Challenges and opportunities in Dire Dawa city, Ethiopia. *Jàmbá: Journal of Disaster Risk Studies, 11*(3), 1–8.
15. IFRC. (2020). Africa: Floods and multiple disasters – Information bulletin. International Federation of Red Cross and Red Crescent Societies.
16. NMA. (2012). Report of DRMFSS-led Flood Task Force comprised of representatives from the National Metrological Agency of Ethiopia, Sectoral Line Ministries, UN agencies, NGOs and donors, under the Disaster Risk Management Technical Working Group (DRMTWG). 2012, National Metreological Agency of Ethiopia.
17. Achamyeleh, K. (2003). WMO/GWP Associated Programme on Flood Management, in Ethiopia: integrated flood management, T.S. Unit. World Meteorological Organization – Global Water Partnership.
18. Vojtek, M., & Vojteková, J. (2016). Flood hazard and flood risk assessment at the local spatial scale: a case study. *Geomatics, Natural Hazards and Risk, 7*(6), 1973–1992.
19. Nasiri, H., Mohd Yusof, M. J., & Mohammad Ali, T. A. (2016). An overview to flood vulnerability assessment methods. *Sustainable Water Resources Management, 2*(3), 331–336.
20. Huang, D., et al. (2012). An assessment of multidimensional flood vulnerability at the provincial scale in China based on the DEA method. *Natural Hazards, 64*(2), 1575–1586.
21. Chang, H., et al. (2021). Assessment of urban flood vulnerability using the social-ecological-technological systems framework in six US cities. *Sustainable Cities and Society, 68*, 102786.
22. Bigi, V., et al. (2021). Flood vulnerability analysis in urban context: A socioeconomic sub-indicators overview. *Climate, 9*(1), 12.
23. Usman Kaoje, I., et al. (2021). Physical flood vulnerability assessment using geospatial indicator-based approach and participatory analytical hierarchy process: A case study in Kota bharu, Malaysia. *Water, 13*(13), 1786.
24. Chan, S. W., et al. (2022). A systematic review of the flood vulnerability using geographic information system. *Heliyon, 8*(3), e09075.
25. ANRS. (2007). Population Size of Bahir Dar City administration, Bahir Dar, B.o.F.a.E. Development.
26. CSA.(1991). The 1984 Population and Housing Census of Ethiopia, C.S. Authority, Editor. Addi Ababa.
27. CSA. (1995). The 1994 Population and Housing Census of Ethiopia, C.S. Authority, Editor.: Addis Ababa.
28. CSA. (2007). The 2007 Population and Housing Census of Ethiopia, S. Authority, Editor. Addis Ababa.
29. Bahir Dar City Structural Plan Preparation Project Office. (2020). Bahir Dar Regio-Polytan City Structure Plan: Population Projection Report. Bahir Dar.
30. Bahir Dar City Structural Plan Preparation Project Office. (2020). An Executive Summary of Study Report of the Bahir Dar Regio-Politan City Structure Plan. Bahir Dar University: Bahir Dar.
31. OpenStreetMap. (2023). OpenTopoMap basemap, accessed through QGIS's quickmap services plugin.
32. Abson, D. J., Dougill, A. J., & Stringer, L. C. (2012). Using principal component analysis for information-rich socio-ecological vulnerability mapping in Southern Africa. *Applied Geography, 35*(1–2), 515–524.
33. Müller, A., Reiter, J., & Weiland, U. (2011). Assessment of urban vulnerability towards floods using an indicator-based approach–a case study for Santiago de Chile. *Natural Hazards and Earth System Sciences, 11*(8), 2107–2123.
34. Atteslander, P., et al. (2008). *Methoden der empirischen Sozialforschung* (p. 359). Erich Schmidt Verlag.
35. Birkmann, J., et al. (2013). Framing vulnerability, risk and societal responses: the MOVE framework. *Natural Hazards, 67*(2), 193–211.

36. Balica, S. F. (2007). Development and application of flood vulnerability indices for various spatial scales.
37. Esri. (n.d). How Hot Spot Analysis (Getis-Ord Gi*) works [cited 2023 12/07/2023]; Available from: https://pro.arcgis.com/en/pro-app/latest/tool-reference/spatial-statistics/h-how-hot-spot-analysis-getis-ord-gi-spatial-stati.htm

Chapter 24
Hydrological Impact of Khat Cultivation in Aba Gerima Watershed, Lake Tana Basin, Ethiopia

Kindie Bitew Worku, Dagnenet Sultan, Nigussie Haregeweyn, Mulatu Liyew Berihun, Menber Berihun Nigatu, and Mequanint Demeke Aynalem

24.1 Introduction

Land degradation in the Ethiopian highlands poses a significant threat to subsistence agriculture, which is the backbone of the economy [1]. The sector faces persistent challenges due to natural resource depletion, soil erosion, and a shift towards commercial crops [2–4].

Khat (*Catha edulis*) is a perennial crop that is extensively cultivated in Ethiopia for its commercial value, ranking second in export value after coffee. It is a flowering evergreen shrub [3] that is chewed for its stimulating effects and is deeply ingrained in the social fabric of certain Ethiopian communities [5]. Khat is harvested multiple times a year for up to 50 years and is grown as a cash crop within a limited height range [6]. Although it offers short-term economic benefits, its cultivation replaces sustainable agroforestry systems [8]. Khat requires minimal labor and inputs and can be intercropped with other food crops like maize and coffee [9]. In the eastern part of Ethiopia, it is planted on hillsides in rows along contour lines [7].

The growth of khat plantations depends on rainfall in terraced mountains or shallow trench/pipe irrigation in flatlands. It adapts well to well-drained soil and various

K. B. Worku (✉) · D. Sultan · M. L. Berihun · M. B. Nigatu
Faculty of Civil and Water Resources Engineering, Bahir Dar Institute of Technology, Bahir Dar University, Bahir Dar, Ethiopia

N. Haregeweyn
International Platform for Dryland Research and Education, Tottori University, Tottori, Japan

M. D. Aynalem
Faculty of Chemical and Food Engineering, Bahir Dar Institute of Technology, Bahir Dar University, Bahir Dar, Ethiopia

S. W. Fanta et al. (eds.), *Sustainable Development Research in Manufacturing, Process Engineering, Green Infrastructure, and Water Resources*, Green Energy and Technology, https://doi.org/10.1007/978-3-031-77339-6_24

climatic conditions, tolerating drought for several months [10]. Unlike coffee or cereals, khat does not have a specific harvest time [7]. Its cultivation is closely tied to the availability of sufficient water, although it contributes to the rapid depletion of groundwater due to its high water requirements [3, 11].

Three percent of Ethiopia's total khat production comes from the Bahir Dar Zuria district, specifically the Aba Gerima watershed [12]. Khat cultivation has become increasingly common in this area, with a significant expansion observed over the years. Satellite image analysis from 2005 to 2016 indicates an annual areal expansion rate of approximately 53% in the Aba Gerima watershed [13].

While khat-related studies in Ethiopia and around the world have mainly emphasized its impact on human health, legal stance, social-economic, cultural, and historical perspectives [3, 9, 14–19], there is a lack of research on the water requirements of khat [5]. Furthermore, no systematic studies have been conducted on the environmental impact of khat plantations, including their effects on runoff and soil conservation. Understanding the hydrological effects of plantation practices is crucial for developing effective strategies to control soil erosion. Therefore, we aim to study surface runoff, soil loss, and moisture in different age categories of khat in the Aba Gerima watershed.

24.2 Methods

24.2.1 Study Area Description

The study was conducted in the Aba Gerima watershed, which is located in the Lake Tana basin of the Amhara National Regional State (ANRS) in Ethiopia (see Fig. 24.1). It is situated 20 km northeast of the regional city of Bahir Dar, with its outlet spanning from 37°29′30″E to 37°31′0″E of longitude and 11°38′30″N to 11°40′0″N of latitude. The altitude of the watershed ranges from 1993 to 2120 meters above sea level, covering an area of approximately 720 hectares. The topography is characterized by flat and gentle slopes, with some undulating areas in the upstream edges.

The study site has been part of the national government's regular extension programs and public-based soil and water conservation interventions, although experiences with externally funded programs have varied greatly [20, 21]. Since 2011, the Aba Gerima watershed has been one of the learning watersheds in Ethiopia where integrated watershed management has been implemented by the Water and Land Resource Center (WLRC), funded by the Swiss Agency for Development and Cooperation (SADC) [4]. A recent study conducted in our study watershed [13] revealed that khat cultivation in the Aba Gerima site has increased over time due to farmers' growing interest in increasing farm income by shifting from cereal-based farming to cash crops.

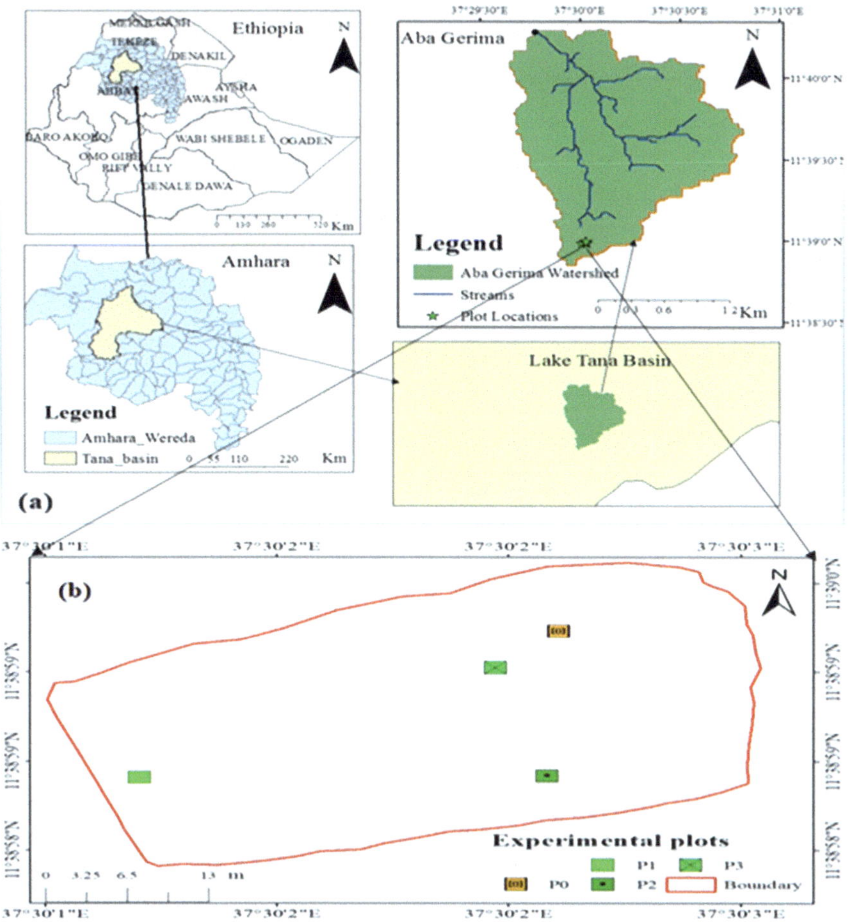

Fig. 24.1 Map of the study area. (**a**) location of Aba Gerima watershed within Ethiopia, Amhara region, and Lake Tana basin. (**b**) location of the experimental plots in the study watershed (P0: control plot, P1: initial stage of khat plot, P2: developmental stage of khat plot, P3: maturity stage of khat plot)

24.2.2 Experimental Setup, Data Collection, and Analysis

Four experimental plots were established for the farmer's khat fields, each measuring six meters in length and three meters in width. These plots were delimited by a 60 cm galvanized sheet, with 15 cm inserted into the ground to prevent lateral runoff. The experimental plots represented different stages of khat plantation: a control plot (bare land, P0), an initial growth stage (2 years old, P1), a developmental stage (6 years old, P2), and a maturity stage (9 years old, P3) (Fig. 24.2).

Fig. 24.2 The runoff plots from the khat field in the Aba Gerima watershed; the photographs were taken in 2018 during the rainy season

Table 24.1 Laboratory and gravimetric soil analysis for the experimental plots

		Grain size distribution (%)								
Plot (code)	Soil class	Sand	Clay	silt	MC (%)	EC (ds/m)	FC (%)	PWP (%)	AWC (%)	Bd (g/cm^3)
P1	Clay	17	56	27	27	0.05	32	21.2	10.8	1.4
P2	Clay loom	39	34	31	39	0.16	32	21.7	10.2	1.3
P3	Clay loom	33	36	31	40	0.1	33	23.2	9.8	1.3
P0	Clay	33	40	27	42	0.06	35	22.6	12.1	1.4

To characterize each site, various soil variables were determined (Table 24.1). Soil texture was analyzed by taking three samples from the top 0 cm of the soil profile at 10 cm intervals using the hydrometric method [22]. The average particle-size distribution was used to characterize the site. Undisturbed soil samples were also collected at 10 cm intervals to a depth of 90 cm using a core sampler with a volume of 100 cm^3. These samples were oven-dried at 105 °C for 24 h and weighed to determine soil bulk density.

All experimental plots were situated on a horizontal slope of approximately 10%. At the lower end of each plot, a 1.5 m^3 pit (1 m deep, 1.5 m long, 1 m wide)

was dug and covered with waterproof geomembrane plastic (0.5 mm thick) to collect daily runoff and sediment data. The design of the runoff collection pits took into account extreme rainfall events, considering anticipated rainfall and runoff coefficients. Due to financial and field constraints, no replicates were included in the experimental plots. However, the age and tree densities of khat, as well as the distribution pattern and slope appearance, were similar among the plots.

To gather data for the study, primary data sources were utilized. Data collection involved general field observation, daily monitoring during the rainy season (including rain, runoff, and sediment data collection), providing water for khat irrigation in the dry season, and using tools such as the Em50 recorder and Decagon EC-5 sensor, core sampler, and auger for soil data collection. Additionally, focus group discussions were conducted to support the study objectives.

24.2.3 Rainfall Measurement and Climate Data

Meteorological data spanning from 2000 to 2019 were collected from the Bahir Dar and Aba Gerima meteorological stations, which were installed on-site by the SATREPS project during the investigation period. The collected data included precipitation, maximum and minimum temperature, relative humidity, sunshine hours, solar radiation, wind speed, and other relevant data required by the FAO-56 Penman-Monteith method. These data were used to calculate the ETO and crop water requirements using the coefficient approach for crop evapotranspiration.

24.2.4 Soil Data Analysis

Field and soil property measurements were conducted during the periods of 2018 and 2019. A total of 24 soil samples were collected and sent to the Amhara Design and Supervision Works Enterprise laboratory for analysis. The analyzed parameters included field capacity (FC), permanent wilting point (PWP), bulk density, and texture. For each depth range (0–30, 30–60, and 60–90 cm), two soil samples were taken (Table 24.1).

The soil's textural composition was determined using the Bouzoukis hydrometer [22] procedure at all depths of the soil moisture sensor. Bulk density was measured by drying the soil samples in an oven at 105 °C for 24 h and then weighing them using a mass balance. The soil moisture content at field capacity and permanent wilting points were determined using a pressure plate apparatus [23, 24], applying pressures of 1/3 bar and 15 bar, respectively. Based on the obtained results, the total available water from the experimental field was calculated using Eq. 24.1.

$$\text{TAW} = \theta\ FC - \theta\ PWP \tag{24.1}$$

Where TAW = Total available water in the root zone (mm/m), θFC = Moisture content (vol. %) at field capacity, and θPWP = Moisture content (vol. %) at permanent wilting point.

24.2.5 Plot Water balance Determination

To investigate possible trends and fluctuations in the khat field of the basin, it is essential to measure the inputs and outputs of water and changes in water storage. This process is known as the water balance. Once the ETc value is determined, the required irrigation water can be calculated by solving the water balance equation. The depletion of water is expressed in Eq. 24.2.

$$ETc = \Delta S + \mathrm{P} + \mathrm{I} \pm \mathrm{D} - \mathrm{R} \tag{24.2}$$

Where ΔS = change in root zone soil moisture storage for a given interval, the inflow and outflow variables required in the water balance equation are measured from the Decagon EC-5 soil moisture sensor and can be either positive or negative, P = Precipitation(mm), I = Irrigation(mm), R = Runoff(mm), taken from average value for the 2 years from experimental plots of the khat field, D = Drainage or Groundwater Loss(mm), since the groundwater loss due to deep percolation is difficult to measure, it is minimized by maintaining the soil moisture condition in the plot at field capacity, which can be either positive or negative in value. A positive D represents downward drainage, while a negative D represents a net gain of water by a capillary rise below the soil profile, ETc = crop evapotranspiration(mm).

24.2.6 Rainfall, Runoff, and Soil Loss Measurement

The study site was equipped with an automatic bucket rain gauge (Hobo Data RG3M Rain Gauge Record, Onset Computer Corporation, Bourne, MA, USA) as well as a manual rain gauge. The tipping bucket rain gauge recorded rain peaks, with each tip representing 0.2 mm of rain. To account for the nature of tropical rain, only events lasting at least 15 min and separated by at least 30 min from other events were considered [25]. These events, along with precipitation data from the tipping bucket rain gauge, were used to determine rain intensity and characterize rainfall in the study area from September 2018 to November 2019.

For rainfall-runoff analysis, manual rain gauge data recorded from July to October was utilized since runoff was measured only daily over extended periods. During this period, a total of 136 daily rainfall events were manually measured, ranging from 12 to 49 mm.

To analyze sediment concentration, runoff collected in the pit was mixed vigorously with a floor brush until all deposited sediment was thoroughly mixed with water. A daily sample was taken each morning at 6:00 AM from the runoff collected in plastic water bottles. The runoff and sediment collector pits were emptied each morning after sampling, and the geomembrane was inspected for leaks.

The depth of runoff for each experimental plot was calculated by dividing the runoff volume by the corresponding plot area while accounting for the direct rainfall on the collecting pit. The runoff depth and sediment concentration were measured and analyzed in the forest laboratory of the BDU Zenzelma campus using Whatman filter paper no. 42. Sediment concentration was determined after filtering and drying the sediment in an oven at 105 °C for 24 h. Daily soil loss for each plot was obtained by multiplying the sediment concentration (in g L^{-1}) by the corresponding total output volume.

The total seasonal soil loss (SL) was calculated by summing the daily soil loss values for the months with runoff-generating rainfall (see Fig. 24.3). Soil loss values were expressed in grams and converted to t ha^{-1}, considering the area of the experimental plot. The control treatment (P0) was considered as having 100% runoff and soil loss, indicating zero reduction. The efficiency of the three treatment plots (representing different ages and tree densities of the khat plant) was analyzed in comparison to the control plots.

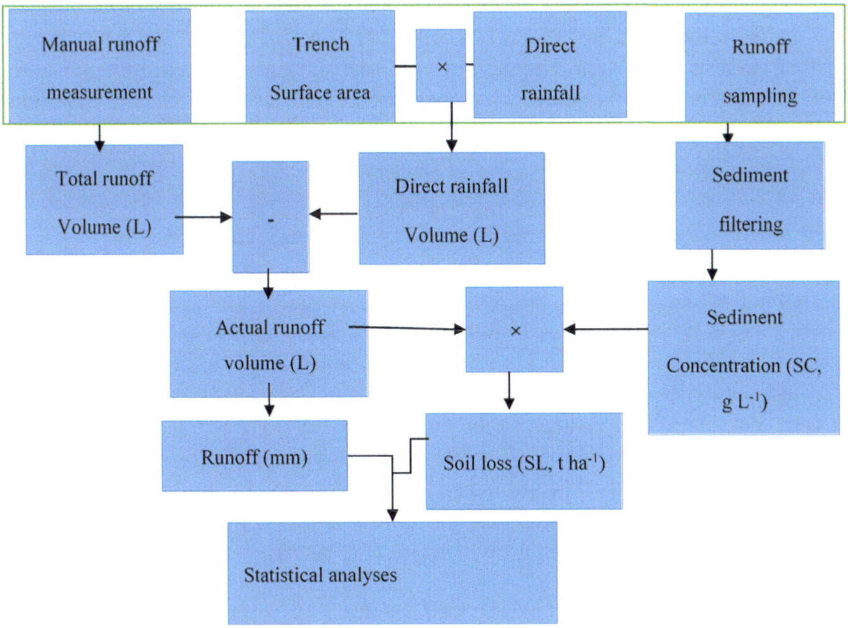

Fig. 24.3 Methodological framework for runoff and soil loss analysis (the arrow indicates simply the flow of the idea)

24.2.7 Moisture Data Measurement

Moisture data was continuously recorded in the field using EC-5 Sensors (Decagon Devices, Pullman, WA) from September 2018 to December 2019. Each of the four experimental plots was equipped with ECH2O probes and EC-5 sensors, along with appropriate data loggers, to enable continuous monitoring of soil moisture.

The Decagon EC-5 sensor utilizes capacity/frequency domain technology to determine the volumetric water content (VWC) by measuring the material's dielectric constant [26]. For this study, EC-5 soil moisture tension sensors were installed separately on the four experimental plots at three different depths (30, 60, and 90 cm) to measure available soil moisture in the khat fields. Port 1, Port 2, and Port 3 were programmed to receive EC-5 sensor readings installed at depths of 30, 60, and 90 cm, respectively, from the ground surface.

Data was recorded at a frequency of 30 min and stored in the Em50 data logger. The data was downloaded through the ECH2O computer software by connecting a personal computer to the software and the wired data logger as needed. The downloaded data was then averaged on a daily basis for further analysis of moisture data.

24.2.8 Reference Evapotranspiration (ETO)

The FAO Penman-Monteith method was used to calculate baseline evapotranspiration for historical data for 19 years (2000–2019) at Bahir Dar and Aba Gerima weather stations (Eq. 24.3). The data includes daily values of maximum and minimum temperature, precipitation, wind speed, relative humidity, and solar radiation.

The Penman-Monteith equation used was:

$$ET_0 = \frac{0.408\Delta\left(Rn - \mathrm{G}\right) + \gamma\frac{900}{\mathrm{T}+273}\mathrm{u2}\left(es - ea\right)}{\Delta + \gamma\left(1 + 0.34\mathrm{u2}\right)} \tag{24.3}$$

Where ETO is reference evapotranspiration [mm day-1], Rn is net radiation at the crop surface [MJ m-2 day-1], G is soil heat flux density [MJ m-2 day-1], es saturation vapor pressure [kPa], ea. actual vapor pressure [kPa], es- ea. saturation vapor pressure deficit [kPa], Δ slope vapor pressure curve [kPa $°C^{-1}$], γ psychometric constant [kPa $°C^{-1}$].

Monthly average daily ETC (mm day-1) was calculated by multiplying the reference monthly average potential evapotranspiration (ETO, mm day-1) by the crop coefficient (Kc) in the different growth stages (Eq. 24.4).

$$ETc = ETo \times Kc \tag{24.4}$$

The duration of the development phases and the corresponding crop coefficient (Kc) for the perennial citrus crop, which has similar characteristics with khat, are the initial phases (150 days, 0.7), the development phase (30 days, 0.65), and the maturity or the final phase values (185 days, 0.64) with cropping index of 0.5 [27].

24.3 Results and Discussions

24.3.1 Effect of Khat on Seasonal Runoff, Runoff Coefficient, and Soil Loss

Figures 24.4 and 24.5 present the amount of runoff and SL observed during the two study seasons (2018 and 2019) in the khat plantation. The overall runoff response for P1, P2, P3, and P0 (from August through October) from the plots for the 2018 to 2019 season varied from 3.04 to 3.41, 2.81 to 3.16, 2.75 to 2.89, and 7.79 to 8.69 mm, respectively.

The soil loss also varied from 0.11 to 0.13, 0.08 to 0.11, 0.06 to 0.07, and 0.34 to 0.27 t ha-1, as shown in Table 24.1.

The daily rainfall that generates runoff and the rate of soil loss for the control plot was higher than that of the khat field plots. We also found that the runoff and rate of soil loss were higher in the initial growth stages than in the development and maturity stages of khat. This is due to the plantation work that softened the soils and the high aggregate stability of the soil during the development and growth stages of maturity compared to the initial growth stages of khat.

During the measuring period of August through October 2018, the total seasonal runoff estimated was 295.89 mm, 115.38 mm, 107.14 mm, and 104.48 mm from P0, P1, P2, and P3, respectively. For the measuring period of June to October 2019, the total seasonal runoff estimated was 852.02 mm, 333.72 mm, 309.39 mm, and 283.27 mm from P0, P1, P2, and P3, respectively. Figure 24.6 shows that the total seasonal soil loss estimated during 2018 was 13.56 t ha-1, 4.32 t ha-1, 3 t ha-1, and 2.45 t ha-1 from P0, P1, P2, and P3· respectively. For the year 2019, it was 26.57 t ha^{-1}, 12.35 t ha^{-1}, 10.78 t ha^{-1}, and 6.78 t ha-1 for P0, P1, P2, and P3, respectively.

The different growth stages of khat significantly affect the runoff and soil loss rate and formation in the Aba Gerima watershed (moist Wayne Dega agro-climatic region). For khat field plots of P1, P2, and P3, the observed reduction in seasonal runoff according to P0 (control plot) was by 61%, 64%, and 65%, respectively, in the measuring season of 2018 and 61%, 64%, and 67%, in the growing year of 2019. The seasonal soil loss was by 54%, 59%, and 74%, respectively, during the measuring season of 2018 and 68%, 78%, and 82% in the measuring season of 2019.

The variation in the value of total seasonal runoff and soil loss for the experimental plots during the two-year analysis periods (2018–2019) is due to the difference in the number of data observed. The highest seasonal soil loss was observed during

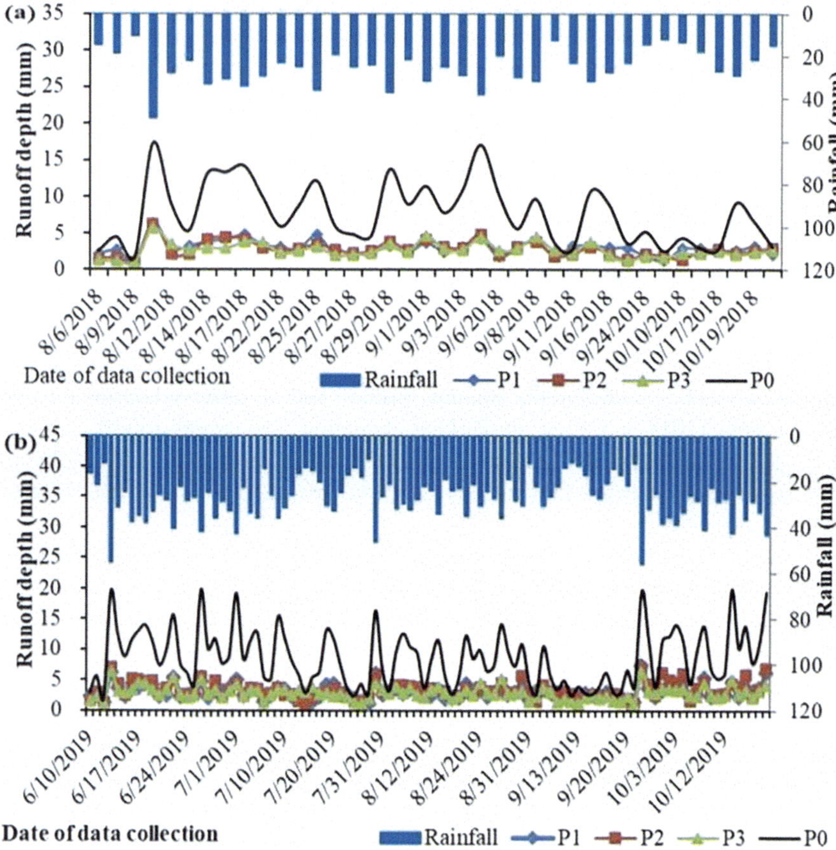

Fig. 24.4 Analysis of seasonal rainfall-runoff relation for experimental plots (**a**), 2018 and (**b**), 2019

the highest runoff generation periods, which means that high soil loss was generated from an experimental plot with high runoff.

In general, the total seasonal runoff depth ranged from 104 to 296 mm in 2018 and 283 to 852 mm in 2019, whereas the total seasonal soil loss ranged from 2.45 to 13.56 t ha-1 in 2018 and 6.81 to 26.57 t ha^{-1} in 2019 from the experimental plots of the Aba Gerima site.

The runoff and soil loss rate for the khat field at Aba Gerima watershed is within the tolerable range as recommended by [2]. For the Upper Blue Nile basin of Ethiopia, the seasonal value of runoff varies from 52 mm to 895 mm, and soil loss rate from 0.07 t ha^{-1} to 39.67 t ha^{-1}. As the age (growth stages) of the khat increases, the efficiency in reducing runoff and soil loss also increases. This is attributed to the well-established and more dense vegetation in the later stages compared to the initial stage (see Fig. 24.2). The runoff from the maturity and developmental growth

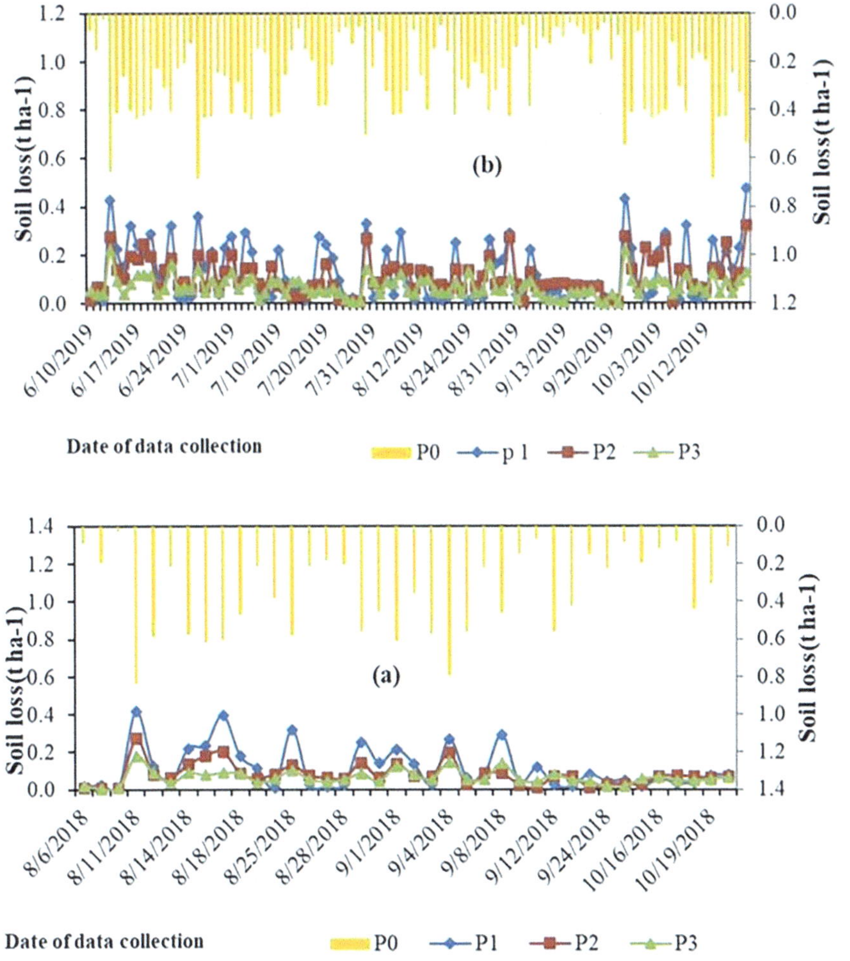

Fig. 24.5 The analysis of soil loss at each experimental plot for the study area in the two measuring seasons (**a**) 2018 and (**b**) 2019

stages of the khat field is low due to the low dry bulk density of the soil (1.3 g cm^{-3}; Table 24.1).

The relationship between rainfall and runoff, as well as runoff and soil loss, was analyzed for each of the experimental plots during the study period (Fig. 24.7). The data for rain runoff and runoff for soil loss are highly correlated with the control plot compared to the khat field plots. This can be attributed to the influence of the khat plantation in minimizing runoff generation and reducing the rate of soil loss. This aligns with the idea that as the age of the khat planted in the agricultural field advances, the effectiveness of runoff and soil loss control increases [28].

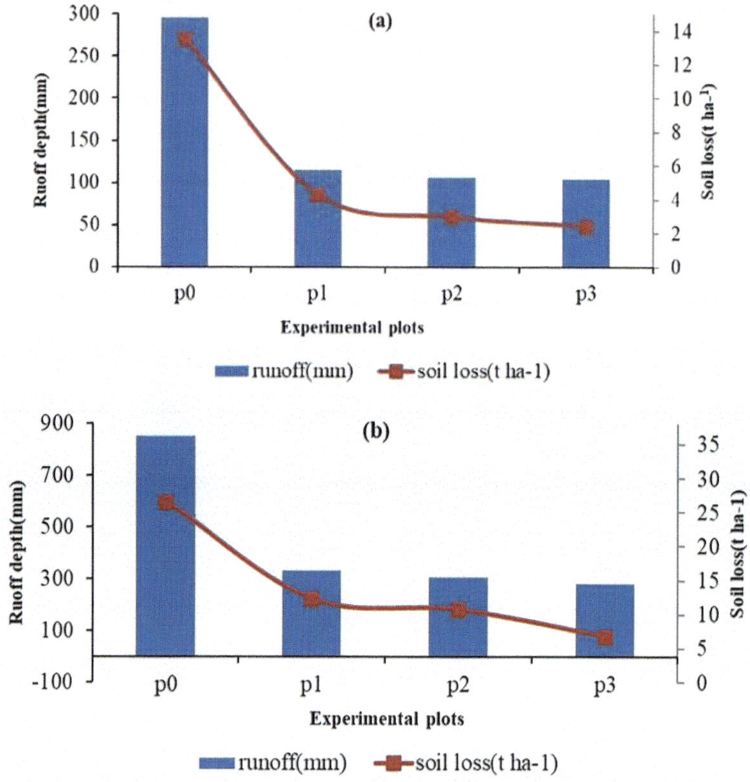

Fig. 24.6 The seasonal runoff and soil loss for an experimental plot in the Aba Gerima watershed (**a**), 2018, and (**b**), 2019

The higher reduction in runoff and soil loss from the khat-treated plots is attributed to the root system of khat, which joins the soil and interlocks with other roots, forming a more solid mass that is less susceptible to runoff and soil loss.

The regression curve fitted to the daily runoff and soil loss, analyzed for the two measuring seasons, is shown in Fig. 24.7. It is observed that khat plantation is more efficient in reducing soil loss than runoff, as indicated by the higher regression coefficient in the soil loss column. This is likely due to the limited storage and infiltration capacity of the khat field for infiltration excess runoff rates. Previous studies in different regions of Ethiopia have also shown that soil and water conservation practices are generally more effective in reducing soil loss than runoff [21, 29–31], which supports the effectiveness of khat plantations in reducing soil loss.

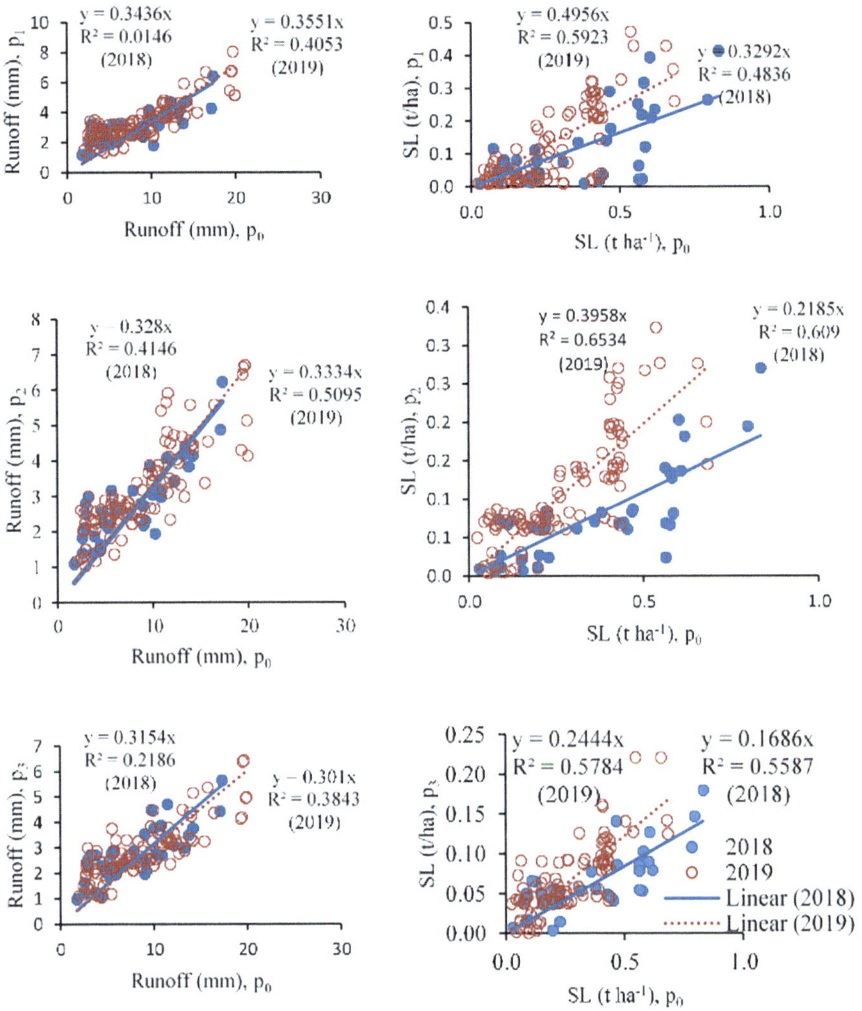

Fig. 24.7 Regression curves fitted to daily runoff (left) and SL (right) data for control (p0) versus khat field (p1, p2, and p3) plots at the Aba Gerima site

24.3.2 Soil Moisture Availability

Vegetation and crops rely heavily on the moisture present at the root level rather than solely on precipitation events [32]. Effective water management and irrigation scheduling require accurate information about local soil moisture conditions [33]. The availability of moisture stored in the soil is crucial for the growth of vegetation [26].

Insufficient water, indicated by soil moisture dropping below the wilting point, can lead to plant stress and survival challenges [34]. Conversely, excessive water can also have detrimental effects. The availability of moisture in the root zone plays a significant role in determining the distribution of plant species [35].

The available soil moisture was found to be high at shallow soil depths (30 cm) and low at effective soil depths (60 cm) across all experimental plots (Fig. 24.8). This can be attributed to the addition of natural fertilizer (compost), the drop of dry leaves of khat, and irrigation during the dry season at the surface of the khat field. The moisture available at 90 cm depth was again high compared to the 60 cm depth, which could be due to the capillary rise of the groundwater.

The monthly average available soil moisture from the experimental plots is shown in Fig. 24.9. It is evident that soil moisture at the control plot is low and shows a linear change between each month compared to the khat field plots. This can be attributed to the effect of plant cover on soil moisture, where plant cover reduces the formation of runoff by storing water in their roots. Among the khat field plots, the average moisture availability from P1 is low compared to P2 and P3 (Fig. 24.9). This could be due to the dense plantation of khat that stores water with low runoff during the rainfall period at P3 or the high runoff occurrence at P1.

The lower moisture availability at P1 can be attributed to the higher water requirement during the initial stage of khat compared to the maturity stage. As a result, the moisture present in the root zone of the soil is extensively utilized by the crop. The EC-5 sensor, used for measuring soil moisture, showed significant results for both soil types (clay loam and clay) in all four plots ($p < 0.001$; Table 24.1).

The performance analysis of the EC-5 sensor revealed that it performed better in P3 compared to P0 and P1 (Fig. 24.10). The available soil moisture was found to be

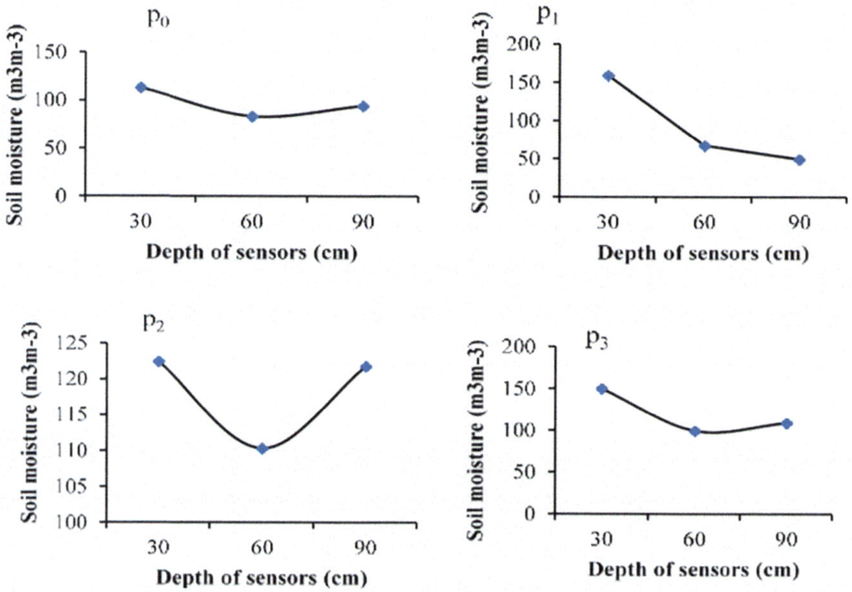

Fig. 24.8 Mean volumetric water content at different depths of each experimental plot

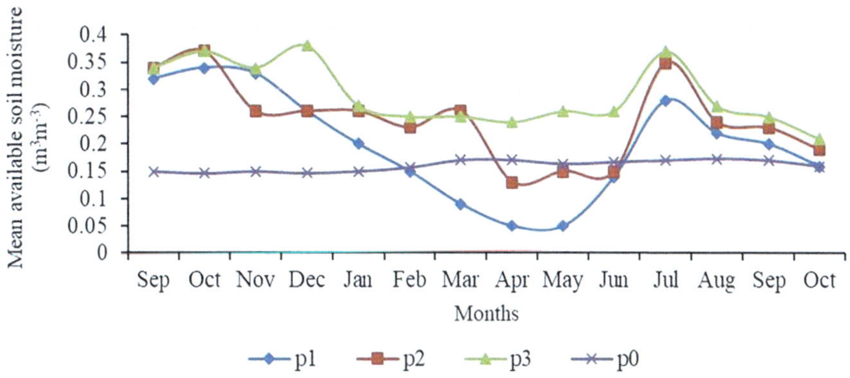

Fig. 24.9 Monthly average soil moisture content for experimental plots

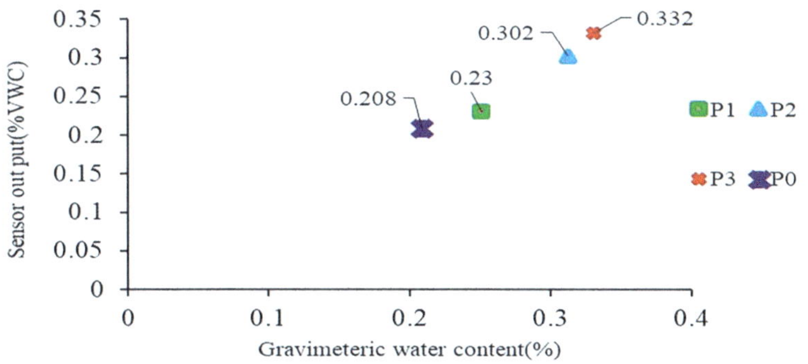

Fig. 24.10 Performance of a decagon EC-5 sensor for the plots

lower in the control plots or plots without khat plantation, highlighting the importance of plantations like khat for soil conservation.

24.3.3 Water Balance in the Khat Field

The average water balance in the Aba Gerima watershed from the khat field was determined by considering the inflow, outflow, and changes in storage parameters for the periods of 2018 and 2019 (Fig. 24.11).

The monthly average values for evapotranspiration (ET), runoff (R), changes in storage (Δs), drainage (D), and precipitation (P) were recorded as 124.3, 151.3, 20.1, 25.5, and 321.3 mm, respectively. The total monthly values for the water balance parameters, including ET, R, Δs, D, and P, were determined as 1865, 2269, 302, 383, and 4819 mm, respectively.

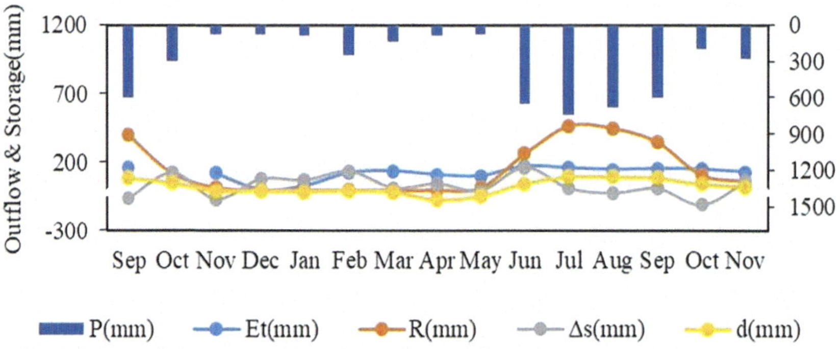

Fig. 24.11 Average monthly water balance from the khat field

24.4 Conclusions

The study investigated the effects of khat plantation on runoff, soil loss, and water balance in the Lake Tana basin by using three different growth stages of khat plots (P1: initial stage; P2: developmental stage; and P3: maturity stage) and one control plot (P0). The results showed that the khat plantation had a significant impact on the hydrology of the Aba Gerima watershed, with strong linkages between growth stages, vegetation structure, and soil conservation activity with their corresponding hydrological responses in the catchment.

The data indicated that seasonal runoff, soil loss, and moisture availability were high at P1 (initial stage) but low at P3 (maturity stage). This is because the plantation of khat improves soil fertility and soil aggregate stability by reducing surface runoff and increasing water infiltration into the soil. Khat cultivation has a profound influence on the hydrological cycle, reducing runoff by 65% and soil loss by 82%. Therefore, the hydrological dynamics of our study basin are greatly influenced by the initial stage rather than the maturity stages of khat crop cultivation.

This knowledge and understanding can help policymakers plan for proper soil and water management in rapidly expanding khat plantations in Ethiopia and other regions.

Acknowledgments We would like to acknowledge the support of the Japan Society for the Promotion of Science KAKENHI program (Grant Number JP25257417) in funding this study. We are grateful to Anteneh Wubet and Agerselam Gualie for their assistance in facilitating our field and laboratory work. We also extend our thanks to the site guards and other individuals who provided valuable support throughout the research process, including plot installation, data collection and processing, and other tasks.

References

1. Wordofa, M. G., Okoyo, E. N., & Erkalo, E. (2020). Factors influencing adoption of improved structural soil and water conservation measures in eastern Ethiopia. *Environmental Systems Research, 9*(1), 1–11.
2. Gebrehiwot, M., Elbakidze, M., Lidestav, G., Sandewall, M., Angelstam, P., & Kassa, H. (2016). From self-subsistence farm production to khat: Driving forces of change in Ethiopian agroforestry home gardens. *Environmental Conservation, 43*(3), 263–272.
3. Mekuria, W. (2018). Public discourse on Khat (Catha edulis) production in Ethiopia. *Journal of Agricultural Extension and Rural Development, 10*(10), 192–201.
4. Sultan, D., et al. (2018). Efficiency of soil and water conservation practices in different agroecological environments in the Upper Blue Nile Basin of Ethiopia. *Journal of Arid Land, 10*(2), 249–263.
5. Atroosh, K. B., & Al-Moayad, M. A. (2012). Water requirements of qat (catha edulis) cultivation in the central highlands of Yemen. *Journal of Scientific Research, 4*(1), 77–77.
6. Ademe, B. W., Coates, J., Dalsgaard, A., Brimer, L., & Lema, T. B. (2017). Khat production and consumption; its implication on land area used for crop production and crop variety production among rural household of Ethiopia. *Journal of Food Security, 5*(4), 148–154.
7. Klingele, R. (1998). *Hararghe farmers on the cross-roads between subsistence & cash economy*. United Nations Development Programme, Emergencies Unit for Ethiopia.
8. Manlosa, A. O., Schultner, J., Dorresteijn, I., & Fischer, J. (2019). Capital asset substitution as a coping strategy: Practices and implications for food security and resilience in southwestern Ethiopia. *Geoforum, 106*, 13–23.
9. Binalfew, T. (2017). The expansion of production, marketing and consumption of chat in Ethiopia. *International Journal, 4*(3), 16–26.
10. Feyisa, T. H., & Aune, J. B. (2003). Effects on the farming system in habro district.
11. Dessie, G. (2013). *Land and water for drugs, cash for food: Khat production and food security in Ethiopia* (p. 60).
12. Birhane, B. W. (2014). The effect of Khat (Catha edulis) chewing on blood pressure among male adult chewers, Bahir Dar, north West Ethiopia. *Science Journal of Public Health, 2*(5), 461.
13. Berihun, M. L., et al. (2019). Exploring land use/land cover changes, drivers and their implications in contrasting agroecological environments of Ethiopia. *Land Use Policy, 87*, 104052.
14. Yahya, A., et al. (2016). Socio-economic and health effects of Khat chewing in Mekelle, Tigray region, Ethiopia. *International Journal of Pharmaceutical Sciences and Research, 8*(1), 11–22.
15. Yimer, A., & Khan, M. (2015). Determination of iron, cobalt, chromium and copper metals in commercially available khat (Catha Edulis Forsk) in Arba Minch, Ethiopia, Int. *Journal of Engineering Research and Applied Science, 5*(1), 66–74.
16. Gebissa, E. (2004). *Leaf of Allah: Khat & agricultural transformation in Harerge, Ethiopia 1875–1991*. Ohio State University Press.
17. Gebissa, E. (2010). Khat in the horn of Africa: Historical perspectives and current trends. *Journal of Ethnopharmacology, 132*(3), 607–614.
18. Kandari, L. S., Yadav, H. R., Thakur, A. K., & Kandari, T. (2014). Chat (Catha edulis): A socio economic crop in Harar region, eastern Ethiopia. *Springerplus, 3*(1), 1–9.
19. Thomas, S., & Williams, T. (2013). Khat (Catha edulis): A systematic review of evidence and literature pertaining to its harms to UK users and society. *Drug Science, Policy and Law, 1*, 2050324513498332.
20. Haregeweyn, N., et al. (2017). Comprehensive assessment of soil erosion risk for better land use planning in river basins: Case study of the Upper Blue Nile River. *Science of the Total Environment, 574*, 95.

21. Sultan, D., et al. (2018). Impact of soil and water conservation interventions on watershed runoff response in a tropical humid Highland of Ethiopia. *Environmental Management, 61*(5), 860–874.
22. Sheldrick, B. H., & Wang, C. (1993). Soil sampling and methods of analysis. In *Particle size distribution. Canadian society of soil science* (pp. 499–511). Lewis Publishers.
23. Gardner, W. R. (1956). Calculation of capillary conductivity from pressure plate outflow data. *Soil Science Society of America Journal, 20*(3), 317–320.
24. Gardner, W. H. (1986). Water content. In *Methods of soil analysis: Physical and mineralogical methods* (Vol. 5, pp. 493–544).
25. Sultan, D., et al. (2017). Analyzing the runoff response to soil and water conservation measures in a tropical humid Ethiopian highland. *Physical Geography, 38*(5), 423–447.
26. Payero, J. O., Nafchi, A. M., Davis, R., & Khalilian, A. (2017). An Arduino-based wireless sensor network for soil moisture monitoring using decagon EC-5 sensors. *Open Journal of Soil Science, 7*(10), 288–300.
27. Setegn, S. G., Chowdary, V. M., Mal, B. C., Yohannes, F., & Kono, Y. (2011). Water balance study and irrigation strategies for sustainable management of a tropical Ethiopian Lake: A case study of Lake Alemaya. *Water Resources Management, 25*, 2081–2107.
28. Casalí, J., Giménez, R., De Santisteban, L., Álvarez-Mozos, J., Mena, J., & de Lersundi, J. D. V. (2009). Determination of long-term erosion rates in vineyards of Navarre (Spain) using botanical benchmarks. *Catena, 78*(1), 12–19.
29. Amare, T., Derebe, A., & Yitaferu, B. (2014). Ecohydrology & hydrobiology combined effect of soil bund with biological soil and water conservation measures in the northwestern Ethiopian highlands. *Integrative Medicine Research, 14*(3), 192–199.
30. Ebabu, K., et al. (2019). Effects of land use and sustainable land management practices on runoff and soil loss in the Upper Blue Nile basin, Ethiopia. *Science of the Total Environment, 648*, 1462–1475.
31. Adimassu, Z., Mekonnen, K., Yirga, C., & Kessler, A. (2014). Effect of soil bunds on runoff, soil and nutrient losses, and crop yield in the central highlands of Ethiopia. *Land Degradation & Development, 25*(6), 554–564.
32. Zotarelli, L., Dukes, M. D., & Morgan, K. T. (2010). *Interpretation of soil moisture content to determine soil field capacity and avoid over-irrigating sandy soils using soil moisture sensors (AE460)*. The University of Florida's Institute of Food and Agricultural Sciences (UF/IFAS). Gainesville, FL.
33. Aguilar, J., Rogers, D., & Kisekka, I. (2015). Irrigation scheduling based on soil moisture sensors and evapotranspiration. *Kansas Agricultural Experiment Station Research Reports, 1*(5), 20.
34. Ilaibi, H., & Al-Sudani, Z. (2018). Study of morphometric properties and water balance using Thornthwaite method in Khanaqin Basin, east of Iraq. *Journal of University of Babylon for Engineering Sciences, 26*(3), 165–175.
35. Kar, D., & Kar, D. (2016). Methodologies of different types of studies. In *Epizootic ulcerative fish disease syndrome* (pp. 187–221).

Chapter 25
Surface Water Quality Assessment of Chilanchil Abay Watershed: The Case of Bahir-Dar City Waste Disposal Site

Dargie Haile and Nigus Gabbiye

25.1 Introduction

Water is the most abundant resource on earth, but only 3% is accessible for human activities, while the remaining is present in the ocean as salt water [1]. It may be available in various forms and quantities, but its use for various purposes is the subject of quality. Of all the environmental concerns that developing countries face, the lack of adequate and clean water remains the most serious problem [2]. Diseases may spring up through water pollution and rapidly spread beyond human expectation because of its flow [3]. Wastes of different types, mostly solid wastes, are the major input of dumpsites/landfills. With respect to the hydrological analysis of groundwater, it flows from areas of higher topography towards areas of lower topography, thereby bringing about the examination of the degradable material that forms leachate and contaminates the surface water of the study area. Properly managed, derelict voids can be reclaimed by a process of sanitary landfilling, ultimately bringing the land back into productive use and providing much of the needed waste disposal site. A reduction in either the quality or quantity of the discharging groundwater can significantly influence surface water quality and the attainment of water quality standards. Surface water and groundwater are, therefore, intimately linked in the water cycle, with many common issues. The protection of surface quality is of paramount importance. The depressions into which solid wastes are often dumped include valleys (abandoned), sites of quarries, excavations, or sometimes a selected

D. Haile (✉)
Department of Chemical Engineering, Institute of Technology, Woldia University, Woldia, Ethiopia

N. Gabbiye
Department of Chemical Engineering, Bahir-Dar Institute of Technology, Bahir-Dar University, Bahir-Dar, Ethiopia

S. W. Fanta et al. (eds.), *Sustainable Development Research in Manufacturing, Process Engineering, Green Infrastructure, and Water Resources*, Green Energy and Technology, https://doi.org/10.1007/978-3-031-77339-6_25

portion within the residential and commercial areas in many urban settlements where the capacity to collect, process, dispose of, or re-use solid waste in a cost-efficient, safe manner is often limited. Increasing urbanization results in an increased generation of waste materials, and landfills/dump sites have become the most convenient way of disposal. Most of the water supply facilities may likely be affected by the generated leachate through waste decomposition from the dumpsites. Generally, one of the oldest and most common methods of municipal waste disposal is dumping/simple landfilling at the selected site. This is most likely to remain the ultimate fate of solid waste since it appears to be one of the cheapest ways. Once dumping the waste to the site, solid waste undergoes various physico-chemical and biological changes. Consequently, the degradation of the organic fraction of the wastes in combination with percolating rainwater leads to the generation of a highly contaminated liquid called leachate [4]. The leachate so produced is a high-strength wastewater with the levels of wastewater parameters, such as chemical oxygen demand (COD), biochemical oxygen demand (BOD), ammonium nitrogen, turbidity, and solids being many times greater than those in municipal wastewaters. Leachate generated owing to moisture release, precipitation, and infiltration is easily increased by soluble or suspended organic matter, ammonium nitrogen, and inorganic ions such as heavy metals and may cause severe environmental hazards if they are not properly collected, treated, and safely disposed of [5]. If not dealt with proper waste management, such kind of dumping sites can cause pollution to surface water (through contaminant transport by flooding, wind, and groundwater movement from the open dump sites). The Bahir-Dar open site is one amongst the improper open dump sites and is located in a place where a number of people are living around. The communities who are living near the disposal site (in a position of downstream and upstream) are using polluted ground and surface water for their day-to-day activities. As a result of this, it becomes a great deal of threat to those communities regarding the water quality aspect. Therefore, the focus of this research was to assess and evaluate the water quality in that watershed, especially near this waste disposal site, to examine its effect on surface water quality.

25.2 Materials and Methods

25.2.1 Descriptions of the Study Area

The Eriamecharia municipal waste disposal site is 5 km from Bahir-Dar city, Ethiopia, near the expressway to Addis Ababa and the TIS Abay waterfall. It forms part of the Sebatamet provincial network. According to the Central Statistics Agency of Ethiopia (CSA 2007 G. C), approximately 6401 people, 3053 females and 3348 males, live around the dump site. Its geographical location is as follows: latitude, 11.54; longitude, 37.38; altitude, 1803 meters at 3 degrees; and elevation above sea level, 1801 meters. The length and width of this irregularly shaped removal site are

384 and 174 m, respectively. This site was not equipped with liners or a leachate sorting system until 10 years ago and was not efficiently planned before being utilized for waste removal/dumping. In addition, no environmental impact assessment was performed before this location was established as a waste disposal site. Trucks and other vehicles from various parts of the city collect the waste, carry it to this site, and dump it in a disorganized manner. The waste is dumped as is without isolation. The base amount of solid waste that is generated from the city and dumped at the site is as follows: private waste 12,610 kg/day, business waste 4202 kg/day, service provider waste 98 kg/day, municipal waste 1044 kg/day, and overall waste 22,774 kg/day (Source: Solid Waste Portrayal and Evaluation from the Bahir-Dar city report, 2007). Currently, the average amount of waste dumped at the site is estimated to be 31,321 kg/day.

25.2.2 Sample Collection, Preservation, and Laboratory Analyses

Water samples were collected from the selected sample areas near the dumpsite called locally Zebir, Abohoy Manekia, and Tikkurit to the laboratory for analysis. Those five sampling points were chosen based on their accessibility and proximity to pollution sources, such as communal sites and cottages. Global positioning system navigator (GPS etrex VISTA HCX) was used to determine the actual positions of the sampling points and referenced to ensure consistency in the sampling points during subsequent sample periods. The sampling points were selected purposefully to include all the target points near the communities, as shown in Fig. 25.1. Taking the sample started during the dry season from the month of March and continued with a wet season up in the month of August from all the five sample points. To assess the water quality, water samples were kept in 1 L polyethylene plastic bottles cleaned with metal-free soap, rinsed with deionized water, soaked with 10% nitric acid for 24 h, and finally rinsed with ultra-pure water. All water samples were stored in an icebox and delivered on the same day to the laboratory, and all samples were kept at a constant temperature of 4 °C to avoid sample deterioration due to the effects of light and temperature until laboratory analysis was taken [6]. Water quality parameters such as temperature, pH, electrical conductivity, total dissolved solids, turbidity, and salinity were measured with a handy YSI pro 30 Multipara meter (model HI 98130 HANNA). Other water quality parameters analyzed include dissolved oxygen (DO) using HI 98193 dissolved oxygen meter, BOD_5 using BOD_5 incubator in the presence of NaOH, COD using COD analysis method in the presence of standard potassium dichromate, sulfuric acid reagent, ferroin indicator and standard ferrous ammonium sulfate titrant and total coliform was done using membrane filter method. Heavy metals (Cr, Pb, and Mn), phosphates (PO_4^{2-}), and nitrate ($NO_3^{-)}$ were determined using the method of Palintest Spectrophotometer

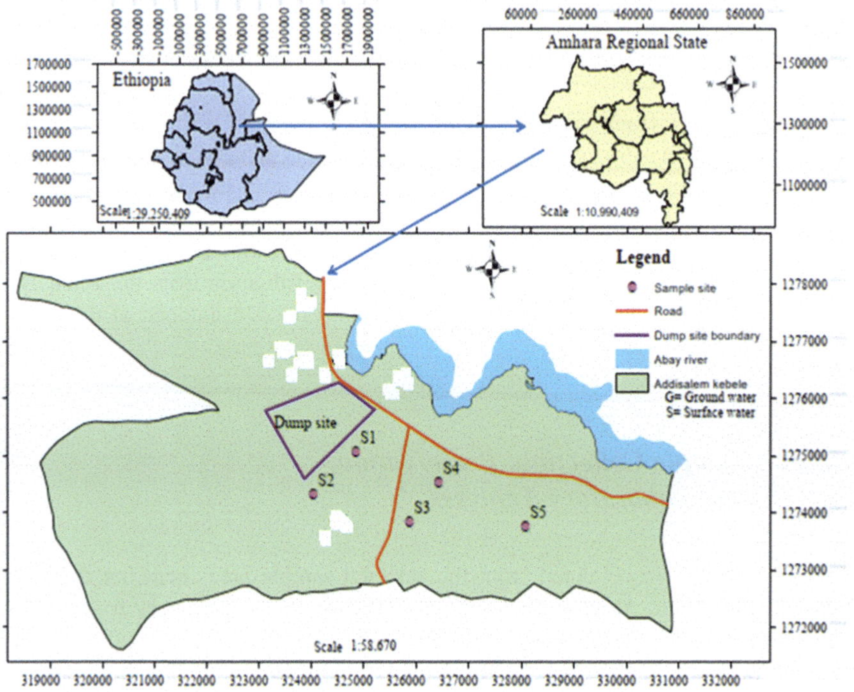

Fig. 25.1 Topographic map of the study area

(WAGTECH 8000). All the methodologies for laboratory analysis were conducted according to the Standard Methods for the Examination of Water and Wastewater [7].

25.3 Results and Discussions

25.3.1 Spatial and Temporal Variation of Surface Water Quality

The results of spatial and temporal variation of surface water physical parameters measured in all five sampling points throughout the study period are presented in Table 25.1. The results are discussed in subsequent sections.

Table 25.1 Seasonal and spatial variation in physical parameters of surface water quality

Sam pling sites	E.C (µS/cm)		Turbidity (NTU)		TDS (mg/L)		Temperature (°C)	
	Dry season	Wet season	Dry season	Wet season	Dry season	Wet season	Dry season	Wet season
SS1	1899 ± 21.18	2192.33 ± 121.37	13.82 ± 1.55	14.58 ± 1.14	1227.86 ± 35.91	1588.67 ± 187.56	26.57 ± 0.85	23.37 ± 0.73
SS2	1929 ± 6.25	2504 ± 118.19	8.59 ± 1.41	27.38 ± 1.23	1234.97 ± 44.83	1539.67 ± 135.77	27.2 ± 0.72	23.93 ± 0.85
SS3	1977.33 ± 5.51	1086 ± 175.92	12.11 ± 1.08	17.18 ± 1.58	1256.67 ± 47.44	814.67 ± 121.33	26.14 ± 0.32	22.53 ± 0.45
SS4	2060.33 ± 25.15	1013.67 ± 171.55	7.12 ± 1.85	16.02 ± 0.77	1338.06 ± 4.82	798.83 ± 163.29	26.7 ± 0.95	23.70 ± 0.75
SS5	2181 ± 25.36	1021.33 ± 79.58	8.77 ± 0.82	24.39 ± 1.04	1413.80 ± 11.57	815.33 ± 111.69	26.37 ± 1.07	23.06 ± 1.82

Where SS stands for sampling site

25.3.1.1 Spatial and Temporal Variation of Physical Characteristics of Surface Water

Electrical Conductivity Electrical conductivity (specific conductance, EC) is a measure of the resistance of a solution to electrical flow, and it is useful in aquatic studies to provide a direct measurement of dissolved ionic matter in the water. As can be seen in Table 25.1, the electrical conductivity of surface water samples during the dry season ranged from1899 ± 21.18 to 2181 ± 25.36 μS/cm with a mean value of 2009.333 ± 16.69 μS/cm and ranged from 1013.67 ± 171.55 to 2192.33 ± 121.37 μS/cm with a mean value of 1563.467 ± 32.33294 μS/cm in a wet season (Table 25.1). The result indicates that both values of the dry and wet seasons were above the WHO's standard limit of 1000 μS/cm of drinkable water, implying that the water is not suitable for drinking. Even though all values were beyond the permissible level, $SS_{2,}$ which is near the dump site in a wet season, is too high (2192.33 ± 121.37 μS/cm) (Table 25.1). SS_2 was found 317 meters below the dump site at the point where surface water flow joins with the leachate, which comes from the dump site during the wet season. Then, this derived waste from the disposal site was suspected to have a significant amount of inorganic dissolved solids, resulting in a salinity problem. The lowest value was found at the sampling point of (SS_1). This was because the location of this sample point was found at the site just before the surface water passes through the dump site. Generally, these higher values of electrical conductivity were caused by the high salinity problem. The values of salinity for all samples were recorded in the range of 0.98 to 1.24put in the dry season and 0.36 to 2.82put in the wet season.

Total Dissolved Solid From this study, the total dissolved solids for the dry season ranged from1227. 87 ± 35. 93 to 1413. 80 ± 11. 57 mg/L with a mean value of 1294. 27 ± 28. 92 mg/L, while in a wet season, it ranged from 798. 83 ± 163. 29 to 1588. 67 ± 187. 56 mg/L (Table 25.1). Both the highest (1588.67 ± 187.56 mg/L) and the lowest (798.83 ± 163.29 mg/L) values were recorded during the wet season. The highest value 1588.67 ± 187.56 mg/L) that was obtained at SS_1 was due to the large amount of sediment load transported from the watershed during the wet season since the water sample SS_1 comes through the settlement of population, a number of wastes are thrown by the community at a number of points during the wet season. The lowest value 798. 83 ± 163. 29 mg/L that was obtained at SS_4 was due to this sampling point was found far from the dumpsite and the source of this SS_4 water sample is locally called Tikurit (independent source which has no interaction with the dump site). Generally, almost all samples of the dry season (Table 25.1 Graphs of Appendix C) were observed above the WHO limit of 500–1000 mg/L for drinking purposes, aquatic life, and irrigation, while in a wet season, only SS_1 and SS_2 (which are the nearest to the waste disposal site) were in the marginal range of maximum permissible limit of 1000 mg/L. On the other hand, the mean value of the water sample in the dry season1294.27 ± 28.92 mg/L was higher than the mean value of the wet season1111 mg/L. The idea of this study (TDS of the dry season was higher than TDS of the wet season) was in line with [8] who investigated the spatial and temporal variation of Jamuna river water quality parameters in Bangladesh. This temporal variation was significant as it was supported by the analysis of variance (one-way ANOVA, Appendix A-1.2) with a P-value of 0.,008, which is less than the alpha value ($\alpha = 0.05$). This significant variation could have been due to the higher temperatures observed during the dry season, which facilitated dissolution, ion exchange capacity, desorption, and weathering processes. Also, during the dry season, the water evaporated, and ion concentrations increased. TDS is the sum of the concentrations of dissolved major ions [9] and, at high concentrations, is referred to as macropollution [9]. These ions are highly mobile and widespread and can be transported in all flow conditions [10], rendering them more complex to monitor than their particulate form [11]. TDS increases the physiological stress and chronic effects in the receiving water body systems [12] and may cause several health problems to living organisms and adversely affect the fertility of soil if used for irrigation purposes. According to WHO, total dissolved solid values for drinking water <500 mg/L is the minimum desirable limit, and 1000 mg/L is the maximum permissible limit. The excess of dissolved solids creates an imbalance due to increased turbidity and causes suffocation to aquatic life, even in the presence of high dissolved oxygen. Water with high residue is normally less suitable and may induce an unfavorable physiological reaction in the transient consumer and even may cause gastrointestinal irritation. Moreover, water containing a high solid concentration may cause constipation effects. Thus, a high level of TDS may aesthetically be unsatisfactory for bathing and washing.

Turbidity Colloidal matter would scatter or absorb light and thus prevent its transmission [18]. From this study, during the dry season, the turbidity of the water sample ranged from 7.12 ± 1.85 to 13.82 ± 1.55 NTU with a mean value of 10.14 ± 1. 34NTU, while in a wet season, it ranged from14.58 ± 1.14 to 27.38 ± 1.23 with a mean value of 19.91 ± 1.15 NTU. Values recorded in the wet season were higher than those of the dry season and EPA's/WHO's standard values (5–25 NTU). This could be caused by higher flow rates during rainfalls that might have carried sediments, clay, silt, significant chemical fertilizer, pesticide runoff from agricultural activities, high suspended matter, or impurities that interfere with the clarity of the water. Moreover, finely divided inorganic and organic matter, soluble colored organic compounds, and other microscopic organisms from the waste disposal site to those sampling points can be transported. The low values of turbidity recorded in the dry season (Table 25.1, graphs of physical parameter Appendix C) may be due to the absence of runoff, recession in flow level, and settlement of silt and clay. This finding of higher values in wet seasons than in the dry season was similar to the findings of those who assess the effects of seasonal variations in physical parameters on the quality of gravity flow water in Kyanamira sub-county, Kabale District, Uganda. The highest value of all those five sample points was recorded in the wet season at SS_2 (27.38 ± 1.23 NTU) due to the flow of finely divided inorganic and organic matter/ high suspended matter or impurities, soluble colored organic compounds from the municipal waste disposal site since this sample point was found at the exit of the dump sit in the downstream direction whereas a minimum value was recorded in the dry season at SS_4. This was due to points that are far apart being exposed for self-purification when they pass through different wetlands. Therefore, SS_4 was found 3 km away from the dump site. Turbidity in the studied area was within WHO guideline values of 5–25 NTU during the dry season. Since the appearance of water with turbidity in this range is usually acceptable to consumers, the probability of microorganisms, typically bacteria, viruses, and protozoa, attached to particulates is minimized. However, the measured results during the wet season (Table 25.1) were above the WHO water quality guideline limit. Furthermore, high levels of turbidity can protect microorganisms from the effects of disinfection, giving rise to a significant chlorine demand and reducing the performance of some disinfection treatment methods. Thus, turbidity could represent a key issue regarding the microbiological quality and disinfection of water.

Temperature The measured temperature of the surface water sample ranges between 26.14 ± 0.32 and 27.2 ± 0.72 $^{\circ C}$ during the dry season with a mean value of 26.59 ± 0.78 °C, whereas in the wet season, it ranges from 22.53 ± 0.45 to 23.93 ± 0.85 °C (Table 25.1, Appendix C) with a mean value 23.32 ± 0.92 Oz. The maximum temperature of 27.2 ± 0.72 as was observed at SS_2 during the dry season, whereas a minimum of 22.53 ± 0.45 was recorded at SS_3 during the wet season. High quantity discharge of sewage, agricultural runoff, and leachate from a heap of municipal solid waste along the watershed and septic tanks close to streams was a significant cause of the change in water temperature [13]. The high-temperature values during the dry season indicate a higher rate of metabolic processes or activities in the water. The mean value (26.59 ± 0.78 Oz) of the temperature shows that the temperatures of the water body were believed to have been influenced by the intensity of sunlight during the dry season. The mean value of the temperature in the rainy season was within the range of WHO's standard values of surface water (12–25 °C) for drinking, bathing, and irrigation purposes. This may be due to the absence/insignificant addition of warm water from point sources, including kitchens and bathrooms. The fluctuations of the mean value of temperature in both seasons show that the temporal variation was significant, as it was observed during the ANOVA- test<0.05 (Appendix A-1). Generally, factors that influence the change of temperature are climatic fluctuations and respond to factors such as season, time of day, air circulation, cloud cover, and flow of water in the natural system. Increased temperature, for example, could cause a change in the species of receiving water body. In addition, oxygen is less soluble in warm water than in cold water, and this could result in serious depletion of oxygen concentration. Many of the physical, biological, and chemical characteristics of surface water are dependent on temperature [14]. The optimal health of aquatic organisms, from microbes to fish, depends on temperature [15]. J. Pankow [15] reported that the temperature of the water also affects the volume of dissolved oxygen it can hold. The rates of bacterial growth, decay of disinfection residual, corrosion rates, taste, odor, and color problems are all affected by water temperature.

25.3.1.2 Spatial and Temporal Variation of Chemical Characteristics of Surface water Quality

The results of spatial and temporal variation of surface water chemical properties measured in all five sampling points throughout the study period are presented in Table 25.2.

Surface Water pH pH indicates the sample acidity/basicity but is actually a measurement of the potential activity of hydrogen ions (H+) in the water sample. From this study, the pH values of the dry season were recorded between 7.21 ± 0.16 and 7.69 ± 0.15 with a mean value of 7.25 ± 0.18 while in the wet season, pH ranged from 8.22 ± 0.32 to 9.05 ± 0.84 (Table 25.2) with the mean value of 8.62 ± 0.64. The result shows that the pH of all the samples of surface water during the dry season was generally good throughout the sampling points. They all conform to the EPA guideline values for drinking water quality 6.5–8.5. Although they all conform to the standard set by the EPA, the surface water pH for SS_5 (7.21 ± 0.16) was lower than all and very close to neutral. This sample point was found 6 km downstream away from the dumpsite, and it was the combination of different water sources, locally called Tikurit and Chilanchil Abay. Then, the reason for the values to be lower than the others was due to the dilution effect of those surface waters and might be due to self-purification while it flows through different natural wetlands before it reaches the sample point (selected sampling point). On the other hand, SS_1 was higher than that of the other sample points. When this (SS_1) surface water runs through the streams or canals, it passes through the discharge of the leather factory at some points, and then the contribution might be from such effluent. In addition, runoff from the communities could also contribute to the increase in the pH of the sample. Based on the analyzed result, the pH values of the wet season were higher than the dry season at all five sample sites. This means the temporal variations of pH value during the study period were significant, as evidenced by the one-way ANOVA test, which results in $P < 0.05$ (Appendix A-2.2). The idea of this study (the pH of the wet season was higher than the pH of the dry season) was in line with [8] who investigated the spatial and temporal variation of Jamuna river water quality parameters in Bangladesh. It is likely that the elevated pH of the wet season might be due to the use of detergents by neighboring populations for washing clothes, vehicles, and utensils, and human contributions to elevated pH primarily in the form of nutrient runoff (most commonly fertilizer), which leads to increased algae growth and higher pH. The maximum pH values were observed at SS_4 (9.05 ± 0.84) during the wet season. This might be due to this that the surface water sampling point was found downstream of the old and unregulated pit latrines of the prison camp. Higher pH values during the rainy season revealed aerobic conditions and lesser anthropogenic sources as compared to the dry season [16]. The pH of water affects the solubility of many toxic and nutritive chemicals; therefore, the availability of these substances to aquatic organisms is affected. For example, as acidity increases, most metals become more water-soluble and more toxic [17]. found out that high or low pH will adversely affect the availability of certain chemicals or nutrients in the

Table 25.2 Seasonal and temporal variation in chemical properties of surface water quality

Sampling sites	pH		TH (mg/L)		DO (mg/L)		BOD (mg/L)		COD (mg/L)	
	Dry season	Wet season	Dry season	Wet season	Dry season	Wet season	Dry season	Wet season	Dry season	Wet season
SS1	7.69 ± 0.0.15	8.22 ± 0.32	248.33 ± 13.2	277.33 ± 9.45	2.1 ± 0.05	2.78 ± 0.17	3.78 ± 0.08	4.14 ± 0.39	72.84 ± 2.11	84 ± 6
SS2	7.31 ± 0.27	8.37 ± 0.57	307.67 ± 14.22	343.3 ± 14.05	1.89 ± 0.07	1.37 ± 0.25	4.13 ± 0.12	4.64 ± 0.12	87.87 ± 2.86	95.07 ± 1.79
SS3	7.25 ± 0.22	8.63 ± 0.63	257.33 ± 16.50	300.67 ± 10.02	3.83 ± 0.19	9.31 ± 0.36	3.38 ± 0.04	3.97 ± 0.17	73.23 ± 3.22	86.10 ± 5.73
SS4	7.40 ± 0.12	9.05 ± 0.84	276.33 ± 8.02	293 ± 5.92	3.1 ± 0.13	8.42 ± 0.50	3.81 ± 0.16	3.98 ± 0.36	78.33 ± 5.97	88.93 ± 4.10
SS5	7.21 ± 0.16	8.82 ± 0.85	247 ± 12.53	271. 3 ± 9.45	4.67 ± 035	9.61 ± 0.19	2.19 ± 0.09	2.71 ± 0.39	71.1 ± 1.9	79.57 ± 2.89

Where SS stands for sampling site

water for use by plants. According to [18], when hydrogen ions are in extreme concentration in water, it is difficult to treat by biological means. If the concentration is not altered before discharge, the wastewater effluent may alter the concentration in the natural waters. According to [19], this parameter is very important in an aquatic environment to many metabolic reactions in microbial cells, which includes energy.

Total Hardness Total hardness is the parameter of water quality used to describe the effect of dissolved minerals (mostly Ca and Mg), determining the suitability of water for domestic, industrial, and drinking purposes attributed to the presence of bicarbonates, sulfates, chloride and nitrates of calcium and magnesium. From the study, the total hardness of all samples of surface water during the dry season ranged from 247 ± 12.53 to 307.67 ± 14.22 mg/L, with a mean value of 267.33 ± 12.89 mg/L, while during the wet season, the total hardness of surface water samples ranged from 271. 3 ± 9.45 to 343.33 ± 14.05 mg/L (Table 25.2), with a mean value of 297.25 ± 9.78 mg/L. From the result, one can see that even though the values of the wet season were higher than those of the dry season, values recorded in both seasons were below the WHO report (2004a) standard value (500 mg $CaCO_3$ mg/L). In this study, the hardness criterion was used to determine the usability of the water supplies understudy for domestic and drinking purposes. The hardness of water for domestic use relates mainly to its reaction with soap. Since soap is precipitated principally by Ca^{2+} and Mg^{2+}, hardness is defined as the sum of the concentrations of these ions expressed as mg/L of $CaCO_3$. Comparing the recorded values with the WHO (2004a) value of less than 500 mg $CaCO_3$/L, surface water bodies in the studied area could generally be described as safe. Thus, surface waters in the studied area and its surrounding watershed area can be used for domestic purposes without any human health concerns in terms of total hardness. The minimum amount of total hardness in the water sample was recorded during the dry season at SS_5 (247 ± 12.53 mg/L). This might be due to fewer minerals suspended in water samples, less weathering of limestone, sedimentary rocks, and calcium-bearing minerals. On the other hand, the maximum amount of total hardness in the water sample was recorded during the wet season at SS_2 (307.67 ± 14.22 mg/L). This high total hardness in the wet season may be ascribed to different detergents dumped to waste disposal sites from the city, agricultural runoff, discharge of domestic sewage, and cloth washing stations through open drains in water bodies may be a significant cause for increasing the value of total hardness in water bodies during the wet season, which puts large amounts of alkaline ions into surface water bodies. This means higher values of total hardness during the wet season as compared to the dry season were in line with the study of [20] that evaluated seasonal variation in water quality parameters of river Mkomon Kwande local government area, Nigeria.

Dissolved Oxygen Dissolved oxygen is one of the most important factors in stream health. Its deficiency directly affects the ecosystem of a water body due to bioaccumulation (how pollutants enter a food chain) and biomagnifications or bioconcentration (the tendency of pollutants to concentrate as they move from one trophic level to the others). The oxygen content in water samples depends on a number of physical, chemical, biological, and microbiological processes. From this study, the amount of dissolved oxygen for all samples of surface water during the dry season ranged from 1.89 ± 0.07 to 4.67 ± 035 mg/L, with the mean values of 3.12 ± 0.16 mg/L, while in the wet season, the amount of dissolved oxygen in the surface water samples ranged from1.37 ± 0.25 to 9.31 ± 0.36 mg/L (Table 25.2), with a mean value of 6.30 ± 0.29 mg/L. The result shows that the content of dissolved oxygen for the surface water samples in the dry season was below the EPA's guideline values for surface water 5 mg/l. This could be due to the increasing salinity of the water sample as it records (0.98–1.2 ppt in the dry season and 0.36–2.82 ppt in the wet season) high amount of nutrients and decomposable organic wastes and slow movement of the surface water since the nature of the area around is very flat. As the amount of dissolved salt in water increases, the amount of oxygen the water can hold decreases, conversely, as water becomes fresh (lower salinity), more oxygen can dissolve into water [21]. Even though all samples of dry season were below the standard limit of EPA, the lowest and the highest values of DO during the dry season were recorded at SS_2 1.89 ± 0.07 mg/L and SS5, 4.67 ± 0.35 mg/L, respectively. This was due to SS_2 being located 317 m downstream from the dump site in the direction of more waste flow towards the surface water. Then, more organic waste was suspected from the dump site, and then, as the amount of organic waste increased, more bacteria would grow, and more oxygen would be used. Therefore, the concentration of DO was dropped. Generally, like [12], who found lower values of DO during the dry season in the assessment of temporal variations of water quality characteristics and their principal sources in the tropical Vellar estuary, southeast coast of India, the mean value of dissolved oxygen in the dry season was 3.12 ± 0.16 mg/L during the time of this study, even lower than wet or sopping season(6.30 ± 0.29 mg/L). This was due to high rates of respiration, organic decomposition, chemical fertilizers, rise in temperature, and rate of photosynthesis. Like dry season, the highest and lowest values of dissolved oxygen were also recorded in the wet season at SS_5 and $SS_{2,}$ respectively. The reason why dissolved oxygen decreases at $SS_{2,}$ which is the nearest of all downstream during the wet season, was more due to the increased input of organic load from the accumulated waste disposal site into the water sample point with the help of rain. On the other way, $SS_{5\ \text{was}}$ found 6 km downstream away from the dumpsite, and it was the combination of different water sources, locally called Tikurit and Chilanchil Abay. Then, the reason for the values to be higher than those of others was due to the dilution effect of those surface waters, and when they were combined, the speed of the movement was increased. In fast-moving streams, rushing water is aerated by bubbles as it churns over space and falls down hundreds of tiny waterfalls. The presence of dissolved oxygen in wastewater according to [12] is desirable because it prevents the formation of harmful odors. Dissolved oxygen is required for the respiration of aerobic microorganisms as well as all other aerobic life forms [22]. However, its solubility in water is generally affected by, among other factors, temperature, turbulence, salinity, and biological processes [23]. According to [12], the DO of water is affected by raw or poorly treated sewage, speed of flow condition, contaminated stormwater discharges, wastewater from human activities, animal feedlots, and natural sources like decaying aquatic plants and animals. Most organic waste means more decomposed and more oxygen being used.

Biological Oxygen Demand From the study, the amount of BOD for all samples of surface water during the dry season ranged from 2.19 ± 0.09 to 4.13 ± 0.12 mg/L, with a mean value of 3.46 ± 0.09 while in a wet season it ranged from2.71 ± 0.39 to 4.64 ± 0.12 mg/L, with a mean value of 89 ± 0.31 mg/L (Table 25.2, Appendix D). In both seasons, the amount of BOD in the water samples was below the EPA's guideline values for surface water. This indicates that the water was moderately good in terms of BOD for different household activities because a low BOD is an indicator of good quality water, while a high BOD indicates polluted water [24]. In water with a BOD level above the EPA's WHO guideline value, the water is considered somewhat polluted because there is usually organic matter present, and bacteria are decomposing this waste. The higher the BOD value, the greater the amount of organic matter or food available for oxygen-consuming bacteria. The average concentration of BOD in the study was 3.46 ± 0.09 mg/L in the dry season and increased to 3.89 ± 0.31 mg/L in the rainy season, which was below the permissible limit of WHO. Biochemical oxygen demand indicated higher values in the downstream positions at (SS_2) than others during the dry and rainy seasons, with values of 4.13 ± 0.12 mg/L and 4.64 ± 0.12 mg/L, respectively. This is expected because only the downstream water receives high organic pollutants through the effluents. The occurrence of this significant spatial variation between and within groups was assured by the result of ANOVA ($P < 0.05$, Appendix A-2.1). Seasonal analysis reveals that BOD values are higher during the rainy season (2.71 ± 0.39 to 4.64 ± 0.12 mg/L) than in the dry season (2.19 ± 0.09 to 4.13 ± 0.12 mg/L). The trend of seasonality in BOD followed that of dissolved oxygen concentration, with higher values during the rainy season than in the dry season. This is similar to [12], who reported that the value of BOD was higher in a wet season (40 to 63 mg/L) than in the dry season (20 to 43 mg/L) in the assessment of the Jamuna river water quality parameters. Biochemical oxygen demand is a measure of the oxygen in the water that is required by aerobic organisms. The degradation of organic materials exerts oxygen tension in the water and increases the biochemical oxygen demand [25]. Water samples with low BOD have low nutrient levels; therefore, much of the oxygen remains in the water. BOD directly affects the amount of dissolved oxygen in the given sample sites. The greater the BOD, the more rapidly oxygen is depleted in the stream. This means less oxygen is available to higher forms of aquatic life. The consequences of high BOD are the same as those of low dissolved oxygen: aquatic organisms become stressed, suffocate, and die. Sources of BOD include leaves and woody debris; dead plants and animals; animal manure; effluents from pulp and paper mills, Wastewater treatment plants, waste disposal sites, feedlots, and food-processing plants; failing septic systems; and urban stormwater runoff [26].

Chemical Oxygen Demand From the study, the chemical oxygen demand of all samples of surface water during the dry season ranged from 78.33 ± 5.97 to 87.87 ± 2.86 mg/L, with a mean value of 76.67 ± 3.21 mg/L, while the chemical oxygen demand of surface water samples in the wet season ranged from 79.57 ± 2.89 to 95.07 ± 1.79 mg/L (Table 25.2, Appendix D), with a mean value of 86.73 ± 4.10 mg/L. In both seasons, the values of COD for the surface water samples were below the EPA/WHO standard value (≤ 150 mg/L). This could probably be due to the fact that most of the wastes that are being generated from various communities are biodegradable. Even though all samples of both seasons were below the standard values, the spatial and seasonal variation of COD obtained at SS_2 (Table 25.4) was slightly higher than that of other values, this could probably be due to the fact that the dump site may contain household chemical products such as detergents, soaps, shampoos, preservatives, dyes, and cleaners. Then, since SS_2 is found within a call /at the hand of the dump site in the downstream direction, a considerable amount of chemicals may reach the sample point without having enough contact time for dilution. But SS_5 has smaller values relative to the others; this could probably be due to the dilution effect; secondly, there are wetlands along the stream banks at this stretch. According to [21], wetlands serve as natural purification media (from SS_2 to $SS_{5,}$ the decrease in COD value could be due to self-purification of the stream within this long stretch of about 6 km). This is in line with Metcalf and Eddy [18], who reported that, as the travel time in the collection system increases, natural purification can take place. In general, the result showed an increase in COD in all sampling points in the study period, but when compared values at the base of the season, the amount of COD in surface water samples during the wet season was higher than that of the dry season (Table 25.2) due to the fact that during the dry season, the reduced water quantity or rainfall result to decreased COD value. This was supported by the statistical analysis of variance (ANOVA) result, which indicates that there was a significant difference in mean COD concentration between dry and wet seasons during the study period ($p < 0.05$, Appendix A-2.2). The chemical oxygen demand is a measure of the oxygen equivalent of the organic matter in a water sample that is susceptible to oxidation by a strong chemical oxidant, such as dichromate. It provides an index to assess the effect that discharged wastes would have on the receiving watershed environment.

25.3.2 Spatial and Temporal Variation in the Measured Values of Nutrients and Microbiological Parameters of Surface Water Quality

The results of spatial and temporal variability of surface water concentrations of nutrients and microbiological parameters measured are presented in Table 25.3.

Dynamics of Nitrate in Surface water Quality Nitrate concentration of surface water samples fluctuated from 0.050 ± 0.002 to 0.303 ± 0.23 mg/L, with a mean value of 0.141 ± 0.068 mg/L, while in the wet season it ranged from 0.316 ± 0.056 to 0.903 ± 0.083 mg /L (Table 25.3, Appendix E.3), with a mean value of 0.601 ± 0.074 mg/L. In both seasons, the concentrations of nitrate at all sampling

Table 25.3 Spatial and temporal variation in concentrations of nutrients and microbiological parameters of surface water in the Chilanchil Abay watershed

Sampling sites	NO_3^- (mg/L)		PO_4^{3-} (mg/L)		TC (CFU/100 ml)	
	Dry season	Wet season	Dry season	Wet season	Dry season	Wet season
SS_1	0.303 ± 0.23	0.42 ± 0.11	0.153 ± 0.031	0.253 ± 0.055	11 ± 2	19.33 ± 3.055
SS_2	0.0664 ± 0.0023	0.903 ± 0.083	0.13 ± 0.02	0.6233 ± 0.051	22 ± 4	34.333 ± 3.512
SS_3	0.0593 ± 0.0021	0.7633 ± 0.106	0.21333 ± 0.0153	0.27 ± 0.0361	16 ± 2.65	20.667 ± 3.512
SS_4	0.225 ± 0.11	0.607 ± 0.021	0.245 ± 0.033	0.363 ± 0.050	21.67 ± 2.082	30.66 ± 3.512
SS_5	0.050 ± 0.002	0.316 ± 0.056	0.127 ± 0.012	0.58 ± 0.046	13.67 ± 2.082	24 ± 3.6

points were well below the EPA guideline value of 10 mg/L. This could probably be due to the absence of DO significantly in the water samples; maybe a smaller amount of nitrate is produced from ammonia as well as the decomposition of food wastes and other sources of protein. Sources of nitrates in water include decaying plant or animal material, agricultural fertilizers, manure, human or animal waste, and domestic sewage. A relatively high concentration of nitrate has been observed at SS_2 (0.903 ± 0.083 mg/L), SS_3 (0.7633 ± 0.106 mg/L), and SS_4 (0.607 ± 0.021 mg/L), respectively, during the wet season due to evaporation, percolating NO_3^- from sources such as decaying plant and animal materials, agricultural fertilizers and domestic sewage, inappropriate and excessive applications of fertilizer, increasing loadings of organic waste disposal from the point and non-point sources. The minimum amount of nitrate in the water samples of the studied area was recorded during the dry season at SS_2 (0.0664 ± 0.0023 mg/L), SS_3 0.0593 ± 0.0021 mg/L and SS_6 (0.050 ± 0.002 mg/L), respectively, as compared with the wet season. According to the WHO water quality guideline limit of 10 mg/L, the water in the studied area would be safe from polluting the existing water bodies in terms of nitrate. This could be due to the naturally occurring nitrifying bacteria fixing the free nitrogen from the atmosphere into nitrate and making it usable by photosynthetic plants around the study area. However, excessive amounts of nitrate can cause water quality problems and accelerate eutrophication.

Dynamics of Phosphate in the Surface Water Quality In this study, the concentration of phosphate in the form of PO_4^{3-} in all samples in the dry season ranged from 0.127 ± 0.012 to 0.245 ± 0.033 mg/L. with a mean value of 0.174 ± 0.023 mg/L, while in a wet season, the concentrations of phosphate vary from 0.253 ± 0.055 to 0.6233 ± 0.051 mg/L (Table 25.3, Appendix E.3), with a mean value of 0.42 ± 0.048 mg/L. Those with laboratory results indicated that PO_4^{3-} concentration for all samples of surface water in the dry and wet seasons were generally above the acceptable limit set by the EPA/WHO prescribed limit of 0.1 mg/L for drinking water. This could be due to the waste disposal site since it contains different municipal wastes like human waste, medicated shampoos, food waste, and cosmetics. This was in line with the work of [12] who stated that phosphate would increase due to an increase in the uses of phosphate-producing items such as cleaning products (soaps, detergents), cosmetics, medicated shampoos, food products, discharge of wastewater with high phosphate levels from the communities, runoff containing soil-bound phosphate, yard waste, runoff from animal feedlots, stormwater, and other sources including human excretions (feces and urine). High concentrations of phosphates indicate the presence of pollution and are largely responsible for eutrophic conditions. Phosphate is essential to the growth of algae and other biological organisms [27]. argued that the illegal way of waste disposal into an area of aquatic environments can result in excessive algae growth, eutrophication, and the depletion of oxygen in water bodies. Based on this study, Bahir-Dar city waste disposal site was found to have an anarchy way of waste disposal site and stands for the contributions of this problem. Even though all values were above the standard limit when compared to the base of the season, the highest values of phosphate were found in wet seasons at sampling points SS_2, SS_5, and SS_4, and the lowest values in dry season, at sampling points SS_5, SS_2, and SS_1 (Table 25.3, Appendix E) respectively. Generally, the phosphate concentration was higher during the rainy season (0.6233 ± 0.051 mg/L) at SS_2 compared to the highest value recorded during the dry season (0.245 ± 0.033 mg/L) at SS_4. This could be attributed to the rising of flooding during the rainy season from the dump site (domestic wastes) and upstream agricultural intensification to the water sample sites, particularly those containing detergents and fertilizer runoff from dump sites and other various sources, while the minimum values recorded in the dry season could be due to evaporation. It could thus be inferred with respect to this parameter (phosphate) that the water sources in the study area might not be suitable for drinking and domestic purposes. This was in line with the findings of several researchers. For example, [12] showed that phosphate concentration in the water was above the limit of the World Health Organization (WHO) standard. Excess phosphate concentration in water gives rise to algal bloom and causes risk to human beings as algae produce toxins, which damage the neurological system and as well as cause oxygen depletion in the water. From the ANOVA (Appendix A-3.2) result, there was a significant temporal variation of phosphate concentration between the wet season and dry season. This notifies that season was one of the main factors that determined the concentration of the measured parameter.

Total Coliform on Surface Water Quality The purpose of the total coliform counts in water bodies was to estimate the number of coliforms in water samples as an index of the magnitude of biological contamination. The total coliform of surface water samples in the dry season ranged from 11 ± 2 to 22 ± 4 CFU/100 mL, while in the wet season, from 19.33 ± 3.055 to 34.333 ± 3.512 CFU/100 mL. On the other hand, Table 25.3 above shows that wet season samples analyzed revealed the maximum presence of total coliform at SS_2 (34.333 ± 3.512 CFU/100 mL), whereas at SS_1 (19.33 ± 3.055 CFU/100 mL) recorded minimum total coliform in the same season. During the dry season, maximum total coliform was recorded at SS_2 (22 ± 4 CFU/100 mL), whereas SS_1 (11 ± 2 CFU/100 mL) recorded minimum total coliform in the same season. From this study, it is revealed that in the studied area, total coliform counts were low in the dry season and high in the wet season. This is due to the flow of effluents from pit latrines, domestic sewage, municipal waste, and decaying organic material from the waste disposal site to the surface water sample points. Besides this, increased water temperature and depletion of dissolved oxygen may contribute to a significant elevation of total coliform contamination in water bodies in the studied area [28]. According to the WHO's permissible level, the average values of total coliform recorded in both seasons in the studied area were above the guideline limit value. Therefore, from the aspect of this parameter, the water is unsuitable for use. The high total coliform count during the rainy season (than the dry season) was in line with [29], who investigated the physico-chemical and bacteriological parameters of Cauvery kollegal water shade in Karnataka. Their results revealed that the rainy season exhibited more counts than the dry season. This high coliform count during the rainy season was attributed to the discharging of domestic wastes containing fecal matter to the water body and open defecation along the sides of the stream bank. Generally, the result of SPSS ANOVA (Appendix A-3.2) shows that the total coliform was significant temporally but not spatially during the study period. This means season plays a dominant role over space due to environmental conditions like rainfall, humidity, temperature, and moisture.

25.3.3 Spatial and Temporal Variation in the Concentrations of Heavy Metals of Surface Water Quality

The results of spatial and temporal variation of surface water concentrations of heavy metals measured in all five sampling sites throughout the study period are presented in Table 25.4.

Manganese Concentrations of manganese for all samples of surface water during dry season varied from 0.00202 ± 0.0026 mg/L to 0.0029 ± 0.0037 mg/L (Table 25.4), with a mean value of 0.00291 ± 0.00326 mg/L, while in a wet season it ranged from 0.0053 ± 0.0067 to 0.023 ± 0.0057 mg/L, with a mean value of 0.012 ± 0.01014 mg/L. As one can see from the results in both dry and wet seasons, the concentrations of manganese were below the EPA/WHO January 2004 standard value (0.5 mg/L).

Chromium From the study, concentrations of chromium in all samples of surface water during the dry season varied from 0.0066 ± 0.0001 to 0.068 ± 0.0015 mg/L (Table 25.4), with a mean value of 0.044 ± 0.18 mg/L, while in a wet season the concentrations of chromium ranged from 0.04 ± 0.001to 0.29 ± 0.053 mg/L, with a mean value of 0.178 ± 0.0311 mg/L. On the other hand, the maximum concentrations of chromium during the dry season were recorded at sampling point 2, sampling point 4, and sampling point 5, with the minimum value at sampling point 1. In the wet season, almost all values were increased and became above WHO/ (January 2004) standard value (0.05 mg/L). This might be that significant amount of chromium through migration of leachate due to rainfall from the dumpsite was exist in the sampling points. Therefore, the water sample in the study area was not recommended for use by people in the aspect of chromium since chromium is one of the most toxic metals that affects the nervous system and has damaging effects on the brain and kidneys. Infants and children who drink water containing chromium in excess of the action level could experience delays in their physical or mental development. The children could have deficits in attention span and learning abilities, while adults who take such waters could easily develop kidney problems or high blood pressure [30].

Lead concentrations of lead in all samples of surface water during dry season were varied from 0.00167 ± 0.000322 mg/L to 0.002233 ± 0.00038 mg/L (Table 25.4), with a mean value of 0.00192 ± 0.00032 mg/L, while in a wet season it ranged from

Table 25.4 Spatial and temporal variation of heavy metals in surface water quality

Sampling sites	Cr (mg/L)		Mn (mg/L)		Pb (mg/L)	
	Dry season	Wet season	Dry season	Wet season	Dry season	Wet season
SS_1	0.0066 ± 0.0001	0.063 ± 0.0086	0.0046 ± 0.0034	0.023 ± 0.0057	0.0017 ± 0.000252	0.009133 ± 0.0112
SS_2	0.0677 ± 0.0015	0.29 ± 0.053	0.0029 ± 0.0037	0.00933 ± 0.0112	0.002233 ± 0.00038	0.00293 ± 0.000252
SS_3	0.0241 ± 0.0311	0.04 ± 0.001	0.0026 ± 0.0034	0.0072 ± 0.0094	0.001967 ± 0.0032	0.00267 ± 0.00031
SS_4	0.064 ± 0.00153	0.2467 ± 0.057	0.00202 ± 0.0026	0.0053 ± 0.0067	0.002067 ± 0.00032	0.0074 ± 0.00922
SS_5	0.056 ± 0.0034	0.25 ± 0.0361	0.00251 ± 0.0033	0.0133 ± 0.0171	0.00167 ± 0.000322	0.00467 ± 0.0055

0.00267 ± 0.00031 to 0.009133 ± 0.0112 mg/L, with a mean value 0.0054 ± 0.0053 mg/L. As one can see from the results in both dry and wet seasons, the concentrations of lead were below the EPA/WHO January 2004 standard value (0.01 mg/L). This might be due to the absence of a significant amount of discarded lead acid batteries, paint, gasoline, and pigments in the waste disposal site that can possibly flow to the surface water samples. Then, it is possible to say that the disposal waste site was not contributing significantly to the increase in lead concentration above the standard values. Even though all values were below the standard limit, the concentration of lead which were obtained during the study period varied from site to site and season to season as compared with each other. Therefore, the maximum concentrations of lead were recorded at sampling point 1, sampling point 4, and sampling point 5, respectively, with the minimum value at sampling point 3 during the wet season and during the dry season the maximum values were recorded orderly at sampling point 2 and 4 with a minimum value at sampling point 5. In the base of the season, the maximum mean value of 0.0054 ± 0.0053 mg/L was obtained in the wet season, whereas a minimum of 0.00192 ± 0.00032 mg/L in the dry season due to the effect of rainfall that forces the pollutants to flow from their source to the surface water sample points.

25.4 Conclusion

The temporal and spatial variations in surface water quality of the Chilanchil Abay watershed were assessed following WHO quality parameters. The analysis of physical-chemical properties, concentrations of heavy metals, and nutrient recorded values show that TH, BOD, COD, NO_3^-, Mn, and Pb were within the acceptable limit during the study time while the remains fluctuated with the seasons.

The total coliform counts detected were above the permissible limits for drinking water at all sample points. The data suggested the importance of greater attention to household contamination, environmental sanitation control, and awareness about water contamination since the open waste disposal site and practice of open defection have actually impacted the watershed of the study area.

From the result of the statistical analysis of variance (one-way ANOVA) performed on the data of surface water related to TDS, BOD, and temperature suggested that all the measured parameters varied significantly among the sites. Besides the temporal variation, values of Cr, Mn, Pb, nitrate, phosphate, COD, pH, total coliform, temperature, and turbidity were significant during the study period (P-value <0.05).

Conflict of Interest The author declares that there is no conflict of interest regarding the publications of this chapter.

Appendices

Appendix A: Graphs of Spatial and Seasonal Variation of Physical Parameters of Surface Water

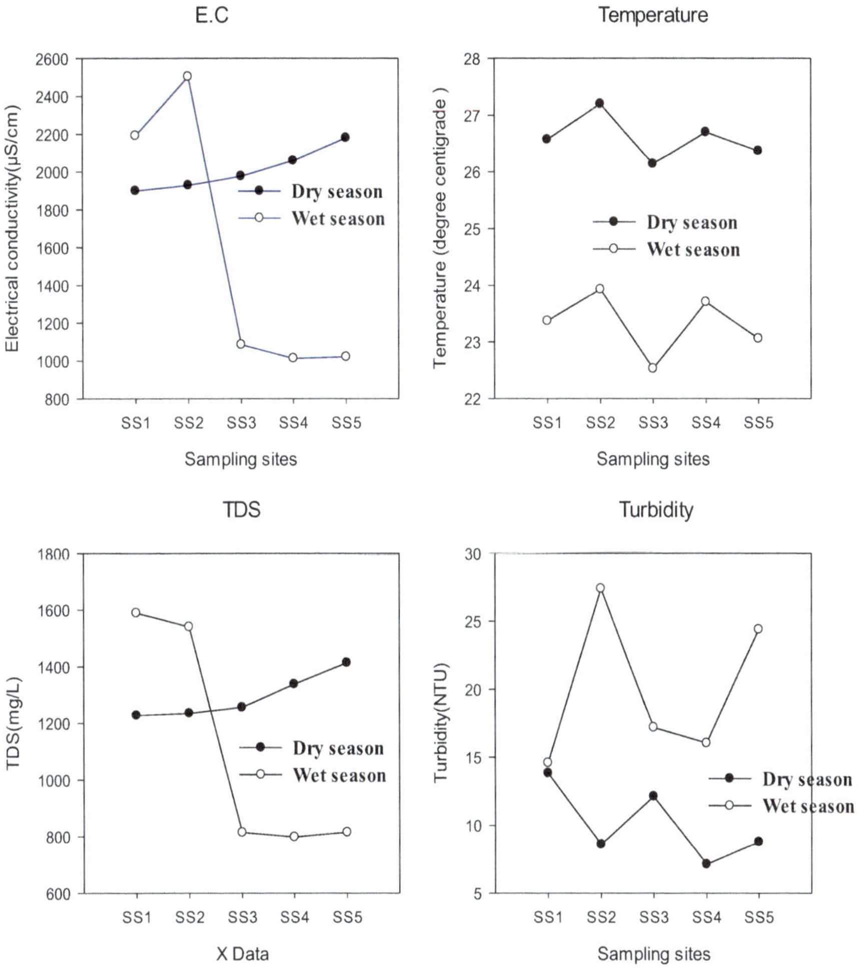

Appendix B: Graphs of Spatial and Seasonal Variation of Chemical Parameters of Surface Water

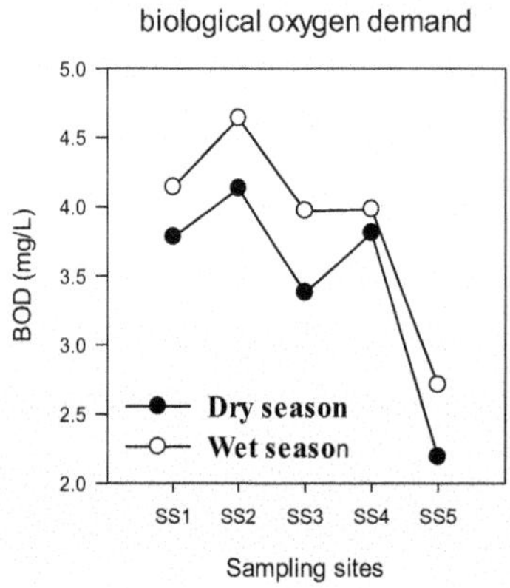

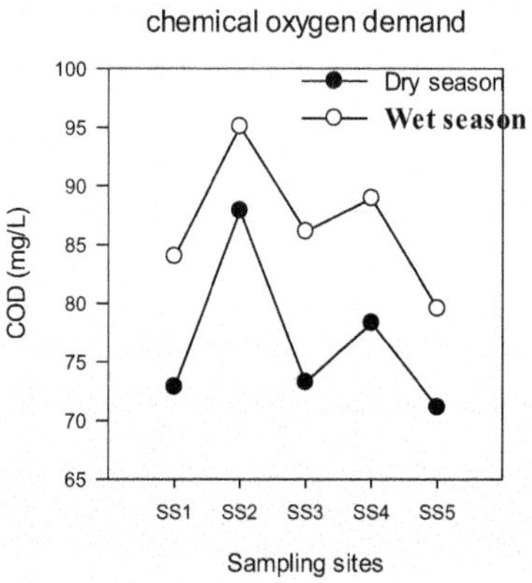

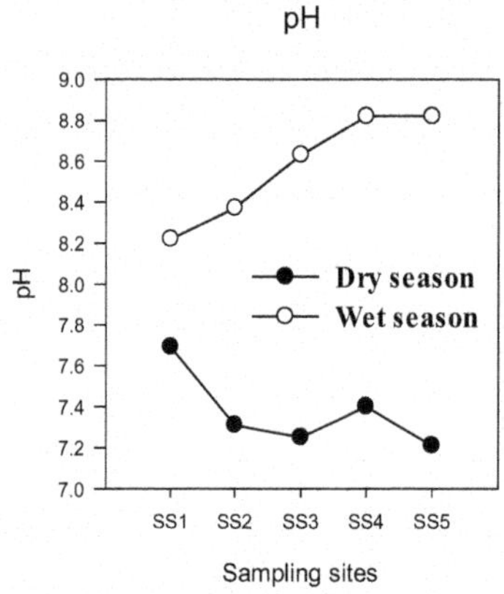

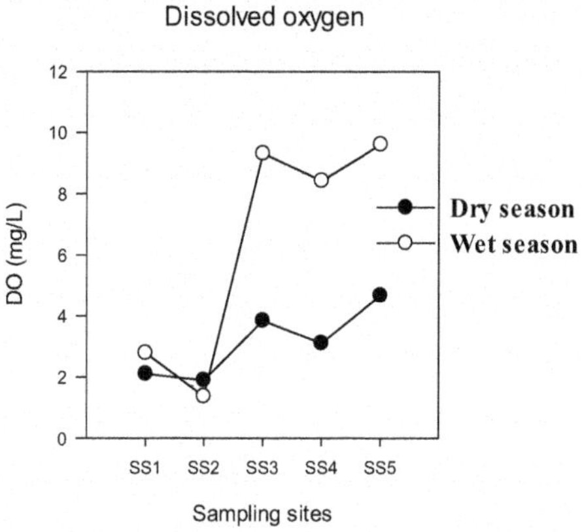

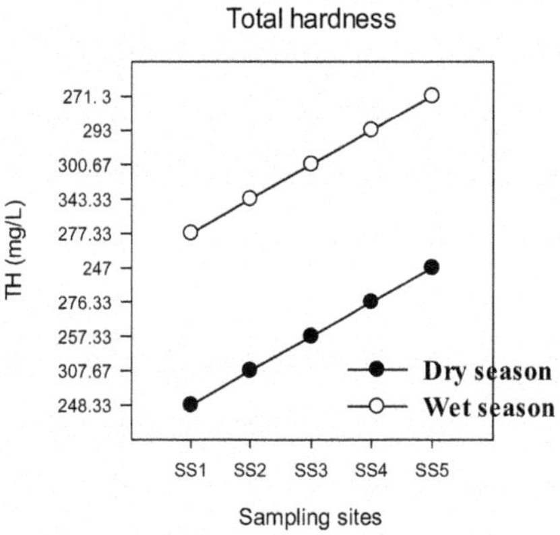

Appendix C: Graphs of Spatial and Seasonal Variation of Nutrients and Biological Parameters of Surface Water

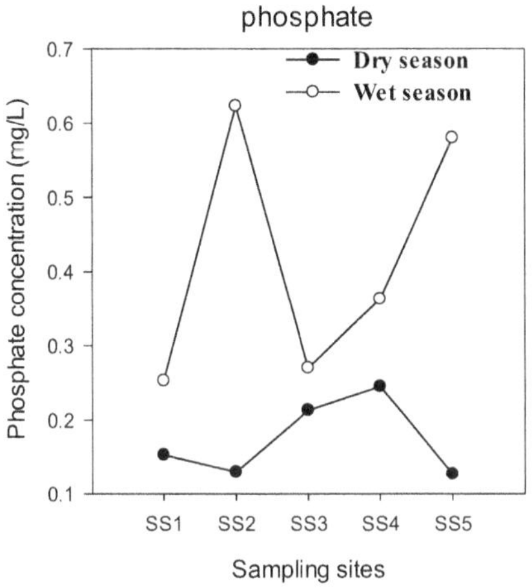

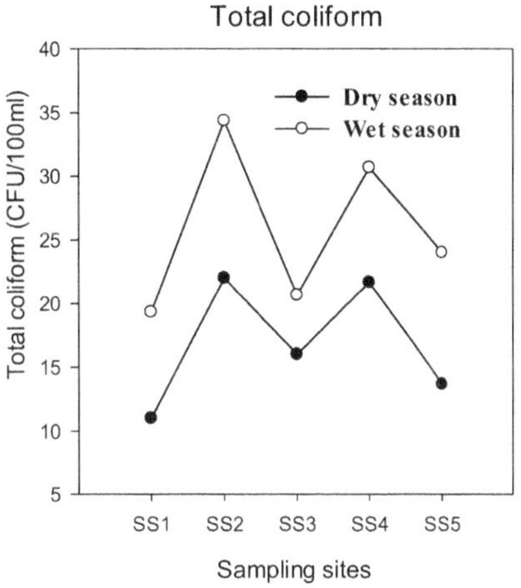

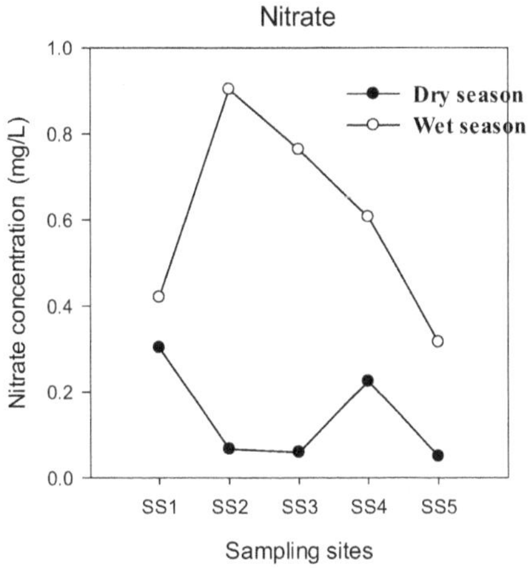

Appendix D: Graphs of Spatial and Seasonal Variation of Heavy Metal Concentration in Surface Water

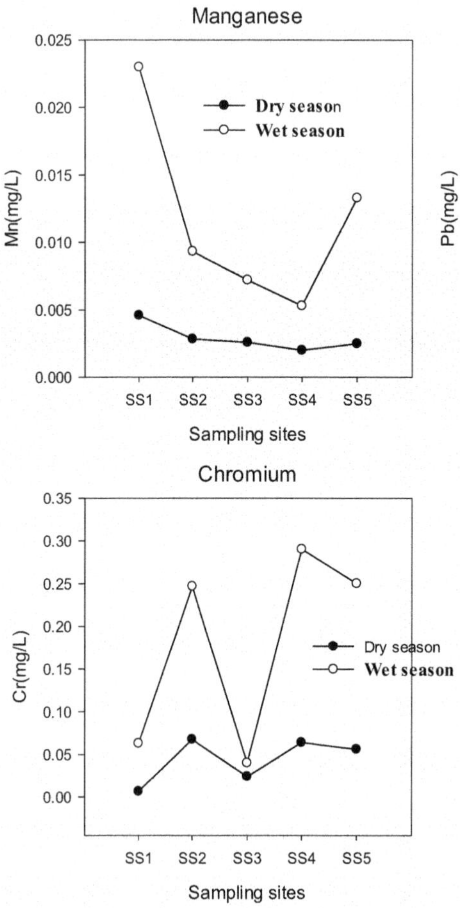

References

1. Love, J., & Luchsinger, V. (2014). Sustainability and water resources. *Journal of Sustainability and Green Business, 2*, 1–12.
2. Markandya, A. (2006). Water quality issues in developing countries. In *Economic development and environmental sustainability: New policy options* (pp. 307–344).
3. Afolayan, O., Ogundele, F., & Odewumi, S. (2012). Hydrological implication of solid waste disposal on groundwater quality in urbanized area of Lagos state, Nigeria. *International Journal of Applied, 2*(5).

4. Kamaruddin, M. A., Yusoff, M. S., Rui, L. M., Isa, A. M., Zawawi, M. H., & Alrozi, R. (2017). An overview of municipal solid waste management and landfill leachate treatment: Malaysia and asian perspectives. *Environmental Science and Pollution Research, 24*(35), 26988–27020.
5. Aziz, H. A. (2015). *Control and treatment of landfill leachate for sanitary waste disposal*. IGI Global.
6. Clesceri, L. S. (Ed.). (1989). *Standard methods for the examination of water and wastewater* (Vol. 17). American Public Health Association.
7. Apha Awwa, W. P. C. F. (1998). *Standard methods for the examination of water and wastewater* (20th ed.). American Public Health Association.
8. Uddin, M., Alam, M., Mobin, M., & Miah, M. (2014). An assessment of the river water quality parameters: A case of jamuna river. *Journal of Environmental Science and Natural Resources, 7*(1), 249–256.
9. Göbel, P., Dierkes, C., & Coldewey, W. (2007). Storm water runoff concentration matrix for urban areas. *Journal of Contaminant Hydrology, 91*(1–2), 26–42.
10. Mitchell, G. (2005). Mapping hazard from urban non-point pollution: A screening model to support sustainable urban drainage planning. *Journal of Environmental Management, 74*(1), 1–9.
11. Viklander, M. (1999). Substances in urban snow. A comparison of the contamination of snow in different parts of the city of luleå, Sweden. *Water, Air, and Soil Pollution, 114*(3–4), 377–394.
12. R. J. B. J. O. M. Mishra. (2023). A. Studies, The effect of eutrophication on drinking water. *4*(1), 7–20.
13. Nkwachukwu, O. I., Chidi, N. I., & Charles, K. O. (2010). Issues of roadside disposal habit of municipal solid waste, environmental impacts and implementation of sound management practices in developing country "Nigeria". *International Journal of Environmental Science and Development, 1*(5), 409–418.
14. Smith, K. (1979). Temperature characteristics of British rivers and the effects of thermal pollution. In *Man's impact on the hydrological cycle un the United Kingdom* (pp. 229–242).
15. Pankow, J. F. (2018). *Aquatic chemistry concepts*. CRC Press.
16. Makwe, E., & Chup, C. (2013). Seasonal variation in physico-chemical properties of groundwater around Karu abattoir. *Ethiopian Journal of Environmental Studies and Management, 6*(5), 489–497.
17. Haines, S. G., & Cleveland, G. (1981). Seasonal variation in properties of five forest soils in Southwest Georgia 1. *Soil Science Society of America Journal, 45*(1), 139–143.
18. Metcalf and Eddy, Inc, Asano, T., Burton, F. L., Leverenz, H., Tsuchihashi, R., & Tchobanoglous, G. (2007). *Water reuse* (pp. 6–15). McGraw-Hill Professional Publishing.
19. Boakye, R. (2012). *Impacts of waste on urban surface water resourcs: Case study of Aboabo stream*. Doctoral dissertation.
20. Ioryue, I. S., Wuana, R. A., & Augustine, A. U. (2018). Seasonal variation in water quality parameters of river Mkomon Kwande local government area, Nigeria. *International Journal of Recent Research in Physics and Chemical Sciences, 5*(1), 42–62.
21. Wilson, P. C. (2010). Water quality notes: Dissolved oxygen1. *Sea, 1000*, 5000.
22. Metcalf, I. (2003). *Wastewater engineering; treatment and reuse*. Mcgraw-Hill.
23. Canadian Council of Ministers of the Environment. (1999). *Canadian environmental quality guidelines* (Vol. 2). Canadian Council of Ministers of the Environment.
24. Kumar, D., Malik, D. S., & Gupta, V. (2018). Seasonal assessment of surface water quality in the middle stretch of river Ganga for suitability of fish and human health. *Journal of Experimental Zoology India, 21*(3), 667–677.
25. Begum, A. (2008). Study on the quality of water in some streams of cauvery river. *Journal of Chemistry, 5*(2), 377–384.
26. Olomukoro, J. O., & Dirisu, A.-R. (2014). Macroinvertebrate community and pollution tolerance index in edion and omodo rivers in derived savannah wetlands in southern Nigeria. *Jordan Journal of Biological Sciences, 147*(1570), 1–6.

27. Tjandraatmadja, G., Pollard, C., Sheedy, C., & Gozukara, Y. (2010). *Sources of contaminants in domestic wastewater: Nutrients and additional elements from household products*. Water for a Healthy Country Flagship Report: CSIRO, Canberra.
28. Kale, V. S. (2016). Consequence of temperature, ph, turbidity and dissolved oxygen water quality parameters. *International Advanced Research Journal in Science, Engineering and Technology, 3*(8), 186–190.
29. Dalpian, G. M, Chelikowsky, J. R. J. P. R. L. (2006). Self-purification in semiconductor nanocrystals. *96*(22), 226802.
30. Water, D. (2000). State of Florida department of environmental protection. *DEP, 22*, 2088.

Index

A

Aba Gerima, 396–399, 402–404, 406, 407, 409, 410
Aerospace components, 49–65, 71–76
Aloe vera gel, 245–247, 249–251, 254
Anaerobic digestion (AD), 298, 306–309, 311, 312, 318, 319, 322–326, 329
Angle of repose, 144, 145, 147–148, 151, 152
Antidandruff, 244, 259–275
Antioxidant activities, 185, 189, 195–200
Aquatic weeds, 297, 306, 308, 309
Awaze, 185–200

B

Bahir Dar, 205, 246, 262, 263, 334, 335, 341–357, 374, 377–391, 396, 399, 402, 414, 415, 428
Berseem and lucerne seeds, 96, 97, 102, 144–152
Biogas production, 309, 311, 318
Biological processes, 299, 308, 424
Biomechanics, 156, 158, 162–164
Biopesticides, 226–240
Birbira seed, 226–230, 233, 235, 238

C

Cage wheels, 33–35, 37, 39–43
Cellulose-recovery, 205, 206, 219, 220
Characterization, 207–209, 227, 230–233, 235, 239, 247–250, 360
Cleaning capacity, 115, 120, 122
Coffee husks, 203–220
Contract times, 341–357
Controlling activities, 345
Customer order decoupling point (CODP), 125–129, 137, 138
Customization capabilities, 132

D

Damage grain, 115, 119–122
Deep learning, 18, 19, 84, 92, 164
Delignification, 216, 219, 220
Delphi method, 6, 8, 10
Digital supply chain, 170, 172
Digital transformation, 129
Digital twin (DT), 125, 126, 129–139, 162
Discharged beer, 243–254
Draft forces, 35, 40–43
Drying, 208, 227, 230, 248, 261, 265, 280, 283, 284, 286, 288, 290, 292, 399, 401

E

E-commerce, 170–181
Edge detection, 18, 19, 21–25, 27
Enabling factors, 2, 6, 7, 10–13
E-supply chain, 170–174, 180
Ethiopia, 36, 111, 175, 185, 205, 226, 260, 297, 334, 342, 360, 377, 395, 414
Extractions, 84, 89, 156, 160, 225–240, 261, 262, 280, 282, 283, 285–293

F

Fault diagnosis, 84, 87–89, 91, 92

S. W. Fanta et al. (eds.), *Sustainable Development Research in Manufacturing, Process Engineering, Green Infrastructure, and Water Resources*, Green Energy and Technology, https://doi.org/10.1007/978-3-031-77339-6

Formulations, 65, 200, 226, 228–229, 233–235, 237–240, 243–254, 259–275, 339

H
Heavy metals, 359–362, 371–374, 414, 415, 430–432, 436
Herbal extracts, 260, 263

K
Khat, 395–410

L
Living labs, 333–339

M
Machine vision, 17, 18, 21, 27
Mass customization, 1–13, 125–129, 132–137
Mean square error (MSE), 25–27
Metal printing, 49
Metering devices, 96–108
Microalgae biomasses, 279–293, 318, 328, 329
Microbial analysis, 191, 248–249, 254, 255
Microwave-hydrothermal process, 203–220
Moisture, 35–37, 112, 114, 115, 117–122, 145, 149, 150, 198, 228, 230, 233, 249, 261, 283, 303, 396, 399, 400, 402, 407–410, 414, 430
Multi-material structure, 48
Musculoskeletal disorders (MSDs), 155, 156, 159, 161–165

N
Neem leaves, 226–230, 233, 235, 238
Nutrient recovery, 309
Nutrient retention, 312

O
One-dimensional convolutional neural network (1D-CNN), 84, 87–92
Optimization, 43, 53, 65, 74, 88, 131, 133, 203–220, 292
Organic hair shampoo, 243–254

P
Perfect models, 19–22, 24–27
Physical parameters, 360, 362, 368–371, 374, 416, 417, 419, 433–434
Physicochemical properties, 197, 227, 233, 245, 249, 250, 254, 374
Pneumatic tyres, 33, 34, 36–43
Polyunsaturated fatty acids (PUFAs), 279–283, 287, 288, 292
Prediction algorithms, 156, 162–165
Pretreatments, 203–220, 280, 283, 284, 290–292, 298–300, 302–312, 318, 323, 324
Process stability, 318, 322–324
Productivity rates, 342, 343, 345–357

R
Rotating machinery, 83, 84, 87, 91, 92
Rubber tracks, 34, 35, 37–40, 42, 43
Runoff, 369, 371, 373, 377, 396–410, 419–421, 423, 425, 428, 429
Rural entrepreneurship, 333–339

S
Seasonal variations, 281, 368, 419, 423, 426, 433–436
Sedimentation, 280, 298–312
Seed damage, 96, 100–102, 106–108
Seed rate, 96, 98, 100–104, 106, 108, 144
Seed spacing, 102, 108
Selamco, 359–374
Separation loss, 115, 119, 120, 122
Shampoo formulations, 244, 246, 247, 249, 252–254
Simple distillation, 229
Small-seeded, 96, 144, 145
Smart supply chain, 172
Soil loss (SL), 396, 400–407, 410
Soxhlet extraction, 226–228
Sphericity, 144, 146, 149–151
SS removal, 301–305, 312
Surface waters, 368, 369, 413, 414, 416–432
Sustainable energy, 337

T
Test models, 20, 21, 23–27
3D Printing, 47, 49, 54, 62, 63, 65, 73, 77

Threshing capacity, 115, 118, 119, 121, 122
Threshing performances, 112, 115, 118–122
Tractive devices, 33–35, 37–43
Tractive performances, 33–43
True density, 144–147, 150–152
Two wheeled tractors, 31–33, 35

U
Upflow anaerobic digesters, 319, 320, 324, 325, 328
Urban flood, 382
Utilizations, 5, 32, 53, 56, 71, 92, 137, 169, 186, 192–194, 292, 298, 318, 329

V
Vulnerability, 66, 378, 380–390

W
Waste disposal sites, 413–432
Water balance, 400, 409, 410
Water hyacinth juice (WHJ), 297–312, 318, 319, 321, 323–326, 328, 329
Watersheds, 396–398, 403, 404, 406, 409, 410, 414, 418, 420, 422, 423, 426, 427, 432
Wearable devices, 156, 157, 159–162, 164, 165
Work item scheduling, 341–357

The manufacturer's authorised representative in the EU is Springer Nature Customer Service Centre GmbH, Europaplatz 3, 69115 Heidelberg, Germany. If you have any concerns regarding our products, please contact ProductSafety@springernature.com

Printed and bound by CPI Group (UK) Ltd, Croydon, CR0 4YY
01/10/2026
02225958-0013